For The Right Reasons

TRUE STORY OF:
7 Deaths + • Treason • USA National Security
• USA National Defence • Billions Stolen, Squandered, Lost
• Planetary Crimes • F.B.I. • NASA
• Human Species Future

STEVEN W. AGEE
(FBI Code Name "Brains")

FOR THE RIGHT REASONS

Manuscript History:

Originally written in 1989 while wearing a licensed concealed gun at Renton Wash., USA. Raw manuscript travelled to Australia, then to New Zealand and was finally updated and completed in New Zealand on 5 January 2017 by Steven W. Agee.
At the time of originally writing this book my life was considered to be in danger from numerous criminals at Morton Thiokol Inc. in Utah, officials of the Mormon Church, Attorney General of the USA - Ed Meese, and probably other failed / misguided high level U.S. Government officials.

Published by Steven W. Agee
ISBN: 978-0-473-38892-8

Date of First Publication is 06/2017
Written By: Steven W. Agee
Publisher: Steven W. Agee
ISBN: 978-0-473-38892-8

Design and Layout by: clarkdesign@xtra.co.nz

Printed by: www.yourbooks.co.nz

Political Incorrectness And Its Ramifications

I am old, now 74. I am ill and I will not live especially long. I am saddened and disgusted by 'Political Correctness'. Political Correctness has wasted vast amounts of earth's assets, vast amounts of societies currencies, has wasted billions of peoples 'time', and has caused untold deaths and in my view is a blight on our planet. Political Incorrectness does not mean bad manners, courtesies, or lack of temperance. Political Incorrectness does not hinder human progress like Political Correctness does. At age 16 I told myself I was not going to waste my life and energy placating the world. I wanted to make effective contributions, and indeed I have. My life has produced many positive results because I refused to be 'politically correct' or always 'go with the flow', or to waste time avoiding 'dealing with problems' for political correctness reasons .

I grew up around cussing and swearing but was a good, 'clean mouthed' lad until I was in the U.S. Navy and was given over to the U.S Marine Corps. The U.S. Marine Corps can operate on less than 20 words, mostly explicative's. When I was in the Vietnam war I found our most cherished explicative's to be woefully inadequate to the sound of bullets going by your head, or the explosion of the IED that still has 70 pieces of shrapnel in me. The Vietnam War was a political and deadly 'war' which was played as an ego game. This book is written in the 'language' of its place and time deliberately - it is *not politicaly correct or diluted, or moderated.* Live with it!

This book is written as close to 'truth and reality' as I can write it's sad, deadly and pivotal history. In no way is it intended to offend anyone unless they are the criminals or failed persons I have identified, etc... I am not perfect, I have made my mistakes, and I am not ashamed to admit them and move on. I do not have to be politically correct and I am not.

If you are offended I am sorry for you. You are absolutely free to read no further, or sue me immediately as you see fit. However by design my total assets on Earth are only seven well worn Italian suits, four pairs of shoes, some shirts and socks and underwear. If you should be stupid enough to sue me I doubt if I would even bother to show up in court unless it was for amusement.

Why I Wrote This Book

Earth is a very small, rare and precious place in our universe. Earth is our only home for now, and will be far into the future. Looking at Earth from planet Saturn from one of our Cassini Satellites here is a photograph of Earth. The photo following is the final argument of how small our world is and how isolated we are. Crimes against our planet and our lives can easily destroy such a small and sensitive world. Planetary crimes cannot be tolerated, we and our earth are very isolated, insignificant and fragile.

The destruction of the Space Shuttle Challenger ranks as one of the most analyzed, talked about and reported events in American History. The explosion and deaths were iconic. Virtually all of the documentation focused on the narrow picture of the disaster itself and what led to it. None has centred on the events since and the unflattering picture of the country's space program they reveal. After the tragedy Americans rightly expected all of their country's enormous technical and human resources would be brought to bear in a redoubled effort to ensure no possible avenue was left open for anything similar to happen again.

I was the first person involved in the post-Challenger Shuttle re-evaluation and redesign to step forward and show how criminal conspiracies, treason and cover-up's thwarted safety and progress. While working at Morton Thiokol I found massive crimes. The months I spent working under cover for the FBI gave me a unique insight into the mindboggling scope, breadth and negative synergy of these crimes. This is the story of lessons from the disaster wilfully going unheeded. It is the story of self - serving organizations and individuals treasonously jeopardizing national security and criminally endangering lives for their own short term gain. It is the story of the creeping ethical desert which has engulfed so much of American government and corporate life. Finally, it is the story of the man who blew the whistle and the heavy price I and my family are still paying for my ethical actions.

Americans support their country's space program, Americans are concerned about National Security, Americans believe in their country should reward only the best efforts. This book, *"For The Right Reasons"* is a clear and honest documentation of criminality and treason long exposed but still unpunished. It will leave Americans shaken and demanding answers *"For The Right Reasons"*.

Now America has retired all of our Space Shuttles. NASA has wasted enough money to have bought (and paid for) substantial nations. Now NASA is looking to non-government Rocket Contractors to build America's newest generation of rockets. Foreign nations are advancing on space exploration and America's space efforts have crumbled into disgrace and decay. Resurrection and resuscitation are in order!

It is a felony to lie to the FBI. So by virtue that I am not in prison and I ended up consulting to the FBI for years you can be assured all of this book is true. Supporting documents are available by the U.S. Freedom Of Information Act, via the various U.S. Agencies.

Steven W. Agee Nov. 2016

NASA, Public Domain

Earth and our Moon are the white speck at lower right of photograph.

We all need to keep proper perspective of our insignificance and fragility. This photo from our Cassini Satellite taken from our near neighbour Saturn shows Earth is hardly a speck of dust in the grand scheme of the universe.

Dedication

"For The Right Reasons" is dedicated to the few persons who had the knowledge, insight, courage, and dedication to pursue the continuance of Free World Space Exploration no matter the dangers or cost to themselves. Also to those who encouraged and supported me on to write this book.

These persons became part of my life and came to know this story as well as I do. Some of them battled fraud, treason, bureaucracy, cover ups, threats, slander, and libel. Their contributions to the welfare of the Free World are incalculable.

Dedicated to the living:

Dr. Stafford and Mrs. Ella Bourke:
Dedicated friends who relentlessly encouraged me to complete and publish this book. They are retired global medical missionaries.

FBI Special Agent Tim Healy:
He pursued Justice against all odds and all obstacles. Now retired from a high level management position at the FBI. He is now president of a substantial corporation.

Edwin McLean, Attorney:
He pursued justice with a vengeance for his clients and for America. He fell victim to DOJ Crimes. Now semi-retired.

Carl Haller
A very supportive friend for many years. He retired from the nuclear industry. He is an electrical engineer and project manager.

Dedicated in Memorium:

Dr. Rema Walker:
She told me who and what I was and she freed me to pursue what I thought was right. She was a retired US. Air Force Colonel, and Psychologist.

Mr. Roger Boisjoly:
Mechanical Engineer and manager who fought to stop the launch of the Space Shuttle Challenger. He was massively destroyed by the U.S. Federal Court System being in the full bloom of its failures and criminality. He was a great man and a great American.

Tony Laine Ph.D.:
Physicist - Witness and Victim.
His outstanding Contributions destroyed him first and then 'killed him"! He was a Great American.

Alan Salzmann:
Former IBM Services Sr. Enterprise Architect
Friend and Confidant who urged me to complete this book.

Acknowledgement And Gratitude For Great American Space Explorers From Planet Earth:

Astronauts of the Challenger Mission:
All Deceased from fraud and corruption at Morton Thiokol, Inc., and NASA

All Dedicated Explorers, Professionals & Experts

- Francis Scobee - Mission Commander
- Michael Smith - Pilot
- Ronald McNair - Mission Specialist
- Ellison Onizuka - Mission Specialist
- Judith Resnik - Mission Specialist
- Gregory Jarvis - Payload Specialist
- Sharon Christa McAuliffe - Payload Specialist, the first teacher to fly in space

Ern Reynolds, Attorney.

"This book reveals how the truth of one whistleblower outweighs the bogus integrity of all the religious leaders and government officeholders arrayed against him."

Ern Reynolds, "Attorney, Washington D.C. Bar"
Expert Whistleblower Attorney.

Table Of Contents

Appendices

Bibliographies

Bib. 1 Massive amounts of this book's documentation are from FBI documents, DOJ documents, and NASA Documents all on file and obtainable from the FBI and the U.S. Dept. of Justice via the Freedom of Information Act. Acquisition of these documents was authorized by a Federal Judge in all cases. Thousands of pages are available.

Bib. 2 News Paper Articles are all Public Domain

Bib. 3 Photographs of persons supplied by Steven W. Agee

Bib. 4. Photographs from Cassini are Public Domain

Bib. 5 Drawings are Public Domain

Photographs

Photograph 1 Cassini Saturn Earth photo NASA, Public Domain
Photograph 2 Ern Reynolds, Attorney
Photograph 3 Challenger Astronauts NASA, Public Domain
Photograph 4 Steven W. Agee (Author)
Photograph 5 Steven W. Agee & Carl Haller (Author)
Photograph 6 Steven W. Agee & / FBI Special Agent Tim Healy
Photograph 7 SRB separation from Space Shuttle, Public Domain
Photograph 8 Rail Road Cars carrying SRB components / explosives
Photograph 9 Paul Edwards OBE (outside back cover)
Photographs 10, 11, 13 SpaceX, WWW, Public Domain
Photograph 14 Yacht

CARTOONS

Cartoons: Public Domain

Authors Biography

Steven W. Agee was born at Ventura, California on Armistice Day 1942. First and only child of Sylva and George Agee Jr.. He was raised in Southern and Central California. He attended grade schools in Madera, and Folsom, California. He went to and graduated from Mt. Whitney High School in Visalia, Calif. He grew up close to family members and was substantially influenced by his maternal grandparents. Their strong work ethics, and their high standards of integrity, their love.

While Steven Agee was in high school he set over sixty swimming records in Central California. He came close to making the U.S. Olympics team in 1960. He also placed very high in the Central Valley Science Fairs in his Sophomore and Junior years. In his Senior year he won the Central California Valley Science Fair and went on to Kansas City to place 4th in the *'11th National Science Fair International'*. In his Junior year (1960) he nearly made the USA Olympic Swimming Team. He was lean and thin then but had not accumulated a lot of his strength yet as he would later in life.

Steven Agee went to the University of Texas at El Paso on Physics Scholarships in 1961. He subsequently graduated with a B.S. in Physics, a second major in Mathematics, and third major in Metallurgical Engineering. He was not a good student as he worked two and three jobs simultaneously to help pay for his university education. He did not have family financial support. There is much more to tell about his education however because of his consulting to 'un-discussable' governmental entities and projects it won't be elaborated on here.

After being absorbed into USA military service against his will (just prior to graduating from UTEP) in 1965 at the height of the Vietnam War he became a U.S. Navy Hospital Corpsman attached to the Fleet Marine Force Reconnaissance Marines during the Vietnam War. He was wounded in action while on a reconnaissance mission near the "Rock Pile" in South Vietnam - I Corps, just south of the DMZ (Demilitarized Zone). He underwent surgery then to remove a number of large shrapnel pieces from both of his legs, and to this day he has over 70 pieces of shrapnel in both hands and both legs. Steven Agee later developed colitis in Vietnam. After years of suffering he ended up having ten operations to cure it. It nearly killed him on several occasions over a period of 45 years. Now at age 74 it is probable that the Vietnam War will be the cause of his death - numerous issues suspected from Agent Orange.

After his university training Agee was involved in both 'smoke stack' and 'high technology' work ever since entering the work force. He worked on numerous space, subsea, industrial, military, communications, and medical technology projects. He is considered a SME (Subject Matter Expert) at salvaging troubled multimillion / multibillion dollar projects. He has been consulting and advising for years and has owned several small businesses. He is a global business traveller and has been in 73 nations on business. Along the way he picked up an MBA and additional education and expertise. He is fluent in international finance. He is a "crisis manager."

There was a major motion picture pending based on this book. He is currently writing two other books - His autobiography, and about his sailing to the South Pacific from Friendship, MD, USA.

The reality portrayed in this book destroyed his life as he had known it and as it would have been. Steven, his wife Fay, and their (then) new daughter Mariel fled the United States as CIA men escorted them to Cairns, Queensland, Australia.

Eight years were wasted by him trying to settle in Australia. He was 'Persona Non-Grata' in Australia. He was hated for his superior education, his directness, and his unwillingness to suck up to Australian moron Prima Donnas. He interviewed for the number two position at Australia's CSIRO, 'Commonwealth Science and Industry Research Organization'. He beat the competition hands down. During the process he evaluated all of CSIRO's Laboratory research projects and made substantial commercial recommendations. His analysis was lauded and then the position went to a copy machine sales manager with a small amount of scientific training because Agee was not a 'registered Labour Voter'.

Later on Agee raised AU$22,000,000.00 to start up the world's leading edge robotics company, *Android Protocol*, in Queensland. An Australian Senator demanded AU$120,000.00 in a brown paper bag to obtain his '100% Tax Exemption stamp' for the investors. At this point Agee wrote off Australia as a 'basket case' of criminality in his view and he packed his family and shifted to New Zealand. He was burned out from the tribulations of his displacement from his former life.

In New Zealand Agee found far more honesty than in Australia, however he was absolutely unemployable due to past high incomes, 'over education' in their eyes, and simply an inability to downgrade to low New Zealand standards, expectations and incomes.

Eventually Agee started commuting to the USA to work. He went to Fort Collins, Colorado and straightened out a GIS (Geographic Information Systems) company. Then on to a new aircraft company building a new four seat twin jet. When he came home it was clear to him his marriage had unravelled and he filed for divorce. While he and Fay had signed off in New Zealand on 'shared custody' of Mariel (11 years) and Evan (9 years) he was frustrated by Fay's incalcitrance over everything. Soon he returned to the USA to create the *Emergency Operations Plan* (usually known as a COP Plan) for the Federal Reserve Board / Bank. When he landed in the USA he was notified Fay had abducted their children to Australia and broken the Hague Convention on Child Abduction. A massive Hague Battle was fought.

Australia was a signatory of the Hague Convention on Child Abduction but clearly had never read it. The battle was not about children's welfare, it was about how much money Australia could steal or usurp from all parties.

While the battles raged he completed his project for the Federal Reserve Bank. Then he set up from scratch the FBI Internet Fraud Complaint Center (FBI IFCC) for / with Special Agent Tim Healy. It was and is a resounding success.

Later he set up from scratch the FBI Terrorist Screening Center (FBI TSC) for / with Tim Healy and Presidential Appointee Donna Bucella (Attorney). Upon initialization and to this day it is the center for global terrorist activities information. It operates far above ordinary Top Secret.

Upon completion of establishment of the remarkable and powerful FBI TSC his health was in decline and he skippered and sailed his 44 ft. sailing yacht from Friendship, Maryland, to Bermuda, all through the Caribbean, then to Panama and onward to the Marquis, Tonga, and his South Pacific Home.

Steven Agee is now retired and on a beach somewhere. He is writing, and also doing scientific developments to replace petrochemical fuel use globally. He is an accomplished wood sculptor, and a disgraceful classical guitar player.

Steven W. Agee
2002, Wash., D.C.

Steven's yacht Mariel

Foreword By Roger Boisjoly

(1 / 8 / 1989)

Roger Boisjoly is the Mechanical Engineer and Manager who tried to stop the launch of the Challenger due to icing conditions.

Many years since the episodes of this book have been a living hell for many of us who were associated with the *space shuttle challenger disaster* which occurred on 28 Jan. 1986.

Many readers will no doubt immediately draw the conclusion of exaggeration of facts by Steven about his personal involvement and observations concerning Morton Thiokol Inc. (MTI), NASA, U.S. Dept. Of Justice (DOJ), and the FBI. I assure you as a result of my very independent involvement and personal observations that my experiences during the past three years actually parallel Steve's.

I have not met Steve, but we have talked on the phone and I find him to be a man driven as I am by the truth and a deep sense of doing right for country and our national space program. For his service and position, we have both been summarily cast aside by the system at MTI, and NASA as know-nothing troublemakers.

I have always felt that my two lawsuits may be too hot to handle politically, but thought that a fair and impartial review of the facts would result – How naive I was to think that above-board fair play would prevail in the arena of every day foul play!

My biggest disappointment has been the failure to bring real accountability to Morton Thiokol and NASA managers for their involvement in the Challenger Disaster and to have a legal accord as a foundation for others to stand upon who come forward for the protection of the health safety and welfare of our citizens against wrongdoing in our industry.

I hope in some small way, and perhaps in a large way this book will alert the public to the foul play system we now operate in and embrace as a nation, when justice is not allowed to freely prevail on the side of right and truth without interface from political and industrial big guns abusing power.

My sincerest best wishes to Steve and his wife Fay for the success of this book and to the enlightenment of those who care about doing what is right.

Roger M. Boisjoly

NOTE: Sometime later Roger Boisjoly and his wife and I met at their home in Phoenix, Arizona. We had in - depth discussions of the massive crimes we had witnessed. Roger was a fine Professional Mechanical Engineer and Manager. He was the 'salt of the earth' and profoundly honest. He and I had collided with what President Dwight D. Eisenhower called the "industrial military complex" being out of control. Now in 2017 the USA "industrial military complex" is one of the strongest forces on earth.

CHAPTER 1

SPECTOR OF AMERICAN AND FREEWORLD DEFEAT

TIME: After the Apollo moon missions to the untold future
Note: {Writing on this book began in 1989}

Subtle innocuous changes can and do change our lives forever. Profound and permanent changes in my life began with a simple phone call from my friend Malcolm Phillips. This phone call eventually changed my career, endangered my life, alienated me from an entire community and groups of local friends, exiled me from America, caused me to seriously question the wisdom of American justice, and brought me to a *"state of shock"* when I realized that *"we the American people"*, have in fact lost the space race. I found myself surrounded by crime and treason.

America has lost the first and probably last Space Race in Earth History. America was far ahead of all contenders for the 'high ground'. Well educated intellectuals had already calculated that the cost in resources and money to go to the Moon and Mars, and beyond were finite and should not be squandered. As a nation we currently have a substantially reduced ability to defend ourselves as a result of losing the Space Race. In the very near future we Americans will probably be *'barred from space'* by our numerous economic and military enemies. The U.S.A. has lost its gambit for many of the best earth orbits for its important satellites (commercial and military). The U.S. has lost a huge portion of its "in place operational satellites" necessary to protect the United States and Free World due to gargantuan crimes. The U.S. can now launch only a small fraction of its commercial and military satellites which were built and available to launch. The U.S. can no longer fix most of its space related problems on earth or in space. The United States national bottom line is simply that the largest "land and space grab" in history or ever to be in the future of human beings is taking place now and the United states is barely a participant. America has fallen by the wayside! This celestial land and space grab encompasses our entire solar system and all of its assets: minerals, energy, positions and orbits for equipment and satellites, new frontiers, military high ground, and the human race's entire volume for future expansion. The current space race into our solar system is well underway. Extensive exploitation of space beyond our solar system cannot occur in the near future. However our solar system holds secrets , assets of extreme value to us earthlings beyond our current imagination. These assets and wonders are available for the taking by the bold, and wise of which America is no longer leader.

Whether we individuals approve or not, man has already become a permanent fixture in space and will continue to inhabit space beyond earth. Space is not easily or pleasantly inhabitable by humans. Space is hostile, lonely, and very dangerous. Nonetheless it is the compelling realm of man's destiny. Even young children understand their destiny lies in space when they play such games as Solar Quest, Lego Land Space System, Space link and Robotix. The United States did not have a person in space from the time of the Challenger Disaster (Jan. 18, 1986) through the next 32 months. The U.S.'s next successful manned space mission was the Space Shuttle Columbia's flight (Space Shuttle flight #26). The Challenger mission (STS 51L otherwise known as flight #25) ended in a tragic explosion during ascent towards orbit. STS is an acronym for Space Transportation System. Seven valued lives were lost, a Space Shuttle Obiter Vehicle was lost, Scientific and Military missions and equipment were lost. In total - a national catastrophic loss of enormous proportions! Seven lives, a four and a half billion dollar STS, a two hundred and fifty million dollar plus payload, thirty two months of lost heavy space launches and the Space Race at that time.

The Russians have had persons in Space frequently since 1971. There is no adequate means to compare national space efforts. However, it is certain that those who 'can go to and return' from space and those who inhabit space will prevail in conquering space. The Russians have inhabited space for some time. The U.S. cannot 'go to or return from' space in a reliable manner, and the U.S.A. has not inhabited space in a long term, meaningful way yet. The U.S.'s attention to space has been sporadic, ill planned, and improperly supported since the last Apollo moon missions. Currently the European Space Agency, the Peoples Republic of China, the Japanese and the Russians all have launch capability. In total 14 nations can hurtle objects into space at this time.

This book is my frustrated effort and moral plea to inform American Citizens, politicians, and the free world of the fact that our space efforts have fallen victim to what is probably the largest and farthest reaching crime in recorded history up to 1990. Aerospace vendors have failed our nation and the posterity of the entire free world miserably. As a nation, the U.S. cannot blame other nations and peoples for our national failures. The problems lie 'within'! The only conceivable solution lies with you, the readers. You are the only force in the universe which can precipitate an acceptable solution to this monolithic 'species planetary disaster'.

For seven months in 1987 I was a friend, 'undercover source', and worker for the F.B.I. I was in the vortex of crimes, lies, and cover-ups. I could not tolerate what I saw, heard, felt, or sensed!. Truth and reality are often frightening, but the true story will now be told – no matter what danger or cost to myself. The personal costs to me have been immense but seemingly genetically mandated.

Hopefully more of the story will be told in the courts, newspapers, radio, TV, movies, press conferences and lecture halls. The problem is bigger than me. The problems and crimes are bigger than NASA. They are bigger than the U.S. Government. The problems and the crimes are as big as the inevitable future expansion and continuance of the human race. The crimes I witnessed can be effectively dealt with only for a short time to come after which no solutions will be attainable for America or any other nation when resources are exhausted. In order to understand how and why the U.S. has lost the Space Race we must consider the following questions and their scope:

- Who is in charge of the United States Space Program?
- Between 1969 and 1972 the U.S. sent 12 men to the moon and returned them to Earth. Why can't the U.S.A. repeat this performance, and why has there been in effect no meaningful space program for years?
- Why is the U.S. currently out of the Space Race? There is rhetoric but little reality.
- What has the U.S. to show for hundreds of billions of dollars spent on space exploration and space operations? What more should it have to show?
- Why isn't America the current leader in Space explorations, habitation, development, and space related research?
- Why did the Space Shuttle Challenger really crash?
- Why is America's Space Station so political?
- Why can't the U.S.A. even get to its International Space station reliably currently?
- Can the U.S.A. recover?
- Why do our governing agencies, such as the U.S.A.F., U.S.N., U.S.M.C., CIA, NSA, DOD, DOJ, FBI, NASA, U.S. Congress, U.S. Senate

the others not cooperate effectively for our national space efforts effectiveness?

- Why had NASA known about numerous and clearly lethal problems on the Space Shuttles for 13+ years prior to Challenger and not effectively responded?
- Why were Morton Thiokol Inc. and NASA in bed?
- Why did Morton Thiokol Inc. cover up over 3,000 catastrophic Hazards identified by Consultants working on the space shuttle's solid rocket motors?
- Why did Morton Thiokol Inc. lie to the President's Commissions on the Challenger Accident many times and what effects did the lies have?
- Why were the Space Shuttle's Solid Rocket Motors sole-sourced to Morton Thiokol Inc. against most of NASA managements wishes?
- How and why did Morton Thiokol Inc. 'threaten' and 'control' its employees?
- Can the U.S. afford for the U.S.A.F. and NASA to compete for funds and equipment?
- Why did NASA avoid, knowingly deceive, and lie to the FBI?
- Why did the D.O.J. avoid prosecuting Morton Thiokol and NASA?

I will answer these and many other pertinent questions so that you, the readers can understand our space related problems and the scope of great crimes and failures that affect us all.

You readers have a vote in the U.S., and in this case a vote in the entire destiny of humanity! Please exercise your vote in any way that you can through understanding. Your vote / voice may correct incredible crimes and errors. Leaders of the U.S. Government have not consummated a U.S. Space Plan which is effectively followed or even respected. The US Congress and Senate have vacillated, our military and corporations have disagreed! You readers probably have more sense and insight than our aforementioned leaders. You need to support space in any appropriate way so your children and subsequent progeny are not *"dirt poor"* for their whole lives; so your progeny are able to share in space related assets which will be available in the near future. Without acting 'NOW' all of our American progeny are faced with eventual containment and domination on Planet Earth! Never before in Earth history have persons such as ourselves been forced to deal with such complex issues of future national development! Our *immediate and continuing* National Futures will be determined by our nations' positions in the Space Race! Can human beings, as creatures, respond to such imminent decisional pressures when the immediate needs appear to be distant rather than imminently critical and urgent? We as a species, a nation, and as a planet society are at the crossroads of our evolution if we like it or not, if we are ready or not. There are far too many of us, a lot of our jobs are being taken over by robots, and global finances and resources are collapsing, wars are raging and several wars are pending.

The choices are ours for only a brief time to come – beyond which all choices may be made for America by its economic or military enemies or mother nature. The world over persons frequently capitulate their decisional options to the whims of fate. Fate is not often kind, desirable, or optimal. "Fate" has historically provided the cruellest of miseries and frequently 'extinction' in the case of 99% of most earth species.

The gargantuan crimes of Morton Thiokol Inc., NASA, and others have provided an opportunity for the enemies of western civilization to 'contain' western civilization on Earth while our enemies domains expand intractably and space forever remains inaccessible to the USA and other second and third world nations. Forever is a long time. A perfect analogy and example is recorded in history when Portugal did not follow up its lead in 'new world exploration'. Portugal treatie'd out of the conquest via the Treaty of Torricelli's and then collapsed. Portugal has remained painfully poor every since – almost 500 years now. The USA will be the new loser if it doesn't get moving 'RIGHT NOW'! Being the biggest and best is not the issue. Freedom and quality of our lives are forever at risk.

Russia's space weapons program is far advanced. The United States Star War's programs were mostly fantasy or trapped on earth or cancelled. Russia's violations of all existing treaties in recorded history are well chronicled. Russia's destruction of U.S. Satellites and space assets is common knowledge within informed military and aerospace circles. While men on earth make global treaties apparently lessening nuclear threats, the balance of power passes to the possessors of the 'high ground'. *The United States had been at war with the USSR in space! Soon it will be with China and others.*

More recently there have been more attacks. Our attacks and theirs are of course kept very secret for obvious reasons. Momentous crimes have been committed and witnessed. These crimes extend beyond the $20+ billion essentially wasted by Morton Thiokol Inc., the $150 billion essentially wasted by NASA on the Space Shuttles, and beyond the unnecessary and tragic loss of valued lives. These crimes are probably the most insidious and far reaching crimes in human history up to 1990. These crimes will remain a permanent stain of the entire remainder of human history. Many nations have been 'sold out' via treason. In this case the entire Free world has been sold out.

Now my true story of death, space exploration, crimes, National Defence, FBI, and Treason to the entire human race. As this true story unfolded souls were sold, deals were cut, crimes raged, justice faltered, and America's Space programs regressed into historical status of waste, failures and ineptitude.

Appendix 1: Soviet Lasers 'Zapping' U.S. Sattelites UPI 24 Jan. 1989

CHAPTER 2

THE STORY BEGINS A NEW CONSULTING CONTRACT

Time: Early Nov. 1989

My phone rang, it was my friend and associate Malcolm Phillips, who lived at Palm Beach, Florida. Though he didn't know it, that single phone call forever changed my life, the lives of thousands of others, the destiny of a major American Corporation, the potency of several government agencies, and ultimately the direction of Earth and human history.

We had probably met a little over a year and a half ago while working on the *Peacekeeper (MX) Missile Weapon System* at Boeing Ballistic Missile facility located at the Boeing Aerospace Facility known as Boeing Plant 2. Boeing Plant 2 is across the street from Boeing Air Field (King County International Airport) at the South end of Seattle, Washington. It is where thousands of B-17's were built during WWII. We sat next to each other at work and were both hired 'aerospace slaves' – i.e. contract engineers / scientists / managers.

Malcolm was an elegantly knowledgeable and capable man. He was Australian born, educated, lean, balding with a fringe of hair left around the edges. In his early fifties, he had the energy of most any two men in their twenties. He was distinguished by his radiant intelligence and a 1 ½ inch diameter skin graft on his right temple. He beat a dangerous melanoma. Malcolm grew up interested in and enthusiastic about aviation and space. As a young air cadet in Australia, he walked away from the crash of a two place trainer aircraft when his instructor demonstrated the ultimate blunder. His instructor and Malcolm ran out of petrol. Malcolm learned early and firsthand the price of aviation mistakes. Later Malcolm obtained a bachelors degree in Mechanical Engineering and did some graduate work. He matured to become a Project Manager on Australian Government Defence Projects. Later he was involved in firing operational missiles on joint American and Australian operations in Australia. Then he worked for the European Space Agency and was on many launch teams sent to launch European missiles in Koru, French Guiana, now known as Surinam.

Eventually, Malcolm migrated to the United States. Malcolm later made a strategic career move to become an 'aerospace contractor' rather than a captive employee. I had made a similar decision some years prior. Captive aerospace employment has continually been on shaky ground since the end of the Apollo moon landing projects. At the end of the NASA Apollo missions most of our national backing and funding was pulled from the U.S. Aerospace industries. Fickle American interests and politics turned away from space exploration to Defence, to social welfare, and frequently to insanity.

Aerospace contracting usually occurs via three routes: Contracting through what is known as a 'Contract Engineering House', having a direct contract (known as a 1099) or self employed contract, or a contract through a tax registered consulting firm authorized to handle Security Clearance related work billings. When Malcolm and I met, we were both working as 'contract engineers'. We shared similar interests and ideas. We found that we worked well together and could do nearly anything technical, political, or managerial between us. We were powerful and synergistic. Whatever came our way technically or managerially we would accomplish. We would tackle 'critical path analysis', 'dog and pony shows' for the USAF Generals, 'Cross check launch system electronics', 'software to hardware interfaces', 'missile systems to ground systems', etc.. It just didn't matter to us. We were both fully competent and matured in the industry and had seen and worked on virtually everything imaginable on earth, in space, or in the oceans, commercial or military. We had both worked on many Classified 'unmentionables'. We nearly always held one or another level of Security Clearance appropriate to the 'Project of the Moment'.

At Boeing Ballistic Missile Division we reported to Mr. Stephen Ashley, the Boeing Aerospace manager for the Peacekeeper Missile System 'Test Group'. Between Steve Ashley and ourselves was Steve Ashley's technical 'right arm', Mel Dybdahl, and a contract engineer Mahalon Hollway. Mel Dybdahl was a brilliant Project Systems Engineering type. He probably knew more about the overall Peacekeeper Missile Weapon System than anyone else in the world at that time. Mel Dybdahl kept track of more information, systems needs, changes, interpretations, and interfaces than any ordinary mortal or engineer can imagine. Mahalon Hollway was a contract employee. Mahalon told us he was an 'Optical Physicist'. We learned he was knowledgeable in rocket Gas Plume toxicology.

Our work area was an enormous room in an aircraft hangar balcony. Well over a thousand people worked in the Peacekeeper missile system there on 'the balcony'. During WWII this is where B-17's were built by the thousands. Most of the area was divided up into what were called 'cubes' by 4 ½ foot high dividers which divided off eight workers typically. Cubes are also known as "Cell Blocks" since they were painted grey. Some of the 'cubes' were full, others only partially manned. Mahalon was our cube's 'lead man' primarily due to the fact he came to the project months ahead of others in our cube.

There were no clear or hard lines of authority when I arrived. Malcolm, Mahalon, myself and others in our cube jumped when any of the associated aerospace contractors, the U.S. Air Force, or the Boeing Hierarchy needed anything. Our job was to satisfy everyone ASAP and bring the Peacekeeper Missile System to life to protect America as soon as possible. We were at the azimuth of the *Cold War.* East and west were postured to annihilate each other - immediately!

Our Department was to test fire the Peacekeeper Weapon System and support live missile test firings. We supported enthusiastically! A mature working environment is, however not without its unique and inevitable problems. The main problem there being Mahalon Hollway. He was two years into 'recovery' from a lifelong alcohol problem. His 'mood swings' ranged from kind and reasonable to irrational every ten minutes or so. Even though we all realized that Mahalon couldn't help himself it was difficult for all of us to tolerate. Mahalon's moods spilled over to frustrate everyone around him. He had the most insatiable ego I had ever witnessed up to then. There was no replacement for the alcohol that had previously fuelled him. I am considered by many to be arrogant, and overconfident, but I wasn't a down payment on Mahalon. We all worked and contributed massively to the MX Missile system. The MX (Peacekeeper) Missile itself was contracted to and built by Martin Marietta. The 'MX Weapon System' was contracted to Boeing and was a 20 billion dollar project. The Weapon System was the lion's share of the funding, i.e. missile silos, special buildings, software, etc..... Millions of components outside of the missile it's self.

Then, a year after completing the MX Missile Project, I answered Malcolm's fateful phone call at my beautiful Seattle apartment. I was pleased to hear from Malcolm. Malcolm, Mahalon, and myself had kept loosely in contact after our work together on the Peacekeeper (MX) missile system was handed over to the USAF. The primary reason we kept in contact was that we were all continuously pursuing consulting contracts and there were essentially none in the U.S. at that time. The Gramm-Rudman Law had just gone into effect for the first time and it literally assassinated the U.S. Aerospace industry. The Gramm - Rudman law cut 15% out of nearly every Federal Agency and department budget in America every year for some years. Reasons for the Gramm-Rudman law were both devastating and numerous. The U.S. Congress didn't like the Gramm-Rudman Law and didn't know what to do with it or about it. Our U.S. military didn't know how to deal with the new concept of curtailed spending and cutbacks. The Gramm-Rudman Law was an anomaly in American History, it ran contrary to irrational spending!

U.S. aerospace contractors and vendors were near the end of the whip while confusion reigned supreme above them. At the tip of the whip were Contract Engineering Persons and Independent Consultants - persons like Malcolm, Mahalon, and myself. Above us contract engineers was the culmination of 'national lack of planning', 'congressional confusion', 'bickering', 'indecision', 'military confusion', 'aerospace corporate fear and confusion', and the usual fraud, deceit and embezzlement issues. What could possibly go wrong? High end contractors such as Malcolm and I generally made 2 1/2 to 3 times what the captive Vice Presidents made when we worked. We were high tech mercenaries. The concluding result was a devastating disappearance of aerospace and high technology contracts of any kind.

Even though there were more military budget items than ever according to some sources, there was a devastating reduction in the number of jobs and a profound rollback in pay rates. During this slowdown huge amounts of 'unclassified' American high technology business emigrated to Europe and Asia.

In the last half of 1985 I couldn't get a contract as a parts cleaner let alone a Senior Engineer, Senior Scientist, or Project Manager, or CEO. So our keeping in touch was a necessity. We appreciated each other's moral support. Who could say which of us would score a contract first and maybe help his associates? We each had our circles of influence and industry insights, and wealth of knowledge. This time it was Malcolm Phillips who scored first.

Malcolm called to tell me he scored a contract. He had just gone to work at Morton Thiokol Inc. at their Wasach Plant near Brigham City, Utah. This was the largest Solid Rocket Motor Plant in the world. He was rightfully proud of himself for finding what was probably the only remaining pocket of contracts in the United States. We talked a bit and he told me about the place. Twenty-seven miles from Brigham City, cold, isolated, grim, and generally undesirable – a far cry from his Florida home. So much for the bad news I thought – Malcolm, can you get me a contract there? I needed to get back to work ASAP! I owed the IRS a new well equipped luxury car! Malcolm said the magic word 'yes'! Now for the best news 'how much are they paying?' "Well you aren't going to like this Steven. The contract is only about half of our normal." Jessuuessss Christ Malcolm! Have you gone out of your mind?? By taking a shit contract like that you will ruin yourself in the industry! Malcolm then explained to me what I should have instinctively guessed – there were considerable reasons for taking this contract. Even at 1/2 normal it was a very good income in the American scheme of incomes at that time.

First and foremost was the fact that Morton Thiokol Inc. (MTI) contracts were the only contracts in the United States at that moment and would be the only ones until the middle of January 1987 at the earliest. Secondly the contracts were funded essentially by order of the Presidents' Commission investigating the Space Shuttle Challenger Accident via the U.S. Congress. Therefore indeed the contracts were funded (unusual after the Gramm-Rudman law). Furthermore, Malcolm agreed that the money per hour was abominable, however he continued to explain that this contract had 'unlimited overtime' at the usual 'time and a half' and his boss, Mr. Kent McKinnon had authorized the unlimited use of overtime as needed for the MTI Solid Rocket Motor System Safety group. The SRM's are the Solid Rocket Motors for the Space Transportation System (STS or Space Shuttle System). It was one of the SRM's which exploded causing the Challenger Disaster and loss of seven lives.

Simple arithmetic demonstrated this contract was immediately worth enough. Additionally it sounded like fun. Why should a lot of overtime be a deterrent when you are an ocean person beached in Utah ice and snow in the winter?

I had spent considerable time at high dollar hourly contracts by that time. Usually the more per hour, the less overtime aerospace companies permit. It you think time and a half for overtime is outrageous at that income consider a lot of companies bill the Government at overhead rates of 200% and 500% of what I was paid. President D. Eisenhower had been correct in fearing the U.S. Industrial Military Complex. The Industrial Military Complex was financially carnivorous!

Malcolm said the MTI personnel were pleasant and that local rentals were reasonable in Brigham City and the surrounding area. OK, Malcolm you have convinced me, and my IRS bill demands me! What do I do to get a contract there at MTI? And so Malcolm innocently set into motion events that would irrevocably change my life and lives of many others.

Malcolm told me he would immediately tell his MTI boss, Kent McKinnon, about me and he would also call Gene Pusey immediately. Mr. Gene Pusey was the manager of ARC Inc. which was an Engineering

Contract Firm in Phoenix, Arizona. Malcolm told me ARC Inc. was apparently the only contract house with a conduit into MTI. Very peculiar I thought? A sole source contract house? I would have to use ARC rather than our regular contract houses. Malcolm was not impressed with ARC for reasons to unfold later, but ARC paid their bills so far as Malcolm knew. After a brief time, either Gene Pusey called me or I called him and we talked briefly.

He said that as soon as he received my resume from either myself or Malcolm he would set up a telephone interview between me and MTI (standard procedure then). Due to pressures from the President's "Rogers Commission" on the Shuttle Challenger Disaster Gene Pusey had an urgent need to fill a number of open slots ASAP. Experts who could make the Solid Rocket Motors safer and more reliable – ASAP.

A few days passed and I talked with Gene Pusey again and he was in receipt of my resume and said he was very impressed. Gene told me that Malcolm and I were both highly overqualified, but he understood we needed contracts and work. Gene immediately set up a telephone interview with MTI. I was supposed to be telephone interviewed by Kent McKinnon, Manager of the MTI System Safety Group regarding the Solid Rocket Motors on the STS.

An hour or two later the call came from MTI. It was not Kent McKinnon, it was his 'right arm', Mr. Kerry Sanofsky who called and interviewed me. Such changes always generate paranoia and worry; is this guy smart or dumb as a pet rock? Can this guy make any decisions? What if we don't like each other? Is my background a threat to his position? Standard interview paranoia! Why worry, I closed effectively on 99% of all interviews I have and why should this be any different? And so, my interview was on...Kerry asked me if I was a good aerospace person and of course I told him the usual – I walked on the water. In three minutes it was settled. I was to be hired, and to come to the project immediately.

The interview was brief, polite, honest, and effective for all concerned. Payroll would be via ARC and Gene Pusey in Phoenix, Arizona, but I was hired by Kerry Sanofsky at MTI directed by MTI, and 100% responsible to MTI, and MTI would hold my Security Clearance Open. Closing out any Security Clearance was / is death to people like me. Kerry Sanofsky of course interviewed me on Kent McKinnon's orders. Kerry and his boss had quickly filled one more prime and necessary slot, and I had an income assured from what was certainly on the very last funded contracts in late 1986. Perhaps I would even be able to pay the last $13,000.00 of my overdue IRS bill before the pitiful game of IRS Hardball War began. I immediately began to pack and plan my trip from Seattle to Brigham City, Utah (wherever that was?).

I decided to just take what would fit into my new Toyota Tercell station wagon. Basically take clothes, and little else. I would have a friend, Miss Janet Brewer keep an eye on my place and keep track of things for me and do my book keeping. In 24 hours I was ready to go and departed. Had to rush, the IRS was depending on me to support America!

As I was leaving the Seattle area I was introspective and frustrated by my career. Was I a success or a failure? If I worked steady I could pull down over $350,000.00 per year which was huge at that time in history. While I was steady as a rock, the stability of the aerospace and high technology industries were erratic. What could be done to stabilize the industry or me? Could an individual have any affect? Not likely! Individuals were more disposable than influential.

The great promises of the sixties had washed out! When I was in high school in the 1950's, and universities in the 1960's there was a non-written contractual promises that 'for students who academically prevailed in the tough aerospace disciplines (Physics, 'Engineering, Mathematics, Chemistry) there would be no end to good jobs, and a bright, interesting, and dynamic future. I was naive enough to buy into the erroneous promise hook line and sinker. Perhaps it was a true promise then, but times had changed and the changes had been accelerating.

While on route from Seattle to Richland, Washington, I clearly remember pondering all of my great successes and failures. Most of the U.S. aerospace companies had short fuses toward their engineers, scientists, and managers like they were experimental animals – 'experiment is over, we don't need them anymore, euthanize them!'. Since the early 1970'S, just after returning from the Vietnam War I noticed that aerospace corporate Vice Presidents started turning over more frequently. Over the years the attrition rate accelerated to the point it is currently difficult to call aerospace corporate offices twice and find the same person at any desk or phone after a few months have passed.

Also observable to the most mundane observer was the reality that aerospace retirement is frequently prevented by most of the corporations "dispatching" their trusted and dedicated employees at about their eighteenth year of employment. Law Suits over this practice have continually raged since 1968 when pre-retirement terminations became standard practice. The American aerospace industry was massive and really did not truly care much about their employees welfare. The aerospace industry worries about projects, and contracts, and profits, not personnel. Well educated employees had never before in American History been so disposable. Aerospace jobs generally pay well, but the long term benefits are spotty, questionable and elusive. Most dangerously they depend on government stability, an elusive and unreliable commodity. Other observations I contemplated were decisive degradations of industry standards. When I entered the Aerospace Industry in the mid- sixties there were typical political problems and wars, however the government and corporations intents seemed relatively pure - "let's do the best job we can, as fast as we can, and as cheap as we can". Then they would move on to the next project. The decay of industry standards I and at least a million others observed up to that time had catastrophically accelerated. Causes of the decay of industry standards are not difficult to sort out. The causes are simply the lack of clear and directed national space and technology programs – no strategic vision. Also the uncertainties about which programs are viable. Should we have B1-B bombers, MX Missiles, death rays? Nano - dollar stupidity over the last 30 years is a new world record in waste and fraud, etc.... The need for the B1-B bomber was simple and academic. The raging stupidities were the political reasons driving Congressional and Senatorial decisions and expenditures. After wasting years to decide to build the B1-B bombers for billions, then to save money it was decided to not have de-icing boots on the wings. A brilliant decision since the aircraft were to be flown into Russia at low altitudes, subsonic over Siberia! How could the wings possibly ice up? One more powerful weapon system castrated by U.S. Congressional morons! Also simultaneously the U.S.A. National educational system had been in severe decline to the point it had plummeted to third world standards. Our world was changing.

The U.S. Congress has played perpetual games with military organizations in search of votes. Confused and frightened military organizations competing for their share of the national budget pie were driven to warring with each other and the U.S. House and Senate! Then of course the House and Senate antagonized the military, which in turn abused the aerospace contractors. Up until approximately 1985, when government contracts were awarded the money was immediately forthcoming, i.e. the U.S. Government's word was good. Then the U.S. economy declined (recession after recession) the U.S. Government started awarding contracts with start dates and then delaying payments. Aerospace contractors finally revolted. The U.S. Government was using Contract Houses as bankers. American aerospace corporations ceased to trust the U.S. Government and came to a new position of 'no pay, no work'! As a result the integration and strength of the aerospace industry unravelled. The U.S. Government paid better for unfinished work and poor performance than for honourable, excellent and timely efforts and products. All of this was economic death to those of us at the high velocity tip of the whip. As a group, we were the best educated and best experienced and broadest spectrumed engineers, and scientists, and managers in the history of the world. We worked very hard, we cared. We were also used and abused. We simply were not in control of our lives or our destinies. Any of a thousand factors above us were more in control than ourselves, or were they in control? Perhaps all of those 'factors and agencies' were 'out of control'? Perhaps I should 'change' industries? I wondered if my aerospace career would improve? Does America want Space Dominance more than illegal drugs? What are America's priorities? The American illegal drug industry was huge, and now, 30 years later it has escalated to $100 Billion Dollars a year.

One industry offers euphoria, the other goes to the sun, the moon, and the stars!. Hell, even the U.S. Sports industry is more important than our space programs. And so I drove on over the mountain passes through the snow, the valleys, and high desert to Richland, Washington initially.

At Richland, Washington I stopped and had lunch with my friends, Carl and Angie Haller, and their children. Carl is an electrical engineer / manager who has worked his way up the totem pole in both electronics high tech and the nuclear industries. He was a Project Manager at the Hanford, Washington Nuclear site. We had a pleasant lunch and visit. He has been a dedicated friend and supporter for years. He has helped me through thick and thin. We briefly discussed our dissatisfaction with 'high tech' confusion and then I went on my way. Carl and I had always seemed to charge each other's intellectual batteries. Carl had the permanent job security I needed and lacked. I had the freedom Carl lacked and craved. Nothing is ever perfect. Later that afternoon and evening I drove on toward Utah. Always before a new contract I wonder what the personnel will be like, what the working conditions will be, what part will I be assigned to, how long will any contract go on, etc.? All the usual contract employee concerns and paranoia. That night I stayed at some small town in the middle of the high desert. The weather was very cold. The next morning was miserable, snowy and ominous. Having always been an ocean person and sun worshiper, I felt lonely and in a foreign land out there in the middle of what I conceived as the high desert. Conflict between anxiety and frustration peaked as I rolled through Twin Falls, Idaho. My car could get nearly 40 miles per gallon. It was the longest stretch of 'snow and nothing' I had ever seen. Much more of 'nothing' than I expected. Feeling like a fool, I arrived at Snowville, Utah 'on fumes' with visions of freezing to death on a Utah Highway 'out of gas'.

I was never so glad to see a gas station before. Snowville had gas and no other facilities. What the hell! Who cared what necessities were executed along a desolate road? No one would notice the 'yellow snow'. From Snowville it was a short hop to Tremonton, and then a few miles to the mysterious metropolis of Brigham City, Utah. I thought Brigham City looked like a nice, quiet, and cold place. Probably a great place to raise children and dogs. It sure was not the epitome of a west coast high tech community. No yuppymobiles (Mercedes or BMW's) in sight. I did not feel physically well and therefore I proceeded directly to acquire a motel room. Gene Pusey of ARC Inc. had suggested for me to use a Best Western motel initially. This motel was reported to give contract engineers a welcome discount. Determining the MTI plant was best accessed from the North end of Brigham City, I examined the Best Western Motel at the north end. It was clean, neat, and cheap and would do for a few days. By the time I unloaded what I needed from the car, it was sunset. From the motel I called Malcolm Phillips who was at work at MTI and would be there for some time. We made arrangements to meet for dinner at a local restaurant that evening.

Later Malcolm showed up and gave me the 'low down' on MTI as he understood them. He told me the MTI check-in procedure, and set me at ease. We had a nice visit over dinner. We decided to carpool together after the first day. However, on each person's first day, there were check in procedures, forms to fill out, safety movies to see, Security Badge photos to be taken, etc.... Also, a later 'start time' and earlier 'quitting time' on the first day. So Malcolm went on his way and I on mine. It was good to see Malcolm again. A friend in a foreign land! I had now arrived innocently at a most unexpected turning point of my life. Little did I know this was the last evening of a normal life that I would know. Now 30 years later I can confirm my life was changed forever! I can also confirm that world history was changed forever.

CHAPTER 3

MTI ORIENTATION

Time: 9 Nov. 1986

On the morning of 9 Nov. 1986 I awoke, showered, dressed respectably in a suit and tie as usual, and proceeded to breakfast near the center of town at J&B's Restaurant. I was feeling notably tired and cold. I couldn't help notice the people at the restaurant were cliquish. I didn't think much about it at the time. After all, I was just a stranger in 'their town'. The breakfast was fine. After breakfast I started toward the great rocket motor plant at the top end of the Great Salt Lake, driving slowly north through town. At the north edge of town was a sign which pointed west. It said "Morton Thiokol Inc." and had an arrow pointing to the west. So I turned west like any good 'worker bee' or 'corporate slave' should. The IRS , U.S. Tax Agency was in luck now!

It was a two lane road and much to my horror the drivers there drove like possessed maniacs. Even worse than me! My God did they roll! As you leave Brigham City you go through the metropolis of Corine at about 65 mph. Even a naive person like me figured out what drove the economy there in the high northern Utah desert. The driving force must be whatever dwells 27 miles down this frozen road! And sure as hell it turned out to be a huge Solid Rocket Motor plant - the Largest Solid Rocket Motor plant in the world.

Half way there was a hot spring gushing steam and vapour into the cold morning air and creating a mystic veil around the spring. Simultaneously the drivers were awakening. Speeds were rising, bumper to bumper space closing! A substantial number of drivers took to passing by driving in the 'oncoming' lane. I told myself 'these people are crazy'! For the last half of the drive you can see structures around the hills that appear to be rocket test areas. At about 26 miles into the snow-covered oblivion of desert you round a bend with hills and there is the Wasach Morton Thiokol Inc. plant. The 'worker bees' drove behind the main buildings to a huge parking lot capable of parking the 6,000 to 8,000 employees who work there. Since it was my first day I was not yet officially a 'worker bee' so I drove up to the front door and parked right next to the rockets and rocket components exhibited in 'Rocket Gardens' in front of the MTI main office buildings and parked in visitor parking. Exiting my car, I looked out over the wasteland and told myself 'what an excellent place for a rocket motor plant'! If this plant blew up, there probably wouldn't even be any killed rabbits or snakes! Couldn't be any rabbits or snakes dumb enough to live in this cold or desolation!

The Great Salt Lake was clearly visible to the south, the Wasach Mountains to the East, hills to the west and north. I turned my back on the desolation and walked through the front door, into an unexpectedly new and dangerous life!

The Morton Thiokol Inc. Reception area was clean and neat. It was even yuppyish, like a California company might be. A few well dressed persons were standing around. I went to the front desk and was greeted by one of two pleasant older women who looked so much alike I honestly thought they were sisters. I told them I was Steven Agee, a Contract Engineer / Scientist reporting to work in their Solid Rocket Motor System Department. A secretary told me they would notify Sharla Rose , would I please wait? I complied.

In about 15 minutes, Sharla Rose came to the lobby, called off my name along with three others and introduced herself to us four. Sharla was a lovely lady, blond, thin, cheerful and clearly quite comfortable with herself and her position. She told us that we were required to see some plant safety movies before anything else. So, off we went. Out the back door of the main lobby, around to the northeast to a trailer filled with offices. We went inside telling the first available person the reason we were there. She led us into a small room to watch a video plant safety movie. I was fascinated by the movie. It told about the flashing lights on buildings, and what the lights meant to anyone in the area. Also about the smoking rules around the rocket propellant and rocket motors. The subject of 'Not Smoking' was thoroughly covered. Light match for cigarette. Boom. No Rocket Plant! Some of the areas required removal of street clothes and required employees to wear 'low static' clothes and shoes. Other areas had strict requirements about turning over cigarettes and lighters and matches when entering an area. It all fascinated me and I revelled in the hassles to 'smokers'. I was pleased smoking was diminishing in the United States. Anything smells better than cigarettes to me. I'm not really concerned about smoker's cancers; it's the smell I object to. Morton Thiokol Inc. of course was concerned about smoking induced explosions. Come to think of it, so was I!

The movie was of industrial importance and I believe a good thing for every employee at any rocket motor / propellant plant to see and ponder. Oh yes, the movie also told of the powerful industrial X-Ray machines at the

MTI Wasach plant. A mistake around these could remove you from the gene pool or as they say "shrink your nads"! Another major aspect of the movie was telling about the explosion / evaporation of the Solid Rocket Motor Casting Building a year before. Twenty two million up in smoke and a narrow escape for 15 very lucky workers. Parts of that plant may be on the moon! That explosion was just one of the many in a long Thiokol history of industrial explosions and employee killings. The most recent killings were seven astronauts. Soon there would be five more at one of Morton Thiokol's MX Missile Rocket Propellant Casting Plants.

Back to Reception. Sharla Rose met us and we were off to fill out a stack of documents and forms. Then we all went down a long hallway for mug shots for preparation of our MTI Plant Security, DOD (Department of Defence) Security, Identification and access badges. While at the badge room, Malcolm walked by and friendly greetings went around. After badging, I can't remember the exact sequence of events. Somewhere along the way Sharla escorted each of us to our office areas. We had to be escorted for a few days prior to our DOD Security Clearances being 'transferred and received' in to Thiokol from our last 'Classified Venue'. When my turn to be escorted came we went down and around an incredibly long corridor, and then out into a trailer complex. En route Sharla told me that MTI would have my Security Clearance in hand very soon. I told Sharla I appreciated that. Under no circumstance could a contractor afford to let their Clearance lapse - much too difficult to re- establish. She knew I would not have agreed to come to MTI if they wouldn't pick up and transfer my Security Clearance. After the Walker Family of the United States Navy spies were rounded up, the requirements for DOD Security Clearances had escalated. In addition to escalated requirements if you ever were off of a DOD Project for a short time your clearance was automatically terminated at that time. Once lost or relinquished it could take an act of God to get it back.

The DIA (Defence Intelligence Agency) and FBI apparently used the Walker Case as a hedge against Gramm-Rudman Government cut backs. Sharla told me there weren't desks for anybody new. Desks and equipment were on a "Steal your own" basis. Since I was here early mine was an easy 'acquisition by possession and occupation', however it had to be defended daily from 'unscrupulous' new contractors wanting to sit down. I asked about a phone and was told we would all have to share for a while. Ultimately I came to learn that about 15 people would share the same two phones for months. U.S. Government contracts standard 'boiler plate' clauses always require workers to be properly supported with desks, chairs, phones, etc.... It appeared to me that Morton Thiokol was trying to save a few bucks there. Much later I would come to understand it was a policy of "screw the contractors & employees and pocket the money."

In a bit, Malcolm arrived to escort me to lunch. There sure as hell wouldn't be any vendor lunches away from the plant on this contract. It was about 22 miles to the nearest restaurant in Corine. The plant cafeteria was at the farthest possible point from our trailer offices. Down what seemed to be a mile of hallways, down flights of stairs, we came to the main offices' communal cafeteria. It was clean, neat and had an entire wall of windows which I would come to cherish in the winter months to come. The food was reasonable and acceptable, but never glamorous. After lunch, we returned to our trailer office and Malcolm told me he was having a difficult time finding pertinent documents at the plant. He had adopted a 'copy or steal it' policy due to the rarity of pertinent solid rocket motor (SRM) documents availability. Malcolm had a one week head start on me and Malcolm admonished me that we should work together to acquire a proper data base for our work. Sounded right to me.

Our task was to develop a System Safely Hazard Analysis on absolutely any existing or potential hazards we could find on the SRM's or their integration into the Space Shuttle (STS, Space Transportation System) as directed by the NASA 5399.4 Document. Since the Space Shuttle Challenger had blown up, killing its seven member crew, and destroying a 4.5 billion dollar Space Shuttle System and a 200 + million dollar payload, we expected there would be some identifiable hazards. We both discussed the importance and urgency of our work. It must be treated with the moral respect it commanded. Numerous lives at stake, billions of dollars of equipment at stake, national defence was at stake. Hell, our careers were at stake!

After all, Thiokol was a 'sole source' contractor for the SRM boosters, and it would take years to develop another source even if NASA wanted to do so. We both knew that nearly all of America's 'space eggs' would have to ride this 'basket' into space. If MTI's rocket boosters would not safely propel the Space Shuttle into space, then the U.S. was out of the space race until adequate rocket boosters would be subsequently developed. Subsequent development would be five years minimum. So far 25 launches and one catastrophe. Also the number of launches was massively behind schedule. At that time Malcolm and I had no doubt we would make intelligent and professional contributions. So it was with optimism and enthusiasm that Malcolm and I tore into our new tasks. Malcolm proceeded to give me the 25 cent tour of the great rocket plant offices and ancillary buildings. There was every imaginable department: rocket nozzle engineering, contract compliance, accounting, drafting, computer support, Project Management, Configuration Management, and many others. The tour was interesting and we covered a lot of ground fast. Ostensibly it looked like a well set up operation. One of our primary attentions was on possible sources of data and support material for flight SRM's. I asked Malcolm where the copies of the contract were kept – he didn't know. No problem, we would find a NASA/MTI contract copy soon and study it. All government and space projects are to be accomplished in accordance with the contract – that is fundamental. An important note here is that personal computers were just becoming available at a very high price at that point in history. So any computing we did at the MTI plant was on a very restricted basis on their mainframe computer. No personal security whatsoever and MTI really did not want additional users on their computers.

Sometime in the late afternoon, I was introduced to Mr. Kerry Sanofsky, the young man who had interviewed and hired me via a telephone interview. He was a fine looking young man, about 31 years old, emotionally mature for his age, very polite and clearly supportive of his (and my) boss, Kent McKinnon. There were cursory social amenities and Kerry apologized for the lack of desks, equipment, telephones and such. He said a lot of equipment was ordered but it took six months to get desks and chairs and such into Thiokol. I thought that was curious. Kerry told us we would meet Kent McKinnon as soon as Kent's schedule of meetings slowed down. Malcolm and I told Kerry 'no problem' we knew what to do and were not dependent on supervision anyway. Kerry Sanofsky was not an Engineer or Scientist, he was a liberal arts paper pusher so our knowledge vastly exceeded his on all rocket matters at hand. Malcolm and I were more used to directing than to being directed. Kerry then went about his business and we went about ours. We extended our fingers to feel the pulse of the Space Shuttle Solid Rocket Booster project.

Leaving the plant that first afternoon I noticed the cold air at that altitude exhausted me. I left at a much earlier hour than Malcolm. There is an 'unwritten law' in consulting (contracting) that says you do not know the job and the location of things well enough on the first day to 'dip into the till' for overtime yet – wait until tomorrow! Besides I needed to find an apartment. On the trip back to Brigham City the drivers were even more crazed. Now they were both 'tired, pissed off, and crazed'. I felt very weary and short of breath.

Both lanes of the two lane roads were used by 'mad men and women' to speed one direction – into town. God help any poor soul going against the traffic. Clearly the local constabulary wouldn't risk their lives on that 27 miles of frozen madness. Back in town I tried to locate a local newspaper. I was shocked to find out it was published once a week, if the publisher was in the mood at all. This week the paper was not published, so I could try next week! This did not drive me to drink, however there were the inevitable explicative's. So ended my first day of my odyssey – innocently and innocuously enough.

CHAPTER 4

SURVEY OF THE SOLID ROCKET MOTORS

Time: 10 Nov. 1986 - Early Dec. 1986

On 'day two' of my Morton Thiokol Inc,. Contract, Malcolm and I car-pooled. We always discussed everything from soup to nuts. Never a dull moment until months later when a creeping exhaustion settled over us. Being my second day, we were off very early in full anticipation of maximum overtime and massive work to do. Malcolm and I never abused overtime. We worked our tails off. We literally did the work of three or four ordinary Engineers each. We were more dedicated than most as we believed in human exploration and expansion into space.

Beginning on day two of my contract I fell into a 91 hour per week routine. We worked seven days a week and the only break we took was when we worked a short day on Sunday – only 10 hours usually. We really had no time to explore the metropolis of Brigham City. Not only did we push our work, we were dedicated to it, and enjoyed it. Our work would affect the entire future of America's and Earth's space exploration efforts. The Space Transportation System (STS) had to fly again soon since it was the cornerstone of America's space program - sooner the better. We loved being part of it and we knew we were the right persons at the right place and time. We were technically a perfect match. Egotistical and arrogant as it sounds, few engineers, scientists, or managers who ever lived had known as much as Malcolm and / or I. How could this be? Simple. A great deal of what we knew was new to our generation and we were the ones who were at the apex of this 'new knowledge'. We were the implements of the new knowledge and sophisticated methods necessary to manage new technologies. Furthermore, we were most dedicated, and had extraordinary breadth and depth of experience. Malcolm's Australian and my American educations and experiences were both at the climax of highest standards – just prior to grade inflation and decline of academic standards in both nations. We were both teachers and students to the newer generation of scientists and engineers. We did not know at that time it would be generations before our standards would be matched again, if ever! Now, 30 years later standards in both nations are in 'free fall'.

Before analyzing any problem one must put the proper foundation under themselves. So, I went to Peggy and began my long quest for pertinent information, data, documents, and Thiokol's SRM technical history. Peggy was a documents acquisition person who was fifty - ish and always enthusiastic and helpful.

Documents I acquired immediately were what are called base line documents; items such as the MTI /NASA contract manuals, acronym dictionaries, primary overview blue prints, flight test data, flight records, and literally hundreds of other types of documents. Just zeroing in on the fine points of how the Solid Rocket Motors for the Space Shuttle functioned in detail required the reading of thousands of pages of diffuse information and data. This reading was comprehensive of mechanical engineering aspects, chemical engineering aspects, physics, rocket propellant chemistries, electronic grounding and electrostatic charge dissipations, lightening effects, shipping parameters, ageing degradations, storage studies, etc., etc.....

It is impossible to estimate to you the readers how many pages I read in the first two weeks at MTI. An overview of the entire SRM booster system had to be acquired with discrete knowledge, narrowing down more and more to very specific interactions of each component in the SRM booster system. Then the interaction of the SRM booster to the External Tank (ET) which is made by Martin Marietta, and then finally the interaction of the SRM to the orbiter systems, and the ground control systems. I located and sized up Morton Thiokol's technical library. It was a 1952 library and I wondered where the new knowledge was? It was then 1986! Hummmmm?????? The old expression 'light'em and rid'em' wasn't really germane anymore.

Here is a drawing of the Space Shuttle (STS) to familiarize you with the various Components. [2]

Appendices 2a, 2b, 2c: Other drawings of the Space Shuttle

Space Shuttle Launch Configuration

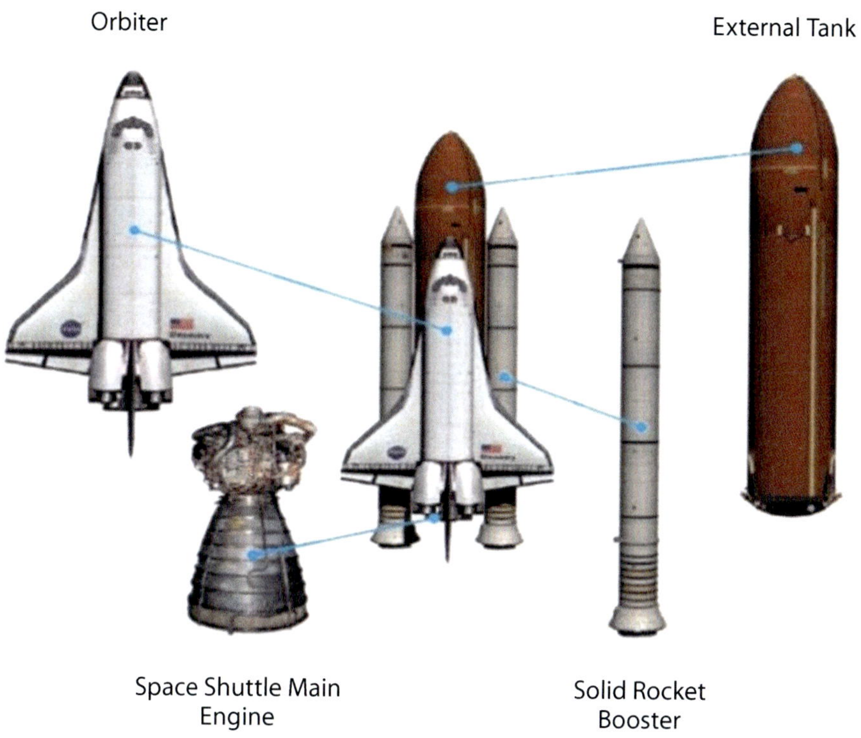

To tell the story I need to explain the various components of the STS. The STS was developed under the auspices of the United States National Aeronautics and Space Agency to be America's and the Free World's most inexpensive and reusable means to and from space.

The STS was the first rocket vehicle designed, developed, and tested by America not flown unmanned prior to final testing with a human being on board. The first time the Space Shuttle flew its 'Primary Test Flight' was enormously dangerous. It carried several astronauts.

The first test flight was enormously risky in every respect. A new system, new concept, new equipment, new software, new contractors, the only systems on board which had been properly flight tested were the crew members. The original crew members were some of the best trained and tested persons in the world. Their performance had been impeccable all of their lives.

We had to understand comprehensively what it took to develop, test, qualify (certify), store, assemble, ship, launch, operate, recover, and refurbish a set of solid rocket motors. Note that the multiple refurbishments were to be done on each one 19 times. Refurbishability of the SRM's was where NASA had planned to save the free world massive costs. We had to evaluate the effects of various conditions: humidity, temperature, acoustical vibrations, aerodynamic forces, rocket propellant thermodynamics, railroad shipment vibrations, variable G-Forces, etc., etc... Propellant for both left and right SRM's had to be accommodated for equal thrust. You get the idea! The crew were brave and trusting souls indeed. We weren't sure about the hardware. We assumed Morton Thiokol knew what they were doing, but our job was to make certain Thiokol hadn't missed anything and even the best companies and people can miss a few things. We had no prejudices and we were absolutely objective and neutral.

Now a description and explanation of the STS System. [3]

The STS was contracted to be reusable 19 times for a total of 20 flights. Originally proponents of the STS claimed and contracted for a launch cost of approximately USD$110,000,000.00 per flight carrying a pay load of 65,000 pounds (29,000 KG) into orbit circa 1989. Also a safety factor of 1.5 at the end of the 20 launch (use) of the cases. Post Challenger Accident payloads were planned to be diminished down to approximately 45,000 lbs.. As payload into orbit decreases the cost per pound into orbit increases dramatically. The contractor team I was on computed that

---

Appendix 3a, 3b, 3c, 3d, 3e.
Explanation and description of the Space Transportation System

the Solid Rocket Boosters never had a 1.5 safety factor for even one launch. First launch was less than a safety factor of 1.0. In other words we eventually determined they were statistically, analytically, and engineering wise unsafe for a single flight!

Appendix 4 addresses more Space Shuttle flight missions and scenarios. [4 & 5]

As you can see from the table, MTI had the contract to develop and build the SRM's. The SRM's would be the first solid rockets in American history to propel humans into orbit. Also, these motors were the largest SRM's ever built in America. These SRM's were designed to function for approximately 120 seconds (two minutes) at which time they would flame down and then flame out, and would be jettisoned from the ET External Tank) and then fall into the Atlantic Ocean via parachute to be recovered and recycled. The SRM's were to be refurbished nineteen times for a total of 20 uses. This refurbishment was to be the focal point of operational cost savings. These SRM Stainless Steel rocket cases were huge and proportionally expensive. Proper functional reuse was supposed to save taxpayers and NASA billions in no time at all. What could go wrong?

The orbiter eventually jettison the E.T. (External Tank) and the orbiter proceeds upward into outer space and then orbits around the earth 'on mission'. NASA's decision to have the external tanks reusable meant they would have to be dropped into a remote ocean area or be lost forever. The STS can only orbit in what are defined as low earth orbits. When launched from the Kennedy Space Center it can not attain proper polar orbits which are the favourites for military 'spook intelligence' hardware. The military prefers polar orbits because then their equipment can observe the entire earth as our planet spins beneath the satellite.

NASA's decision to have MTI's Solid Rocket Boosters (SRB's) recoverable meant they would fall back to earth, enormous parachutes would open, and the SRB's would fall into the Atlantic Ocean. The SRB's would then be recovered, inspected, refurbished, and be available for reuse at a later time (in theory at least). As it turned out frequently the SRM's would be swinging from the parachute and some get massively bashed up on contact with ocean waves. The damage was what is called 'oil canning'. Repairs to 'oil canning' always reduce strength.

Appendix 4a, 4b, 4c, 4d, 4e, 4f, 4g, 4h: Space Shuttle Flight Missions and Scenarios
Appendix 5a, 5b, 5c: List of Principal Space Shuttle Contractors for discrete components of the NASA STS Project

All of my 'base line' reading required that we academically dissect all of the SRB components functions in discrete detail. No one told us to do this. From experience, Malcolm, I, and others knew this was the best way to do the job. We were looking for any errors in design, any faults in concept, any indiscretions conceivable which would precipitate loss of the STS Vehicle, the mission, or lives. We even examined the scientific and engineering mathematics and physics of the systems. Our findings were to be prepared and presented on *'Hazard Analysis Forms'.* Our job was to execute Hazard Analysis on the Solid Rocket Motors, (SRM's). Our marching orders were mandated in the *Nasa 5400.3 Safety And Reliability, And Quality Assurance Document* and by U.S. Congressional oversight. We were specifically directed by the U.S. Congress to report any and all findings. As I studied the thousands of engineering documents and the ten's of thousands of pages, it became clear to me that almost everything on Solid Rocket Motors would cause a catastrophic hazard if there was any malfunction. Most SRM Components were critical. In many devices catastrophic hazards only occur due to infrequent or massive Malfunctions. On the SRM's nearly any minuscule malfunction either directly or indirectly in a short period of time (seconds) would precipitate a total malfunction. As this severe criticality unfolded before my eyes, others (Malcolm Phillips, Gerald Woods, and others) were discovering (confirming) the same thing.

Mr. Gerald Woods (contract engineer from Phoenix, Arizona) was assigned to study the Solid Rocket Propellant and its relationship to the SRM System and specifically the propellant loading of SRM cases and case liners. Malcolm and I talked frequently about the critical nature of each and every component of the SRM. So did Gerald, he was a bright meticulous plodder who studied, and read, and studied ... he seemed to know a lot of propellant chemistry. Gerald found plenty wrong and asked a lot of compelling questions MTI engineers and scientists couldn't answer. Furthermore they proved to be not much interested.... He was sincerely concerned for Space Shuttle Propellent Safety issues. Gerald knew the sad state of our industry. Gerald told our boss, Kent McKinnon, that he was going 'head to head' with Morton Thiokol higher management. Kent McKinnon was very supportive of Gerald's work. Gerald meticulously documented everything.

The more we read, the more documents we knew we needed to locate. It soon became clear that learning the SRM system in and of itself was inadequate. We had to delve into the SRB (Solid Rocket Booster,) then into the ET which the SRB's attached to, then into the orbiter, then into the launch pad interfaces, and finally into the launch control systems and translation of old systems to newest. Many people are amazed at how complex the SRM's are. The forces, and extremes these SRM's had to endure was mind boggling.

The SRB's themselves are 154 feet long, 13 feet in diameter, and weigh 1,300,000 pounds at launch. SRM's are lit off by an 'Ignition propellant train' initiation device (igniter) which releases 68,000 lbs. of thrust in the first ½ second of operation. It also shoots a flame over 100 feet and lights off the entire length of rocket propellant (top to bottom). Once the SRM's are lit off they irrevocably release 2,650,000 pounds of thrust each over a 120 second period (two minutes). There are two SRM's required for each launch (or STS), therefore each launch requires the accumulative thrust from the two SRM's or about 5, 300,000 pounds of SRM thrust, plus the three main engines on the obiter (liquid propellant) which are fed from the liquid propellants in the external tank. Enormously powerful and dangerous rockets!

If any malfunctions occur in the STS which cause imbalanced ignition, or non simultaneous ignition between the left and right SRM, the entire system will break up and blow up on the launch pad. If one SRM does not ignite properly and another does the SRM system and STS will explode on the launch pad. If the systems SRM'S burn at an uneven rate the STS will be not be steerable and will have to be destroyed by the Range Safety Officer even if Astronauts were on board at that time. If an SRM flames out at a substantially different rate from the other SRM the STS will not be controllable. It is absolutely imperative that the SRM's work perfectly, simultaneously, and homogeneously. In order to assure an even burn rate and an even flame, the SRM segments are cast in left and right matched segment pairs. Every effort needs to be made to make certain that the consistency and all other aspects of the SRM cast pair of segments are as identical as possible. No unmatched pairs allowed.

Solid Rocket Motor propellant begins as a liquid mixture of rubber-like compounds, bonding agents, curing chemicals, and aluminium dust. It is poured into a large mixing vat, stirred until thoroughly mixed, and then poured (cast) into SRM Rocket Case Segments. Then the propellant is cured over time with pressure and heat. Since the propellant is not a stable chemical compound, but is a mixture of mixtures no two batches are predictably homogenous or equally dependable, or of equal explosive force. Just as I had delved into thousands of systems prior to working on the STS I was amazed how the same old laws of physics always apply. The simple fundamental truths of the universe never change. The laws of physics are the gospel of the universe. Malcolm and I probably absorbed more information in a week's time during this period than most men and women absorb during a university education. Our ability to absorb this information came from years of experience and knowing what questions to ask and knowing what principles applied. Neither of us were ever stumped for very long. Every time something looks simple there are always red flags because ultimately we found 'nothing is truly simple'.

We were driven to search for the final truths in the SRM system because we knew that lives and National Security would depend on the SRM reliability. Our future as space and high technology experts would be affected by our own work here at MTI. Hell, even the Bedouin in North Africa and the future of his progeny were depending on our work if they knew it or not. And, of course, they did not.

Even today, 30 years later very few understand the escalating pressures of human expansion into the solar system territory. Earth now supports 7 1/2 billion people and every day another 257,000 are born and around 94,000 die. Simple arithmetic, planet earth cannot support us much longer. Earth is nearly out of both room and resources.

Earth is but a spec of dust and water existing in the 'Goldilocks Zone' around our star - the Sun! The 'Goldilocks Zone' is the area where life as we know it can exist.

CHAPTER 5:

FIRST PROBLEMS

Time: Nov. 11, 1986 to Early December 1986

My third day at Morton Thiokol Inc,. (Nov. 11, 1986) was also my 44th birthday. I just aged a year. It was not a particularly remarkable birthday, but then none of mine are because I never much cared about them. I read several hundred pages on the 11th. The document stacks around my desk were growing. One stack for the obiter, one stack for the SRM propellant cases, one stack for the SRM Electrical system, one for the nozzle system – stacks, everywhere stacks!! No shelves or tables to put them on of course. So, there were labelled stacks on the floor circularly around our desks. I felt like I was doing high-tech analysis around an aboriginal campfire - "situation preposterous!"

On this day Kent McKinnon, the manager of System Safety, finally had time to meet and talk with me for a few minutes. He was about 40, a very pleasant and civil man. Polite and sincere. Helpful and professional in every way. He apologized profusely to me for his being continually absorbed in meetings, however one of the President's Commissions on the Space Shuttle Accident was having meetings in Utah and he was preparing for them. Additionally there were several other governmental commissions from Washington D.C., and NASA which were interrogating Morton Thiokol's prior and current performance on the SRM's and related systems in light of the Challenger Accident. For the most part Kent appeared to be consumed by government and Thiokol drones wasting his time. He was too consumed in meetings to do his System Safety job until after normal working hours. This has happened to everyone in our industry at one time or another. Liberal Arts degree'd or undegree'd people wasting the time of Science and Engineering degree'd people because they don't know better. Kent McKinnon was clearly a harried man. He had recently been appointed as manager of System Safety for the SRM's. During our brief conversation I forced my resume on him because I sensed he had not had time to examine mine or any others. I informed him that several of the new people including myself had extremely diverse and strong backgrounds. He was glad and relieved to hear it. He said he needed all of the help he could get! He promised to pay attention to which persons would work on which parts of the SRM system. Then he was off to more meetings at a dead run! I felt sorry for him. I could see the drones and steers were busy jumping him through corporate hoops.

Malcolm and I were just beginning to see the tip of a 'data acquisition problem'. Obtaining proper documents was tough. Even though there was a documents vault, we contract engineers had 'no access'. Subsequent time at MTI demonstrated it took an act of God to get any 'Challenger related documents or films or photographs' from the vault. There were document files, but their resources were limited, full of holes, their assets scattered throughout MTI or lost. Unknown problems lurked here. Even though I had already procured hundreds of documents it was clear thousands more were needed. Malcolm and I were astounded that the MTI in-house personnel had never researched the SRM System as we were currently. We soon would be working about 90 hours a week digging into the SRM systems. The MTI personnel worked a normal 40 hours per week and had no time for digging since they were consumed with perpetual staff meetings, elk hunting and other activities. Could it be these MTI staff System Safety personnel didn't know the discrete elements of the SRM system? Hummmm?? At their present rate of departmental consumption of staff activities they would never have time to come up to speed. In a matter of a few days we were technically ahead of them. They did not know what documents to ask for. They did not know what questions to ask. Perhaps they hadn't learned very much over the last 13 years? How long had the most crucial employees been on site? Deer and Elk hunting seemed to be the pressing issues among MTI discussions while the permanents literally left their phones off of the hooks! A startling revelation. Saving Astronauts' lives was not pre-eminent to the MTI permanents.

As is my custom I started picking up the telephone and talking to people around the plant and searching for appropriate experts and meetings to attend. Data acquisition by any means, anywhere, anytime! These TI (technical interchange) meetings and discussions shed light on related subjects and guided me to specific problems and documents I could use, and documenting search activities would soon plunge me and others into a lot of trouble with frightened Thiokol management.

Perhaps this is the right time to explain that while I was located at the Morton Thiokol Solid Rocket Plant I did not directly serve MTI. Nor did I serve ARC Contracting House which paid me. The team I was on literally answered to the U.S. Congress, and specifically the Rogers Commission appointed by the President George H. Bush to get the STS returned to space safely again. It was a somewhat unusual symbiotic relationship between all parties. What type of symbiosis?? At most Corporations System Safety is considered a pain in the ass, however it is respected as a necessary and vital corporate asset, even an insurance if you will. In good corporations, designs are based on safety considerations first and foremost. In greedy and terrible corporations System Safety is used as an excuse, is twisted, and frequently is a covering of 'warm and fuzzy' in an effort to hide grim realities and dangers. System Safety Departments are to find and list ALL POSSIBLE PROBLEMS, from most severe to most innocuous. Then typically the top 10 to 20% of the most dangerous are addressed. It is always too expensive to fix 100% of all problems discovered.

System Safety persons are the soul and conscience of Project Management, Design Engineers, contract requirements, end user safety, and contract compliance. The entire 'working system' (whatever it is) must first be safe and secondly there must be assurance that the system works reliably. Typically safety is further considered continually throughout the typical evolution of design, development, test, life cycle and end use. System Safety covers the breadth of systems technologies and all aspects of systems functions and ultimately applications. There is a spectrum of System Safety methods just as there is an analogous spectrum of devices and applications.

Immediately, as I started attending meetings or talking to people on the telephone about System Safety business it became blatantly clear to me that 'System Safety' at MTI was a dirty word. Some of the MTI personnel were polite and courteous, but made their feelings of disdain for System Safety crystal clear. MTI engineers and scientists couldn't be bothered with such mundane and unnecessary things as someone to look over their shoulder and examine their work and review it with them! Others at MTI were flat out rude, discourteous, and frankly not interested in safety. Some said they did not understand System Safety which was at least honest, though ignorant! Many of those who were disdainful were poor engineers or totally unqualified as it turned out. Some were liberal arts degree'd or non-degree'd persons. Most MTI personnel placated System Safety persons and continued to do whatever they wanted to do! As in all corporations there were some profoundly arrogant managers at MTI. Some of the arrogance was liberal arts ignorance, some ignorant opinions, some was incorrect engineers arrogance, and some of it was Mormon religious arrogance, i.e. God will take care of any and all problems.

System Safety is the department which has crucial signatory vote on acceptability of design and tests, as they affect personnel, equipment, and the mission and lives. In most corporations Systems Safety personnel come from a variety of technical analytical backgrounds. An academic mixture is typically scientific, or engineering, with a thorough knowledge of multiple disciplines, and procedures. Safety degree'd professionals are not usually the technical members of the System Safety team for the simple reason that safety professionals are not trained adequately in engineering or science to identify many insidious problems. Non - engineering System Safety personnel's primary responsibilities and strengths lie in understanding and executing the documentation inside the company and preparing the findings into final presentation form for interfacing government agencies and associated aerospace contractors (ASCONS). Many well qualified technocrats who do System Safety work are weak on System Safety "lingo" and long on engineering and technical insights. The professionals with System Safety degrees are typically short on technical education and deep insights and long on 'format, lingo and presentation formats'. Safety degree persons are usually strong in insurance 'loss control' work. So a comprehensive team is composed of both types. In our modern world, and where well run and well integrated it is System Safety that finalizes the overall safety of any project and system. On my third day at MTI I was learning from Malcolm and others that the MTI System Safety group had for years been from one to three System Safety individuals in the MTI Space Division. MTI's System Safety had been simply 'rubber stamping' anything an MTI engineer, or technician, or manager placed in front of them.

Personnel doing the System Safety analysis were simply unable to handle the volume of work previously ignored prior to our team's arrival and in most cases previous personnel were not qualified to do any or much of the analytical work legitimately required. I met those who had been doing the work formally, and their knowledge was often childishly and dangerously limited.

In the past Thiokol System Safety simply 'rubber stamped' anything they were told to rubber stamp. Often without even reading it in many cases. This situation was common knowledge. This practice vastly increased profits and vastly speeded up design and development time.

Just prior to my arrival at the MTI System Safety group Kent McKinnon was given the task by MTI management to put a proper System Safety team together which would in fact do the required analysis - or so he thought. He was first told what his task was and then he was immediately stripped of any ability to accomplish the task. Every obstacle imaginable was placed in Kent's way. Malcolm, Gerald Woods, and myself, were immediately frustrated. We and others experienced 'controlled and managed Thiokol hostilities' when we attended meetings. Immediately we felt 'across the board' resistance's from the MTI personnel and management. We knew there was a limit to how far we could push them (for their own good and for America's). With the exception of Kent McKinnon and his staff of permanent employee's which included Kerry Sanofsky, Phil Greenhalgh, Ron Larson and Lance Heller (new on board) we were 'contract lepers' at Thiokol even though we were paid much more than MTI Vice Presidents. The permanent System Safety staff pretended to be too busy to bother with us and they were not technical enough to understand us. They were consumed with internal meetings and jumping through various corporate formalities to pay any attention to us. The permanent Thiokol System Safety staff were glad we were there and told us so frequently. They had just realised they might be held accountable for seven astronauts' deaths. We contractors could be bold, and we were. We had no Thiokol careers to protect. Honesty would not hurt us as it could a permanent employee – or so we thought! As a group we were not politically correct at that time, emerging 'Political Correctness' was a communicable virus and spreading.

Simultaneously, when I heard of a scheduled meeting I would go to it and usually find no one else in attendance. I asked the others in System Safety about this and they said they were experiencing the same resistance and undisciplined meeting schedules every day. There was a uniform lack of concern and lack of discipline by MTI staff. How in hell did Thiokol ever get a SRM together? A serious concern?

Often we contractors found scheduled meetings would be cancelled by the MTI insiders. System Safety, being of no importance to other MTI departments meant there was no reason for MTI personnel to notify System Safety of meeting cancellations, or schedules for that matter.

Then on other occasions we would go to scheduled meetings and MTI personnel would simply stand up and say something like 'due to lack of interest we are not holding this meeting today', and they walked out! Clearly whatever problems there were could wait or be ignored! Other times the meetings were brief, inconclusive, and innocuous – always undisciplined. Frequently it was poor engineering or science. Malcolm, Gerald and myself were growing more disturbed. Personnel at meetings would wander in and out, interrupt, change subjects, tell jokes, and were especially delighted to tell 'deer and elk hunting tales'. We contractors joked about the occasional Mormon prayer meetings' - no fooling! Most of MTI meetings were far more social and theocratic than working. Every corporation has some loose meetings. At MTI 90%++ were 'loose' on the strong side, 'never attended' on the weak side. Beyond irresponsible and already seven Astronauts dead!

For reader insight I will tell you about a few meetings. One of my favourites was the Electrostatic Discharge meeting. The purpose of the ESD team meeting was to discuss electrostatic discharge (ESD) in the SRM system, the lightening protection system, and all important electrical grounding of the SRM. This was a very explosive subject in reality. One small spark, and approximately 1,100,000 pounds of rocket fuel can go off in one SRM, or in a railroad car carrying a SRM segment (approximately ¼ of a SRM) could explode en-route resulting in deaths from an explosion first and secondly from toxic fumes for miles downwind. Such an explosion could literally waste a small town and the toxic fumes from such an explosion could kill thousands in a large city, a possible Bhopal, India, or perhaps an Atlanta, Georgia!

It took me two or three weeks to find out if there was an ESD meeting at all - I had only heard rumours. When I verified the "ESD TEAM's" existence I couldn't find any one who could tell me 'where or when", just that it existed. So, I kindly asked Kent McKinnon. After a few days he told me to go see Jed Williams (another contract person). I found Jed in the Systems Integration Group and he said he was continually trying to get the ESD team together, but there wasn't much MTI interest! He gave me the schedule and an ESD pep talk. He was preaching to the choir. The meeting was scheduled once a week. Jed and I were the only ones there frequently.

Until the week I left MTI, the ESD meetings were a seriously dangerous bad joke. No one cared! Jed worked hard to notify every one. Jed literally begged them to come. He sent memos, he called them, he cornered them in hallways. Did MTI really know the rockets could be set off by the smallest spark or electrical charge? Were MTI personnel restricted from attending? I identified a lot of serious SRM electrical grounding and ESD problems and I wanted to talk – technical interchange (TI). I viewed the STS & SRM system to be in serious danger from ANY electrostatic discharge.

The 'Safe and Arm Team' meeting was another interesting example of irregular meetings. The 'good old boys' and the contract personnel, those who knew and infidels who didn't! At these meetings there were jokes, stories, a limited amount of disciplined discussion and a world class amount of wanderings, comings and goings. At nearly every meeting some MTI employee would say something pertinent like "NASA shipped us 300 defective NSI's (NASA Standard Initiators) and they arrived today". Then the manager running the meeting would smile, look at me, then around the table, and say "System Safety is here fellas – let's talk about this later"! And so via exclusion it took me a week to verify that 300 NSI's were truly defective, and were truly supplied by NASA. How could NASA be that lax and errant? If one initiation failed the STS would explode at launch on the pad!

NSI's are 'NASA Standard Initiators' which are spark devices that set off the SRM's. If these were used on a rocket they might or might not fire. If used on a set of rockets they could have non-identical ignition. Clearly System Safety shouldn't be concerned about such a minor problem as exploding an SRM. The explosion would destroy the STS, its launch facility, and take more lives. Such actions gave new meaning to 'corporate sheep'. NASA had erred royally. Thiokol caught NASA's error that time, but MTI was covering up for NASA rather than developing safeguards against ultimate errors and failures. I was very pissed off about the MTI secrecy and hiding facts. My feelings about NASA were on 'alert'. Thiokol and we contractors were paid to surface and fix problems.

Many of us in System Safety wanted to watch testing, look over new data, and correlate it with old data, but it was made inordinately tedious beyond description. Tests were not run on accurate schedules, Project Offices were literally not in touch with Test Offices. Test data was generally not librated – it was in someone's bottom drawer and they forgot they put it there. Much of it was locked up. Only rarely would they willingly share a look at it. The only well run meetings I ever attended at MTI were Kent McKinnon's. He was professional and realistic. He was not a technocrat, he was a System Safety integrator. He scheduled, he was organized, he followed a disciplined agenda, he kept his team informed, and best of all he gave total support to all of us at all times and in every way he was allowed to. Kent was a team player and he well understood we were all miles ahead of him technically - as it was supposed to be. He was in no way egotistical.

The first problems I had at Thiokol can be summarized as causing a sense of fear and suppression. Both Thiokol and Mormon exclusion were 'wearing'. We non-Mormon Contract Engineers and Scientists were rapidly dismissed by most Thiokol personnel as 'to be tolerated only'! In no way did I go to Utah with prejudice or preconceived ideas about Mormons. The numerous Mormons I had met and worked with over the years were wonderful family people, super workers, and team players – patriots!

What I immediately ran into at MTI was something different and quite unexpected. We outsiders were mostly surprised by our lack of acceptance. Personally I felt I was in the presence of either an evil conspiracy or evil negligence. I could not then pinpoint the source of the problems. We contractors wanted to do a good job, but we felt unaccepted, uncomfortable, and therefore insecure. The last time I had felt like this was when my parents sent me off to Catholic School at age 5 and I wasn't a Catholic. The Nuns beat the shit out of me and the other non-Catholic students for that miserable and wasted two year period of my life.

Mormon doctrine elevates its believers. Each of the elevated regards himself as in the world, but not of the world. The insidious consequence of this belief is that whatever escaped doctrinal mention is discounted – infidels, scruples, engineering facts. The disconnect is not only theoretical, but descriptive, prescriptive, and applied. It is the problem of trying to hold mutually inconsistent feelings, at the same time. Intellectually correctable errors persist because they are doctrinally warranted. Jungian psychoanalysis recognised this trait as 'pure thinking'. Mature adulthood is forever put off, because reaching it provokes an irreversible spiritual crisis. Thus there is tremendous community pressure to never fully face anything they dismissed as 'of the world'. Imagination and creativity are made subversive. All tendencies to self-correct are stifled. This is the hope MTI floated in. There is no groupthink in the world that qualitatively matches Mormon behaviour in my experience. At that time in history the Mormon Church claimed to have approximately 8,000,000 members and was in most nations. Mormonism was and still is growing rapidly.

CHAPTER 6

SOLID ROCKET MOTOR TECHNOLOGY

Time: December 1986

Regressing to the first three or four days at MTI I was simply on a roll digging into documents, reading, calculating, asking questions, immersing my mind. My mind was accumulating a matrix and array of solid rocket booster knowledge. My mind was entering the SRM Systems physical relationship with the universe, the Space Shuttle Systems, and with human beings. About this time Kent McKinnon called a meeting with his entire System Safety staff.

Kent related that there were numerous frustrating problems and that he was very glad to have all of us consultants/contractors on board. Kent's core staff of three had been inadequate. Kent related that he had read all of our resumes and was overwhelmed and pleased with the quality of people he now had at his disposal. He had never before had such talent at his disposal. Maybe the Gramm-Rudman Law was a good thing? Kent said he felt he should work for us! In fact he asked us consultants why he was so lucky? We 'point blank' explained to him the reasons was that there were no other contracts in the United States since the Space Shuttle Challenger had blown up and the Gramm-Rudman Law had come into effect. As politely and discretely as possible I explained to Kent the fact that all of the contractors there would be off on new contracts that paid a lot more money (per hour) right after the new year if our Morton Thiokol base hourly income did not come up to snuff! Kent admitted he was well aware that MTI paid very poorly and that he would look into arranging us more appropriate money so he could hang on to us, but we all knew that would be tough. MTI was nationally famous for being 'Aerospace K-Mart' for poor pay. High tech persons come to the desert to hunt and fish, not to get rich. Kent understood that this Consultants nearly always go where they get paid the most or liked the projects best. I cannot speak for all of the consultants, but for Malcolm and myself the pay scale at MTI for contractors 'regular time' was disgusting, but the only activity in America at the moment. For everyone of us it was a financial 'filler' until 'NEXT' contract. Interesting, and important, and contributory yes, profitable but embarrassing to our usual incomes!

Kent intended to divide the System Safety work up amongst our department. In the next meeting or two it was agreed that Malcolm would work on the SRM Ground Support Equipment (GSE), and that I would work on the solid rocket motors nozzle systems, Gerald Woods would work on the Solid Rocket Propellant, Lance Heller, a new MTI 'permanent' hire, on the SRM case insulation and case liner materials, Kerry Sanofsky on the SRM System Safety Plan, Phil Greenhalgh on the something or other, Rakesh and Arun (the Indians) on the SRM cases and joints, Wayne Tilsey on the SRM cases, etc. etc... We all had a piece of the system. We all had to communicate and immerse in Technical Interchange (TI). All in all it was a good start.

Malcolm was very pleased that he was assigned the ground support equipment. As he realized that those who went 'head-to-head' with MTI on any aspect of flight hardware we're going to be immersed in a lot of nasty unpleasant fights and arguments. Malcolm's choice was very wise and I believe he saved himself a great deal of trouble.

Malcolm was manoeuvring far into the future. Malcolm was from Australia and had a different view of most things. At that time I didn't realize nearly all Australians have 'genetic permanent shackle mark's on their wrists.' Sometimes a joke, sometimes the truth.

All of the rest of us were destined to fight unpleasant wars related to SRM flight hardware. We were focusing on the fact the MTI hardware was generally 'crap' design and 'crap' hardware assembly with 'crap' materials. Why else would MTI management maintain 'there are no problems so fervently? NASA was interested in ground support equipment, but focused more on flight hardware. Either could cause fatalities.

In my seemingly infinite lack of wisdom I was pleased to be assigned to the SRM nozzles. I had already detected a number of problems related to the nozzles which I wished to look further into. Furthermore, I was extremely interested in the nozzles. At this time it was unknown to me if these problems were serious or not, but they should be examined. I sensed the obvious problems were not being examined. The Solid Rocket Motor nozzles (the back end of the rocket) are where all of the extremely hot gasses blast out of the rocket. Also the nozzles are moved by computerized hydraulic controls which steer the rockets. They are made of woven layers of carbon cloth and epoxies which were bonded together in such a way as to burn away slowly (ablated) and thus carry the extreme heat away from the nozzles themselves. If the nozzles break or ablate too fast there is a catastrophic explosion and / or no steering. Gerald Woods was the obvious natural candidate to work on the SRM propellant. He had been studying it intensively for three months before Malcolm and I arrived. Currently, Gerald Woods probably knew as much about the solid rocket propellant as anyone in the world and had the freshest look at the comprehensive technology. He had studied diligently and had excellent overviews. Gerald had already found a number of Catastrophic Hazards and was in the process of writing them up in 'Hazard Analysis form'.

At this time our System Safety Group had simply identified the tip of the iceberg. How big was the iceberg? The MTI permanent staff had done very little work prior to the November and Dec. 1986 arrival of contractor personnel. Over the last 14 years MTI had produced some HA's (Hazard Analysis) for NASA delivery. We all read them, and all concurred they were innocuous crap offering no insight or depth or of understanding of clear and present dangers. In fact they were 'busy work' and 'whitewash'. They were technically illiterate.

Malcolm and I sensed NASA paid best for 'no problems'. System Safety work at Morton Thiokol prior to our work was indisputably minimal or zilch. Had NASA received its money's worth?

So, with our new assignments, Kent and his permanent staff simply stood back and let us rip into our work. Kent's System Safety staff knew that we knew what we were doing. They also knew they couldn't keep up with us. They worked 40 hours per week minus elk and deer hunting discussions, we worked 90+ hours per week. They attended many meetings and were active in departmental busy work, we focused all of our time to our area of SRM technical concerns. All of the permanent System Safety staff including Kent McKinnon knew they all could learn a lot from us. The permanent staff were complaining frequently and just as bitterly about the resistance they were encountering from MTI personnel in other departments. We contractors were complaining more frequently and just as bitterly about resistance we were all encountering in the MTI corporate system. Now that we were assigned specific tasks we zeroed in on appropriate meetings and started attending them when and where possible. No simple task!

Malcolm and I spent a lot of time lovingly helping Kerry Sanofsky and Phil Greenhalgh through their technical and political problems. They were stressed and asked for and appreciated the help several times a week early in the project. We became their mentors. Malcolm and others had solicited Kent McKinnon's help to find their 'necessary' meetings. Kent and I had a quiet talk in the corner one day and he said, "Steven, the fellow who owns the nozzle team meeting is not a nice person. His name is John Sucher, and he comes from the MTI Project Management Office. He is not an engineer. He is not a project manager, he is essentially a gross encounter of the worst kind! He is a 'blatant rooster type' who will raise unholy hell to have his way and he is not driven by knowledge and facts. Kent instructed me to use caution and discretion at this meeting because System Safety would certainly not be welcome there. Kent wanted me to try and overcome John Suchers and the Project Management Offices resistance. Kent had previously been barred from attending some of the key President's Commission meetings on the Challenger SRM's by John Sucher who told Kent that he would "physically restrain" Kent from one of the meetings if Kent showed up. How's that for a good corporation safety attitude with seven astronauts dead and billions wasted?

I am hardened and experienced in corporate insanity, but this was too much even for me. I was PISSED OFF! Kent also told me that John Sucher had bought a huge amount of nozzle manufacturing materials in an unauthorized purchase which had caused a lot of manufacturing / assembly problems. Very irregular, both Kent and I thought, something System Safety should look into no doubt. How, when, what, why?

Kent requested for me to use all of my management dignity and skill to try to gain John Sucher's and MTI management's confidence so there wouldn't be war and hell to pay for attending their meetings. Kent also asked me to pass this information and concern along to the others in our group. In the next 48 hours I had quite a few quiet meetings over cups of coffee. A word here, a word there...

I was always well dressed and courteous. On the occasion of my first 'nozzle team meeting' appearance and demeanour were not pertinent! The meeting was held in a conference room dedicated to space shuttle activities. The conference room had a long table and a projection system at one end. A clean, pleasant room. As I entered the room I was introduced to a number of persons and they were polite and courteous to me. About then Mr. John Sucher entered the room. It was a gross arrival of the worst kind! As John Sucher entered the door he was screaming, yelling, and clearly had the rudest and most sullen of demeanours possible. I had no opportunity to introduce myself to him because he was screaming and yelling all through the entire nozzle team meeting. Finally he left screaming, yelling and grumbling. He was especially abusive to Saukeen who was a knowledgeable scheduling person. Saukeen was an especially pleasant lady. His management style was 'wounded chimpanzee' mentality.

Experience told me I wouldn't have to look far to find enormous problems in the nozzle system. Any product unfortunate enough to have this sort of person for a Project Manager wouldn't likely be in good shape. Rudeness and inappropriate behaviour are what that class of persons substitute for knowledge, confidence, and know how! Perhaps I had just witnessed the reason for MTI's SRM rocket nozzles being referred to as "REMBRANTS" – each a unique artistic creation! No two rocket nozzles were even close to identical. It was common knowledge at MTI. No one could learn what's going on or manage a project by bitching, and screaming. Listening is often helpful and informative - A good starting place. After returning from the meeting I told Kent about the pitiful meeting and he said it sounded like' business as usual'. Kent indicated that he had previously heard there were problems with the nozzle but he had not had the time to delve into them nor was he qualified.

This was Nov. 1986, 25 STS launches and one disastrous failure with loss of all on board. Kent was very enthusiastic to bring all problems forward and have them fixed as best possible prior to the next shuttle flight (flight 26). After having attended such a meeting as the 'nozzle team meeting' I felt extreme anxiety. I realized that probably my job was in danger by virtue of being assigned to the nozzle system. One part of me wanted to escape the obvious troubles looming ahead and another part of me wanted to go straight forward and 'tangle' with the problems and if necessary the people. I am not famous for being a wimpy person. Most of my successes have come from tearing straight into the guts of problems and obstacles. Playing games and avoiding issues causes me insomnia. I hadn't had insomnia for years! The primary reason I am a consultant is so I can face issues rather than hide from them as most castrated corporate personnel are encouraged, or more often forced to do.

Kent made gestures and motions indicating the telephones were probably tapped and that a lot of persons in our work area were 'stooles' for various MTI managers. I had sensed and wondered if there wasn't an MTI Gestapo throughout the MTI Wasach Plant. I had no proof of a pervasive Gestapo, but I sensed it from my second day on site. Was the resident Gestapo MTI or was it Theocratic? It didn't matter. Any way you slice it protective corporate Gestapo's are dangerous. Now Kent simply confirmed my suspicions and his fears. I sensed and feared an incredibly incestuous work environment and staff than I had ever encountered before in my career. My internal resolve was that I would do one hell of a job on this nozzle system and anything else they gave me regardless of who liked it or not. That's just how I have always worked, for the job more than the employer. Projects and jobs are usually important. Employees and also employers are usually only a necessity and transient encumbrance. I would have been diligent even if Kent McKinnon hadn't treated me right himself. However, I must say Kent was extremely open, enthusiastic, and of proper frame of mind to deal with all legitimate SRM problems. I was beginning to get a glimmer of an image that

Kent was one of the very few people at MTI who had a proper attitude and bearing towards solving SRM problems prior to the next launch. He was MTI's "Lone Stranger"!

Kent was the lone prophetic MTI 'Voice in the Wilderness'. Kent was in deep shit and his boat could sink. After all, when John Scucher threatened to physically restrain Kent from attending government meetings Kent found no relief or support from other MTI managers such as his boss, Don Thompson, or Russ Rasmussen, or Carver Kennedy (VP), etc. MTI was apparently a 'gorilla society'.

The morning after I attended my first nozzle team meeting Kent McKinnon pulled me aside and asked what went on in the nozzle team meeting yesterday? I told him the details. He said we have some serious problems. So I asked what they were. He said, "Well Steve, John Sucher had a screaming fit after the nozzle team meeting that I had attended. John Sucher had asked someone who I was and where I was from and when he heard I was from System Safety he went 'berserk'. John Sucher and others from MTI's Project Management Office caught Kent in the hallway a few minutes ago and told Kent they did not want any System Safety people at his meetings ever again PERIOD! I asked Kent if I should continue to attend the meetings and he said SURE! Have your wits about you because they are clearly trying to keep us from the nozzle problems. Kent McKinnon was embarrassed by Sucher's activities. He was also worried about what other repercussions he would have to endure from any / all directions. We both sensed trouble was brewing and crimes were underway and many lives were in danger. I could only have great respect for Kent at this time. More research, more reading, more blueprints, more calculations, more discussions. Finally, after a couple of weeks of research I had identified some serious unresolved problems with SRM nozzles. My first macro view of the nozzle problems was that all of the nozzles were different from each other and they were not built to one standard configuration, dimensions, weight, or strength. They really were 'flawed *Rembrandts*' so to speak. Secondly, the nozzles were of completely different designs. There were test motors, developmental motors, flight motors, and various other hermaphrodites at multi - millions each. Toys for the boys at MTI. So, one fateful day I asked around to find out if MTI had engineering drawing trees for the SRM nozzles or a central location at MTI for engineering drawing trees and 'interface control documents'.

Malcolm and others told me there were 'drawing trees' here and there but they 'were not to be trusted!' Technical components drawing trees are like family trees. They show what set of drawings go together to make a subunit and what subunits make a larger unit, and what larger units make up the whole or vice versa depending on the system preferred by a company – from the bottom up or the top down. These also show the evolution of engineering thinking. Malcolm and others indicated to me that when an engineering drawing is marked 'final' at MTI it truly did not mean final. Huge numbers of drawings of discreet nozzle components were marked 'final'. Peculiar thing – nearly all of the final drawings were different! How had NASA allowed this to happen? Was NASA asleep or possibly Elk Hunting? The natural processes of 'Engineering Change Paper' had obviously run amuck at both MTI and at NASA.

At all other companies I have worked for final truly means "FINAL" – the final, accepted, 'cast in concrete' drawing to be used for fabrication and lives trusted to it. If in time it is superseded it is marked boldly as superseded. At any rate no one seemed to know where to find nozzle drawing trees and drawings, so I discussed the problem with Kent McKinnon who directed me to talk with Jerry Fonnesbeck. I gathered up my note pad and went off to the nozzle engineering area and located Jerry Fonnesbeck.

Jerry was a 35ish type engineer who had been working on nozzles for some time. Jerry was having a meeting at his desk in an open office area when I arrived. I stayed until he finished his meeting and then explained to him who I was and what I was there for. He apologized to me profusely, stating that there were bits and pieces of drawing trees. There were no direct lineage of drawing trees and no traceable evolution of nozzle design from the existing drawing trees. Helter Skelter! For the SRM nozzle such a compilation of drawing trees was on his list of things to do someday. I thought to myself, "about time, you're 13 years into the project"! I asked for copies of what Jerry had. Jerry said he would acquire the available drawings for me in a day's time. When I returned and asked the next day I was not talking to the same Jerry Fonnesbeck. Jerry had miraculously changed!

I was talking to a man who had been told to be courteous but definitely not to cooperate. Jerry was in the demeanour 'Yes, Mr. Agee', 'No Mr. 'Agee', etc ... Jerry furnished me with four sets of drawing trees which were helter skelter, chaos and confusion! I asked him when the final series of drawing trees would be in order and Jerry said he did not know, but it was on his agenda. I asked him to let me know when the final drawing trees would be available and he promised he would notify me. As I was walking from the nozzle design area back down the corridor to my desk in the trailers I was thinking to myself that this man had been counselled by John Sucher who clearly owned this man's job and perhaps his balls and his soul. I remembered the basic corporate rule –"When you have them by the balls their hearts and minds will follow". Fonnesbeck's heart and mind were following.

After returning to my desk I sat down and spent more than a day going over these drawing trees. I was shocked and appalled! Kent told me he was now getting the 'Yes sir", "No sir" treatment from Sucher, Fonnesbeck, and others as well. MTI scuttlebutt had filtered back to Kent that I had caused another substantial upheaval in MTI"s Project Management Office. Kent's boss, Don Thompson, had been in MTI's Project Management Office for years and was reported to still be effectively subordinate to it. Thompson's heart and mind also followed. So it was that Don Thompson was angry at both Kent and I both according to Kent. It was becoming Don Thompson's 'state of mind'. In the hallways Don Thompson appeared 'GLAZED' and 'CATATONIC' - 'OBLIVIOUS'!

My insights into nozzle problems were accelerating concurrent to my ability to acquire documents, drawings, and test reports. Approximately a week prior to Thanksgiving in 1986 I called Kent McKinnon aside and told him that I had newly discovered very serious problems in the nozzle that were shocking and astounding and would be fatal. I explained my findings to him briefly and he said "write them up and get them to me as soon as you can" and he would try to set up a conference with Morley Cox. Morley was an MTI manager who came up through the ranks as a rocket nozzle designer. Kent felt Morley was possibly more enlightened than most MTI managers and was certainly less likely to ignore or suppress problems than his boss Don Thompson.

Kent told me he did not want my findings suppressed by Don Thompson or anyone else. Kent wanted to grease my pending discussion with Morley Cox so that when I talked to him, he would be aware of my background and track record. Kent didn't want Morley to initially think I was a typical 'contract engineer' that Morton Thiokol usually acquired due to low pay. Kent told me that if we went to his boss the problem would be immediately suppressed and hidden from NASA and all others. At that particular moment Don Thompson worked for Morley Cox. Kent further explained that Don Thompson killed all negative findings out of System Safety dead on sight. So it was official. Don Thompson's technical expertise was as an independent U.S. IRS Tax Preparer. He had no technical training at all.

Kent was beginning to open up his terrible insecure situation to me. Soon I had another very serious discussion with Kent along the lines of "Kent, I am getting the feeling that 'your position' and the MTI System Safety Department are untenable"? Kent replied that his position was 'untenable'. I said well, if it is untenable we need to sit down and plot some plans for what we can all do to go around the Morton Thiokol and NASA obstacles and to see what can be done to bring problems to light and solutions to bear. Kent said he would be pleased to have any and all the help he could get since he was not getting any support from any of his MTI superiors or peers. None of them were willing to deal with the emerging SRM problems and guilt from the 7 deaths. In fact Kent was receiving all kinds of flack from the entire echelon of persons above him at MTI. His activities were actually suppressed by Don Thompson, Russ Rasmussen, John Elwell (V.P.), Ed Dorsey, John Therkill, John Sucher, Al McKinsey and numerous others.

Kent and I agreed that in a day or two we would have further discussions on how to 'un-suppress' SRM problems so we could save some Astronauts lives and return America to Space Operations soonest. Kent nodded at the phone and held his finger to his lips as if to say shhhhhhhhh! Kent was clearly frightened and fearful of the MTI phones, Management, and Security. Kent's actions were not casual or joking – he was worried and informed of MTI eavesdropping. I sensed I should also be cautious. This project was not a matter of a job, it was the Global Space Race, High Ground race to National and Solar System dominance. Most aerospace and high tech companies where Federal Security Clearances are required have either sporadic or total eavesdropping. Frequently it is done by disabled persons such as those in wheel chairs from remote locations. The sensitivity of the U.S. Government eavesdropping is frequently related to the level of security required - Secret, Top Secret, TS SCI, others, depending. However many U.S. Corporations maintain separate eavesdropping operations just to cover their internal secrets, legitimate or otherwise.

I committed Malcolm and myself to help Kent as much as possible. I shouldn't have done that! It was very presumptuous and dangerous of me! Kent told me he certainly recognized Malcolm's rocket experience, however, he wanted to be very selective of whose help he received. Kent said it was nearly impossible to keep any secrets inside MTI!. MTI had a 'corporate security system' and a 'religious security system'. The religious security system was the more dangerous. These worked together synergistically. I asked Kent if that was because of the isolation of the MTI Wasach Plant or due to the fact that most of the employees were Mormons? Kent (a Mormon) replied "all of the above and more!" Kent was a practicing Mormon and had done missionary work in Australia. He was practical, pragmatic, and intensely honest. He was a bit inflexible due to his wife being overwhelmingly tied to the Mormon Church. He told me a fascinating story about MTI's Wasach Plant. He said there is an imaginary line somewhere between Brigham City and the rocket plant. On the Brigham City side of the line (mostly Mormons) were fine persons who loved family, God and Country. He said once they crossed the line to the MTI plant side they smoked, conservatives cussed, pleasant pillars of local communities became corporate tyrants. Kent could not explain this sociological anomaly of 'crossing the line'.

Later I came to see his observations were indeed valid. We then ended our meeting and went about our separate business. Two lonely men in a quandary and a dangerous space. How could we best get America's Space Shuttle safely flying again?? If we could not prevail, then who would?

As I went about my business I felt sad for Kent McKinnon – the 'designated MTI sacrificial lamb'. I envisioned three scenarios.

SCENARIO ONE;
Kent does as he is ordered, SRM Safety is suppressed, a SRM fails, Kent hangs high, takes all of the MTI blame and probably goes to jail in disgrace – MTI and NASA walk free.

SCENARIO TWO;
Kent walks out the door and never looks back, MTI replaces him with the next 'anointed sacrificial lamb'.

SCENARIO THREE;
Kent 'tangles' with MTI management and is fired and a castrated corporate sheep ultimately replaces him. All three scenarios concluded in the devastation the American Space Program, loss of missions, further loss of equipment and lives. Loss of future aspirations for all living and yet unborn Americans.

Kent McKinnon was in very deep shit!. America was in deep shit! The free world was in deep shit! What irony! What power a few immorals can have! What moral demands can befall an unsuspecting person suddenly! Kent had the weight of the free world on his shoulders! He knew it, he felt the load. He understood the consequences and he was afraid and shaken. History was about to be made as a religiously dominated group unilaterally disarmed the United States and the free western world!

CHAPTER 7

SRM NOZZLE PROBLEMS

Time: December 1986

As was my usual practice at that time I hand-wrote a rough draft of the memo Kent asked for on my extensive list of rocket nozzle problems and sent it off for double space rough typing so I could further refine the memo later. There were some persons allowed to use the MTI computer, but not contractors! This was the unique time of historical change when the world was transitioning from hand written and then typed to computerized. Not nearly enough computer terminals to go around and of course MTI didn't want us to use them and speed up our findings and operations. I am sure nearly every contractor and scientist there could type and most would leave MTI typists for dead in accuracy and speed - however none of us had the time after doing our daily research.

This memo addressed SRM nozzle problems I had identified to date. I was astonished how rapidly I could identify the problems and then I realized it wasn't because I was any great thinker; it was because these problems were so blatant and fundamental it was shocking. After I hand wrote my memo I took it to Wendy Anderson, one of Kent's shared secretaries. I had previously noticed that Wendy and Erma (Kent's other secretary) were very close friends as well as secretaries to Kent and several other MTI managers. I gave the memo directly to Wendy for 'rough draft' and expected it back the same day or early the next day. Since it was ultimately headed to Morley Cox it should be professional quality. Morley Cox was Don Thompson's boss at that time.

Midmorning the next day Kent caught me in a hallway and said we must have a meeting! I said fine. So, he indicated that he was afraid to talk in his office (he gestured that it could be bugged) so we went into another trailer and found an unoccupied and isolated meeting room. We closed the door and sat down at the conference table and I asked him what was up and he said "Plenty"! Kent looked terrorized! I thought to myself "oh, oh – the troubles begin, or rather continue, or perhaps are over?"

Kent said that my rough draft memo for him on the nozzle problems had been intercepted by his boss Don Thompson. Wendy, Erma, and others were supervised by Mrs. Sylvia Rollins. Sylvia was a pleasant gal. Either she (Sylvia) gave the memo to Don Thompson or, Don found it – I really don't know? Sylvia claimed she was helping Wendy and Erma when Don Thompson supposedly grabbed it. Also, supposedly, Don Thompson was looking for it! How could he have known about it? MTI's grapevine was second to none. Damn; I should have typed it myself, but there was also a shortage of typewriters and how much time did I have anyway? I was already exhausted. Kent reported that Don Thompson turned red in the face and was jumping mad, and in fact was furious that I, Steven Agee, knew so much about the SRM Nozzle and its menu of very dangerous problems! (Does something smell amiss?) Don explained simply to Kent that he (Kent) didn't understand! Neither did Steven Agee! Everything relating to SRM nozzles was "under control" and that it was not Kent's concern to cause problems. It was Kent's' job to keep 'those contractors under control'! Don threatened Kent to shape up or ship out! Don told Kent that it was up to Kent if he fired Steven Agee or not, but it would certainly go better with Kent if he let Agee go today on the spot! Kent told Don that he was not going to let me go because Steven Agee was doing an excellent job. In fact Agee was doing exactly what he was hired to do and what NASA and the U.S. Congress directed him to do!

Don then ordered Kent to remove me from working on the rocket nozzle and I was to be reassigned to an innocuous part of the SRM System immediately. Kent and Don discussed what was the most innocuous part of the SRM System and it was decided that the "Safe and Arm Device" and the SRM "ignition train" were innocuous. The initiator and igniters train system would be good parts to assign Agee to since there were no known problems in such tried and true system. Twenty five STS's launched, (300 failed NASA Igniters) and Kent had no choice but to agree to my reassignment! Kent told Don he would reassign me immediately. To his credit Kent nearly had to sell his soul to preserve my job. Clearly Don Thompson had beat up on Kent severely and wrongfully. Law suit material really. At this point I should reiterate that Don Thompson was not a scientist or an engineer. He was a local Brigham City Mormon Church Bishop, and he made his living prior and during his Thiokol period as a local IRS Tax Accountant. His position of Trust and System Safety was based on his reliably paying his Mormon Church Tithe! How could anyone be bad if they paid their 10% regularly? Here were Kent and I in an isolated conference room with the door locked behind us. Kent, was telling me that by doing a good job I have gotten him a hell of lot of trouble (nearly fired) and myself almost fired. We both affirmed that *No Good Deed goes Unpunished!* I was not surprised by any of it. Over the years I have seen and heard nearly everything. By this time it was clear to me that MTI was a sick and criminal corporation and that corporate tyranny reigned paramount. Morton Thiokol Corporation was extremely dangerous to Free World Security and human life. Secretaries Wendy and Erma were both worthless and dangerous as secretaries.

I was angry about my typing being sent on up rather than returned to me for review and correction. To say I was pissed off would be a vast understatement. I was so mad I could have jumped, run down the hall and beat the shit out of Don Thompson! Seven dead and Don Thompson will not face reality and work toward safety and reliability, or towards USA Space Superiority? I asked myself how could Don Thompson or any other dumb SOB could do such a cover-up and maintain a family or sleep at night? In many ways it might have been better if I would have run down the hall and beat the shit out of him, i.e. stopped him. There would have been satisfaction. Over years of consulting I have become accustomed to every type of dimwit imaginable: educated dimwits, uneducated dimwits, and low down incorrigibles. I was angry to the core of my genes. My inner spirit cried "foul"! Then a more serious thought entered my mind - who was Don Thompson owned by other than the Mormon Church and Thiokol?

Rather than getting mad and embarrassing Kent McKinnon, I just sat there placidly and said, "Well, so what next Kent?" Kent said "First of all, you have to take a very low profile immediately". If you go to any meetings, you have to go quietly and ask very few questions and literally "fall out of sight of all MTI management". "You are clearly being watched." In fact Kent thought I should not go to any meetings prior to the Christmas shutdown and then quietly reappear after the first of the year. Another thing that Kent said must be done is that I must immediately have a "Safe and Arm" document on my desk so that when someone is looking over my shoulder they clearly see "Safe and Arm" documents. Kent assured me that both Don Thompson, upper management and MTI Security would be looking over our shoulders and through our desks. Kent explained that the MTI Security Department was not for protecting National Security as required, but was clearly enforcing internal MTI tyranny.

Don Thompson ordered Kent that only "Safe and Arm" and "ignition" documents were to be on my desk and that I was not be seen with SRM nozzle documents. Kent said the MTI management feelings were that the safe and arm device and propellant train were totally innocuous and safe systems. Kent went on- there hadn't been any talent around which had truly examined these systems. Kent felt there were no doubt plenty of problems on the propellant train and Kent *directed* me to find them if they indeed existed! Was Don Thompson telling Kent not to find problems in the "Safe and Arm Device" Kent and I had no doubt.

Kent looked most forlorn and said "Steven I apologize to you for myself, for my company and for my bosses"! He said this entire uprising has been a travesty and it could not possibly be more inappropriate than it has been. Kent said "I'm begging you to stay in my group and work on the SRM problems". Kent said, "never has there been a man more appropriately at the right place and time than you Steve".

Kent said Malcolm and these other contract people were very intelligent and very professional. Kent then said " but you alone looked out over the entire situation and saw all of the technical and corporate problems in their full intricate matrix. Kent said he had never met a man who saw so much as quickly as I did. Kent said for the sake of the SRM program, and the sake of the Space Shuttle Program he would do anything he could do to keep me on this project in any capacity. He said even though I was assigned to the safe and arm device and the ignition propellant train, he wanted me to keep a sharp eye on all of the other systems and alert him to other problems I could identify so that none would be inadvertently missed. We both remembered that NASA and Congress had ordered ALL Hazards to be reported.

Kent then asked me if I would stay at MTI or if I was just going to walk out the door? Kent said he would not be mad if I just walked out the door since both he and I knew MTI's actions were disgusting and dangerous. I told him that frankly that I and the IRS needed the money right now. I felt a deep personal commitment to America's space program. I previously had a deep commitment before I knew there were MTI corporate and hardware problems. My commitment was deeper now that I knew how dangerous MTI was to the U.S. Space Program. I felt it would be wise to just go on working at MTI until they threw me out the door or leave at a later date of my own choosing. My 9th gene so to speak spoke quietly to me,. It said *"I Am Surrounded By Treason!"* Kent then discussed MTI phones being repeatedly and randomly bugged. Basically, corporate tyranny by the MTI Security Office, Kent's bosses, and a group of MTI managers who were working around the clock to keep all Systems Safety problems and reliability problems suppressed and hidden from NASA, Presidential Commissions, news media, insurance investigators, astronauts' families and others. Kent said this was his opinion and it troubled him.

He further said there were a number of things which could happen in the very near future which could alter the current suppression and he felt my 'help and insight' would be instrumental in opening up the MTI SRM problems to NASA and Congressional visibility.

So in the end we agreed I would go low profile, keep my eyes open, and Kent and I would seldom be seen together at all since it would be bad for Kent's reputation to associate with anyone who had found anything wrong with an MTI design, process, procedure, or activity. After all, 'Don Thompson and MTI knew best!' Steven Agee needed to be treated as a leper! Like hell, MTI knew best! Morton Thiokol didn't know its ass from a hole in the ground and Morton Thiokol didn't care to do even minimal work or about the lives of Space Shuttle workers or astronauts. Not one of us had seen a copy of the NASA / MTI Contract yet, but I didn't have to see the contract to know safety suppression was a violation of any NASA or government contract. All of a sudden it hit me. The seven Challenger astronauts deaths were not by accident. Their seven deaths were *Negligent Homicides!*

This all happened in late Nov. 1986 and I was so mad that I couldn't see straight!. 'Shoot outs' in the Vietnam war didn't get me this riled up! In Vietnam I had a 'sudden cure' for my enemies and their threats. When I saw Don Thompson in the hallways, I had to suppress my Vietnam commando training! A few slow sloppy killings were in order. Additional stress to me at this time was double pneumonia. About two days before I wrote my 'nozzle memo' breathing became difficult. So I asked a local Brigham City physician, Dr. Bunderson, to order a late evening X-ray. The next day I called him and he said I had double pneumonia. I went in to see him the following day and he prescribed antibiotics. I was too busy to slow down my 15 hour days. Therefore I did not experience a startling recovery,. Cold, snow, and corporate suppression were not contributing favourably to my quick recovery either. I was lonely and worried. So much work needed to be done. The best thing that happened to me in Utah was Dr. Bunderson's bill - $12.00. I wrote him a note and congratulated him for being the last honest physician in America! Not only were his prices honest, he took time to listen, and he was very good. I salute you Dr. Bunderson! You were a friend in need at a dark hour of my life. *Respect to you Dr. Bunderson!*

CHAPTER 8

CLOSE CALL

TIME: Mid-December 1986

By now I had arranged interviews between MTI and Mahalon Hollway in Seattle via ARC in Phoenix. Soon Malcolm told me Mahalon was set to work at MTI. Both Malcolm and I informed Kent McKinnon and the System Safety staff that Mahalon would be a good man to put on test documents, procedures, and specifically rocket Gas Plume toxicology. Kent McKinnon assured us he could talk to Mahalon when he arrived. Kent said they would probably assign him to evaluate MTI's past and present test program performance. New contract engineer arrivals were coming in from all over America. There were two Indian fellows from the east coast with tremendous credentials, but their phone interviews left the erroneous impression they were weak in spoken English. Also MTI was concerned these persons couldn't be kept unless they were paid substantially more than MTI was offering. Kent talked to me about them and I said, "Well, I don't know these engineers, and I can't recommend them, however, you might tell them about your efforts to get more money after the first of the year because if you don't get more money, not only will you not likely have the Indians, you will not have any of the rest of us. Kent took my suggestion. He really appreciated us and wanted to keep us. We we're not asking for too much – we just wanted our normal going rates.

Just before the Christmas holidays, Mahalon Hollway reported into work. Mahalon was glad to be back at work since he was low on money. Malcolm and I gave Mahalon the 25 cent tour and set him in motion. Mahalon had meetings with Kent McKinnon and Kerry Sanofsky. Mahalon was directed by Kent McKinnon to begin looking over the MTI Test Documents and doing research into evaluating MTI's testing effectiveness on the SRM Systems.

Discussions of more money prompted Kent McKinnon to have meetings with Sharla Rose. Sharla and Kent sat down and mapped out their strategy to try and secure us contractors a more appropriate base income. Something that wouldn't embarrass us. It was a hot subject when your government kills everyone's general incomes via a vehicle such as the Gramm Rudman Law. We would need it in a flash if 'overtime' was curtailed. Sharla

Rose had quietly become Malcolm's and my confidant. She had recently been promoted to be in charge of all MTI contract personnel. So, at this time she had to look after 68 contract (consultants) with many more coming. Sharla Rose told us she was having wars with a MTI procurement officer, Mr. John Gardner. She said he was in bed with ARC, the contract engineering house we came in through. No other vendor would be considered. The following is Kent McKinnon's letter regarding salary increases.

MORTON THIOKOL INC.
Space Division
Interoffice Memo

11 December 1986
L322-FY87-M32

TO: R. G. Gardner - Procurement

FROM: D. K. McKinnon, Supervisor
SRM System Safety

SUBJECT: Compensation for Job Shoppers

Due to the recent down turn in the Aerospace Industry across the country, we have obtained the services of several very sharp Job Shoppers which normally work for a great deal more money than we offered them. This is fine for now, but it is my understanding that after the first of the year several large Aerospace Companies are beginning some large new contracts and all these men could easily leave us.

As I said they are very skilled and talented. I would like to suggest that you evaluate the possibility of these men being transferred from the current rate to the new rate as provided by TWR 6376, Job Description #37, System Safety Analyst. Their names are Steve Agee, Malon Hollaway, Malcom Phillips, Wayne Tilse, and Joel Henry. These men deserve full compensation for their efforts and talents.

I feel that if we do not take this action now, we will probably lose these people and lose the benefit of the knowledge they have gained in the last several months.

R. G. Gardner
11 December 1986
Page 2

The Redesign/Re-evaluation Hazard Analysis is heavily weighted toward utilizing Job Shopper talent. Any efforts on your part to help us retain these people would be greatly appreciated.

D. Kent McKinnon

D. K. McKinnon

Concurrence:

D. E. Thompson

D. Thompson, Manager
System Assurance

As you can see this letter was for the purpose of keeping good people on board. Even if Don Thompson did not want good people on board he and MTI were directed to keep good people on board by both the President's (Rogers) Commission on the Space Shuttle Challenger accident and the Rodney (NASA) Committee. Don Thompson knew that he and his bosses would eventually have to answer to these commissions, not MTI. MTI at least had to have consultants in place or lose billions a year. Appear to perform or perish.

No work product had yet progressed past our department, that was yet another matter. Meetings were then set between Sharla Rose and her bosses, and Don Thompson approved and signed off (if he liked it or not!). On at least two occasions beginning about this time and subsequently later, Don Thompson came running into the System Safety Department and told Kent or Kerry Sanofsky that MTI and NASA were on a teleconference and that NASA demanded synoptic resumes of the Systems Safety Personnel ASAP! It sounded like NASA was accosting MTI for our resumes. Sharla Rose talked with an MTI procurement officer, Mr. Jim Gardner who said that MTI wasn't going to pay the contractors any more money period! If the contractors didn't like, they could be replaced! He had a questionable 13+ year relationship with Mr. Gene Pusey, President of ARC Inc. It was between Mr. Pusey and Mr. Gardner that ARC had essentially been a de facto 'sole source' vendor for approximately 13 years I learned from several sources. This set off more bells in my mind. Sharla Rose, Malcolm Phillips and myself had quite detailed discussions on the subject of ARC Inc.. ARC was in fact criminal. Sharla said she had nothing but trouble with ARC Inc. Sharla indicated to us that the jobs open at MTI were in fact not being filled by ARC but were being filled by referral from persons like myself and Malcolm. Our referrals in turn referred other contractors into MTI. Some came to MTI and some didn't. Sharla was trying to expand contractor acquisition to more than just ARC.

ARC was not even advertising to fill the numerous open positions at that time so far as Sharla could determine. When the good contractors were headed in they should have come through a contract house of their own choice. There was currently no choice other than ARC Inc.. So, of course ARC took a slice of everyone's contract. Gross violations of US Gov. Contract Law at that time. There were approximately 68 contractors on board and escalating. That volume of contractors would cost a minimum of one quarter million dollars per week. Any slice of that would add up. ARC provided substantially less support, services, and perks to contractors than normal contract houses. ARC profits had to be accordingly escalated.

Competition and public bidding were the cornerstones of U.S. Government Procurement Regulations. For some reason, MTI out in the middle of the northern Utah desert had circumvented the general competitive spirit, regulations, and law for a long time. I personally did not know for certain that ARC and MTI had any criminal conspiracy or involvement at that time, however, the fact they were the sole source vendor on the SRM contract was illegal. Or was ARC sole source? Was there someone else who supplied "a body or two" who essentially legalized the entire slanted operation? MTI and NASA 'window dressing'?

Sharla related other ARC problems to Malcolm and I which were driving her crazy. Her problems included ARC's going to management above her head and settling the issues behind her back even though it was her assigned job task to manage all of the contract personnel. Sharla was coming to conclude she was MTI's "token woman" in this part of their operations. Again, the MTI methods of management was emerging – "this job is yours but you will do it our way or else!" Also known as a 'hidden agenda'?

Sharla came and talked to Malcolm and me frequently because she, in a most sincere and excellent way, was trying to elevate the quality of contractors acquired, treat the contractors properly and to supply the best possible contractors for the dollar to MTI. Sharla was a "right" person in all ways. She had not yet learned that ""No Good Deed Goes Unpunished" I told Sharla that I would be happy to write her a memo listing all of the virtues and services she could look for when selecting a contract house and give her a list of some of the better contract houses I and others had done business with. She was delighted and said she would be looking forward to receiving the memo. The letter was written over the next few days informally and was given to her handwritten since it was not an official internal or external memo.

I had no personal bones to pick with ARC; however, the quality and scope of their services to their contractors was definitely limited and restricted compared to the services I was accustomed to. The only two good things I could say in defence of ARC were that (1) they did put me on a contract and (2) they did pay their payroll with few errors. Only twice while at MTI did I have payroll errors and that was more likely an MTI secretarial failure to send in the time cards. A second error was improper deductions of Utah personal income tax (a too large deduction of course). Other contractors experienced more serious errors. Several contractors at MTI packed up and left when ARC changed pay rates to lower amounts than they were hired for. I did not know these contractors personally; however I heard from totally reliable persons and Sharla Rose that these persons were very angry and would probably sue MTI and ARC. They already had sent letters to MTI and ARC from their attorneys. These contractors had come to work at an agreed hourly rate. After a short time ARC called them up and lowered their rates for MTI convenience and profit. The contractors thought it was somehow a violation of their work contract. I wonder why?

I gave some serious thought to MTI having an apparent sole source arrangement with ARC. Every aerospace vendor and contracting company in the industry would like to 'sole source' for simplicity, cost savings, and for incestuous gains. However, there were a number of government restrictions which usually prevent sole sourcing from happening. Usually government contract "boiler plate" says there shall be no sole sources, and multiple bids are required. Usually government DCAS inspectors who look at their books must verify if there "was" or "was not" sole sourcing. DCAS (Defence Contract Administration Services) had an office and representatives at MTI Wasach Plant. I wondered if they had ever looked at MTI"s sole sourcing from vendors really closely? In this case NASA should have been looking over the shoulder of MTI and prevented anything close to sole sourcing. I wondered quietly in the back of my mind how could such a thing could go on for so long? Then I had an epiphany!

Why be concerned about "small potatoes?" ARC being a sole source when "big potatoes" MTI was a sole source on a number of huge government programs. Come to think of it MTI was America's "Sole Source" for solid fuelled rockets into orbit and space research. I told several of the contract houses it was Sharlas' intent to get a list out which would be a contract firm bidders list and she was going to give them all a fair chance if she was allowed to by the MTI hierarchy. Sharla was a fair and honest soul. She fought the good fight. Each and every one of the contract house managers, Frank Pegg (CDI), Jim Kalember (Volt), Hal Costello (RHO), and Oscar Cervels laughed hysterically. One of these managers told me he had been trying to get into MTI for 13+ years and was perpetually beat out by what he described as "ARC INC. and the Mormon connection". I found out nearly a year later Gene Pusey wasn't a Mormon – his first wife was. She had been the initial Mormon connection between ARC and MTI. My friend thought it would be a cold day in hell that he would ever receive any business from MTI. He told me the conspiracy was infamous in the aerospace contracting industry! Similar to Martin Marietta at Vandenberg. I had earlier in my career been at Vandenberg and truly Martin Marietta was also ensconced with a single suppler.

Another, who had only been trying to sell to MTI for 10 years, said that he would sure like to see MTI become a fair place to do business! He knew a lot of contract personnel went to work there, but he had never been able to get MTI to talk to him. All of the contract houses had similar comments. All were amused at the possibility of honest bidding taking place at MTI INC. The entire industry knew business wasn't Kosher at MTI – wonder if DCAS or NASA did? Just what did DCAS and NASA do in their offices at MTI?

About this time I made a special effort to meet Dr. Leland Davis. He was a physicist and MTI's local Electro-Static Discharge Expert. When we met I asked questions and he was congenial, but his attitude was "run along sonny". He was deceptive. He and I both knew it. I suspected problems lurked here. Why be deceptive if there wasn't something to hide? Dr. Davis and I talked about the electrical grounding of the SRM's for a while. I went away laughing. Dr. Davis did not express science and engineering. He preached the 'MTI Company Line' - *"There Are No Problems"!*

I talked with Sharla Rose about my personal problem related to my nozzle memo and she told me it was "business as usual at MTI". Sharla was keenly aware that Kent McKinnon was fighting an uphill battle against forces beyond his control. Sharla knew 95% of the incredible evils at MTI. She admired Kent and would support Kent all she could – a helpful word here, and word there. She was diligently and quietly supporting Kent in her own way every day to obtain the best personnel for Kent to prevail in the end. Sharla knew Kent had proper intention, did good work, and she likewise understood the 'where and why for' of MTI persons trying to suppress SRM safety and engineering problems. Though the guilt of the Challenger explosion was not hers she carried the burden as did many of the honest and concerned MTI employees.

Simultaneously, from several sources I learned that the chief nozzle designer, Mr. Jim Lemere, was about to quit his job over MTI dangerous practices. Jim Lemere was angry about MTI Procurement buying poor materials for use in fabricating of the SRM nozzles. Jim Lemere was reportedly also unhappy with nozzle design problems and "cure cycle" issues. All of northern Utah Knew Jim Lemere was pissed off at dangerous Thiokol practices.

I soon found out that there was a war going on between MTI and NASA over what was the best cure cycle for the SRM nozzle manufacturing process. I sensed no one knew the correct answers. Sharla did not know all of the details nor could be expected to since she was not a technical person. She was a public relations person and scheduling type. Sharla talked to Malcolm and me numerous times hoping that we would be able to pull MTI's System Safety Department together.

Sharla, like many of the worker bees at MTI did not understand the entire problem but certainly had good intentions. She hoped MTI would be successful in its efforts to supply NASA good, safe SRM's. Sharla and many others at MTI had pride in their work and loved to be an important part of the USA space efforts. There seemed to be one group with a pride motive, and others with a profit motive. You could identify each type in a flash. On any project anywhere the fastest ways to save money is simply to CUT SAFETY! Sharla recommended that I go have a long talk with John Lemere on the SRM nozzle problems I had identified and was concerned about. I told her that I "could not" at this time since I was in the "low profile mode" necessary for my long term survival. Kent McKinnon also felt I should talk with John Lemere when the time and situation was appropriate, but not while I was under current MTI management scrutiny and pressure. Maybe soon things at MTI would improve! There seemed to be hope. Kent McKinnon had a plan and he was optimistic for a breakthrough. My technical reading went on and on until I was sick of it. Complexities of the 'Safe and Arm Device' and the propellant ignition train were just beginning to come together in my mind. Also, about this time the President's Commission returned to MTI for another meeting on the SRM's status. MTI went into hysterical meetings and overtime.

This meeting of one of the President's 'Rodgers Commission' was held in Salt Lake City. Kent McKinnon, Don Thompson, and a number of MTI managers were there to answer questions from the Commission Investigators. Kent McKinnon was about to have a stroke. He had been instructed by his superiors (Ed Dorsey, Don Thompson, Russ Rasmussen) pointedly about what "he could and could not say". Kent was literally in a hollow eyed sweat over the pressure and possible complicity to MTI crimes! Kent was outraged and who could blame him. I honestly do not believe Kent could have been made to say other than the truth at gun point. However he certainly felt the pressure of his career and future being essentially destroyed at gun point. He told me he probably should just blurt out his problems to the committee and let the cookies crumble where they may. Kent's situation was terminal. Where else could he work in Utah, the land of Zion? This quiet well meaning man was at the center of one of mankind's historical crossroads towards future success or failure.

When Kent returned from MTI's meeting with the President's Commission meeting he said it had not all gone bad and that some members of the President's Commission had "slam dunked" some of the MTI managers! And, furthermore, somebody on the commission had stood up and said they did not believe all problems from MTI System Safety were getting through to the commission. So the Commission was to some degree insightful if not completely effective and timely. When the President's Commission left, the local tension calmed down some. Kent said there had been MTI lies told to the Commission, but he didn't' tell them. Kent was angry and near the end of his rope! I wondered to myself –"is MTI trying to grind Kent down and then replace him with a conforming castrated corporate sheep type?" Perhaps someone who couldn't recognize a problem if it bit him on the ass? MTI would be shut down for the Christmas Holidays from approx. Dec. 19th, 1986 to Jan. 5, 1987. Under normal conditions the consultants such as ourselves would not be there working through the Christmas Holidays. However due to the urgent needs of NASA to launch the Space Shuttle again soon we were scheduled to be there through the holidays as was a small core of MTI permanent personnel. All companies have Christmas festivities for Christmas, but MTI seemed unique to me. This was not a company this was more of a close internal family festivity which went on for weeks. Food, drink, friends, the good life in the high bleak Utah desert and the land of Zion (whatever that means to a Mormon).

What Morton Thiokol lacked in good pay for its permanent staff it compensated for in family spirit and camaraderie for Mormons. Don't forget you could shoot an elk out in the parking lot as you left work if you needed fresh meat. Hummmmm – a disgusting consultant / contractor could easily be confused with an elk?

CHAPTER 9

LOW PROFILE OR ELSE

Time: Late Dec. 1986

The last few days before Christmas (1986) were not especially productive for either contract or permanent MTI employees. All over the plant there were traditional Christmas parties and festivities since this was a strongly Mormon community. The family ties were closer than in communities of random religions or random corporate workers. These Utah people literally lived in the same community together, worked in the same few businesses together, hunted elk together, and their lives were entwined far more than ordinary employees of ordinary American corporations. There was also a hidden remnant of illegal polygamy unspoken but closely lurking.

I thoroughly enjoyed all of the yummmmmm yummmmm's and edibles around the different departments as I wandered around the plant – low profile of course! Kent McKinnon was in a friendly and kind mood towards all of his departmental employees. I sensed that Kent envisioned that there were massive wars looming in the immediate future, and he wanted to have a nice Christmas before he lost the pending wars and his job. I believe he did find respite during these holidays and I was glad for him. I felt pain and immense respect for the man. He was alone and in agony. Even though there were many festivities, we continued with our necessary reading and research. I had come to adopt Malcolm's "acquisition mentality" – if I found it, I either copied it or acquired it for our System Safety department. Where ever I went my radar scanned for documentation on SRM systems. The same acquisition mentally remained pervasive throughout the System Safety Department.

I was personally pleased to see the MTI permanents on their way home for holidays so they were out of our way. Now I could get down to more concerted studying of the huge and dangerous SRM's systems without their politics and interruptions. Just prior to the Morton Thiokol Corporate personnel leaving for holidays, I had a list of about 200 documents to be pulled from the documents room. Huffing and puffing, with double pneumonia, I took my documents list down to one of the ladies who worked there, gave it to her and asked her to pull everything on the list and she promptly informed me that I would have to pull them myself! She and her crew only pulled lists of five documents or less! This woman had time to hobnob all over the plant 10 hours a day, time to try and "put the make" on any man that showed up near her area but couldn't fit her job description in during working hours! I was pissed off! I am certainly not too proud to pull documents from files. In fact I had already pulled hundreds. It was a terrible abuse of my cost per time allocation on a regular basis. This gal probably made 1/20th of my salary and the tax payers weren't well served if I took time to search for and pull documents.

The enormous amount of time spent looking for data and other documents around MTI rather than in the documents area was more than a full time job under any conditions. At MTI, component or full up rocket tests are frequently run, a test report written, and then it was kept in the test engineer's / scientist's desk. Some results are published, some filed, many lost. We actually found cases where tests were scheduled, supposedly run and no one could confess to having seen the test, the results and / or data. Most peculiar?

Again where was NASA and DCAS? Queer indeed? I was polite to her originally but politeness did not overcome her scroll of excuses. So, I talked to Kent McKinnon and told him I thought I or he should go have a quiet talk with her bosses and Kent agreed – with the usual comments for me to keep a "low profile".

I set up meetings with the managers directly over the MTI documents crypt and vault personnel and went down there and politely told them I was dying of double pneumonia and not trying to cause the third world war, but was trying to be cost effective and that over the holidays I had to read and analyze these documents and thousands of pages more.

These gentlemen thanked me for coming to them rather than escalating higher up the totem pole. They assured me I had come to the right place and they assured me I would have my documents "real soon". Within 24 hours I had all documents on my list that could be found and the 45ish documents gal had a new and improved attitude. It turns my stomach to have to be involved in such low class crap! Generally speaking, personnel at the documents window were helpful and this was the only time I ever had a serious problem with them up to that date. The really serious problem was a lack of comprehensive, integrated, detailed and congruent documents. Another problem of deep concern was that MTI had locked up huge amounts of documents after the Challenger Accident in their vault and didn't want them seen. In fact they wouldn't even provide a list. Do you suppose they might have been incriminating? Another big problem was lack of documents control. There were no comprehensive document records and no way to know what was available and where it was? No log of document comings and goings. About a month later I found a document locator Directory which even helpful Peggy and her boss didn't know about. They were tickled to death to have it discovered and shocked to find out of its existence. This was an excellent document locator for certain types of documents, worthless for others. Still, far better than nothing at all. I made copies of this document and gave one to all System Safety staff. Kent was glad to see his copy. Shocked actually. A great find.

Finally, the big day came and 99% of the MTI permanent staff disappeared into Christmas Holiday oblivion – down the cold lonely road to their homes. I was personally relieved to see them go because it meant I could get more work done. Also we felt more at ease acquiring office materials and equipment under the "procure your own system". Also, Don Thompson and associates weren't around to fire anyone for doing their jobs!

Two weeks prior to Christmas Vacation NASA officials jumped Kent McKinnon through hoops severely by changing the "form and format" of our Hazard Analysis documents. In fact NASA could not even make up its mind which documents should be submitted to NASA. In other words NASA, after 13 years on the contract had not decided what System Safety "deliverables" to NASA should be!! Outrageous incompetency! Additionally, NASA couldn't decide what formats, which routes of submittal, or who the documents would be submitted to. NASA couldn't even decide if our Hazard Analysis, a normal product of System Safety, should go to the Marshall Space Flight Center (MSFC), the Kennedy Space Center (KSC), or the Johnson Space Center (JSC)! Who was in charge at NASA? Who ran the MTI contract? Where was the NASA / MTI contract? Where was DCAS? Damn good questions! Did the cart push the horse or were the cart and the horse disconnected and on separate farms? NO! The cart pushed the horse off the cliff! Kent McKinnon was diligently trying to force professional decisions in a reasonable manner. Kent didn't fail, Morton Thiokol and NASA failed initially, then others joined them!

Oh yes, about the elusive MTI / NASA contract! The entire System Safety staff had diligently looked for the NASA / MTI contract ever since my third day at MTI, but could not find a copy. We asked a lot of people and they diligently "knew nothing". Finally, I cornered Malcolm Phillips and Kent McKinnon and asked Kent to find a copy of the contract for us. We emphatically explained its importance to our doing a credible analysis and it was the baseline of everything Kent's System Safety Department was about.

Malcolm and I were seasoned project managers and our careers evolved around forcing projects to conform to contract requirements and standards. Kent understood and said he would see if he could run a copy down. He confided he had never seen a copy of the contract. The contract's absence was a curiosity to him. It was astounding to Malcolm, me, and others.

When Kent McKinnon queried Don Thompson for a copy of the NASA / MTI contract Don told Kent to "forget it"! As the mere manager of System Safety Kent was not allowed to see copies of the contract nor were the other departmental managers! Clearly we working slaves were restricted. Don Thompson was a sad case. Don Thompson admitted freely to Kent McKinnon that he didn't know what System Safety really was or should do. Kent was then asked to explain System Safety to Don Thompson. Don was of course manager of the Space Shuttle's Solid Rocket Motor's safety at MTI! A Brigham City Tax Accountant!

A document called the MTI 3600 document was then being fabricated by MTI to be the cornerstone of our work in lieu of a contract. I was astonished. The 3600 document was not yet NASA approved after 14 years and in fact the document was just being written. Some preliminary versions of the 3600 document were floating around. *Different versions??* The 3600 document was already being prepared to replace the 3300 document. The 3300 document called out some performance specifications and parameters seemingly specifically and diligently, but incomplete. The word around MTI was that the 3600 was loose and vague, thus easier to comply with than a detailed contract. Larger cracks for details and safety to fall through so to speak. Everywhere I have ever been previously, the contract copies are readily available for all. How can you build and design a system to criteria you haven't seen? Very irregular! Were we going to have to engineer the updated SRM to a "moving target" called the 3600 document? What in the hell did MTI use for the last 13 1/2 years?

How naive I should think a government contract (NASA is USA government) would be clarified in such a short time span – a mere 13.5 years. This was a leadership problem, and a government problem. Was NASA's contract under control? Clearly Not!! NASA was becoming 'highly suspect' to most Reliability and System Safety personnel

I was getting glimpses of NASA confusion and incompetence. At that time, I figured NASA confusion was simply a nuance derived from the Challenger Accident, and the Rogers Commission investigation which certainly pressured everyone in the space industry. How wrong I was! How frustrated Kent McKinnon and others were! Confusion, chaos, incompetence, and deceit turned out to be the life blood of NASA. The contract should have been 'rock solid' at time of issuance thirteen point five years prior.

Malcolm Phillips was finding his first major catastrophic hazards on the SRM systems. One of his first hazards was the startling fact that a large number of the SRM ground support equipment (handling devices) were not electrically grounded, and in fact, had never been! To make a long and dangerous story short, if there is a transient electrical current, or spark, an unscheduled, unlicensed, and damned inconvenient explosion occurs! A gross explosion of the worst kind! Such an explosion could kill astronauts, but more likely a few hundred NASA and contract employees. During this period Don Thompson seemed to be raising an inordinate amount of hell with Kent on a number of subjects. It was rumoured that there were similar problems in the Reliability Department next door to System Safety which was run by Mr. Keith Dixon. Don Thompson was also reportedly "controlling" the Reliability Department. Apparently Don Thompson's designated agenda was to prevent any technical problems or questions from occurring from any source. Don said it best – "You don't understand! It's not a problem! There are no problems"! The MTI Mantra ....

The Reliability Department was responsible for preparing FEMA's which are Failure Evaluation Modes Analysis. Two of the FEMA people sat next to Malcolm and me. The four of us had excellent dynamic technical interchange about our work all of the time. We helped each other and as a group we were synergistic. One of them was a profoundly capable electronic engineer, Mr. Frank Howell. He had identified a number of very serious problems. Frank was like a demon. He was also doing a lot of work on and thinking about Shuttle Crew Egress (escape). He worked in an environment of consistent disappointment regarding MTI's past engineering analyses and studies. A small portion of the available historical record was elegant, most were disgusting. MTI was a D+ to a C- corporation at the very best in our opinions. Most of us realized that under no condition would we bet our own lives on MTI's substandard engineering work. None of us would fly on the space shuttles. I held a U.S.A. 'Commercial Pilots Certificate' and flew a lot of different planes at various times. The Space Shuttle simply had too many 'issues' to be considered reasonably safe.

I was often amused as I sized up other engineers and scientists working on solid rocket motors. After they had been around for a few weeks I would casually ask if they would ride the space shuttle? If they answered "sure I would" then I knew I was confronting an extremely poor technocrat, or a politically motivated corporate sheep. If they had the knowledge and balls to say "No fucking way!" I considered them to be aware, wise and thus respected. A large body of us in the System Safety department thought the solid rocket motors were 'projectile trash cans' - not reliable rocket motors.

CHAPTER 10

HOLIDAY CHAOS

Time: Late Dec. 1986

Now that the Christmas vacation was on for most of the MTI employees, the hallways were 95% empty. This absence included most of the MTI Security 'Gestapo'. The cafeteria was closed so we had to bring bag and / or microwave lunches. All of us felt a much freer hand at running through offices and work areas, gathering up documents and looking at charts on the walls here and there to find out what was really going on or what failed to be done. This is the only contract I have ever had where information was so hard to come by. Some of our discoveries were astounding!

Ours was not a particularly covert activity, just something we felt we had to do to zero in on our specific areas of concern since the MTI jungle drums were so quiet. The Indians and Malcolm were the most aggressive in perusing the other offices and areas. The American Space Program depended on it.

During this Christmas Holiday period, I was determined to pin down my first formal Hazard Analysis and have some Hazard Analysis forms filled out and ready to go forward to NASA when the permanent employees returned on 5 Jan. 1987. The following was the Hazard Analysis format which was used at that time with the exception that the form we used ran from left to right and not top down: There had been many formats.

As time passed, the formats changed over and over and over ad nauseaum. The hazard numbering system changed dozens of times. Definitions of what each column was to do in the format changed weekly. In fact, after Christmas, it became clear there would be so many changes that nothing could become final. This wasn't simple confusion , it was a way of keeping Hazard Analysis from ever being published in the opinions of a lot of us on the System Safety team. Some of the changes were absurd – especially the ones from Don Thompson. An example – one day he wanted us to swap the *Safety Requirements* and the *Hazard Control* columns thus making both useless. It was a simple matter of say "ah, here is something I do not agree with or believe is an inadequate design, condition or fix". A liberal arts mind in an engineering company. Whatever the problem is, it

HAZARD ANALYSIS FORM:

Prepared by:
Approved by:
Sub System:
Project Engineer:
Hazard Number:
Mission Phase:
Date:

HAZARD CONDITION: 1.
2.
3. Etc.

HAZARD CAUSE: 1.
2.
3. Etc.

HAZARD EFFECT: 1.
2.
3. Etc.

SAFETY REQUIREMENTS: 1.
2.
3. Etc.

HAZARD CONTROL: 1.
2.
3. Etc.

One of many changing formats

needs to be described in some sort of an accepted and approved way. The universally accepted contract method was via Hazard Analysis. There were diverse opinions concerning how thorough Hazard Analysis (HA's) should be done and how they should be prepared and presented. MTI opinions were based on cost to MTI, and MTI exposure - not on safety.

Kent McKinnon asked me to develop a format that he could show to the Rodney Commission, the Rogers Commission, MTI management, and others in his hope of locking in a format. This format was to be how System Safety felt the HA should be researched and derived. So I set to the task of drawing a step by step procedure of how I envisioned a thorough Hazard Analysis should be prepared for the MTI System Safety department taking all local MTI factors into consideration One contractor had a lap top, and MTI had a mainframe with only a few stations. The internet had not become a public forum at this time in history.

After I wrote what I call the base line analysis procedure, I brought all system personnel in, one at a time, told them what I had done and "could they in turn add to it?" In every case, these experts had one or more contributory points. Then, after I had extracted something from everybody, I recycled everybody through one last time and we picked a few more overlooked odds and ends. The final System Safety Departmental format was sound and comprehensive. I had the HA "flow sheets" computer drawn and ready for Kent McKinnon when he returned from Christmas Vacation. He was very pleased. He had a "hard copy" in his hand and was off to get NASA and MTI formats pinned down, solidified, and finalized.

General Joel Henry, a retired two star U.S. Air Force General typed it up. He was a contractor just like Malcolm and me. He was a very expensive typist. Simultaneously, I continued into the 'Safe and Arm Device' technology and SRM (Solid Rocket Motor) Ignition System technology. During this period I prepared a substantial number of HA's in that 'currently approved form and numbering system'. I prepared a HA form just as it should be. Some HA's were scientifically complex, others were extraordinarily simple in nature. A few simple but important examples of hand written Hazard Analysis attempting to comply with daily changes are in Appendices 6A, 6B, 6C, 6D, 6E and 6F.[6]

There are varying formats possible. Catastrophic Hazards are the most severe and are defined by NASA and the U.S. Air Force as discrepancies causing loss of life, loss of equipment, and loss of mission. Any group of

---

Appendix 6a, 6b, 6c, 6d, 6e, 6f, 6g, 6h: Hand Written Hazard Analysis
{Note, the messiness was caused by massively changing formats and more..... }

engineers could argue over the possibility or probability of any discrepancies or defects occurring. There are legitimate differences of opinion. Sometimes all parties to the discussion are not completely informed, and of course they are from various technical and discipline mixes. However, I can assure you that any rational management would conclude the profound necessity for examining every known problem as completely as humanly practical. NASA knew that and had originally ordered everything examined. NASA safety directives directed MTI to examine and divulge every possible hazard, no matter how innocuous. Were NASA's actions in keeping with its specifications and design? Now I had my teeth on the bone better and was shaking the meat off! I was effectively doing what I was hired to do in spite of obstacles. I was doing what I do best – I was thinking far into the future, the culmination of important and complex engineering, scientific, operational, and management events - even historical events.

All the contractors and myself had numerous discussions about our sense and feeling of a low MTI technical threshold. It seemed to us that many MTI personnel were far more "liberal arts like" than they were "technocrat" types. Bit by bit, interesting facts began to unfold, such as:

- Many of the MTI project managers were really former high school teachers and not engineers or technocrats.

- MTI project managers could be generalized as high school teachers who were Mormons in good standing and who had paid their 'tithes' up to date. The man in charge of SRM reliability, Mr. Keith Dixon, had been an auto -body teacher at the local high school before he was brought to MTI and became SRM Reliability Manager! No background in rockets whatsoever! All of us were shocked at such revelations.

We simply felt it was important for us to work around their lack of knowledge and experience and enlighten them. Enlighten them in polite, courteous, and constructive ways. We would work with anybody - even criminals and children!

One afternoon I was talking to an MTI guard who was bored and lonesome at his desk. I asked him if the SRM Cases or railroad cars were physically guarded on a regular basis. He told me they used to be, but due to expenses MTI had curtailed most rail road car guarding. I asked him if the railroad cars were escorted in transit. He told me no. I was shocked. A rocket segment loaded with rocket propellant is an awesome amount of explosive power. The explosion would produce an awesome amount of toxic gas. When I dug into this I found a lot of seriously Catastrophic Hazards.

The guard unknowingly set me on paths which led ultimately to discoveries such as the SRM Cases loaded with rocket propellant were inadequately or not at all electrically grounded on rail road cars as they crossed the U.S.A.. Vulnerability of the Cases to terrorists was revealed. A lack of any adequate safeguards was revealed. Some of the rail road cars went from Utah to Vandenberg for storage and then crossed the nation to Cape Kennedy! Serious exposure of entire towns to 'mass exterminations'!

Malcolm and I felt that we should have a Christmas party and invite some of the local MTI people to it. After all, many of them were delightful people, and MTI was putting a lot of money in our pockets (against their will). We worked such long hours and so hard we had no social life at this time. Malcolm and I have both had a sense that we should try to maintain social amenities even though we were a long way from our homes. So we agreed we would gather at my newly leased apartment in Brigham City. We would invite the contract engineers, landladies and a number of nicer MTI people from System Safety and Reliability. Our landladies were wonderful people. They had extended their best hospitality to us and were anxious to meet our famous General Joel Henry (US Air Force fighter pilot, retired).

General Henry had been consulting to Boeing in Seattle for years. A retired U.S. Air Force General, a former fighter pilot, Joel was caught in a squeeze at Boeing and was temporarily out of the door as were Malcolm, Mahalon, and myself thanks to the Gramm Rudman Law. So in mid-December he came on contract to Morton Thiokol. Malcolm, Mahalon and I had all worked with him on the MX (Peacekeeper) ballistic missile system. Joel was an extraordinarily jovial and pleasant fellow. He had fought his wars and now sought peace. He was a high technology public relations grandpa type. He was not a technocrat. Our landladies had heard us always speak well of General Henry. Not many generals live in Brigham City. Malcolm and I had casual concern if MTI Mormon folks would come to a party with us non-Mormon infidels? Who cared – we were going to throw a party anyway!

We needed a party. Mahalon was sick and frustrated over very poor MTI testing of the SRM components and systems. Wayne Tilsey, (another contractor from Seattle) was at his limits of frustration regarding suppression of System Safety activities. I had double pneumonia and did not feel well. I was also under incredible stress, knowing MTI upper management wanted me out of my job. I was in pain from Thiokol suppression and we had hardly begun.

Malcolm had been our appointed "invitation document artist". Malcolm wrote up a 8 ½" X 11" cartoon type of invitation. It was distinctive, memorable and appropriate. Invitations were passed out to nearly forty people and posted on local bulletin boards. A map was included – how could you get lost in Brigham City it was so small? Easier than you might think! For all of Utah's high taxes, there was profound absence of street lights and signs. The fateful party night came, and the party was held at my place - Apt. 1, 105 W. 700S. Mrs. Deloris Hill, my landlady and Mrs. Patty Vanderstein, General Henry's landlady came and then there was Malcolm, Gerald Woods, Mahalon, General Henry, myself, Sharla Rose, and her husband. Wayne Tilsey and a few others. I was on the mend from my pneumonia, General Henry and Wayne had terrible colds. Our worst fears regarding most of the Mormon folks not attending were dead correct. Other than Sharla Rose none of the MTI Mormons came. What the hell! We tried. We knew infidels didn't pull much weight in the land of Zion.

The entire System Safety permanent personnel including secretaries and others had been invited and did not show. We enjoyed the party just fine without them. Fewer attendees provided more fried chicken for Malcolm and myself for days after the party. God knows we didn't have much time to cook.

The party, small as it was, was interesting, fun, and informative. Sharla Rose's husband was a pleasant fellow. I liked Sharla and her husband very much. The landladies almost talked General Joel Henry to death. General Joel Henry had a terrible cold, but was in a jovial spirit. Malcolm and I were in a good mood and Mahalon was on his best behaviour. Many pleasant discussions on everything from soup to nuts! As is to be expected in any group of socializing "worker bees" the discussions finally turned to work at MTI. Gerald Woods, Malcolm and I all later commented that we observed the same thing. Everyone was "horrified" by MTI practices, incompetence, and suppression. "Everyone " wanted it fixed! No one knew what to do.

Looking back I believe it was on this night my eventual course to the FBI was charted. I was "emotionally disturbed" by the dangerous, inept, and apparently (to me and numerous others) fraudulent practices of MTI. Yes *fraudulent*! The entire group discussed fraud, crime, and MTI raking in billions! The total lack of ethics was the most prominent thought. I was angry and ashamed to be part of it. I leaned on the door jam listening to all of them and concluded I was not even half informed of how serious the MTI and NASA failures and crimes were! My vision had been limited. I needed to expand my vision and knowledge. Shouldn't everybody?

During the Christmas Holidays, there were almost no NASA, DCAS, or USAF personnel at the MTI plant. After the hubbub caused by the Rodney Commission and the Rogers Commissions the commotions temporarily died down and we were immersed in our technical work. General Joel Henry, Mahalon Hollway, Malcolm, myself, and Wayne Tilsey each thrashed around in our areas of expertise. Wayne Tilsey was crazed by shoddy structural and metallurgical analysis work he found on the SRM cases and joints. Wayne and Malcolm talked in great depth about the very poor condition and work done on what is called the "Flight Loads Books". The" Flight Load Books" were in abysmal condition and they notified the rest of us.

Mahalon Hollway ran up and down the hallways crazed, ranting and raving about either "no test documents", or "pitiful quality test documents"! At this time Mahalon Hollway felt that MTI test documents were the worst he had ever seen in his 35+ years of aerospace work. "Stupid S.O.B.'s" he ranted and raved. We were all in accord, and nearly everyone was ranting and raving. We were becoming concerned if we were being set up to take the blame for the next astronauts deaths if everyone ignored our results.

General Henry had been a jet fighter pilot and had held high strategic planning positions in the U.S. Air Force. For years he had consulted at very high income levels to Boeing Aerospace which just recently rolled him and all other contract personnel out the door during a budget purge and bloodletting. General Henry anticipated Boeing would roll him back in the door after the first of January 1987. He was not clear on how and when he would go back to Boeing. He was always a gentleman, but he was hurt that Boeing had temporarily "sacked him". Temporary no doubt for all of us. His value to Boeing was not his technical prowess at all, it was his membership in the 'USAF Good Old Boys Club' right to the top of the Pentagon. Boeing used General Henry as a 'DOD political access tool'.

During the Christmas period Wayne Tilsey was writing documents as were Malcolm and I. Mahalon Hollway hand wrote a number of papers and essentially General Henry was used as our 'pool typist'. General Joel Henry was no doubt the highest paid typist in the world (approx. $14,000.00 per month). Later on Eric Nalder of the Seattle Times would refer to General Henry as a *"Two Star Typist"* as opposed to a *"Two Star General"*

Wayne Tilsey had determined that the structural characteristics of the SRM cases were terrible. In fact Wayne calculated the SRM was structurally unfit for one flight. He thought the existing SRM structure did not meet the required safety factors of 1.4 after 20 launches, and in fact fell below 1.0 for its first flight. *In other words, the rocket case probably was not safe for a single flight, let alone an additional 19 flights.* Wayne determined the MTI SRM field joints O-Ring design was very poor. Everyone agreed except for MTI and NASA.

I should mention here that I never met a single person, permanent

MTI or contract employee, engineer, or scientist, who thought the MTI "O – Ring" field joint of the SRM Cases was a sound and desirable engineering solution. Every single scientist, engineer and technical person I talked to was afraid of these joints and openly hated MTI's design. Every engineer at the plant had a better design, drawings, sketches, and calculations on different field designs that floated around on scratch papers and napkins everywhere.

Dr. Tony Laine had come to consult in early January 1987. He was a Ph.D. in Physics. He was nice, sincere, and went to work looking for hazards and he appeared to not have any negative thoughts about MTI. He was there to do a professional job. He didn't need to look long to find problems. He immediately found numerous problems with the nozzles and other SRM components. He was shocked and dazed by what he found. He did not take his findings quietly or sitting down. Soon he could be found in meetings, the men's room, the hallways discussing how "fucked up" MTI designs were, "what dumb SOB's" MTI engineers were, what criminals MTI managers must be! Dr. Laine caught on real fast and didn't take it well because he understood the political / historical ramifications without being told. He was 'his own man'. This went on for weeks. Dr. Tony Laine scared Kent McKinnon and pissed off MTI in general. Tony was honest, and sincere, but he really brought unwelcome MTI pressure and focus to Kent. Tony was so far ahead of most MTI engineers they couldn't even talk to each other. Tony was eccentric. He had a runaway ego, but there was no question he was highly competent. A short time before Tony came to MTI he had been consulting to the CIA for $13,000.00 a week which was 'off the planet' income at that time. He had also held an 'off the planet Security Clearance'.

Wayne Tilsey had become disenthralled with clannish, cliquish, separation and isolation of System Safety at MTI! He was frustrated and angry though he remained a gentleman. Mahalon Hollway was so upset with the total lack of overview on the test side of things for the solid rocket boosters, he wrote a pointed memo to the MTI Hierarchy.

None of us had a competitive mentality or feelings that 'us outsiders' were against' them 'insiders'. We consultants had respect for the project and we worked very hard through our Christmas non-vacation. Our System Safety group literally read a mountain of documents. We did mathematical analyses, held group discussions on problems to be examined, and in general did most decisively excellent work. We were all wondering why this work hadn't een done years ago? It seemed as if the SRM's had been built, used, and were now being engineered – and the engineering wasn't looking flash to us! The status quo was ugly! SRM Segments (four per side or per SRM, 8 per STS) were 'oil canned'; (bent) from bashing into the Atlantic Ocean on swinging parachute re -entry. MTI personnel did not clean the dirt, or the straw, or the bird shit from 'O-ring' grooves when assembling them etc. ad nauseam.

We each had developed a long laundry list of more documents to acquire, discussions to open, and questions to ask. When the "permanents" came back on Jan. 5th we were going to put the cattle prod to them. Wayne Tilsey was anxious to get back to Boeing Aerospace in Seattle and General Henry was in a quandary as to how long before Boeing would pull him back. So we went on through the Christmas Holidays raiding, working, cussing, and discussing. We were surrounded by a cold hostile environment, we were lonely and isolated. One day a frozen water pipe broke and flooded the men's rooms – very exciting stuff! Most excitement we had for two weeks. Outside the buildings was pseudo Siberia.

In many ways this was a great time. It was an intellectual probing into a system of subtle complexity and extreme importance. There were didactic discussions, probing, and it was very stimulating! Superiority of the Western World military depended on the STS in many more ways than the public ever conceived of. Our task was very hard work. We all talked about the SRM Boosters Rockets coming to work, over lunch, and on the way home. We were immersed in the 'cold war' and the future of everything. Our group findings to date frightened me! How could we tell a large American Corporation (NASA), the USAF and the United States Government their primary heavy lift rocket launch system was launched with two 'TRASH CANS' (SRM's) per shot? America's Greyhound busses to space were proving to be extraordinarily expensive, were now proven to be unreliable, and as a system were bound to fail America and the Free Western World.

NASA's original estimated cost of $150,000,000.00 per launch was conservative. Perhaps that would be defensible if the equipment was reliable and of high quality – it was neither. Furthermore, after the Challenger Accident the pay loads were reduced to approximately 40.000 lbs when the contract called for 65,000 lb pay loads. It turned out that $250,000,000.00 per launch was more realistic.

I flew over to Denver to visit old friends for three days. Frank and Shirley Abshier were senior geologists I attended university with. I barely made it out due to terrible weather. A brief and enjoyable interlude. We visited and talked of our careers and world problems.

For geologists their problem was that each geological commodity acquisition escalated and collapsed. Gold up, then collapse. Uranium was the element of the moment then. All of our careers were similar. The difference was they lived and died on 'commodities' and I lived nd died on 'projects'. We all pondered the new world economics we lived in. A new debt economy, engineered to be unstable it seemed. 'Globalization' and the new 'Neo-Liberal Capitalism' had arrived.

CHAPTER 11

A NEW YEAR

Time: Early Jan. 1987

Early in the morning of 5 Jan. 1987, we all headed for work and we could tell from the highway madness between Brigham City and the MTI Wasach Plant that we were back to business (madness) as usual. Kent McKinnon had his staff in a whirlwind of "start the new year" meetings, preparations, and the whole plant was abuzz. A new year, new budgets, etc., the usual necessities. We contractors had stayed in high gear so we hadn't broken stride and had no trouble continuing. Furthermore, we were not encumbered by so many mandatory and non-productive meetings as the permanent staff. At most companies there are a lot of TI (Technical Interchange) meetings. At MTI they were EHS's (Elk Hunting Stories) meetings, etc. Anything but rocket business.

In fact we hated to have the permanents back to get in our way. During the holidays Frank Howell had written two reports on Space Shuttle Crew Egress. How the Crew could escape out of the Space Shuttle.

Frank Howell was now trying to find an MTI management type to read his work and to discuss and push his life – saving ideas on up the MTI totem pole for ultimate submission to NASA. The Reliability side of the house was in fever pitch, preparing some final documents for NASA meetings. System Safety was in high pitch also because there were a lot of diverse and serious SRM Hazards emerging from everyone in our group. Shortly after Jan. 5th, Wayne Tilsey appeared to be close to a nervous breakdown in frustration over what he referred to as the 'Dumb son-of-a-bitch – engineers' at MTI. He had a lot of company. He had demonstrated quite clear mathematics, physics, and engineering from MTI"s own documents that the SRM Cases (skins) were no damn good! Questionable stainless steel alloy, questionable corrosion susceptibility, unpredictable fracture characteristics, etc.. Wayne had also shown that the "O – Ring" joint was insane. That was no revelation – everyone except NASA understood that. However, note that some time earlier NASA awarded MTI a $3,000,000.00 engineering excellence award for MTI's stupid and ludicrous O-ring design.

New Year expectations for our effective pay raise on 5 Jan 1987 culminated. For Malcolm, Mahalon, the Indians, General Henry, and myself the new rate of pay improved. With overtime it ended up much more acceptable.

At that time in history our pay was very good in the scheme of U.S. incomes, but low for top Engineers and Scientists at the time. We would rather make the money in less time, but at least we were working and had good income and we were the lucky ones during the Gramm Rudman 'Dark Ages'! The majority of contract engineers in the U.S. were still out of work, and genuinely without much visible hope.

It was universally accepted that the "O – Ring' joint was no good, but NASA and MTI had made up their minds "that's how it's going to be" so the entire MTI Corporation shut its mouth and did as they were told. Clearly a decision by non-engineering personnel. General Henry was frustrated by many things and continued pushing to re – establish his Boeing contract. His approach was to leave MTI rather than fix them. He had a house and friends in Seattle and he missed them. Also he had a home and wife in Florida, but he had to work as he had been fleeced out of all of his retirement money somehow.

On 5 Jan. 1987, Mahalon Hollway came to work with a startling new attitude. He had a new and different "modus operandi"! He was not the same man I had known four days previously. Mahalon was literally running up and down the hallway glad handing and slapping backs and whispering sweet political coup d'état in MTI managers ears. Malcolm and I speculated about what Mahalon was up to. Soon, it became apparent. Mahalon Hollway had in some way appointed himself to soon replace Kent McKinnon as System Safety manager, in his dreams at least. Mahalon started acting abrasive, antagonistic, and condescending towards Kent and it became clear that Mahalon was trying to make his bed with Don Thompson. He literally "hovered" around Don Thompson's area with "that certain smile", encouraging words, a deadly dagger! Mahalon's vision was that he could save Don Thompson and MTI! He was launching his gambit.

I noticed on Jan. 5th, Mahalon had "cooled off to me" being polite and pleasant to me as though I had been dismissed. That wasn't particularly strange demeanour from Mahalon when he was on an egotistical roll. His moods and mentality would swing with the wind just like most other recovering alcoholics. Clearly his appreciation for his new contract had worn off. It wasn't he who was appreciative to have a job, it was clearly MTI who was lucky to have Mahalon in his view.

System Safety personnel were all stirred up at this time because Kent McKinnon had the usual MTI management forces coming down on him from above like a ton of shit. Kent was trying to determine what NASA really wanted, was trying to determine what our System Safety group policies and practices really should be, was trying to determine how the HA's should be sorted out with the Reliability Analysis prior to the next shuttle launch. MTI at this time came under additional pressure from new law suits from astronaut's families. NASA people were pouring in the door and walking up and down the hallways but they were never at any of the appropriate meetings and they never talked to anybody lower than Don Thompson except on rare occasions. We contract engineers discussed how Don Thompson and NASA seemed to be such buddies! Perhaps NASA officials needed their tax's prepared? Everyone in System Safety was looking optimistically toward the possibility that we would make a political breakthrough and start pumping papers into NASA to get the SRM repaired and prepared for the next Shuttle launch. Time was running out. We were filled with anticipation about the promised computer link between NASA and all of the Space Shuttle Contractors. This link had been promised many times, but never happened while I was at MTI. Optimism and frustrations were everywhere! Would the long promised and contracted computer data link between MTI, NASA, and other STS contractors be installed so we could bring SRM Hazards and FMEA's on line with NASA. Marshall Space Flight Centre. Kennedy Space Center, Johnson Space Centre, and the other SRM Contractors? Dr. Tony Laine suggested to Kent McKinnon, in a full System Safety staff meeting, that we all lease personal computers ourselves and bypass the secretarial log jam.

Everyone thought that was a great idea, and good for all of us. Kent McKinnon thought it was a great idea. Kent proposed this to Don Thompson. Don Thompson denied our request and offer immediately. Note that at this time in history most high school computer buffs were modem - ed into each other, but so far NASA couldn't swing it with its associated aerospace contractors (ASCONS). Why in hell couldn't NASA arrange it? The entire US.A. could be hooked up in one working day for pocket change. Could it be that NASA and MTI didn't want their problems "on line"? Were the problems too many, too big, too expensive, too frightening, too unmanageable, too embarrassing? Oh silly me! After all, *There were no problems!*

CHAPTER 12

ALIENATION

Time: Early Jan. 1987

Sometime in the first week of Jan. 1987 I handed in my first group of prepared SRM HA's. All of them were deadly catastrophic concerns. I handed them into Kent who was going to look them over and tell me if alterations in format were necessary to put them into the NASA system, (whatever that was at the moment). Simultaneously, I was continuing my 'propellant ignition train' and' Safe And Arm' device' studies and preparing more catastrophic HA's. Also at this time, Kent and I agreed it was probably time I resumed attending some meetings again – low key and quietly of course. ' Casper the Ghost' if possible, or even a 'cockroach on the wall'.

When I started attending the 'Safe and Arm' and 'Igniters' team meetings, I did not find the extreme prejudice and restrictions that I previously experienced at the 'Nozzle' team meetings. These managers were more polite, they were never discourteous to me, they simply placated me in the Don Thompson's mandated style. I was in no position to twist their nuts and they counted on it. They also knew I was technically miles ahead of them. They had to tolerate my knowledge and I had to tolerate them! None of us were comfortable. They had been warned and instructed. It was clear from day one they would not discuss certain important, negative, or pertinent issues in front of me. I found there had been quite a number of 'NASA Waivers and Deviations' over the years on the 'ignition train' equipment. Specifically, they would not discuss any major problems unless they couldn't avoid it. This attitude prevailed for the next few months until I left MTI. I didn't actively fight this as I understood the actions process better than them. After all, they were mostly northern Utah 'rocket motor sheep'. Their hearts and minds were in tow. I and others easily worked around them. Someone at one of these meetings would say "well there is problem such and such" and one of the manages would say "hold it"! "System Safety is here, let's discuss this at another place and time – HEE HEE." Then they would smile at me and joke and say, "Oh, just joking"! It was no joke, they knew it and I knew it. They tolerated my presence and again it was clear MTI management orchestrated obstruction of NASA, the US Congress, and Western World Security. Under normal conditions if any company personnel ever treated me like that I would drag them by their balls screaming and yelling straight to the President or

Vice-President of the company. The shit would then hit the fan and this information restriction crap would stop. I couldn't do it at MTI and expect any meaningful back up. Remember – make no waves! Kent McKinnon's ability to back me up was severely limited by his superiors restrictions and threats placed on him which were increasing every day.

I could not afford to have a knock down and drag out with MTI people after what had resulted from the nozzle meetings. I decided to go about my technical investigative research via other routes. The other routes were (A) to obtain all the documents I could get, (B) to quietly have low key discussions over a cup of coffee here and there and squeeze information out of engineers, managers, and other people where I could. I worked on a personal "one to one" basis, negotiating more information out of these people. Typical of these were Doug Bullard (Design Engineer for the Safe and Arm Device) and Pat McClosky (Design Engineer for the Igniter). We talked, they shared their knowledge and frustrations and helped as best they could. Both Pat and Doug were very nice, but also very junior considering the fact that any failures of their equipment would be catastrophic and deadly.

The American public and NASA were not getting their money's worth as long as all problems were not discussed above board by qualified persons. Many technical problems appeared to be serious when viewed by one person and then turn out to be no problem when examined by multiple disciplines and departments. On the other hand sometimes multiple disciplines examination illuminate and elevate an apparently simple problem to a better understood complex deadly problem. Suppression illuminates nothing! On numerous occasions Kent related there had been "such and such" problems and I would think to myself, I'll bet Kent missed some clear and simple solutions, or even obscure and difficult solutions In every case he replied "Oh yes". I tried 'such and such' and in a particular way and MTI management beat him down more in this manner or that! Kent left no reasonable leaf unturned and never missed a stroke. I was impressed. Some of the System Safety personnel were getting the feeling that Kent wasn't a strong enough manager for his job. I was thinking to myself, who in hell is strong enough for his job? A rusty piece of pipe couldn't take the beating!

Kent left no stone unturned, had exceeded his authority in efforts to circumvent political problem areas, problem persons, and to bring out the SRM problems but no matter what he did he was cut off at the pass by MTI hierarchy! It was crystal clear to me that Kent's time at MTI was now very short whether he liked it or not! I knew from prior consulting experience that once the corporate flock became alarmed about someone who was trying to be honest and above board the end was nigh. "NO GOOD DEED GOES

UNPUNISHED". Kent's time and no doubt my time, were clearly short at this point as events unravelled. Madame Guillotine was in free fall and there was a waiting line. I was near the front of the queue!

Soon I cornered Kent and said "Kent, I have high level management friends at several Aerospace Corps. who are always looking for good permanent employees. Would you like me to send some of your papers off with a letter of introduction? Kent grinned from ear to ear and said "that would be lovely". I'll have my resume here tomorrow". And so his papers were there next day. I made some copies and attached my letters of recommendation to them, and dispatched them to various aerospace company executives I knew. When I dispatched them, I was very careful not to use the MTI mail drop since I was afraid of it. The MTI mail drop was just outside MTI"s Security door. I assumed MTI security would just as soon tap the outgoing or incoming mail, as simply as they would tap a phone or suppress negative findings. I guess I had come to the point where I didn't like or trust MTI and my feelings were now irrevocable.

Malcolm's efforts to dispatch his first group of 42 HA's off to NASA was amusing and ominous to watch. Malcolm, a most diligent and excellent document producer, was literally running up and down the hallways. I feared he might take gasoline and 'light the chair off' under a couple of our incompetent near brain dead secretaries. By this time, Malcolm, myself and others had hundreds of pages passed into the secretaries for typing and nearly nothing usable came back. What came back was abominable beyond any description I can provide to you. Some of the atrocities returned by Wendy and Erma were typed so poorly we could not recognize them as our own documents. A lot of our submissions were found in their trash cans which we looked into after their early daily departures.

Every single contract person and permanent was outraged. Some said "if Morton Thiokol wants bullshit then give them bullshit!" It pays the same! Mutiny was in the air. A number of documents returned by Wendy and Erma would have in the order of 30 typo's per page and we could find where they had typed maybe two paragraphs from our work and then three paragraphs' from someone else's work into the same document, never proofed it, and returned it to us. Junior High School students did better work unsupervised. A number of us complained to Kent that these two gals were disgusting and a serious bottleneck for the whole department. Kent told us that they were handpicked by Don Thompson and really, ultimately, answered to Don Thompson via Sylvia Rollins. Kent promised us he would talk to them and try and get them to pay better attention. I assume he tried. He was not optimistic for improvement. He went on further to say the problem was pervasive throughout MTI. Standards were very low in the northern Utah desert! So, it wasn't just System Safety and Reliability under 'failure by secretaries', it was a corporate wide practice.

Quite a number of us were beginning to understand the nepotistic and theocratic employee matrix at MTI. None of us liked what we saw. It was a sure bet any of the permanents we talked to were either supplied by the Mormon Church, or local family relationships. Certainly there were exceptions of different religious background, and some people who worked at MTI did not have other family members there. There are exceptions in every dominant group. The nepotistic and theocratic mix initially amused us, then it wore on us, and ultimately it disgusted us. These people were dangerous.

Some of the contractors / consultants were Utah Mormons. They had been into MTI time after time after time and they knew how MTI worked. They were used to the low pay (much lower than mine), they were courteous to "go with the flow", and they knew that little or nothing was expected of them. They were there for the money and then they would be gone. They were alienated, but they weren't really concerned about it because they weren't going to be around very long. There was then left a small group of highly concerned and diligent contractors and consultants. They were especially lonely, frustrated, and alienated. They fully understood the historical consequences and impact of their actions, MTI failures, NASA failures, and American failures. They needed their incomes but they were there *For The Right Reasons.*

CHAPTER 13

SAFETY RELIABILITY AND QUALITY ASSURANCE

Time: Mid - January 1987

Our entire group of consultants at System Safety were continuing to precipitate escalating MTI anger over MTI's clear cut suppression of known problems. Stories of suppressions and irregularities could be heard in the hallways, the restrooms, the cafeteria, the car pool – anywhere anytime. MTI management had been discussing that Kent might be made Director of the NEW SR & QA'S ("Safety, Reliability and Quality Assurance") organization's System Safety group. The NEW SR & QA organization was ordered by the President's Commissions. We anticipated Kent might at least have a final say and power over the new System Safety organization as directed by the Rogers Commission (The Presidents Commission on the Space Shuttle Challenger Accident) and the Rodney Commission.

This new SR & QA group was to encompass all aspects of SRM safety and to have autonomy to report only to a MTI VP and high ranking NASA officials. Creation of the new SR & QA group appeared to be the final last ditch effort by NASA to bring the SRM's to a safe condition at the order of the Rogers Commission. Currently the SR & QA organization at MTI was impotent, incompetent, and answered to nearly a dozen non-safety and non-technical persons. MTI changed their organization chart frequently so responsibilities appeared to change frequently, i.e. become opaque!

In fact Kent told several of us that the new organization was to ensure total, honest treatment of all identified problems. Most of us could see, smell, and feel the MTI hierarchy resistance to a proper new and autonomous SR & QA process. I had an 'a priori' certainty of MTI fear of the new organization. I wanted the new organization to work, but I was dubious. If Kent was to inherit "final say" and power over the new SR & QA organization, we would supposedly have no further problems forwarding our findings to NASA. All of us could and would then forward our findings on to NASA and other appropriate agencies. I, and others had fears that Kent would not prevail and be made manager of the new organization. Why should we expect Kent to prevail? How could Kent prevail when all of the management above him and around him were doing everything in their power to keep any documents from leaving the systems safety department or the reliability department? Thiokol management was not in search of excellence. Also, several persons including Mahalon Hollway were now campaigning to take Kent's position. There was a bad smell in the hallways.

Kent and I had another talk about a list of MTI personnel which suppressed System Safety findings. The suppression team if you will. No matter which way Kent turned, he was surrounded by overt and covert suppression. If Kent was suppressed first, then the rest of us were subsequently suppressed. The Free World's space program was crumbling before nepotism and theocracy. Frank Howell and others on the reliability side of safety were suffering their own analogous pain and suppressions. Frank and other reliability experts had been called into major meetings to review their work prior to compiling a NASA deliverable document called a CIL (Critical Items List) on the SRM's. Specific findings they discussed and reported during those meetings were stopped dead in their tracks by MTI. Frank Howell told us that he would bring up problems and the reliability management would typically just look around the room and say "we're not going to deal with that problem ourselves – omit it". That would be the end of it, "THERE WAS NO PROBLEM"! Frank tried to fight a couple of them and got himself into serious strife. Frank Howell was a fighter and a very learned and diligent engineer. However, as he led the charge against MTI methods of reliability suppression he felt the heat just like I did.

He was a top professional and kept the best balance possible. He openly felt Morton Thiokol was a terrible corporation and in fact criminal. Many of the others amongst us 'went with the flow'. Several times I thought Frank Howell was going to lose his job. Frank stayed perpetually pissed off. He was older, and sage.

During several System Safety meetings Kent related quite candidly and with dignity all of the political and suppressive problems he was having with his various managers. Kent was not complaining for the sake of complaining – he was giving us all a "heads up" warning so we wouldn't go over the cliff thinking we had any MTI support. He assured us all that *"we had no MTI management support!"*

It was Kent's foremost hope and battle to prevail . Kent loved Utah. Kent and his family were Mormon and had strong ties to the Utah area. Kent was an emotionally mature individual. Kent believed in the Space Shuttle and America's need for it, and the exploration of space. Kent was a quiet American Patriot. Kent put America before his personal Mormon, and Corporate pressures. A sparkling diamond in the Utah desert. A man with no future at MTI, probably no future in Utah.

Finally, around the 10th to 12th of Jan. 1987, the announcement was made that Kent McKinnon was to the be the new System Safety Manager under the new SR & QA manager. The manager of the entire new and larger SR & QA System Safety organization was now Ross Bowman. Points of confidence to mention here are that during the previous few months both Don Thompson and Ross Bowman had asked Kent, "Kent what is System Safety? What does it do or should it do? How should it operate?" The day the announcement was made, Kent was at a management meeting being held at an MTI management retreat in Ogden Canyon.

Kent related to us that hands were shaken all around, his back was slapped, and everyone congratulated him for being so successful and for being appointed manager of the New SR & QA's System Safety organization. Kent was pleased with the outcome and I believe this was the first time Kent truly thought he could really prevail in forwarding the SRM problems to NASA and others. We worker bees now saw the first glimmer of light in the tunnel. We had Catastrophic Hazards in our desk drawers. In fact our drawers were overflowing. We were poised for the dash to the other end of the tunnel – fulfilment of our efforts, making the returning of the Space Shuttle to space safely possible. According to the NASA and Roger's Commission scheme of business, MTI's new SR & QA organization was to be wholly independent from internal MTI pressures and was to answer directly to NASA and one MTI Vice President – all SRM problems were to be surfaced as a matter of NASA direction and government contract law! No more games and suppression! The back log of Hazards at this point in time was several hundred as a minimum, more likely approaching a thousand. Also a point of note is that none of us had yet seen or found a copy of the MTI / NASA contract.

Simultaneously, a number of new contract engineers and scientists were coming into both System Safety and Reliability. Of course, there were precious few desks and chairs to go round. A contract engineer who wanted to sit down had to exhibit ruthless stealth when he stole what he needed. A lot of nice guys stood around for days writing on file cabinet tops or such. Malcolm and I talked quite a lot about a Reliability Engineer, in his expensive blue pin striped suit who had an old broken down chair and a cardboard box as a desk for weeks.

The next day after Kent's appointment to his new position Kent had a number of activities underway and was quite busy as were his entire staff. We consultants were waiting for changes and final directions for preparation of our deliverable Hazard Analyses in whatever format would be final. The word was out from Kent about noon that there would be a mid or late – afternoon meeting in one of the trailer conference rooms to go over our new SR & QA, and System Safety Organization. Everyone in the department was to attend. We all enthusiastically had high hopes for really leaping forward from here on - back to shuttle launches. We wouldn't have to live in fear anymore! What a relief! How sweet it would be to ultimately do a great job in spite of Don Thompson and other unreasonable MTI managers.

About three in the afternoon the meeting convened and the entire staff was keenly in attendance. Kent showed up to lead the meeting and this was a shocker! Kent announced, yes indeed he was appointed in charge of the NEW SR & QA System Safety Organization. That was the good news. Then he announced he had just been told by Don Thompson that even though there was an organizational chart change, "nothing had in effect changed"!

The MTI System Safety operation would run as it always had, would be restricted as it always had been if not more so, and Kent would continue to have to "JUMP" when Don Thompson, Russ Rasmussen, John Elwell, Carver Kennedy, or any other key MTI management said "JUMP". Kent was visibly shaken and upset by this disastrous turn of events. Kent was 'dead meat' and he knew it. Many of us knew we were the epicentre of a massive 'Class A Felony, Fraud', and in my case I realized it was TREASON against the United States. While Kent was telling us the bad news Mahalon Hollway was jumping up and down, grinning and smiling, and blatantly acting as he was now Kent's boss – what was this about? A small minded and disgusting performance by Mahalon – a child in the sand box. Then Kent had to suffer the indignity of announcing that all of his work would be examined and vetted and approved by Mahalon Hollway as per Don Thompson's orders. HOLY SHIT! Don Thompson had nailed Kent's balls to the wall! Kent's fate was sealed. Don Thompson, a good Mormon Bishop fucked Kent, the U.S. Government, NASA, the Free Western World, and children unborn! History in the making. Don Thompson had humiliated Kent, castrated his efforts, and had obviously broken an array of serious laws. Thompson was deep into 'felony' territory - about 40 years worth.

NASA thought they were funding a functioning, autonomous SR & QA organization. What NASA really got was even worse than the same old shit! The *free worlds* cornerstone was now completely shattered! Kent told us the Rodney Commission was due at MTI soon and he would see what he could work quietly to cut out all of his immediate supervision and position himself for forwarding Hazards. AND, there Mahalon sat! It was written in Kent's and Mahalon's face there would be no improvement. He thanked us all very much for all of our good work and he told us to keep it up. I was thinking 'for what'? Kent guaranteed his personal support and told us no hazard was to go unchronicled. Kent's eyes and words were empty. He had lost the battles and the war and his soul was drained. The security of America and the Free World had just been dealt a decisive blow by Mormon Corporate Criminals. If the Russians or the Chinese had done it then it would have been an act of war! At this meeting another interesting sociological phenomenon occurred. Mahalon Hollway, all of a sudden was above Kent McKinnon in his actions and demeanour, and in the MTI hierarchy. Mahalon was looking down on Kent as though Kent was a worm about to be used for fish bait! Kent is not an especially competitive person and certainly not a game player – he was a serious professional. *"No Good Deed Goes Unpunished"!*

Kent was a person who saw his place in the scheme of things and operated as effectively as he could. Mahalon Hollway, was a man who saw himself as a consultant to God and couldn't be satisfied with his position even if he were God! Mahalon could not control his massive alcoholic personality or manias, he had always been out of control but he had just tipped into virtual madness. This was the first time I had noticed Mahalon being literally hostile as well as rude to Kent in public rather than just being high handed. Mahalon's prior rudeness was previously more covert. I and others suspected as a result of this meeting that Mahalon was definitely making his move for Kent's job, or possibly higher? We suspected that Kent would be slipped out of MTI quietly at MTI discretion. Kent was sure to be fired one day when he would not aid in cover ups as MTI directed. We were additionally wondering whether Mahalon was the "chosen one" to replace Kent? I suspected that Kent's last day was marked on Don Thompson's calendar and so did Kent. All of a sudden Mahalon Hollway's whisperings were in our offices and MTI hallways. Whisperings such as "MTI doesn't need all of those Goddamn System Safety people", "no one here knows what they're doing except me", "Don Thompson, you are in trouble – I'll save you", etc. ad nauseam! Mahalon was now positioned to make Kent's life miserable minute by minute and he proceeded to do so with pleasure. Condescension, harassment, stupid arguments, and antagonism, put - downs for all to witness. Kent could no longer truly serve MTI. He stayed for his family's sake.

It was with heavy heart that we all left that meeting, realizing that we were at the focal point of tremendous and lethal crimes and political wars. No heroes were going to get decorated in this war! More likely they would get their balls cut off! Soon after the meeting was over, there was whispering all over our department about what our fate would be -

probably dismissal by firing squad? We were all certain that MTI could not eliminate all of us from the scene without having to answer to NASA, the Rogers Commission and the Rodney Commission for their actions. However it looked like everything we did was just going to cause bloody hell and fill the trashcans!

Also some of us questioned the Rodney Commission - did the Rodney Commission come to Utah to ski and hunt elk? Maybe all of the above. Without sudden and violent overhaul, America's viable space programs were over. At this turn of events I could almost look around our work area and determine who was going to press ahead identifying problems and who was not. Most of them were of the opinion that if they want bullshit, feed them bullshit – ask 'em how much, what colour, where they want it, and when it's to be delivered.

I knew that some of these people would fight for what was right for America and some would simply condescend. Then I realized a direct analogy would determine who would be working at MTI the longest on contract. I knew I was unable to condescend.

Kent's meeting upset me so bad that I sat down and had a talk with Kent very soon after our disappointing departmental meeting. I said "Kent, what you're looking at here is a crime and furthermore what you are looking at is the fact you have been 'chosen and selected' by MTI to be their 'sacrificial lamb' for the next crashed Space Shuttle if you are here long enough to see the next launch."

If a Thiokol SRM caused the next STS crash, and if you do what these MTI managers are asking you, then you will go to prison for committing criminal acts, probably multiple murder and essentially treason!. "Your job is clearly to take this team of people, find hazards, and forward hazards to NASA. You and I both know damn well Don Thompson, his entourage and associates, are not going to stand up and be counted for not forwarding hazards. They're going to point the finger at you Kent, and you are going to be first burned and then jailed". Kent said, "You are exactly correct Steven. I am the appointed MTI sacrificial lamb".

I said "Well Kent, we have to do something about this. Do you think that there is any chance in hell we will be able to get the documents out of MTI and to the Rogers Commission or the Rodney Commission, or any others NASA group without you being fired?" Kent said, "I don't know, - I doubt it."

I told Kent that in my unsolicited opinion the FBI should be notified that America didn't have a Space Shuttle, and therefore an American Space Program. The USAF, the National Reconnaissance Agency, and the CIA could not continue to perform their space and defence related activities. Kent said he agreed it was a good idea to inform the FBI but he was afraid of the ramifications and repercussions to himself and his family from Thiokol and the Mormons. I told Kent I was afraid to, but there were certain times in life, if we like it or not, if it is personally inconvenient or not, when we have to do what we can *For The Right Reasons.*

The question here was not a matter that MTI managers were basically crooks and probably stealing millions or billions of dollars and giving a poor product, but rather that MTI had knowingly and willingly put the USA out of the space business and therefore the *Space Race*. Furthermore, MTI through its internal activities, was jeopardizing our nation's ability to defend itself via satellites necessary for military and National Security operations. Kent agreed with me that the scope of the problem was phenomenally large. Awesome! He had mixed emotions that maybe some small people at MTI had become immersed into "things" that were beyond their scope. I told Kent it may be over their heads, but it doesn't matter if it is or it isn't over their heads. This problem has to be fixed! Kent and I agreed to wait a few days to see how the cookies crumbled. Neither of us thought the cookies were going to crumble favourably.

During the next few days enormous pressures were applied to the System Safety department from all directions. Nobody could get anything typed because the two young women, Wendy and Erma, negated all reasonable efforts to have any work typed. Nobody could get any decision from NASA or MTI management on what format was to be used for deliverable hazard documents. Some eventually found the MTI documents they needed. The lack of other documents caused a lot of frustrations. Kent was looking terrible and was under terminal pressure from both MTI and Mahalon Hollway. MTI supervision above him had a gonad grip on Kent. Kent was torn!

Mahalon Hollway was making me sorry I had ever helped him come to MTI and that I had ever known him. He embarrassed me! Mahalon was treating Kent as though he was a "low life son of a bitch". All of a sudden with no announcement from MTI management Mahalon was literally acting as though he had the blessing of MTI management to treat Kent and the rest of us like dirt. Were these actions planned? Had his alcoholic personality run amok or was he authorized and directed to his bad behaviour?

Had he gone mad? Were these actions planned or helter skelter? Mahalon was now giving Kerry Sanofsky, Phil Greenhalgh, and Lance Heller a bad time as well? The most disgusting aspect of it all was that Mahalon was obviously enjoying it! Mahalon wanted Kent to kiss his ass! Kent was a man with a number of dedicated supporters in his own department but amongst MTI management he was still a lonely, isolated, sinking ship. Mahalon did absolutely no visible work for weeks. Kerry Sanofsky and Phil Greenhalgh were not pleased with Mahalon's "high handed" actions. They were mad about his lack of work. While others worked, Mahalon sat at his desk or stood in the "smokers" hallway glowering at everyone, but never worked while others read thousands of pages of documents, and calculated massive equations. Mahalon smoked and smoked and smoked! Then, he would tell one or another he didn't have to see any documents. He already knew it all! There were a lot of snickers. Mahalon had some tough clashes with very competent engineers and scientists during this time. Mahalon was supposedly an Optical Physicist by training but he had been around Inter - Continental Ballistic Missiles some. He was substantially knowledgeable about rocket propellant and rocket gas plum toxicities. Mahalon was vicious towards the Indian engineers, Arun Murthi and Rakesh Chopra. They were both magnificently educated and trained, first in India, then England and most recently in America. He also tried to put down Dr. Tony Laine. In all three cases it was 'no contest'. All three made a fool of him technically. Mahalon reacted to his technocratic defeats viciously.

My impressions were that I was looking at *Treason* and dozens of other crimes right in front of my eyes. I must go to the FBI and blow MTI wide open soon or someone else must! I was frustrated, angry, and fearful. Enormous impact on the United States military security would result if this situation continued at MTI and NASA. Who was the more dangerous, MTI or NASA? Even if the FBI and NASA knew then, the fixes could be years consummating. Could the problem be fixed? MTI could be 'shit – canned' in a heartbeat, but what about NASA? What do you do with a government organization that has failed? Had NASA failed or had its funds been cut to the point it couldn't function? No doubt about it, NASA had failed! NASA had a small army at Morton Thiokol and they didn't know day from night, they didn't care. It was a pitiful spectre. After thinking it over I also realized replacing MTI would also be a marathon and years in the process. *What a catastrophe!*

Again, Kent and I had a quiet meeting somewhere and I told him that I didn't see any alternative but to bring in the FBI to do whatever was necessary to get the ball rolling to salvage the Space Shuttle and the STS. By this time it was becoming clearer to Kent that his System Safety personnel were finding enormous numbers of problems in a broad spectrum of the SRM systems and components. All these SRM problems in and of themselves were catastrophic, however each of them needed to examined by MTI, NASA and the other engineering organizations including probably the Air Force and in some cases by the National Research Council. Kent said he was concerned about lawsuits from the sort of MTI suppression that had been levied on others in the past. He knew it was MTI's style to intimidate employees and he had heard of MTI suing several former employees. So I asked Kent, "well, what if I go to the FBI, put the red flag up to them, and they come and see you? Will you then co-operate with them under that circumstance?"

Kent said, "oh yes, of course", that would take the burden off of him for initiating the action!" Here at last was an MTI employee who put his country before himself! Before money even! A great man! An American.

So I told him I would probably be doing something along these lines very soon. Kent said he really didn't want to do something like this, but he didn't know what else to do. Every possible line of communications that could have been followed had been exercised and had been nipped in the bud by MTI management. Kent and I were cornered. Where did the pressure come from? What genetic difference forced us to drastic risk and action while the corporate sheep grazed? A mystery to me.

Something unknown to me made us different. I could not be a sheep. Neither could Kent. This sequence of events brought me to the terrible and horrifying knowledge that I was about to be sucked into something that I really didn't want to be sucked into. If I liked it or not, I was the right man at the right place and the right time and I knew it. Sometimes in life you simply have to walk up to the edge of the cliff and step over it.

I looked around the System Safety Department and realized that it was unlikely anyone else would go forward first. Trying to be wise in my decision of what to do was frustrating and frightening. I gave careful thought to what I was getting into but I was not crystal clear as to where best wisdom lurked? I realized that the situation was going to blow up like a bomb once the FBI found out about the MTI and NASA problems and that my life would probably be changed as result of my actions. I did not know a great deal about the Mormon powers in peoples' lives. The Mormons in Utah seemed provincial to me and I was distinctively afraid of physical harm from them. I didn't trust the Mormons because I didn't understand them. I didn't perceive Mormonism as a religion, but rather as a global business! I feared the Mormons or MTI would 'waste me' as a "dangerous infidel". I believed them capable of it from then to now.

I was concerned about MTI having billions of dollars available to them. They could become quite nasty if they found any FBI affiliations. Also I realized Utah had two very dangerous Mormon politicians. Senator Orrin

Hatch and Senator Jake Garns Considered by many to be amongst America's most corrupt, powerful, and dangerous. People get killed over the last cigarette and the last swig of wine. Would these folks kill me over a billion and a half dollars a year and their access to the USA Federal Pork Barrel? Then I pondered more about the Mormon connection? MTI is not all Mormon, however a high percentage of their personnel were Mormons. Would the Mormons do something rash to me aside from MTI? I couldn't have guessed then that later I would be just as concerned about NASA doing something rash to me. Then, later yet, I worried about Motion Salt who owned MTI and even later about the U.S. Department of Justice, and then other American Government Agencies. If any of these groups 'wasted' me it would no doubt be for a good cause in their sight – money and profit! "That son-of –a – bitch Agee is dangerous!" The truth hurts and may not go unpunished! That bastard Agee should not have slammed the Pork Barrel lid on all of those well placed 'trotters'!

I knew that the FBI, the U.S. Department of Justice, and all the people who would become involved directly or indirectly would probably become irate. There would, no doubt be U.S. Federal Court actions, Congressional and Senate Hearings, and God only knew what else. Perhaps media scandals? I thought to myself that I would, of course, ask the FBI to keep me out of all the court cases and legal actions, but down deep I knew that was unlikely. When the FBI needed you they can easily yank anyone's chain, for the benefit of America. You are only a subpoena away. The FBI and DOJ can tap anyone they want, and we live in a very small world.

In a criminal case this serious it was almost ludicrous for me to think I could avoid being involved in what would be inevitable court actions. What should I do? Once I blow the whistle to the FBI there would be *"no going back"!* They open a case, your name goes into the 302 file, and it becomes a permanent legal record, and the case moves forward or is closed. If the case is closed all of the documents then fall under the Freedom of Information Act and can be easily obtained by nearly anyone. Also the data goes to the FBI Data Warehouse where it lives forever. By opening Pandora's box I would have to be on tap whenever the FBI needed or wanted me from that time. I knew that could be demanding on my time, even inconvenient. I tried to consider what this would do to my career and life, but could not see the future clearly. The possibilities were that it wouldn't affect my career at all, or possibly it could completely terminate my aerospace consulting career. In my opinion and the opinion of many others most aerospace companies 'raid' government contracts whenever they can and each of them have their special way of making a little extra money – they call it profit.

Some say this is stealing money, some say it is smart business, some would stand up and say if they didn't do it they couldn't exist on typical small margins of profit allowed by the U.S. Federal Government on most contracts (normally 1 to 15% net at that time). Commercial business cannot live on small profits – in fact commercial business at that time could not exist if it didn't have somewhere in the vicinity of 35% gross profit.

So the U.S. Government asks a lot of its contractors and vendors on one hand to work for nearly nothing, and is open to rape and pillage on the other hand. The contract between NASA and MTI still frustrated everyone. If I had access to proper contract documents, I would probably read it and in a flash put my finger on what the specific crimes were, or so I thought. Maybe NASA had issued hermaphrodite contracts which would allow for this crap at MTI to continue? I did not consider that as likely. After thinking about these and many other things which would 'fall out" if I went to the FBI, it really boiled down to the fact that it didn't matter if my life was endangered, or if it ruined my career. Fate and destiny had placed me in a box. I only had one moral escape route., I had to step over the cliff and notify the FBI. If I walked away my conscience would have suffered for eternity!

Now I would move forward and make my FBI connection. Under normal conditions I would pick up my office telephone and call the FBI. Since the MTI phones were no doubt bugged, and since I was surrounded by MTI personnel, a simple call was a bad idea. I was surrounded by protective and paranoid Mormons I didn't trust. I had to think of something more clever and covert to make contact with the FBI. Additionally there was concern that maybe the entire FBI office in Salt Lake city was full of Mormons sympathetic to MTI? What Mormon or Utah worker would want to put the only large non-government employer in Northern Utah out of business? An improper investigation could be an enormous disaster to me personally and more importantly for the U.S. Space Program and the Free World. Still I realized the need to exercise due caution and wisdom. I was untrained and unprepared for what appeared to be coming and I knew it. It didn't matter. All I could do was the best I could. Would that be good enough?

It would be wrong not to mention here that I gave some thought to just simply going "public" there and then. It occurred to me along the way that a press conference would put MTI and NASA on the "straight and narrow" instantaneously. Going public would certainly forfeit my job and certainly my career, possibly my life. If I just simply took everything I knew public at this time to a major newspaper such as the L.A. Times, or the Washington Post, I knew not only the FBI, but a dozen other agencies would be building a fire under MTI post haste, or would they?

What if I went public and the publicity medium malfunctioned or only did a poor job? Would the public soon lose interest? The evening news continually reaffirms the news media's capacity for errors and inaccuracies. Furthermore, how could the failure of an American Corporation or Space Programs possibly compete with a Sunday football game? A foolish thought! Also, 'if it bleeds, it leads!'

In my mind I suspected that I might be causing a travesty of justice by doing this for the simple reason that once these matters go public there is a period of time before agencies such as the FBI and others can move. During that period of time, a great deal of information and documents can be run through the shredder or be swept under the rug, or buried with the bodies, and a number of people can be forced into MTI's way of thinking by the usual corporate twisting of their 'nads'. If I had gone public I could only speculate what would have happened inside and outside MTI. However, if I had it to do over again, I would give far more consideration to just going public than I did at that time. Many reasons and also times and technologies have changed over the last 30 years. Then journalists practiced some journalism, now journalists are confused and think they should run the world. At that time I completely believed in the FBI, the DOJ, and our American way. As you will see, I have changed a lot since 1989. We live in an entirely different world now, a world that is changing at the speed of global communications - instantly.

I considered some other methods of bringing these SRM safety issues to light, but none of them struck me as strong enough solutions to have adequate merit. I considered a long letter to some congressman, I considered a letter to the President of the United States, I considered forcing a face to face sit-down with NASA. All of these groups might be self seeking so I finally ignored them in frustration and in favour of the American legal system and the FBI.

If the relationship between MTI and NASA was that of normal government agencies and prime contractors, I probably would have simply just sat down and talked to a NASA Inspector General, or DECAS, or a similar group. Since there had been previous problems between NASA and MTI employees complaints I opted not to go that route. NASA couldn't be trusted and had failed to protect sources and helpful informants after the challenger disaster. NASA had an irresponsible track record in every regard.

After the Space Shuttle Challenger blew up, NASA opened a "HOT LINE" to Washington D.C. from MTI guaranteeing that no-one calling problems in on this line would be revealed to MTI. Any callers were to remain anonymous. As soon as the hot line was put into operation a number of concerned MTI personnel used it and were summarily burned! As soon as people would reveal problems in their disciplines to NASA, NASA would then beat up on MTI managers over the specific complaints. NASA beating up on MTI was all well and good, except MTI then said amongst themselves "who had access to that information"? One of the MTI managers would say "him or her", and they would all jump up, run down the hallway, grab "him or her" and run them through the meat grinder. This reportedly happened several times. I was told about this by several MTI permanents. I don't know any of the victims or if this is only well informed hearsay from actual frightened MTI witnesses.

So NASA did respond to their own hotline, however they did it in such an unprofessional way that its success was limited and negated. After several persons at MTI were severely damaged by NASA and MTI word went around MTI in short order. *"No Good Deed Goes Unpunished"!* Everyone stopped using the hot line. The hotline was available to me as well as others when I initially went to MTI. Daily discussion, usually a joke, was "let's call the NASA hotline – shouldn't take more than 10 minutes to become available for a good job."

The only reasonable fix I saw was to put the problems at MTI and NASA into FBI and DOJ hands. I wondered, with so many people at MTI (approximately 7000 at that time) why hadn't someone already gone to the FBI? Maybe it will help – maybe the FBI is already *working this problem and none of us know it!* Perhaps the FBI could simply use some more help? I could only speculate?

Soon I came to expect the FBI had to be 'on to MTI and NASA' and I sincerely hoped they were. I didn't want to be the central focal point of a case. Even though I considered the FBI to be the experts in these matters, I knew the FBI could bite me as well as any other group of civil servants. FBI standards are high and that agency usually gets the "pick of the litter", but even so there are the occasional "retards"!

Sometimes you just have to step over the cliff – life ain't perfect! I stepped over the cliff! It was a long way down and as you are reading this I am probably still in 'free fall' 30 years later! Now it is 2017 and I have not put it all behind me. I never will as I can't.

CHAPTER 14

FBI CONTACT

Time: Mid-January 1987

I was at peace and settled in my mind about the need for the FBI to enter the picture. I was frustrated and anxious to get to the FBI and tell them what was going on. Once over a decision I couldn't move fast enough. National Security was wasting. My awareness was seething. My logic had derived the fourth criteria for obligatory Whistleblowing. There was an immediate need, I knew the details, I had the means to remedy the problems, and no one else was likely to help. America was bleeding!

I took a few days to formulate how to get to the FBI without setting off a Mormon riot at the local FBI office or at MTI or in Brigham City. Was the Utah FBI office loaded with Mormons? The reason I thought that there might be a lot of Mormons in the Salt Lake City office was simply that probably every Mormon in the FBI was trying o get back to the Salt Lake area soonest. Return to Zion! So I finally concluded a plan to contact an FBI Agent friend of mine in Southern California at the Ventura, California FBI Office. Having made that decision, I then had to figure out how to contact my agent friend – Special Agent Bill Pemberton.

I could absolutely not use the telephones at MTI and call the FBI office without expecting a hell of a lot of trouble (or to be shot dead) in short order. People are killed daily for small change, MTI was stealing billions! I thought about taking a day off work and going into town but I had so much work to do that I did not feel taking off was appropriate either.

I considered writing a letter to my agent friend in Ventura, but that wasn't such a good idea because I really didn't have the time to write him an extensive letter to explain a complex situation. Perhaps I could drive to Corine (22 miles) and make a call during my lunch period from a pay phone there? This FBI connection wouldn't be as simple as it sounded. Most of the people in Corine were supported by MTI either directly or indirectly and it's not the sort of tiny town that would have good secure telephone booths. This option was not comfortable to me. Why should I think Special Agent Bill Pemberton would be at his office during my lunch break? He was a very busy man, Hummmmmmmmmm.....

Finally, one day, in total frustration and with resolution, I called a new friend of mine who lived in Brigham City. Even though I had not spent much time in Brigham City, I had met five or six groups of especially nice people. When I called my new friend, I asked him to come to the MTI plant and meet me in the parking lot at 2:00 p.m.. He said he would be happy to and asked me what was up. I said I have a special project that I just had to set in motion. I can't get anything done working these tremendously long hours here, so perhaps he would be willing to help me. He replied, "certainly!"

At 2:00 pm that afternoon, I met with my friend out in the parking lot to the north of the main MTI office entrance. Feeling scared and exposed I climbed into the car and asked him if he could keep his mouth shut? I knew he could, and he assured me he could, and I never questioned him for a moment. I said to him, "here is a note; I want you to call this FBI agent, Bill Pemberton and read it to him." The note said in effect that this was a message from Steven Agee and would Bill please call me at home at a late hour in the evening in the next few days and I would explain to him about a substantial group of National Defence related crimes in progress. My friend acknowledged that he understood, would keep his mouth shut, and would take care of it immediately. My friend then drove back to Brigham City. As I was going back into the MTI main office building, I realized I was in for a lot of trouble whatever it was. From that moment on I became nervous and paranoid about the potential of MTI management or the local Mormons killing me. "It was no doubt an elk hunting accident or an assisted suicide!" Shot in the back only three times! Shotgun of course!

As I walked back down the long hallway, I realized that life from here on out was going to get tough. I was irrevocably over the cliff and was experiencing the nausea of "free fall"! It might be that I would just have one simple meeting with the FBI, tell them the problem, and they takeover or they may want me to help them? I didn't know and the uncertainty frustrated me. I resolved walking down the long hallway that I would start acquiring every document I could lay hands on that would help the FBI understand the scope and diversity of MTI problems, crimes, and cover-ups.

In my mind I started making a shopping list of documents which would educate the FBI as to what a Solid Rocket Motor was and how it worked. Then secondly, types of documents which would show the suppression of Safety and Reliability Documents and Flight Safety personnel's activities, and third 'outright evidence' such as 'signatures'. How would I explain to the FBI the difference between normal corporate bullshit and safety analysis and document suppression? I realized I had my work cut out for me. I could look like a fool if I didn't package all of my knowledge and accusations properly. After all, most of the FBI were Accountants and Attorneys and very few of them if any were Scientists or Engineers at that time in history. What if the FBI sent a retard? It could happen! It took a couple of days for my friend in Brigham City to make contact with the Ventura FBI for me. My friend in Brigham City was a Mormon but I trusted him anyway. In his past life he had been a U.S. Navy officer and weatherman for 1/2 of the planet. He seemed to be of the same cut as Kent McKinnon. He was retired military. Two or three nights later, late at night, I received a phone call from FBI Special Agent Bill Pemberton in Ventura. Bill was a very capable and bright fellow. Bill said, "What's up Steve?". I told him, "You ain't gonna believe this!". I had met Bill Pemberton when he was a case Agent on a B1-B Bomber crime I had uncovered a couple of years earlier. Also because of my longstanding Security Clearances the FBI had a substantial file on me.

I explained to Special Agent Bill Pemberton in brief blunt terms that MTI in my opinion was completely suppressing System Safety and Reliability of the SRM's. MTI's suppression would affect the safety and reliability of subsequent Space Shuttle launches. I told him that MTI had been directed by the President's Commission on The Challenger accident and the Rodney Commission to have open lines of communication to NASA and these lines of communication did not in fact, exist. I told Special Agent Bill Pemberton the President's Commission had ordered an independent SR, System Safety, and QA MTI investigation but that really was a sham. I told him quite a large amount of money was involved, billions! He asked me what the motive of the crime was, and I said I could only assume that it was illegal profit - Build crap, *take the profits and run!* Go through the motions.

I do not remember the totality of the conversation, but Bill was a very bright fellow who had effectively worked aerospace and white collar crime for some time and had a very quick grasp of what is going on. After I explained to him MTI's indiscretions as I saw them, he asked why I was calling him rather than the Salt Lake FBI Office? I told him, "oh yes! I almost forgot, the vast majority of personnel at MTI are Mormons and frankly I'm afraid of them because the obvious suppressors of information and liars to the government are Mormons." In fact a bunch of them are Mormon Bishops. There are two things you must do for me Bill if you forward this and start a case in the FBI system:

1. Whatever FBI Agent is assigned to this, must be a pleasant and bright fellow, because I am working too damn hard to put up with a bunch of Bullshit from some dumb SOB. Bill understood.

2. You have to promise me that this case will not be dropped into the hands of a Mormon FBI Agent in Salt Lake, City. Bill understood and not only agreed,but was adamant that he very much agreed with my thinking.

I told Bill point blank that I was afraid of being 'wasted' by a zealous Mormon who might just do me in to maintain the integrity of the Church and the pipeline of money from the USA National Treasury to the Thiokol and Mormon Coffers. They would kill me for my own and for God's own good! A lot of Mormon Bishops were the apparent MTI criminals. Catholic fanatics, Islamic Terrorists, and Irish Protestants "waste" people on a regular basis – why should I think any different of Mormons? Bill said he didn't think I'd be wasted if it was found out by a Mormon and that I shouldn't be too uptight over the possibility. Easy for him to say! However, he related that there was a serious problem with Mormons within the FBI itself. Some Mormon Agents and their dealings with Mormon suspects and criminals had been contaminated. I asked him if he could elaborate on that and he added he couldn't go into it, but there had been occasional problems and it was even worse for the CIA with Mormons.

He assured me he understood my feelings and where I was coming from on the subject of Mormons and therefore it would take him a few days to locate the proper agent to contact me and he would honour my requests. He said he would contact the FBI's Washington, D.C. office and work it back from there rather than call straight into the Salt Lake office. Bill thanked me for contacting him and he felt from what I had told him the issues were damn serious and needed looking into immediately. "Woe is me" I thought! Freefall uncertainty - where will I land?? Nausea!

So FBI contact was now established and my basic concerns were then a matter of FBI record. Why me God? OOooooh Shit! My career is probably going to take a hit, my income is probably going to be eradicated, and there may be a 'hit man' looking for me soon. *No matter, I geneticaly had to go do what I was doing*. Our American Space Program had to be saved.

About 2 days later, 14 Jan. 1987 in the late evening after I returned home from work there was a knock at my back door. I knew without question a new chapter in my life would begin when I opened the door.

I opened the door, a young man, very friendly, clean cut, wearing a sweater, blue jeans, and a jacket (with a bulge under it), introduced himself as Tim Healy, Special Agent, FBI. I invited him in and offered him a cup of coffee, coke, or beer – whatever in the hell he wanted. I told him I thought we were going to be there a while tonight, and he said, he had a feeling that we had a lot to talk about. He went through the usual formalities of showing his Badge and Credentials (Creds). I could just simply tell by looking at this man and meeting him in the first few seconds that he was no ordinary fellow. He had a radiant smile, clean cut good looks and his intelligence beamed through his entire being at you when he talked. My FBI friend in California, Bill Pemberton, had obviously done well. We sat down at the kitchen table and I said well, where do we start? Tim Healy told me he had read Bill Pemberton's telex and he was shocked and astounded! He had been assigned to it and discussed it with FBI No. 1 man in Salt Lake City and they had agreed on the necessity to meet me immediately and find out what was going on in detail. So then he asked me to give him the rundown on what was going on at MTI as I saw it.

I explained to Tim that I had acquired for him the old but fairly accurate manual on Solid Rocket Motor technology so that he could read it and understand what the Solid Rocket Motors were all about. The manual was about 8 years old and the program was 13 1/2 years old but very little had changed. These manuals had been out of print for about 8 years, but I and others had Xeroxed our own copies from the one last relic dredged up at MTI by Malcolm Phillips. Tim told me that he had been flying Harrier Jets up until a year before and that pleased me. First of all there are no stupid Navy Pilots, secondly, only the cream of the crop of Naval Aviators ends up flying Harriers for the U.S. Marine Corps.. The 'pilot load' of flying a Harrier was at that time roughly estimated to be 300% of flying a normal jet fighter. Harriers then were difficult and dangerous. They are improved now. This man was not just a pilot, he was a great pilot! I also held a U.S. Commercial Pilots Certificate for some time and I was impressed! Later I found out that not only was he a great pilot by virtue that he flew a Harrier, but he was a Harrier "Top Gun" to boot! The cream of the cream. Additionally, Tim had been in charge of the squadron of Harrier's operations and maintenance, so he understood corrosion, general electronics and much to do with military aircraft electronics and weapons systems.

Military jets are not so different from rocket systems. Tim was not a Scientist or Engineer, he had a University Degree in History and Spanish, but he had broad overview knowledge. Harrier jets are actually more complex than the Solid Rocket Motors on the Space Shuttle. Tim knew about military specifications, and he had previously preached Flight and Aircraft System Safety in the U.S. Marine Air Corps..

I told Tim that I had acquired an acronym list for him. Additionally there was a dozen or so other documents that I acquired in anticipation of this meeting which were simply educational in nature. These would help him and others to learn about SRM's and their relations the Space Shuttle systems. With these documents there was also a stack of safety system documents and other pertinent documents for him to begin assessing and digesting. In that first batch of documents was an easy 50 years of prison time.

I then sat back, looked over a cup of coffee, and had a long detailed talk with Tim about everything I could think of at that time that I envisioned to be wrong at MTI and NASA relating to the SRM's and MTI. I detailed to Tim the fact that numerous Systems Safety people had identified a large number of Catastrophic Hazards on the SRM's. First we could not get them in the final form for delivery because NASA and/or MTI changed its mind weekly about what they required and how it should be prepared. I then told him that wasn't the criminal activity in my estimation, however, it was certainly an obstacle to the safety of the U.S. Space Program.

I then related to him the fact that a tremendous number of Catastrophic Hazards were being found and that our boss Kent McKinnon had done everything short of kill people to try and get the documents out of the MTI System Safety department and on to NASA. I explained to Tim the fact that I was almost fired some weeks ago for being effective. I explained the whole scenario to him. He was not pleased. Tim was like a sponge and soaked all of this up and then we discussed the fact that there was a new SR & QA organization as per NASA and Rodney Commission and Rogers Commission requirements, however, the new SR & QA organization was castrated and fraudulent from conception. MTI never even gave it a try!

Everything MTI said about the new SR & QA organization to the Commissions appeared to be "False Statements to the Government" - each statement a crime. Then I explained to Tim that every government contract incurred some "waivers" and "deviations" as the normal course of development and business legitimately. The SRM project had hundreds if not thousands of waivers and deviations. I had witnessed MTI meetings where their solutions to ALL problems were 'deviations and waivers", rarely 'fixes' or 'repairs'. MTI required deviations and waivers as an addict needs a "FIX". NASA freely gave MTI their FIX! Deviations and Waivers are much cheaper than real fixes after all.

However they are deviations from reliability criteria and performance to say the least. Tim literally had a glazed look on his face and he was astounded by the scope of what I was telling him. America's Space Program was fucked!

Then we came to the point which I guess is common with most FBI investigations. Tim asked me point blank, "What are your motives for coming to the FBI Steve"? At first the question surprised me and I told him I hadn't thought about it in those terms. Tim said "well, some people call us because they want money, some people call us because they want to 'do somebody in', who did them harm, and there were thousands of other reasons. Tim told me this is a standard question the FBI always asks.

I had to think a moment, but not long. I told Tim that the entire aerospace industry had been experiencing a lot of ups and downs due to the Gramm-Rudman Law. Due to the USA's lack of a firm commitment to space programs, and due to hassling between civilian and military space programs. I told him that this was probably the natural course of history for any nation entering space oriented activities to experience in my opinion. MTI was, in effect, and in fact, unilaterally negating all of the United State's space efforts! I further explained to Tim that MTI was the sole source supplier for the SRM's for the Space Shuttles. Also there was a body of evidence that MTI was causing problems on MX missile launches, America's new ICBM project, and other systems. Then I explained to him that with the best people in the world it would take 2 to 4 years conservatively to build other SRM's via new designs using a new contractor. An enormous new expenditure in facilities. Therefore, in my opinion MTI had the entire U.S. Military and civilian space efforts of the United States by the 'balls'!

I explained to Tim that MTI was supplying crap to the US government and the US's prime space lift vehicle simply was unreliable presently and would continue to be unreliable for some time into the future.

I explained to Tim that MTI was getting a billion and a half dollars a year in contracts from the U.S. Government. I did not know how the division of contracts between the Trident, Polaris, tactical missile motors and others was, but as for the space shuttle, I guaranteed him that the US was buying crap. I explained to him that the US's larger satellites had no other path into space other than via the Space Shuttle at that time.

Most of President Reagan's Star Wars Program and NASA's Space Station had no other route to space at that time. We did not have other heavy life vehicles and we had committed our National Defence efforts in space primarily to the STS. As a premier world power America was in deep trouble. Russia had several heavy lift rockets. The USA was entangled in the Cold War against Russia. Tim's eyes literally glazed over. Tim was pissed! I told Tim I called the FBI to alert them *"For The Right Reasons!"* I can't remember how long we talked but it was a long time and a lot of coffee went over the lips. Tim and I hit it off well from day one and he indicated to me that they would probably like my help and that he would be talking to FBI No. 1 in Salt Lake City the next morning regarding these matters. {FBI No. 1 in Salt Lake managed 5 states at that time.}

We discussed a few basic ground rules about what we should do about not being seen together – and other security measures. It was early in the morning when Tim left – documents under arm, a glazed look still on his eyes. Not only had I entered a new phase in my life, but Tim had entered a pivotal phase in his also. Where would it all lead? I was tired, frustrated, scared, and committed - *For The Right Reasons*. I was involved in something that was much greater than any one individual. It would lead where it led – over the cliff!

Simultaneously, as I contacted the FBI there were a number of new players arriving at MTI System Safety. These players were, Mr. Kelly Giles, Mr. Tom Bradley, Mr. Horace Hinkle, Mr. Jim Taylor, Mr. George Ave, Mr. Dale Donalds, Mr. Fred Broadhacker, and, Mr. Donald Jonacek and others. Most with Master's Degrees in science, engineering, chemistry, etc... Generally a learned accumulation, i.e. a diverse 'Brain Trust'.

I was frustrated that these new players were going to enter a game that they probably could never win. Certainly it would be financially beneficial to them and I did not begrudge them financial gain as long as their work counted. However, for the national good of the U.S. Space Program. I had my doubts that much of their work would ever be forwarded to NASA or other appropriate agencies unless radical changes took place. Back in December 1986 two Indian Engineers came on board the MTI System Safety Dept.. These two were Mr. Arun Murthi and Mr. Rakesh Chopra. They were the two best trained engineers I had ever worked with. They were trained in India, the U.K., and the United States. They were not myopic technical persons – Broad in scope, deep in their understanding. Also in late January Dr. Tony Laine came on board our System Safety Group. He was an interesting fellow. He was a Ph.D. in Physics, gregarious, cheerful, friendly. He was technically sound and his technical grasp was sudden / instantaneous. He had possibly lived with too many computers over the years. His computer and technical skills were acute, his social couth proportionally diminished. I liked Tony. He was a willing soul. Tony and the Indians (as we all called them) were very heavy weight technical analysts . Others came - some weak, some powerful.

Fear and paranoia swept over me. It is very lonely and uncertain over the cliff in free fall. Pandora's Box was well and truly opened, and it's ugly venomous contents weren't yet completely revealed.

CHAPTER 15

FBI OPENS CASE AGAINST MTI

Time: January 15, 1987 - Jan. 30, 1987

On 15 January, 1987 Special Agent Tim Healy opened an "FBI Case (No. 207C – 77), Morton Thiokol Inc.. Fraud Against Government, Office of Origin – Salt Lake City, Utah OOSU." A few days after my initial contact with Tim we had another meeting at my apartment. Our scheduling put him at my place just after I arrived, home from work at MTI (approximately 8:30 PM.) I arrived, then Tim arrived. I told Tim I was hungry and let's run down and get a hamburger and then come on back and we could talk while I ate. When we came back Tim told me that the Salt Lake City FBI office had "gone off like a bomb" over what I had told him. He told me I could rest assured that No. 1, Bob {Bear} Bryant had restricted it from the in-office Mormons until further notice. The fact that the Utah FBI Special Agent In Charge (FBI SAIC) had restricted the case from Mormon FBI Agents for now made me feel a lot better because I was on insecure ground where Mormons were concerned at that time. From my position Mormons and Morton Thiokol had to be considered one of the same – *very dangerous.*

I could not tolerate additional frustrations. During this meeting I delivered another substantial load of documents to Tim and we discussed my acquisition of these documents. Tim asked me how I had removed these documents from the MTI plant and I told him "any way I could"! I told him the only documents I had taken out were strictly the ones I was forking over to the FBI via Tim himself. Also on occasion the MTI guards would look in my brief case when I left work, but they would never touch anyone's person, therefore some of the documents I brought out via my brief case, most documents were brought out on my person. None of these documents were Department of Defence or Department of Energy 'Classified Documents'. At this time I had no predication into heavy document acquisition for the FBI and was able to simply put a handful here and there into my coat lapel pockets. Tim told me that the FBI managers in Salt Lake City had filed special papers giving me permission on behalf of the U.S. Government and the FBI to acquire papers and documents for the FBI Case against MTI. Since the SRM was a government project all of the documents technically belonged to the U.S. Government. A U.S. Federal Judge had authorized my permissions to remove the documents.

Tim said the filing of these special papers was the U.S. Government's way of sprinkling "holy water" on my document acquisition and was an important issue and he asked me if I had acquired anything which was Department of Defence 'Classified'. I told him I had acquired no Classified Documents. I paid attention to that because I knew God couldn't help anyone who took Classified DOD or DOE Documents without authorization. It should be noted here that nothing about the SRM work by MTI was currently DOD Classified. Certain aspects of the STS were classified, but not MTI's part. "Company Sensitive or Confidential" was quite another matter of course. Tim looked over this evenings set of documents and he was essentially crazed! Tim was a very brilliant young man and he could see in a flash the trend and tendencies of what was happening at MTI. Tim told me that FBI SAIC (Local No. 1 Bob Bryant and FBI local No. 2 (Vernon Kohl) in Salt Lake were hot to meet me at the earliest possible date. I was to rest assured that the FBI and the US. Department of Justice were extremely appreciative of my efforts in coming forward and bringing these problems to their attention. Tim asked me if I was willing to continue helping the FBI? What could I say? I asked Tim what he thought would happen?

Tim said he wasn't sure. Tim said there were a number of possibilities. One possibility was that the FBI could go to MTI management and say "knock off the Bull Shit or else!!" This approach had been used in some cases and was sometimes effective. Another possibility was that the FBI could essentially seize all documents and records at MTI and put MTI out of business in an hour's time, making them unable to conduct business. Yet another possibility was simply the long arduous process of criminal prosecutions. Another possibility was that the FBI might not be able to make a case – however Tim didn't feel closing the FBI's case was a viable alternative. Time would tell.....

I asked Tim what the exact crimes were in legal terms. Tim gave me quite a list of possibilities, many of which were things I hadn't thought about and didn't particularly understand. I had some limited training and experience in business law, but no training or experience in criminal law. Criminal and business law are worlds apart frequently. Therefore, I was surprised by the possibilities that Tim and the other FBI Agents were exploring.

I asked Tim about *"treason"*. I told him it was my feeling that what was really happening here was treason against American and Western World interests. I felt the entire future of the USA in space was being squandered. Necessary deployments of reconnaissance satellites for the protection of the USA were terminated or drastically postponed by unacceptable MTI actions. Was that in fact not "TREASON"? Tim said he didn't know right off the top and he would have to dig into the issue of TREASON. Tim explained to me that he was a new agent and had only been with the FBI about a year. He was technically called a "pup". Tim figured the reason he had this case was because he flew Harrier Jets and knew more about aerospace than anybody else at the Salt Lake City FBI office.

Tim also explained to me that he actually operated out of the Ogden office, but all of the FBI Agents in Utah answered to No. 1 in Salt Lake City – Mr. 'Bear' Bryant. During this meeting Tim set me up with special phone numbers where I could get in touch with him any time of the day or night. These are called "One Call Phone Numbers". You call one of these numbers, somebody answers and they listen but don't say much. When you call a "One Call Number" you then tell them what you have in mind. I told Tim again that I was concerned about my personal safety and he assured me that he didn't think that would be a problem. By having these numbers I had FBI help as close as possible under my circumstances. Tim also gave me a special mailing address for mailing documents from MTI to the FBI via a business address. I had explained to Tim that a number of documents were far too big and bulky to "sneak out" of there under my coat, stuffed down the back of my pants, or hiden in my brief case. Tim asked me how I could get these documents out of MTI and I told him that the only I could think of for the very large one was to simply mail them out via MTI's internal mail system.

Tim asked me how I felt about mailing documents? I told him it scared and frustrated me. I told Tim that the MTI mail deposit in the main building was next door to security. We then had an in-depth discussion about what Kent McKinnon had told me about MTI Security, and what others had told me about it. Everyone I talked to at MTI quietly about the MTI Security Department mentioned to me that they were afraid of them. On the other hand, I told Tim that thousands of documents a day went through that mail room and I didn't really believe security could look at each and every one of them being mailed. Tim thought the mail would be fine. I didn't know for sure? I used it, but it scared the hell out of me.

Tim told me that was fine because I didn't think in a million years the FBI could really direct me as to what would help them until they much more thoroughly understood the problems and the technology involved. Tim told me he was now studying the documents I had given him. We had some "training" talks along these lines – just zeroing in on how the SRM's work. I could not have had a better pupil! What a shame MTI managers and engineers didn't understand as well as this FBI Agent! He was so smart it made me proud to work with him.

Tim and I had several long discussions about the politically sensitive " Mormon problem!" Tim and the entire FBI had to be outrageously careful here. Religious prejudice is a serious and sudden way to lose any court case in America. Furthermore the FBI in particular had to be very careful to not appear prejudiced in any way as an American Governmental law enforcement organization. The FBI case had to be made only on legal grounds, merit and the law standard demands and criteria of the American Constitution.

This book is not U.S. constitutionally restricted so I can and will speak openly in truth and with accuracy. I was not and am not prejudiced, but I am pragmatic. There was a lot of Mormonism woven into these criminal cases. The FBI can never say that, even though they knew all about it. Political Correctness of the time. The FBI had to work around the Mormon issues. I am not religiously prejudiced, I defend everyone's right of choice and belief. If Thiokol's rocket plant was in Mississippi the criminals would probably have been Southern Baptists. In Boston probably Catholics. However, the plant was near Brigham City, Utah. The FBI planned far ahead and worked hard to avoid religious issues so that religious issues would not spoil their pending cases. During this visit, Tim asked me if I had any information at all which tied MTI to Dr. James Fletcher (the Director of NASA)? I told him I did not. This question seemed to crop up from FBI Agents time after time. Contemporary newspapers had numerous articles that Dr. James Fletcher was under substantial pressure for what the newspapers called "the Mormon Connection" and "the Mormon Mafia". I believe this meeting with Tim went to 2:30 AM. Since I would be back at work at 7:00 AM the same morning this is the point where I started physically going downhill fast. It would be several years before I recovered at all. The fact is I massively damaged myself during this period. Now 30 years later I am a physical wreck with broad spectrum serious medical issues. In fact there is no way I can last very long.

When Tim and I were together talking about the problems, the documents, and the personalities at MTI there was never enough time! There was too much to talk about and go over. I doubted to myself that any two people could ever cover it completely.

During the workdays at this time there was tremendous pressure on Kent McKinnon and he was rapidly escalating downhill physically and emotionally. One day I noticed he was in an unusually good frame of mind and I asked him what was up and he said, "thank you, thank you" for contacting those aerospace companies on his behalf. He said "I received an application and a phone call from high level managers at Boeing and it sounded as though they had some very serious interest". I told him that I thought Boeing was probably the best run large corporation in the world. Kent said he was eagerly filling out the application and had his fingers crossed. I wished him well. At this time, I was on a heavy roll preparing Catastrophic Hazards and I continued to prepare them. Most of the equipment I studied on the SRM Systems were literally 'crap'.

Poorest designs, poorest procedures, poorest electrical grounding, poorest drawing control, poorest systems integration. In brief MTI's SRM's were disgusting and I proved it to the FBI beyond any possible argument. The FBI was not pleased!

Steven Agee (left), FBI Special Agent Tim Healy (right)

CHAPTER 16

FBI ON THE MOVE

TIME: 20 Jan. to 1 Feb. 1987

I was asking a very simple question about the SRM's 'Safe and Arm Device' and the SRM 'Ignition Train'. A simple and fundamental question:

"Were these SRM systems properly electrically grounded so as to not ignite prematurely or fail to ignite?"

Life had been difficult enough without having asked this fundamental question! I sat back in my chair, hung blue prints on the wall, charts on the wall, drank coffee, and thought " no doubt these systems are properly grounded". How could they possibly not be grounded? *Wrong*! I jumped clear out of my chair when I saw certain things that apparently had fallen through MTI designers and engineers cracks. I could not believe my eyes and I said to myself "no, no, surely this thing's grounded and I just don't know enough about it yet - have they hidden something?".

I started digging deeper and deeper. The deeper I dug the more it became obvious to me that there were enormously complex and serious electrical grounding failures in the SRM's, the Safe and Arm Device, and the Initiator. The 'igniter' and its electrical relationships between components were deadly *Catastrophic Issues!* On these items I openly wrote Hazard Analysis covering their SRM electrical grounding systems failures.

A new player had entered the scene just after Christmas, Mr. Tom Bradley. Tom was an older fellow who had just resigned or retired as manager of System Safety at Ford Aerospace in Phoenix, Arizona. He had known Kent for years. Kent had a difficult time receiving approvals for enough money to bring Tom Bradley to MTI even though he wanted Tom onboard.

Kent, Don Thompson, and Sharla Rose had received approvals and obtained our raises, Malcolm, myself, and some others had a base salary which was good in its day. If we each eliminated only one catastrophic problem we were a bargain for MTI, NASA, and the USA. So with an expanded budget (thanks to me) Kent had been able to coax Tom Bradley and the Indian Engineers into MTI's System Safety Group. Tom and I had an extensive talk a few days later about these electrical problems and he said that I was exactly correct in his opinion and that my issues were well taken in his opinion. Tom was not ever close to my technical expertise, but he was a System Safety Department professional 'paper pusher' in his career.

The Indian engineers all looked over my findings eventually and all agreed I had identified serious deadly problems – Malcolm also. You will see later these are pivotal points. Kent was frightened and afraid with good reason. Kent was in pain to make MTI's solid rocket motor's as good as they could be. Kent tried everything imaginable to fix the problems at MTI from within the MTI Corporate System. Sadly no one was going to succeed from within the MTI system. So in exasperation he wrote the following letter to Congresswoman Joyce McDivitt on Dec. 26, 1986.

"Dec. 26, 1986

I was very glad to make your acquaintance at the Morton Thiokol presentation.

The reason I am writing to you is because I may have made a serious error in judgement by expressing my personal feelings on the safety situation at Thiokol. I was not given a chance to say anything at the Rodney visit, and I was cautioned several days before your visit by Mr. Don Thompson that if I said things that I felt were the case in your scheduled meeting that I just as well start looking for a job, because they (MTI management) would get rid of me. This was said in front of three of the engineers who were sitting around listening to the discussion.

I was also called into a meeting with Don Thompson and Mr. Elwell in which I was specifically told that I could not continue to be a problem to them because I did not want to be a part of giving a false impression to the Presidential Ad Hoc Committee.

I was told to think about it until after the presentation was over. The day Following the presentation, I was offered to be reassigned or remain in my position. I told him I wanted to remain. (I would have gladly transferred prior to the presentation to avoid being expected to be silent or say that everything was going fine in the meeting) but once all of the problems started to come out in the meeting, I could not restrain myself and allow them to say things and give the impression that things were going fine

I was allowed to retain my position, but I now believe they intend to fulfil the implied promise and I believe I must take this original advice to start looking for a job.

You mentioned that you are going to be getting additional help in your jobs and you appear to be associated with several fine System Safety professionals who might know of some opportunities. Anything you could do to be of assistance would be really appreciated.

Please forgive my forwardness, but I believe I may need some help to find another job quickly. Please understand that I am an honest man. Needless to say, I could really be in difficulty should word of this letter get back to Thiokol.
Respectfully,
(signed) Kent McKinnon "

For obvious reasons Kent was afraid to mail this letter. Then on 10 Jan. 1987 Kent was even more desperate and he wrote the following letter:

"1 / 10 /87

When I first wrote this letter after your visit I was afraid I might be out on the street looking for a job very soon.

I have struggled over whether to send this letter to you for several weeks. It appears that they (MTI) may not remove me immediately, but I now feel that I must leave Thiokol before I become a scapegoat.

NASA is expecting me to re-evaluate all of the analysis and testing that was required and the SRM Hazards. Our re-evaluating efforts indicate that many of the findings and procedures were not very good. What makes this bad is because Thiokol management doesn't really want me to identify anything that might make the company look bad.

If I try to satisfy NASA, program management could do several things to make my situation worse or escort me out the door. If I try to satisfy program management in order to protect my job, it will not be long before NASA will complain that Thiokol is not doing the job well and program management could solve that problem by getting rid of me and saying that I was the problem. I can almost see it coming. It's a catch twenty - two that I don't think can be avoid with the current organizational structure.

It may seem strange to you that I am worried about this, but if you could see the pressure that management is now under to try and look good to the world and to meet the recovery schedule. It would make the pressure around here prior to the 'incident' seem insignificant.

The mood and climate around here is such that everyone knows that Al McDonalds and his fellow engineers were protected by National Attention and even Congressional support, but heaven help the next person to speak out in opposition to management.

Sincerely,
(signed) D. Kent McKinnon

After he wrote it he was concerned for his personal safety if he mailed the letters.

Rather than mail them he showed me these letters and asked me to hand carry them to the FBI to be forwarded officially. Kent hoped that maybe the FBI could intervene in some way and force proper safety efforts from Thiokol. This turned out to be a mistake also. If they had of been mailed their content would have been supported by Federal Postal Law! The FBI was frustrated they hadn't been mailed.

When SA Tim Healy read this last letter he jumped clear out of his chair. He had two 'smoking gun' letters in his hands, and God knows the FBI loves 'smoking guns'!

Soon after this time the FBI had their first meeting with Kent McKinnon. Tim wanted to know if I thought there would be any problems if he talked to Kent McKinnon. I told Tim that Kent was actually expecting to be contacted by the FBI soon. Tim knew Kent was not comfortable coming to the FBI on his own. Tim said he would call Kent and set up some meetings. I gave Kent's numbers and advised caution since Kent was under tremendous pressure, scrutiny and frustration at MTI, at church, and at home. I had uncomfortable questions about Kent's Mormon wife? Tim said he understood and that it is not uncommon. I do not remember the details about that issue. It was not best for Tim to call Kent. I believe the problem was Kent was afraid to talk to Tim at Kent's home. Kent thought of the place and time. An extremely quiet and confidential Japanese restaurant in Salt Lake City.

The attending FBI Agents were both shocked and impressed. Kent had turned them on to a bunch of items I hadn't known about, and therefore the scope of the investigation was expanding. My allegations were verified completely. They asked Kent is Agee worth a damn" and Kent told them he thought "Agee had been sent by God!" Kent confirmed to the FBI and NASA that I was extraordinary competent the FBI later told me. Also about this time, Dr. Tony Laine was digging in and he had found a number of Rocket Plume Gas, and electrical / electronic problems with the SRM under certain launch conditions. Dr. Laine was writing Hazard Analysis at a rapid rate on a wide range of MTI SRM discrepancies. Malcolm Phillips now had well over 100 Catastrophic Hazards prepared.

Other scientists and engineers such as Gerald Woods had a substantial stack of HA's prepared. Mahalon Hollway had written letters condemning the MTI test programs in Dec. 1986. General Henry had written letters of concerns via Kent McKinnon to MTI internal offices, and Wayne Tilsey had written two learned documents which flat out stated that the SRM rocket propellant cases "were no damn good" for a single launch, let alone 20! The Indians (Rakish Chopra and Arun Murthi) were now taking hold and were clearly on board and accelerating. The Indians' minds were the best and quickest. Awesome in fact. Approximately 1000 Catastrophic Hazards had been identified by this time. In a few months there would be 3000 HA, most catastrophic!

Tom Bradley now embarked on preparing a new and improved System Safety Plan for MTI under Kent McKinnon's direction. Kent McKinnon had directed Tom to review the existing System Safety Plan and prepare a new one as he was not satisfied with MTI"s existing plan. Kent wanted a better SRM System Safety Plan in place. As usual Kent wanted the right things included in the plan *For The Right Reasons.* I never heard the total details of what Kent thought was wrong with the old plan. When I read it my impression was that it was an old cursory document rather than a viable modern working document. I sensed a new one would be better than an upgrade. I was never clear what Kerry Sanofsky's contributions were? He may have been too inexperienced to provide upgrades in fact?

Somewhere along in here I started having a lot of quiet discussion with even more technocrats. The FBI had mentioned to me not to change how I acted because FBI experience had taught them that when people in my position (FBI Undercover Source) change their behaviour or personality, it is usually instantly noticed by the insecure and guilty. By now the FBI had asked me if I would be willing to co-operate with them and help them dig into a broad range of emerging problems and of course I felt there was no alternative. So, every day, I was looking for the normal documents I expected to show other problems which would need to be surfaced. When I needed to take a break from reading or calculating I would go make copies of other people's Hazard Analyses, memos, or letters if I thought they would aid the FBI to support their cases. MTI didn't have enough copy machines to go around. The machines they had were worked to death. Waiting lines were common. Discussions while waiting in line were informative and fascinating.

Every day on the way home during this period I was looking much more 'heavy set' than I had on arrival. I always wore a business suit or sport coat and over that in the cold weather a heavy leather trench coat – my own personal spy coat. For years friends had teased me about it, but now I really put my Spy Coat to work. Every time I left work my leather trench coat was covering documents of one type or another. This was the first time in my life I had ever pilfered anything from a corporation and I hoped it would be my last.

Acquiring documents just about gave me a coronary every time I went by the MTI Security Guards. I was not accustomed to "criminal" or "spy" behaviour. I come from an entrenched WASP background. The guards were generally "good old boys" but there were clearly some who usurped power and were on dangerous 'power trips". Then there was the minority who simply realized they had a serious and important job to execute. It was important to be friends with the 'good old boys' and appear to be above suspicion. with the 'good old boys'. A discussion of the latest deer or elk they had just skinned out last week was always a good subject in Utah.

The "power trippers" needed a little stroking to affirm their importance. Those doing a good and proper job were best dealt with by enthusiastically helping them. Open our briefcases before they asked – etc.. I was afraid of the guards. They carried guns, and I did not. In my insecurity I felt certain some of them were MTI zealots and must be considered 'armed and dangerous' to a person. It would take more than a few "planted" consultants to get flowers to grow in that God forsaken cold part of the Utah desert. Some of the guards were ordered to go through selected desks every pm or early am. No one was trusted. Everyone under suspicion. About this time the following article appeared which was solid proof that MTI messed up more than just the Space Shuttle.[8]

Now there were more frequent FBI meetings in the evenings. Usually the meetings were at my apartment. Tim and I would grab a bite at a local fast food place and go eat at my house. My God did Tim love fast food, and he is a connoisseur of hamburgers. Tim was working on documents and "ongoing problems" day by day as they unveiled themselves. Some of the documents I supplied to the FBI were more highly interesting to the FBI than others. Some I thought were horrific didn't cause much of a stir at the FBI, but some I felt were innocuous were considered 'thermonuclear' to the FBI. I could never accurately anticipate which were best for the FBI. There were plenty of irregularities to pick from!

One night during this period, Tim told me he would like for me to come to a meeting in Ogden. FBI No. 1 for Utah and four other states would like to meet me. I told Tim I was fairly confident they had dug into my background quite a bit by now and Tim grinned and nodded his head. Tim said they probably knew more about me than I did. Oh Shit! After all, we all

---

Appendix 8a & 8b: Thiokol docked $ 1 Million A Month For 'Unsatisfactory' Work on MX By Norman Black, AP Military Writer think we have our own secrets – or do we?? The women's panties I hung on the flag pole at University of Texas at El Paso (UTEP), tales of the Juarez , Mexico bars, and numerous other assorted indiscretions!

I was actually glad to hear someone knew about me because sometimes I wonder about myself. I asked Tim if the FBI thought I was telling them the "straight skinny"? He winked and nodded his head that there wasn't any question of validity in their minds. They felt that making a criminal case at that point was almost irrelevant compared to the needs of the nation having another Space Shuttle operable. I was impressed with these FBI people and their attitudes at this time. They had their heads screwed on straight and apparently did most things *For The Right Reasons*.

A few days to a week later, instead of stopping at home in the evening, I rolled straight south, passed Brigham City, and on to the Ogden downtown area and met Tim. On route I was stopped and ticketed for speeding. Hell, I was only doing 85 mph or so, maybe a hundred? The 'highwayman' was kind and only wrote me up for 64 mph and lectured me. I was really pissed. Between work and the FBI I had to schedule a piss two days ahead, and here I was being delayed and getting a speeding ticket to boot! Jeeessssssuuuuuusssss Christ!!!

I parked the car at a McDonalds restaurant in North Ogden. Tim showed up a minute or two later and I jumped into his car and he cleared the area to make sure I wasn't being followed by MTI or who knows! I did not know my way around Ogden, so I was kind of lost as Tim whizzed around the area doing his 'spook evasive manoeuvres' as he was trained to do. Finally we drive to a large quiet restaurant with outstanding privacy.

'Bear' Bryant was sort of short, casually dressed, very polite and friendly. I knew in an instant he was not a 'hostile' to me and his mind was 'razor sharp'. We went to a comfortable high backed booth and sat and talked on several subjects. Then over dinner we got into the diverse issues of MTI and NASA crimes as I saw them. He asked a lot of questions and I had definitive answers for each and every one of them. After the plates were taken away he ramped up and consolidated the crimes into groups in a casual way while he and Tim shot information back and forth. Then I pulled out of my 'spy coat' the day's haul of pertinent documents from MTI.

Healy and Bryant looked them over in an organized and professional manor. Then suddenly Bryant jumped right out of his seat and screamed, "That's Fucking Treason!" Fortunately the restaurant was nearly empty, otherwise they all would have heard him. I was shocked, but then I had long ago deducted both MTI and NASA were into treason so it was not news to me. Even Tim Healy was shocked by 'Bear Bryant's outburst.

Then 'Bear' sat down and reached into his pocket and came up with a note book. He started writing instructions and lists. He said " Tomorrow I will put a 'poison pill' into the file so the President of the United States cannot close down this case.' He had Healy writing his new instructions, and he was one very pissed off FBI SAIC (Special Agent In Charge). He was now the FBI on the war path! I was internally pleased, I felt I had won the battles and the war at that time. Little did I know or understand ....

Soon after that very successful meeting I was to meet FBI No. 2, Mr. Vernon Kohl . Again I parked the car at a McDonald's restaurant in North Ogden and Tim repeated his usual 'clearing the area' manoeuvres. While he cleared the area I could get 10 minutes sleep, I was desperate. We got out of the car, proceeded upstairs to the 4th floor, down the hall and into the FBI offices. I am always amused by the sterility of government offices. Tim's office had a nice clean working area, a desk, a few pretty pictures on the wall, but somehow the place looked like criminals belonged there. I met Mr. Vernon Kohl who was FBI number 2 in Utah. Vernon Kohl was a very courteous and bright fellow. I was tired and he knew it. I suspected he was also tired. We simply put two chairs out in the middle of the room, sat down face to face and went to work on MTI and NASA crimes. He had a lot of questions, and I had a lot of answers. No question went unanswered. I knew by the end of this questioning that my credibility was not in question in these offices. Vernon told me that there was no question that this case was going to be pursued to the maximum. Number 2 was really pissed off at MTI just as Number 1 was. He understood the scope and consequences of the problem.

Either after that meeting or the next Tim and I discussed the FBI 'bugging' Thiokol offices. I am not free to go into that discussion, actions, and inevitabilities of this line of thought. Tim and number 2 informed me that they would like me to talk with NASA officials very soon. I told them that would be fine, I had no qualms and, in fact, it was about time. However, I only wanted to do so in FBI presence because it was more and more clear to me and everyone else at MTI System Safety, that MTI and NASA were in bed together. At this time, I really did not understand the extent of the NASA / MTI incestuous relationship or I would not have allowed the meeting to include me. The forthcoming NASA meeting was to be a disaster to all parties.

At this evening meeting with No. 2, I also asked quite a few questions and received quite a few answers. I could tell No. 2 was hot to trot on this case and I had no question in my mind that the entire forces necessary from the FBI were unleashed. I also delivered another bundle of documents. I noticed at this meeting, Tim was already miles ahead of No. 2 and probably everybody else in the FBI in understanding the SRM problems and MTI and NASA's emersion into crimes and failures.

Tim had been doing his homework, and had come on line as fast as any human being could possibly do it. Tim had university degrees in History and Spanish from a Catholic University and he was very short on scientific knowledge or training with the exception of his being a US Navy Aviator and having flown Harrier Jets which gave him some limited practical science and engineering. After that meeting I drove back north to Brigham City, Utah. It was about 3:30 am and I was very tired. These frequent night and early am meetings were wearing me out on top of my twelve to fourteen hour work days, i.e. my 90 + hour work weeks. I could have slowed up my total hours of work but then I would have been out of timing with other contractors and engineers working long hours of overtime. Also it could have brought attention to myself. Not to forget that the US IRS and I both needed more money. I had just bought the IRS a really nice new small car via my 'late $13,000.00' tax demand from them. Also there was an unlimited amount of work to do on MTI"s SRM's. God and Country needed the work to be done. My work was simply more important to me. The SRM's were in such pitiful shape quite a group of us felt anxiety, heartburn, and fear over them. We wanted America's Space Shuttle to fly – soon and safely.

CHAPTER 17

FBI AND THE NASA INSPECTOR GENERAL'S

Time: Early Feb. 1987

Before long I found myself headed to Ogden again late at night after finishing work at the Great Rocket Plant in the desert. This time I would be meeting Special Agent Tim Healy (Case Agent), Special Agent Dave Latin {FBI No. 2}, Mr. Lou X, and Mr. Jim Y from the NASA Inspector General's Office. These Inspector General's officers had flown in for FBI meetings when the FBI had notified NASA in Washington, D.C. that MTI was under investigation. As usual, I arrived in Ogden at a McDonalds Restaurant, stood out it front in the ice and snow, and was picked up by Tim and zipped off double checking if I was followed and then to the Federal Building basement parking lot.

Tim told me these guys were kind of in a rush, and he didn't think the meeting would go very long. I told him that surprised me - I thought they would probably want to talk all night if they were doing their jobs. Oh well, what the hell, they're probably the usual government drones, I thought. Tim and I kind of chuckled as we entered the FBI office – I met Lou and Jim. Lou was a 56ish gentleman, quiet in nature and clearly a concerned and dedicated civil servant type. Jim was an ex-FBI agent (he had been in the FBI for a couple of years and had been involved in laboratory matters and then he had switched to the NASA Inspector General's Office from the FBI.) Now looking back I think that means he was a failed FBI Agent looking for a 'Port in a Storm'. I don't remember the details of what was discussed and how it was discussed, but they asked me quite a list of questions, and I answered them quickly and to the point. I was not short on answers. Somewhere quite far into the meeting, I noticed we had spent some serious time there, and were going to spend substantially more. I realized that maybe they were paying a lot more attention than they thought they were going to prior to meeting me. I asked these IG's "What do you think about all of this?"

Pained looks went around the room and back and forth and finally Lou looked at me dead in the eyes and said "Steve, I can't believe what you are telling us. I believe that you believe it, but I cannot believe that Thiokol would withhold safety information! I couldn't believe MTI and NASA were in bed as criminals! If MTI is holding this information from NASA this is an extremely serious matter!" I thought *"No Shit Sherlock"!*

I looked Lou back straight in the eyes and in a bold mood and voice told him, *"You Are Dead Wrong"*. I said, "not only are things at MTI as bad as I am telling you they are, but in fact they are clearly one hell of a lot worse!" "The only reason I can't show you how bad the situation really is is because I am restricted from going to the main meetings and I have to keep a low profile because of Don Thompson and others wanting me fired." "And what are Don Thompson's credentials relating to science, engineering, rockets, space vehicles - *Don Thompson is a fucking local Mormon bishop tax accountant!* There are dozens of other witnesses Lou! " Lou was not antagonistic or condescending in any way. He was simply stating his honest ignorant opinion that he honestly found it hard to believe that MTI could be that fucked up or criminal. I asked Jim what he thought about it and he told me that these were very serious accusations and that they would certainly be looked into. However he also stated he did not feel MTI could possibly do anything this seriously wrong. I was astounded and I looked at Tim and could see he was not pleased. I said, "well, if you fellows don't believe it, what about the fact that only a few hazards have been processed and handed over to NASA in the LAST 13 YEARS and all of those were CRAP and non-catastrophic up until this new group of contractors arrived and identified 'real issues'?"

Well, they all looked around the room and shuffled their feet and glances went back and forth. It was clear to me that all present were frustrated, concerned, and were going to have to give some thought as to what to do next. Concerned for different reasons and different agendas I figured this was business as usual with government drones and kept a pleasant tone with them even though I was internally ready to physically attack the IG's for being so stupid, ignorant, and dangerous to United States interests.

I was mad at them and their complacency frustrated me. Their attitude of, "We are NASA and we should know what we are doing really reminded me of MTI"s attitude of "We know what we are doing!" WRONG!! To hell with them both – I'm Steven Agee and I know what I'm doing and I'm telling both groups *"You are incompetent and dangerous criminals!"*

They asked if anybody else had observed these same problems and issues? Tim replied "yes, he had met with Kent McKinnon and Kent had verified everything to Tim and Tim would / could schedule a meeting Kent and the NASA IG's for the next day if they wanted" Jim and Lou said they would like to meet with Kent to sort this out for themselves. Before they left they asked for me to briefly tell them "what would help to fix Morton Thiokol, and the Solid Rocket Motors and NASA?"

I emphatically admonished these two NASA IG's to stop allowing their NASA Contract Officers to hand out 'Contract Deviations' and 'Contract Waivers' like candy canes at the doctors office. I was very specific with the IG's about how MTI's solution to any and all problems was – 'let's get a *WAIVER* and / or a *DEVIATION* so MTI didn't have to fix their problems and could just pocket the money. I asked them why NASA was so generous? These two were 'clueless' as to why! The FBI and I knew why, profits of course.

The undertone of their voices was 'Oh shit' this asshole Agee can't be right'! So, later Tim drove me back to the McDonalds's restaurant. I jumped into my car and drove north to Brigham City exhausted. On my way back I wondered to myself, "are these guys from NASA the right guys?" Are these IG's in bed with MTI? Are these guys going to try and make NASA look good or are they going to really try to work the problems! Were these guys government drones in fact? Will these guys cause me trouble and hurt me? Are these guys scheduled to retire from NASA and go to work for MTI - a not so uncommon practice!" Tim hadn't said much on the way back, but I knew he had a lot of thoughts on the subject. The same thoughts as me – *Outrage! The stupid NASA ignorant bastards!!!*

Before we departed they told Tim and myself that their NASA Inspector General's Office had a contractor / spy which they were going to send into the Thiokol Wasach plant to look into my allegations. I would never find out who, when or how many. Now years later I wonder if there ever were any such?

I had a very uneasy feeling at this point that probably NASA and MTI were far more in bed than any of us had previously sensed. My blood ran cold as I thought about the ramifications of MTI and NASA's mutual symbiosis. What if these NASA IG's went to MTI and they all sat down and decided to cover their asses? I might be discredited? I could be blackballed, more likely a hunting accident . . . I was ill at ease, *Very Uncomfortable, Profoundly Insecure.* It appeared to me that NASA, the rape victim, loved it!

The next day at work I told Kent that Tim and NASA IG's wanted to talk to him. I don't remember the details but it was all put together for Kent to meet Tim and the NASA IG's at a quiet Japanese restaurant in Salt Lake City.

Kent was afraid to meet with anybody around Bountiful, Utah, Ogden, Utah or in any populated venue. Closer to MTI was more difficult to keep secret. I think Kent would have preferred the meeting to be in Ethiopia. I didn't blame him. Kent was afraid of NASA insanity as he was afraid of MTI crimes and insanities. He was not misguided.

CHAPTER 18

TREACHERY

Time: Feb. 1987

Kent's meeting with NASA and the FBI was on and I held my breath with anticipation about the outcome. Kent was nervous, frustrated, and scared. The FBI was "eager beaver", and the NASA IG's were incredulous. Kent had his heavy duty meeting with the FBI and the NASA IG's officers. I had been invited, but desperately needed the sleep instead. Also I didn't want to cramp anyone's style. When I talked to Tim next he was referring to NASA in different terms than previously –instead of referring to "high ranking NASA Officials" it had now been reduced to those "dumb Sons-O-Bitches"! Seemed right to me - how could I argue with Tim on this??

I sensed the FBI and NASA were going over their own cliff as a result of typical inter-agency skirmishes. I wondered how many U.S. Government Agencies were involved. Certainly the National Reconnaissance Office (the super secret U.S. Space Agency), the DOD, and no doubt others. I surmised their problems to be that the FBI knew there were problems, and NASA couldn't dare consider an admission of contract inattention for billions of dollars and lives. Such an admission would make NASA a party to seven recent deaths. NASA had law suits against it from several of the deceased astronauts families. NASA could not accept any blame for non-compliance without admitting its own liabilities and being assigned all or a portion of the blame and subsequent bills.

As I understand it the Kent McKinnon meeting with the FBI and NASA was very successful and no matter how naive or unwilling, NASA had to face up to the realities that there were serious problems with the MTI SRM's. Tim told me that Lou and Jim had returned to the Marshall Space Flight Center (MSFC) at Huntsville, Alabama and would be reporting to Dr. James Fletcher, Director of NASA. Tim spent some time himself wondering if these NASA IG's could be trusted . Tim assured me there was no question of my secrecy and security being respected by NASA. Outwardly I said "GREAT", down deep my stomach churned. In the back of my mind I don't think I ever really believed it. What else could Tim tell me – NASA IG's were jerks and couldn't be trusted? Tim was a good man and knew when to bullshit his 'source'. He and the FBI couldn't afford to tell the truth either.

Part of his job was to put me at ease. Were the NASA IG's a pack of dumb bastards? Would these NASA IG's place NASA's interests above my interests or America's national interests? What would a NASA government drone do? Some government drones could be trusted, some couldn't be. I did not trust NASA.

For the NASA IG's office to tell Dr. James Fletcher (a Mormon) there were problems at MTI meant they were admitting a problem had crept up both on the NASA IG's office and NASA's project office, and the Mormon 'slot machines'. It also meant Dr. Fletcher was at the very least under a cloud. In no way was Tim covering for anybody. I believe at this time he really felt that NASA had some minuscule degree of integrity. Tim related to me that the meeting with Kent was very direct and NASA had quite a few questions to ask Kent which Kent dispatched quickly and accurately. Kent set NASA to rest without question that there had been a great deal of lying going on between MTI and NASA. MTI lied, and NASA bought it. Astronauts and taxpayers paid for it, Astronauts died for it! Western world supremacy was about to collapse from it. At this point NASA could not have any doubts that the President's Commission and the Rodney Commission had been lied to. NASA had now seen the tip of the iceberg and Pandora's box at MTI was clearly opened. So at this point I had achieved a 'partial win with NASA'. The vision inside was ugly! It had taken all of my energy to get this far. I was 'SHATTERED!'

Tim was not certain exactly what would take place from this point forward with NASA. He envisioned that NASA and the FBI would conduct some type of joint effort to somehow correct obvious destructive and synergistic problems. I believe NASA had meetings with FBI No. 1, Bob Bryant, before they returned to MSFC (Marshall Space Flight Center). Subsequent FBI and NASA liaison discussions were anticipated imminently.

Kent McKinnon felt quite good about the meeting. He thought a burden was lifted from his and my shoulders. Kent McKinnon and I had fought the good fight and had lost on all fronts up to this point.

Now the FBI could pick up the pace. I don't believe Kent or any other man on earth in his position could have prevailed against the forces brought against him. Kent only had three alternatives. First – Find a new job and leave MTI immediately. Second – be fired and then find a new job afterwards if he could. Third – shut up and play the role of a castrated MTI Corporate sheep. I really believe MTI prefers castrated sheep and God knows they had plenty of them grazing at their Thiokol Wasach Plant.

Morton Thiokol is the only corporation I have ever seen where numerous workers sit around with their telephones off the hook so they won't be called and they won't have to work. Phone calls were especially unwelcome if they interrupted elk hunting adventure stories.

Sometime in here System Safety was able to get a few documents typed. A gal was borrowed from another department. Her name was something like Miss. VR (let's say 'Very Reliable'). She was a smart, cheerful, helpful gal in her 40's. She was an honest bright spot in a dark cold desert. Besides all else she could actually type and she was even willing! During this time I was perpetually exhausted. I had exceeded my ability to recover. One night I asked VR to go to dinner and much to my surprise she said yes. Even though I desperately needed rest and sleep, I also needed a friend, a personal life, and to eat dinner.

We drove over the mountain one evening and had dinner at a restaurant at Logan, Utah. VR was a lovely soul. She was bright and direct, knowledgeable, and uncomplicated. I was so tired I asked her to drive. Dinner was O.K.. Visiting was wonderful. I needed the break. My double pneumonia was really killing me.

Over dinner I learned a lot about Mormonism, about MTI skulduggery, and about the northern Utah area. VR and I spent bits and pieces of time together over the next few months. VR became my trusted friend and confidant, but I also knew the dangers to her and myself. Exhausted and scared people do not always act rationally. Letting VR be my friend and confidant was actually unkind and dangerous to her. It was a risk to myself and the FBI's Cases. Ours was a dangerous relationship, but then anyone who was my friend during that period of time was subject to innate dangers.

CHAPTER 19

A GOOD MAN FALLS

Time: Feb. 1988

My work load in the System Safety Department was escalating rapidly. Sometime along here, we consultants, contractors, slaves were given more detailed assignments which were essentially the same component systems we had all been working on – however there were two parts to the assignment now. One part of the assignment was to evaluate the new improved design (redesign) after the Challenger Accident and implementations. The Other part was re-evaluation of the system pre-Challenger. These two divisions, past designs and present designs, sounded innocuous enough, however they were entirely different in technical scope, and thinking. Old unsound systems versus new unsound enhanced systems.

By this time, the Indians, Malcolm, General Henry, Tom Bradley, Gerald Woods, myself and others were feeling enormous resistance while trying to acquire documents of all types which we knew existed. We needed them and wanted more pertinent documents. You can't evaluate and assess what you don't have. MTI management had locked away large numbers of key documents from even their own MTI staff. They ranged from numerous government agencies documents, photographs, detailed excellent photographs, taken during the Challenger launch sequence which we were not allowed to see! Ummmm, a picture is often worth a thousand words.

Official word at MTI was that if we could go talk to a particular manager long enough, we might get permission to look at them, but we were probably wasting our breath wanting to look at the stills and motion pictures as that was viewed as 'a suspicious act' and not tolerated by MTI 'watch dog' paranoia. Since Don Thompson wanted me fired there was no use in my asking. MTI didn't want anybody looking at those photographs who might come up with 'culpability conclusions', take them to the press or anything else.

After all, MTI had already made up its mind and was already under a bunch of lawsuits and media pressure. Any new findings at MTI which would increase MTI liabilities were dangerous to them. Personally, I thought this was a travesty because all of us looking at Challenger data would have had a slightly different technical background and points of view. I believe nearly everyone from System Safety could have offered contributory engineering insights after examining these photographs and data. No one could justify MTI locking data, photos, documents and historical records up. We were there to fix / repair, not to cause law suits and trouble. We were there to save lives. The photographic records were owned by the U.S. Government and should have been public record or at least controlled by the government. Additionally MTI locked up telemetry launch data, test data and other flight data. Who knows, if we had access to all of the facts, data, and photographs perhaps in some way we could have exonerated MTI? What was unknown or unfound we didn't know? The Indians, both great souls, fought several bloody wars trying to loosen up some of these documents for their own areas of analysis. Aroun and Rakesh were angry and crazed by the resistance they encountered. Oh hell, let's face it – they were 'fucking, Goddamn mad!'

Tom Bradley threw up his hands and basically quit fighting. Dr. Tony Laine was so upset by his findings he actually seemed to mentally slip over the edge - he was crazed! In just a few weeks several were reduced to an attitude of "give them whatever they want" in their work. Their emotions were battered. I knew there were batches of documents on the SRM nozzles I would love to get my hands on that were locked up. There was no hope of me even asking for them. MTI considered me to be a 'dangerous person' and in retrospect they weren't wrong! NASA was more clearly more complicit by the day.

About this time Dr. Laine identified a bunch of incredibly dangerous hazards. Dr. Laine was crazed by his astute findings and calculations! Dr. Laine was a very nice person, and he had an overt learned personality. He was short on social skills, and he lacked corporate tact. Translating that to 'he didn't suffer fools very well' and he 'was not politically correct', and he was allergic to both technical and managerial bullshit. Despite any personality deviations he may have had, he was sincerely interested and dedicated to launching a safe Space Shuttle. Dr. Laine was absolutely a team player if and when there was a real team. Dr. Laine was not a castrated corporate sheep. He would stand face to face with anybody and defend his point of view. He would also capitulate politely when he lost a point, he was honest, he was flexible, professional, and pragmatic.

By now Kelly Giles had been assigned to the nozzle work. I do not know if he was selected by Kent or if Kent was told to assign Kelly Giles to the nozzles. Kelly was a very nice fellow. He described himself to be a "Jack Mormon" which roughly equates to a "fallen Baptist" or a "failed Catholic". That is, a Mormon who does not practice all of the practices of their 'One True Religion', but , of course, will always be a Mormon. Kelly Giles shared an office with Mahalon Hollway. Mahalon's alcoholic ego could not allow such an important person as himself to sit in a common area, so he found an empty, partitioned office-like area and moved himself into it. Sort of an unused broom closet. Possession by occupation! Mahalon found it best to put this desk in the center of the room so he could sit there glaring out at everyone who went by the door to or from the System Safety department. He 'dominated' this room, which was perfect for the self appointed 'fruit inspector' and smoker. Kelly was technically weak and had great admiration for Mahalon. Mahalon would bullshit Kelly and Kelly sadly 'kowtowed' to Mahalon.

I watched Kelly Giles' process of working on the nozzles and quickly understood that in a million years he wouldn't understand enough engineering to really understand the details of SRM nozzles. He was an uninspired sort of engineer with little insight. Kelly was not a bad person, he just simply condescended to the status quo at MTI. He was a survivor. To put it bluntly, the nozzle was in less than adequate analytical hands. In reality, no one person was adequate to handle all technical aspects and ramifications of the nozzle. A team of approximately 20 diverse experts working around the clock would have taken ages to quantify proper parameters. However MTI felt that since 'there were no problems' on the nozzles, then one or two third rate engineers part time were a generous contingent! Standing at the men's room urinals we heard from some MTI workers the nozzles were known in house as 'Leonardo's' since each one was a piece of unique art, sculpture, not a product of manufacturing standards. In the end Tony locked on to my concerns regarding electrical grounding and electrical currents moving in the ignited rockets' gas plumes. That is the gasses from the Solid Rocket Motors and also the Shuttle liquid fuelled motors creating electrical flow. Tony reached into his bag of papers and pulled out his paper on "Rocket Gas Plume Dynamics" which he had written a couple of years before! MTI generally and Don Thompson specifically did not tolerate Tony's findings well. Don Thompson fired Tony one afternoon and Tony wrote a letter in a 'fit of rage' right after he was wrongfully fired as follows –

To: Mr. Thirkill, Vice President and General Manager MTI
From: Dr. Anthony W. Laine, System Safety
Subject: Overall Summary

Dear All,

It has been told to myself that I am grossly over qualified for the position that I am in. The above, I personally reply that it is going to take persons far more qualified than myself to evaluate all of the Safety and Hazards on the SRM redesign to do a very thorough job.

The President of the United States, the Rogers Commission, NRC, and NASA publicly stated that they want the very best that can be found to evaluate the SRM problems. How can anyone be over qualified, when this is what the above, and the whole United States public demand?

Evidently, there are some persons here at MTI that do not feel the same, as the above paragraph. So, there are problems, nothing that cannot be worked out. Stating that there are no problems in certain areas, is like sweeping them under a rug and hope that they will go away. They will not go away, they will only compound.

A very grave concern to others, as well as myself is this. Our Safety, and Hazard worksheets that are completed are mysteriously disappearing. Does someone not want these concerns to come out? Each and every component on the SRM redesign must be evaluated for Safety and Hazard concerns. Also each and every interface must be evaluated for safety and Hazard concerns also.

I brought up Plasma concerns, it was told to me that 'there are no Plasma problems'. To the above myth as I call it. The Pershing, MX, and other missiles had Plasma problems. It caused chunks of metal to come off, also arcing the o-rings. Why is the SRM any different? It is not.

Nowhere have I found in documents, that MTI has been addressed. This can be a very grave Safety and Hazard concern, because of the field levels. Also just cutting, and pasting the Failure Mode Effects Analysis (FMEA) sheets to become Safety and Hazard analysis sheets is not the complete task. All Safety and Hazard concerns must be evaluated. A very thorough and complete Safety Analysis must be done on the SRM.

Furthermore, I tried to save MTI many, many millions of dollars, by getting in Nuclear test data. This would be on components, materials, Single Event Upset (SEU). This is available at no cost. All I received for my kind intentions from MTI personnel is dislike. I also tried to give to MTI my own personal books which are copyrighted by myself. Again, the same treatment.

The above was to be handled with dignity, pride and total honour. It is all extremely honest. I gave to you a list of numbers to call, so you could verify this.

In closing I personally state, that there is no person who is overqualified for any position that he chooses to do. For instance, this fall I shall be returning to the University. I shall be a Medical student. The best talent for any, and all positions should be gotten.

You may Sir, check with Mr. Kent McKinnon about the disappearing Safety and Hazard Analysis worksheets.

Sincerely,
{signed} Dr. Anthony W. Laine

Tony told me he was going up to the VP's office for a meeting in a few hours. He was red in the face and absolutely outraged at having been fired. Tony didn't understand *"No Good Deed Goes Unpunished!"* He was again summarily fired again at the afternoon meeting in a most improper manner. Illegal in the context of the historical time. John Thirkill (V.P.) tried to get Tony to fall into line, Tony refused to turn his back on deadly SRM nozzle gas plume problems. Thompson, an MTI attorney, and others were present and witnesses. Kent McKinnon told me later he was both shocked and horrified! An obscene corporate encounter of the worst kind! Laws were broken, Dr. Laine was sacked!

It was roughly about this time that Mahalon had turned completely vicious against Kent McKinnon. Mahalon was so rude and overt in his political gamesmanship towards Kent McKinnon it was all I could do to restrain myself from physically attacking Mahalon. Mahalon embarrassed and angered everyone. Mahalon told me one day that he thought he could become an MTI Director! At the mere thought his eyes gleamed - such aspirations! I thought to myself, 'Isn't that just appropriate! What a marriage that would make.' Mahalon had dropped his rocket test-related activities for the most part and was now aligning himself with upper MTI management. He told them such things as "I saved the Apollo Program", "I saved the MX Missile program" and now I'll save you Don Thompson, etc., etc., ad nauseam........

There were over a hundred witnesses and 2/3rds of them talked about Mahalon's runaway ego and alcoholic personality. Our entire System Safety department was disgusted with Mahalon and also feared the comprehensive damage he would probably do. Word was about MTI that Mahalon was 'drinking' again. Oh shit! I personally don't know if this was true. I however wouldn't be surprised. Mahalon had a total change of personality about this time and it certainly was not for the best. Mahalon was no longer on the roller coaster, - he had gone 100 % sour. Kent was damn mad about Mahalon's actions. Mahalon was technically a subordinate contractor, not an MTI employee. Mahalon's main wedge against Kent McKinnon was that Mahalon had bedded with Don Thompson. I imagined Don was giving Mahalon the OK to take certain investigative actions for Don Thompson and furthermore had clearly anointed Mahalon as 'his eyes and ears' inside the System Safety department. Don Thompson was the "right stuff" for Hitler's Gestapo or the U.S. Internal Revenue Service. Both Kent and Kerry Sanofsky talked to Don Thompson requesting Mahalon be fired – no dice! Don loved Mahalon. They were an evil symbiosis.

Mahalon and Sanofsky had special assignments from Don Thompson to investigate Dr. Tony Laine. Probably me and others in System Safety persons too. Mahalon had also absolutely hated Dr. Laine from day one because when Mahalon delivered one of his impromptu "high tech" soliloquies Dr. Laine casually told Mahalon he was wrong. Dr. Laine was right, Mahalon was pervasively wrong. Mahalon's fragile ego could not survive such blatant and direct rebuttal. Now Mahalon was in his full glory gathering information in the proper Gestapo and MTI manner as well as making it up as he went along. Mahalon was working his plan to become an MTI Director. Mahalon was not working on scientific or engineering problems any longer.

I wondered if Mahalon was amazingly going to become a Mormon? Enquiring minds wanted to know?

Don Thompson informed Kent McKinnon that Mahalon, with his superior background, training, and experience would be looking over Kent's shoulder, 'helping Kent', approving Kent's work, and keeping Don Thompson informed. I can't say who's background was technically best, but it was obvious which of the two was a professional and who did everything *For The Right Reasons*. Everything Kent 'the employee' did now had to be approved by Mahalon 'the contractor' now! Was this any way to run a department, corporation, personnel, or a pig farm? What bullshit! Watching Morton Thiokol try to build Solid Rocket Boosters was like watching an animation of George Orwell's book, "ANIMAL FARM"! A couple of scared System Safety members thought the apparent coming change in System Safety managers was good. They liked Kent but thought a stronger personality was required.

Those in favour of Mahalon were not informed completely of Kent's thorough and professional efforts. System Safety staff were insecure and generally frightened. After all, if you lost your job at this time in American history you were adrift without a paddle in the northern Utah desert. About the same time, Kent was called out of town to an interview with one of the aerospace corporations I had connected him with. Kent was frustrated, but very optimistic and I certainly did everything I could to help push this interview. Kent no doubt pleaded his case before God. I'm not a Mormon, but I put in a few kind words for him. Kent bought a new suit and flew out of town with "high hopes". There were 3 or 4 days of heavy duty interviews, a physical, and the newly required drug screening. Kent came back from the interview three feet off of the ground. He had scored, and the money being offered was a pleasing number to him. The position was probably one of the best jobs conceivable to him and one of the top companies in the world. I felt like I had pulled a great coup because Kent was in enormous trouble at MTI and our System Safety Department hearts were bleeding for the man.

Since Kent was not a castrated MTI corporate sheep, he could never be safe or prevail at MTI. MTI wanted him to be disposable and save Don via any coercive means. You may wonder why I didn't seek a permanent position. Very simply, I made 2 to 4 times as much as a consultant or contractor over a permanent.

Almost simultaneously, I was walking down the long glass corridor on the upper deck of the main MTI office building. As I looked out in to the parking lot through this enormous glass wall, I saw Don Thompson sitting in a car in the parking lot with Mahalon Hollway! A clearly *covert meeting.* Mahalon was clearly agitated, waving his arms around, and talking ferociously to Don. No doubt "saving Don". My heart sank. I knew right then and there that Mahalon had done more than simply "get in bed" with Don Thompson if they were talking like this. Actually I thought he was a *treasonous* son-of –a –bitch and so did some members of the FBI.

Mahalon's actions over the last few weeks had indicated to everyone that he planned on having a high position over all of us soon and furthermore we were all cannon fodder to him and his ego. I felt betrayed by Mahalon, hell I was betrayed. We were all betrayed. America's Space Program was betrayed! I was ashamed of myself for having brought Mahalon to MTI. I could blame no one but myself.

The intricacies of life's little twists are sometimes astounding! As usual – *No Good Deed Goes Unpunished!* Within 48 hours of my seeing Don Thompson and Mahalon Hollway covertly in the car together in the parking lot in agitated discussion, Mahalon talking, Don listening, I realized that Mahalon was pulling the same crap here as he had tried at Boeing on the MX Missile also known as the Peace Keeper Missile Project. The only difference being that at Boeing one manager or the another would jerk Mahalon's leash every week or two.

Now Mahalon was unleashed, surrounded by less than average managers and less than average permanent technocrats, and he was "on a roll". Mahalon was now a sheep dog running the MTI sheep. He had a personal charisma and a domineering alcoholic personality that was indeed powerful to insecure, frightened, and unknowledgeable people. The only person willing to tell Mahalon to buzz off when necessary were the Indians, Dr. Tony Laine, and myself and at different times we did.

There were hilarious scenes with the two Indian Engineers mimicking Mahalon's animalistic shoulder rolling rituals. Mahalon knew the Indians thought he was a total *ignorant ass hole* and he had trouble dealing with it. No sense of humour so to speak.

Now word spread through our System Safety department that Mahalon had proposed to Don Thompson and other MTI managers that he could save a fortune and "save MTI" while placating NASA. The word on the street, the hallways, and the men's room was that Mahalon's plan was that since 'there were no new hazards', the few old write ups were all that was needed. MTI could let the rest of the System Safety Department go and just keep one or two enlightened MTI souls around under Mahalon's directions". What a fantastic plan I thought!

Another obvious advantage was that MTI would pay Mahalon a lot more. A lot of department members were plenty pissed over this, but I really didn't think Mahalon could pull it off.

Sharlet Rose was stressed by the trouble Mahalon was stirring up among both contractors and permanents, No question in my mind that MTI would love to cut costs, grab the profits, deliver minimally designed crap, but the Rogers Commission and Rodney Commission probably wouldn't' stand for it. Through this period NASA was nowhere to be seen for weeks and weeks.? I viewed Mahalon's "try for it" as a desperate move on his part – doomed to failure, but nonetheless a serious inconvenience for all concerned. Mahalon's final obstacles of course was that Mahalon was not one of God's Chosen People? An opportunistic conversion to Mormonism should have been possible if he could do without the alcohol and 3 packs of cigarettes a day.

Much later when the Seattle Times did an eight and a half page article on me and my efforts they would learn that *Mahalon proved to not have any College or University degree - Mahalon was a fraud and now it was obvious he had slipped through the FBI's Security checking system!* It's not entirely unheard of. I should note here that it was unlikely that either Mahalon or any of the identified criminals qualified for DOD Security Clearances ever again.

The learned and capable looked straight through Mahalon and many pushed him aside. Why had the contracting house who owned Mahalon not known he was a fraud? Why had the FBI and National Security Services not known? MTI took what they needed from Mahalon and that was the end of it. On the MX Missile System work at Boeing, there had been a few incompetent people and Mahalon worked them like a sheepdog works the sheep. Mahalon ran the stupid to the left, the ignorant to the right, he herded the rest down the hallways. Others on the MX Missile project were smart, and they kept him "in tow".

Now Mahalon was doing the same thing at MTI and I concluded

that this was simply Mahalon's way of shuffling through life. "Listen to me, I'll show you how to save yourselves and make you look wonderful via any sleazy means and in turn you must take care of me!" Mahalon was not without attributes – he knew a lot about aerospace testing procedures and documents, test procedures, and rocket gas toxicology, etc.. Unfortunately he was out of control and little or none of his time was now attributable to repairing the SRM's systems. Within 48 hours, I felt the sting of Mahalon and Don's incestuous liaison. Kent McKinnon called me aside and said " he was terrorized". I asked "What's up Kent?"

Kent explained that Mahalon had gone to Don Thompson and told Don that there were a number of problems in System Safety, and that Kent was not good enough to run System Safety. Don told Kent that Mahalon could show Don how to save huge amounts of money, transfer funds from unneeded consultants into better uses at MTI such as profit – all the while saving the Safety Program! I told Kent that was Mahalon's typical narcissistic approach. Kent said the unfortunate part is that Don Thompson bought it! In fact Don Thompson loved it! This made sense since Don Thompson didn't really even know what System Safety was about since it wasn't part of his Tax Accountant training. Kent McKinnon would never scream *"Oh Shit"*, so I did!

I longed for the days back in the Vietnam jungles where we simply shot our problems. Life was so simple in a real war zone. Don Thompson was no technocrat, he clearly wasn't a manager, he was a Mormon Bishop / Tax accountant and castrated corporate sheep. He was grasping for any escape pod and he had crawled into the wrong one. Sometimes we would see Don Thompson come out of meetings with a glazed look and he wouldn't even acknowledge his workers. He would be in a state of shock. Evidentially, Don's superiors had just ruined his day again. Don ruined Kent's days, weeks, and his sanity. Kent had more than adequate grounds for a massive lawsuit against MTI and I told him so, even though I knew his Mormon wife would prevent it.

Our probable new boss will probably be Mahalon and we'll all be run out the door if the egotistical maniac has his own way. Then Kent McKinnon went on to say that wasn't the worst of it! He said "Steve, Mahalon told Don that he thinks you might have gone to the FBI. Mahalon and Don Thompson's informers here at MTI have seen you at the Xerox machines copying huge amounts of documents." Kent asked if I had copied quite a few documents and I told him it was thousands of pages now, some for the FBI and most for analysis. Kent told me he didn't want to know the details but he felt I should go into very low key operations again. Kent also told me I had his blessing to take whatever I needed for the FBI. Just that I should avoid being caught and Kent didn't want to know about it piece by piece.

Kent frequently directed me to specific documents and information. I was so God Damn'd mad I was ready to run down the hall and attack Mahalon and Don. Mahalon for his own personal gain and greed was sitting there watching all of this crime go on, knowingly trying to hinder any corrective activity – *Treason*. Don was just a dumb lower level criminal,

Mahalon was potentially far more malignant. Over the Christmas break, Mahalon had ranted and raved over MTI stupidity, lack of MTI concern for lives, inept test methods and records, criminal negligence, but now he was an MTI "team player" striving to become a Director. What more did I need to know about Mahalon? Nothing! He was an enemy of the Free World. I asked Kent for more details on what Don had said to him so that I could pass it on to Special Agent Tim Healy. Kent said that when he had asked Don more about 'this and that' Don had asked Kent "are you dealing with the FBI?" This question put the fear of God into Kent and Kent of course replied in the negative. Special Agent Tim Healy had trained Kent and me to "lie, lie, and deny" if confronted with accusations or facts about doing work for the FBI. This was for our own personal safety. After all, we weren't worth much to the FBI dead.

I talked about Mahalon with Miss VR. She hated Mahalon's guts and told me plenty about the troubles he caused throughout MTI, most of which I had not otherwise heard about. His maligning had spread much further than I realized. In fact VR was an incredible source of deep–rooted diverse information. I didn't spend time with her for the information, her information was a bonus to some companionship. VR was a nice friend.

I asked Kent if this is the end for me at MTI? I don't know about you Kent, but they are probably going to throw me out of here on my head if they believe Mahalon! Kent said he didn't think they would throw me out on "hearsay"" but I had better keep my visibility lower than ever. Let others push, shove, and fight for now. Mahalon still had a serious credibility gap – he was not a Mormon with his Tithe ' paid up in full'. He was therefore an 'infidel'. Furthermore, Kent and I decided we should hardly ever be seen together and should hardly ever speak in the hallways. How's that for corporate Bull Shit! Others were experiencing the same sort of thing at MTI. There was a continuing feeling of oppression and suppression in darkest corners of MTI. Very few said what they thought in open hallways or to persons of unknown sympathies.

I did not sleep that night and wondered if I should simply corner Mahalon Hollway and beat the shit out of him? I was up to it! I talked to Tim about it on the phone and he asked me how Mahalon knew or guessed about my current FBI activities? Tim was alarmed and startled by events rapidly unfolding. I had not discussed my activities in any way with Mahalon since he was such a loose cannon. I did not trust or respect Mahalon except as a rocket test document and a rocket plume toxicology purveyor. There was one other person in the company at this time who knew at least vaguely what I was up to other than Kent and that was Malcolm Phillips. Malcolm Phillips was instrumental in helping me acquire many key documents, and helping me get documents out of the plant. I trusted MP with my life at this time. Malcolm didn't know enough about what was going on to really be capable of tipping Mahalon off, but then you never know?

I believe what tipped Mahalon off was the fact that I had been shooting a tremendous number of documents copies at a copy machine right in front of Mahalon's office. Since Mahalon did not work other than stroll up and down the hallways looking sinister, slap backs, and bullshit, he had a lot of time on his hands to smoke a lot of cigarettes and glare out of his strategically located closet door. He glowered at everybody as they went about their work as though he was the appointed fruit inspector. I'm sure he felt his job at System Safety was *"fruit inspection"*.

Kerry Sanofsky came and had a quiet talk with me in Feb. 1987 about how he should or could handle Mahalon. Kerry and his associates were not pleased with Mahalon's lack of work, lack of respect, and general insubordination. I wondered why he didn't simply fire Mahalon for lack of performance and insubordination? What held him back? Evidentally Kent and Kerry were somehow obstructed from firing Mahalon by Don Thompson and others. Strange?

I had unabashedly, on numerous occasions, simply gone to the copy machine with a stack of documents an inch thick and run copies for my research and / or the FBI. Additionally I made copies in other areas of the plant to 'spread copying around'. Different copy machines in different locations ran certain types of documents better and faster, and were easier to access. Sometimes I had to look over ten acres of offices to find a copy machine that worked. MTI simply was not set up to do business. Business was secondary to profits.

When I worked on the MX Missile System at Boeing, Mahalon and I discussed one time that I had previously been contributory to the FBI on a 27 million dollar fraud perpetrated by an aerospace contractor on B1-B Bombers. It turned out that Mahalon told Don Thompson that he knew I had a previous FBI relationship in the past and that when I was on the MX missile group at Boeing I was no doubt working with the FBI. Not true. There was evidence that Mahalon had recommended to Boeing to get rid of me, I was a danger to a certain manager who was later busted by Boeing for incompetence, botched time cards, and a cocaine ring in his department. And now apparently Mahalon told Don Thompson it would be wise to be rid of me because he was sure I was with the FBI again. There were abundant witnesses to Mahalon's actions. I had no FBI relationships ever when at Boeing. I knew the Boeing manager in question was insecure, but not at all about the botched time cards, or the local cocaine dealer. I was disappointed to find I had missed these activities - would have been entertaining.

As a matter of concern, I politely had talks with Kerry Sanofsky and Phil Greenhalgh. I told them they should try and get good jobs with a good company somewhere soonest. I really hoped they would save themselves. I offered to help them. They also were anchored in the Salt Lake area. Mormon wives usually have a death hold over their husbands. Even if the husband wants to shift to a different area for better jobs or incomes, Mormon wives are trained and dedicated to the Mormon Faith and its home grounds in Utah. Sexual bribery is one of their basic tools. It was pitiful to me. I would later learn from the FBI that Utah had the highest rate of 'White Collar Crime' per capita in the USA and also the highest divorce rate per capita of any state. I was surprised.

Up until the 14th of Jan. 1987 I had no active relationships with any government agencies while at MTI. Mahalon saw the MTI crimes, cussed and screamed about them, and then made every effort to participate and in fact to become a 'leader' at MTI. Now Mahalon had joined what I considered a criminal team and was shuffling for power and position within their auspices, I cannot say if Mahalon actually committed any crimes beyond misrepresentation, but he certainly aided and abetted as he could for his own gain. He was too egocentric to be concerned about MTI's gains. Only the FBI and I discussed the parameters of "obstruction of Justice" by Mahalon, by NASA, and several others. Mahalon was under the FBI microscope. I was angry and scared. Kent McKinnon was angry and scared. Don Thompson was probably more angry and more scared. Mahalon was elated at attaining for himself a seemingly more secure position at MTI though he was still a non-Mormon contractor. The FBI had the red flag up that there might be a leak and now there might be a problem in their Case against MTI. Would or could Mahalon obstruct FBI / DOJ and thus justice? Concurrently the FBI and NASA were having serious antagonistic discussions. The NASA Inspector General's Officers who had met with me had reportedly gone face to face with Dr. James Fletcher, NASA Director. Dr. Fletcher was informed of FBI Source allegations (mine and Kent's).

Dr. James Fletcher told the FBI he would not interfere. As I understood it Dr. Fletcher was standing aside and the FBI and NASA IG's were to have a free hand. Investigations were called for. Tim couldn't tell me much, but this is what I pieced together at the time. When I heard Dr. Fletcher knew who I was "jolted by mixed emotions"! After my pro and con emotions slugged it out the 'con' emotions had a clear fifteen round decision. What could Dr. Fletcher try and do to me? Would NASA officials reveal me to MTI officials? Would NASA try to protect MTI in order to protect itself (NASA) from charges of neglect. Dr. Fletcher was reportedly responsible for giving MTI the original SRM contract against everybody's advice. Was he in cahoots with MTI? Dr. Fletcher was known to be a very high ranking Mormon and he had been previously accused of being the "Mormon connection to MTI"! My guts told me I was in deep Mormon criminal shit! Also Dr. Fletcher was reputed to be in bed with Senators Orrin Hatch and Jake Garns. What could possibly go wrong?

CHAPTER 20

FBI RECRUITING AND TAPE RECORDING

Time: January 1987 - April 1987

Simultaneously, I was setting up dinners with various MTI people for the purpose of ether scoping them out to see if they would help the FBI in their efforts, or tape record information they may have been witness to, or documentable crimes. There were a lot of these dinners with a lot of 'Persons of Interest' from all disciplines of the Thiokol Wasach Plant. Most never had any idea the FBI was at the table or listening in from close by.

Sometime in here Don Thompson published his "control the contractors'" letter. Don tried to threaten us by using our overtime as a weapon. It didn't work. In fact it provided an opportunity for us to test Kerry Sanofsky, Kent's possible heir apparent. When our hours were cut back one of the contractors asked Kerry Sanofsky to take money to come to work and supervise Sunday overtime. Kerry took it. This was a grand opportunity for the FBI as Sanofsky was now 'dirty'. Though MTI personnel were around during general working time they never really supervised us. We left them for dead in terms of diligence, knowledge, education, and experience. Kerry was sadly naive!

Some of the dinners were completely successful in identifying viable FBI Sources. Others were successful in identifying those who would be untrustworthy, negative, or dangerous sources. My covert tape recording sessions went superbly. One of these tape recording sessions was on a target Mr. WI (we'll just say 'well informed'). I will use this as an example of the many intricacies of tape recording court-worthy information. Recording court usable information is not a simple as you and I suspect from watching TV crime programs. High anxiety! Complexities, entrapment precautions, mortal fucking fear of being caught 'wired and listening', noise levels in restaurants etc... Hollywood has no concept of reality.

After work on the appointed evening in question I zoomed from MTI straight to Ogden. I was tired, scared, lonely and determined. I parked at a hotel parking lot and was picked up almost immediately by Tim. Tim had prepared a bunch of papers and had been granted court authorization to do this and other recordings. I had to sign some of the court documents. No problems. I think it was something to do with the disposition of my body if I was found out and shot?

Tim escorted me off to the basement of the Federal Office Building in Ogden. There we went into the usual secure garage. In the basement is a locker room for government employees and we simply went in there and I took off my coat and shirt and started strapping on these Swiss manufactured electronic devices. It was the tape recorder which was in a shoulder holster kind of arrangement, held under one arm pit with a wire down a sleeve. Then wired around to a switch on the other sleeve of my shirt. On other occasions we used other types – it just depended what was called for. No telling where or how this equipment could be worn. Just as I was half wired some other government worker came down so I had to run into a toilet stall and hide.

Just because it is a government building doesn't mean the whole damn world had to know who I was and that I was being wired even if they were some long lost government employee. I had extremely high anxiety this night and it took a few minutes to get 'wired up'. I was physically and emotionally pushed due to my incredible schedule. There was also a thin transmitter which dropped into one of my shirt pockets. It is thin and I guess quite sensitive. Tim had been trained to use this equipment but I could tell he wasn't completely at ease with it yet. It was both sensitive and delicate. Such recording events are very serious events for FBI Agents. I think it is what they live for. There are a lot of things to consider when using such refined equipment. Will I be found out and be shot? Probably! "Can I turn the on and off switch on the recorder without drawing any attention to myself?" Doubtful! What if I think I turned it on but fucked up? Probably! With the stress of wearing the equipment and exhaustion I was seriously worried how I would act and talk to the targets? I have never felt like an actor. Oh Shit, what will I say and do and ask and how ....? Fucked if I know!! These are no small matters. At least they were no small matters to me that evening.

Tim had demonstrated to me the transmitting and recording sensitivity parameters. I am not free to explain the sensitivity parameters and they induced stressful considerations to me, the user. Once we attached the recording device adequately, we pushed some buttons to check them out and everything checked out – so we then headed back to the hotel for this evenings recording session. Tim and I used a number of hotel lobbies, motel rooms, and restaurants for our numerous meetings. Due to the tremendous hours I worked we had to have our meeting over a meal because there weren't any hours left in a day for me to eat otherwise. Eighty five to 95 hours per week for Thiokol, another 20 hours per week with the FBI left me in desperate condition. We soon arrived back at the Ogden Hilton Hotel. Tim instructed me that when I met this person (tonight's target), that Tim would be within listening range (a trade secret). We would all listen.

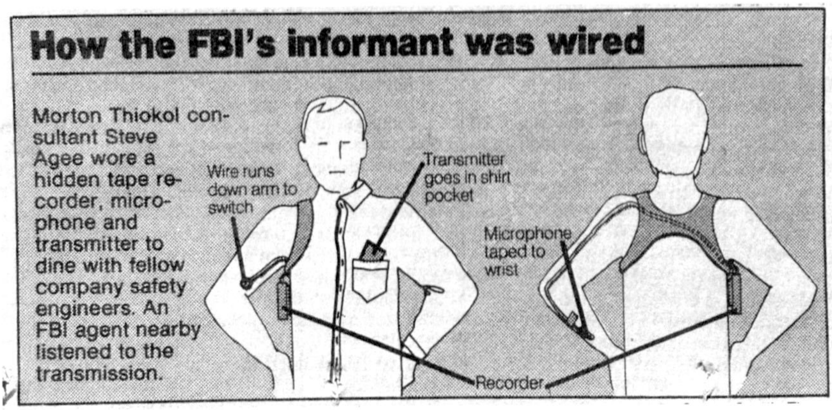

At about two minutes prior to the appointment time for the source, "Well Informed", to arrive I found a quiet hallway and walked down it nervously and fidgeting, reached under my suit coat, fumbled the switch, and turned on the recorder. "Oh God let it be on!" Then in my isolation I made a public announcement over the system which I assumed (prayed) Tim would monitor, that this device is on and should be working and we're all set to go, or something along those lines. I walked back to the lobby and less than a minute later Source WI (Well Informed) showed up at the front door. Source WI and I shook hands and went through the usual social platitudes and headed to the restaurant down at the other end of the Hilton Hotel Lobby. We talked and talked about everything from soup to nuts and I was somewhat embarrassed by portions of our discussions. WI was a most excellent and dedicated person, smart, articulate, hard working, however, source WI hated Mormons so bad it was ludicrous. I now had a tape recording of the most seething indictment of the Mormon Religion that had probably ever been recorded in known history! WI would have 'strung up' Joseph Smith himself. Source 'WI' went on and on about the raving stupid bastards at MTI – it was just incredible. Names, dates, examples, crimes! What MTI manager or person had fucked who, what positions, brand of grease, and number of strokes! As we ate, talked through the evening, the waiter came over and asked me if I was Steve Agee, and I nearly had a heart attack! 'FUUUUuuccccckkkkk!' Who would ask for me by name at a time like this? I said yes, lub-dub, lub-dub, and he said there was a phone call for me – *Oh My God ... What The Hell Now?* This added more stress than I was up to. I was literally weak at the knees. I went over to the cashiers' phone and answered it – it was Tim sitting up in his office at FBI Ogden head quarters. Tim pleasantly and politely gave me instructions to better refine my wired discussions. I felt like I was in the 'electric chair', source 'WI', was at the restaurant booth looking at me, and I was holding the FBI in my left hand! My heart was pounding, knees about to buckle. *Fear, Concern, Perform!* I said "well, Jesus Christ, I don't know what to do" and then he gave me more pointed questions how to discuss this and that and to spend less time listening to anti-Mormon rhetoric. So, back to my dinner I went. Oh shit! Why me God? Maybe being a castrated corporate sheep ain't so bad after all? I could give it a try! I already had ten new crimes out of 'WI' and Tim wanted *More - Faster* before the tape was 'set on fire'.

A few minutes later I looked over my shoulder and 'low and behold' there was Tim sitting at the other end of the restaurant, quietly gulping down a hamburger! Kind of hunched down – he had an ear piece plugged in and was just sitting there enjoying his hamburger and one hell of a hot covert conversation. For an FBI case agent this was probably an orgasmic conversation. Dripping, drooling crimes! Lies, cover-ups, suppressions, threats, *oh, it was wonderful!* I realized from my watch, I only had about ½ hour left to go on the tape and there were still a thousand questions to ask this most excellent person over dinner. I realized this person was a key player for certain. WI covered a broad range of MTI departments, crimes, and programs other than the Solid Rocket Boosters. WI knew about MX missiles, small ICBM's, Tactical Missiles, Polaris Missiles, and other MTI contracts. WI was a good man at the epicentre of a malignant corporation and time in history. He was trying to leave MTI. I was also about to have a stroke from the pressures of handling this recording equipment which I was most unaccustomed to, and uncomfortable with at that time. The dinner finally ended and I shook hands with my source and he left the building and I went out the other side of the building. The evenings stress had 'wasted' me.

A few minutes later I was walking dazed in the parking lot and Tim drove up and picked me up. Tim had really been on the move that night. Tim had not been able to hear all of the conversation. I had made a number of classical mistakes which were very common when inexperienced people did this sort of covert recording. Tim listed all of my mistakes and told me it didn't matter who was doing this type of recording, there were always glitches. Proper spook recording took practice and nobody ever got it perfect and that what we 'got' was plenty good enough! Tim felt that we had obtained a lot of astounding information from this recording. What Tim didn't realize was that the prime issues of information he was listening for, he hadn't heard because he was either between buildings or otherwise distracted. This is typical of a crime related recording session.

I returned home to my apartment tired, frustrated, alone. I felt I was doing the right thing, but its toll on me was profound. My reconnaissance missions in Vietnam were walks in the jungle, this was emotionally crushing. My Vietnam problems were relatively simple and our orders were crystal clear. If the problem had 'slant eyes' we shot it. If it out-numbered us we ran away while shooting. If it cornered or threatened us we called in an airstrike or artillery and 'wasted it'. That was the simplicity of Vietnam, this was Utah. In this Utah based war for *Outer Space* stealth, cool, and control were only a down payment!

Gerald Woods was making a major effort to bail out of MTI and go back to Phoenix and to work with his favourite prime contractor. I put feelers out for contracts. I had mixed emotions between leaving MTI and staying and seeing through to the end with the FBI. What to do? The FBI work at MTI was a "national necessity" in my opinion and in the FBI's opinion. We were in this together if we liked it or not. I was free but I wasn't free. All would be revealed in time.

NASA representatives at MTI were now in a flurry of activity. Even though they appeared to be active we all wondered what in hell they actually did? Uniformly the System Safety departmental opinion was "not much!" System Safety had moved from one trailer into another trailer and my new office space was just outside the door of a small general conference room. NASA representatives were now frequently meeting with Thiokol Reliability Department on certain deliverable documents. The deliverable documents were final SRM FMEA's (Failure Mode and Engineering Analysis) and CIL's (Critical Items Lists). Screaming, yelling, and high tension continued for nearly a week. Thiokol reliability personnel would go in and lay their work out on the table and NASA would quite frankly kick their MTI asses right around the room. I was thinking 'this can't all be bad'! About time NASA woke up!

It was the one and only time I have ever heard NASA really angry at MTI or anyone else. I asked myself "is this happening because of something I may have done?" Have I put the flag up with NASA against Thiokol or is this an actual course of NASA events? God knew there were problems in Thiokol Reliability! Such ass kicking's should have been the natural course of events – I will never know what really happened or drove this chain of events. Maybe NASA had sobered up and come to their senses?

One day the manager of Reliability, Keith Dixon, was called into the conference room and NASA said they wanted to know about this, that and other things. Remember Keith was a high school auto-body repair teacher. Keith said "I can't tell you about it - Don Thompson or higher authority will have to! NASA literally screamed and yelled at the man and told him they were here for answers and that the people who could give answers were to attend these meetings! Then NASA went on to ask why this and that wasn't prepared and Keith told NASA it was because his secretaries had been sick and weren't around lately. The same two 'morons' that System Safety was stuck with as well!. NASA then explained to this naive Thiokol Manager that they didn't give a damn if the secretary was sick or not! There were other secretaries weren't there? I thought to myself that this is classically how Thiokol and some other companies do not perform effectively on government contracts, and frequently get away with it – they bull shit through it, pass the buck, and go in circles. I was tickled to death to see NASA coming down with a heavy hand at least this once. I secretly hoped NASA would do the same when they found out System Safety Hazard Analysis were withheld from them as well.

NASA was at long last getting involved with System Safety matters also! NASA had changed the numbering system to be used on our Hazard Analysis sheets about fifteen times since I had been at MTI. You see, liberal arts educations and inadequate engineers can make great work out of 'how to number' things, and thus avoid engineering and science facts and realities. Many of the NASA employees were liberal arts trained civil servant paper pushers and not engineers or scientists. It was clear to me that NASA didn't have the faintest idea what it wanted from Hazard Analysis and we were a year after the Challenger Disaster. Kent McKinnon felt that NASA's FEMA expertise was much stronger than NASA's System Safety experience. Sixty days after the Christmas holidays there were far more NASA personnel moving into the MTI operation. We saw them in the hallways, we saw them in their offices, we saw them in their own meetings, and in the men's rooms. We never saw NASA in our working meetings until the end of March. NASA personnel finally started showing up at the regular workers meetings where the work is classically done. Always before that NASA would be found only in management meetings where very little was classically decided.

There was suddenly quite a buzz around the MTI Wasach Plant about electrostatic discharge (ESD). I had written a bunch of Catastrophic Hazards by this time on electrostatic discharge and electrical grounding problems on the Solid Rocket Boosters. However, at gun point I couldn't get anybody at MTI outside of System Safety really pay any attention. I couldn't get anyone at the electrostatic discharge meeting to pay attention! Everybody talked about it, but nobody did anything about it. I had cornered the MTI testing department and they ignored me.

I understood that there were some new studies being run by Dr. Leland Davis up on the hill at the labs which would further and better characterize at the electrical ignition properties of the SRM Propellant.

Refined studies were good, however bottom line is one spark and the rocket plant, transport systems or rockets blew up, deaths occurred, billions lost. This was welcome news because the electrical and magnetic properties and values of SRM propellant seemed to be "floating" values and uncertain when we dug into the "test and electromagnetic characterization" literature and studies on file at Thiokol. Knowing how much of a spark it takes to set off the SRM propellant seemed fundamental to all of us in System Safety. Thirteen and a half years had floated by.

What was the electrical flow (ESD) caused from a mosquito crash landing verses a cockroach crash landing? How much induced electrical current flow came from a radar antenna pointed at the SRM's? Suppose improperly grounded static electricity accumulated from the wind blowing over an SRM on the launch pad? Electrical flow arose from galvanic corrosion between numerous metals in the SRM's and the acidic atmosphere caused from the encased rocket propellant. Sea salt in the atmosphere at launch sites? BOOM!

The electrostatic discharge meeting was still being unattended and seemed to be in general ill repute at Thiokol - sort of like syphilis. I was getting the creepy feeling that maybe NASA, after having talked to me, the FBI, Kent McKinnon, and others had come back to Thiokol and said "get hot" on electrostatic discharge and grounding or else! Just as NASA had previously responded to informative telephone calls from well meaning MTI workers and ineptly dumped the problems back in MTI's lap getting the workers promptly into trouble! Was I in trouble? Damn right I was in trouble! I had delivered a bundle of Dr. Tony Laine's Hazard Analyses to the FBI. The FBI had interest in Dr. Tony Laine, but due to his personality being offensive when confronted by stupidity at Thiokol and some Mormons the FBI and I had predicted Tony's days at Thiokol were severely limited. We predicted he would be canned very soon. We were definitely correct on that one. We were all surprised however that he was canned in such an improper and unsophisticated manner and *For The Wrong Reasons* in front of an audience of witnesses. Dr. Tony Laine had roamed the Thiokol hallways rumbling and grumbling about the "Stupid, SOB's"!

How could he have survived? He was of course correct. Tony had fought long and hard to get Thiokol permanent staff to listen to him. He procured a lot of non-Thiokol documents from other arenas of the American Space Programs. A peripheral and sadly amusing peripheral true story is that MTI turned him into the FBI for National Security violations! MTI games and lies again. Tony determined the MTI library was ancient history (I used it and he was correct). Tony of his own volition ordered modern documents from his friends in high places in top laboratories all over the country and from the CIA. His associates sent him most of the data and documents requested, they were unclassified, it was legal, and it was a 'kind deed'. MTI went crazy over it. Tony was actually under house arrest for a while by MTI Security after some 'unclassified' nuclear electronic documents arrived. MTI was ignorant, and out of line, but we must remember that *"No Good Deed Goes Unpunished"*. No wonder Tony finally sued them for the works! Come to think of it – so did I! Eventually our two separate cases were to be conjoined and end up an astounding historical event.

At this time Malcolm had a huge stack of Hazard Analyses written and nowhere to go with them. Malcolm and I were spending enormous amounts of time now changing the numbering system on our identified and analyzed Hazards, every time MTI or NASA changed to a new system (every 3 to 7 days). It may sound very simple to the reader for us to change numbers on our Hazard documents . If you have ten or a thousand Hazards and this numbering changes, you simply cross off the number and write in a new number. *No, Wrong!* It was not that simple! Not even close. In the first part of a week Hazard numbering systems could go from left to right, then from top to bottom, then from the middle out, etc, etc.. Pages altered from portrait to landscape and back. Each number meant something else after each change. The changes were made more complex by the additional changes of what went into each column, and what was meant by what number in various columns? Confused, damn right! The changes were ludicrous and profoundly stupid high liberal arts and corporate criminal stupidity and crimes. Flight phase definitions changed frequently and therefore the start and end points of our analysis.

Sometimes the changing of the definition of a flight phase caused only a labelling problem, other times it caused a huge amount or re-work. Kent was trying to stabilize the numbering systems but Don Thompson forced change after change. None of us were clear whether higher MTI or NASA Management were the driving force behind some or all of these changes? It was clearly more than simple stupidity. An aggregation of great minds could not have encumbered our System Safety Department more. I asked Kent what was going on? He was exercising all possible efforts to stabilize the changes, but Kent could not stop Don Thompson and Mahalon. This was not business as usual by any industry standards. The FBI had at least daily updates on the changes, on occasions hourly. The atmosphere was clear that anyone who stood up and told them how stupid the situation was would be fired on the spot. After all - there were no problems at MTI! The changes were unnecessary, ignorant and disgusting. Confused and angry worker bees, consultants, dedicated employees, all made change after change to our documents. It would have been a big task with a computer, but without one it was overwhelming.

Our computer access continued to be highly restricted. The massive changes were outrageously expensive and when we were changing documents from one form and number to yet another we weren't really making any progress toward finding faults. We had become modern scribes. We were no longer System Safety persons – merely calligraphers. you could have heard the departmental grumbling and bitching two miles away. Concern for the American Space Program, patriotism is the only reason any one stayed. Some left.

The FBI and I viewed these constant changes as delay tactics by MTI. More and more it appeared resistance to changes were driven by NASA. I saw Don Thompson effectively capitalizing on NASA confusion to slow us down. Some changes required a piece of a system to be broken into subcomponents or grouped into larger components, and this would ruin the entire test of a written Hazard so that it might become two hazards or two might become one. Chaos reigned! The "opportunistic" System Safety department personnel quickly observed that none of us ever turned in our HA's, so by not producing any properly numbered, completed, acceptable HA's MTI's efforts of renumbering them was completely successful for a while! NASA never received any HA's for ages. No one should think that NASA received any HA when they were caught up to the newest format. It made no difference. The Hazards were still captive to the Thiokol System Safety Department. Somewhere along this time Thiokol issues a new schedule that showed our System Safety products were not due to NASA until October 1987. There was so much work to do between now and October in order to get everything analyzed the task was awesome, probably not accomplishable. In the back of my mind, I thought how convenient. Rather than NASA not being diligent, and MTI being delinquent delivering documents, NASA just revises the schedule to a much later delivery date! Very clever NASA. I have no way of knowing if this course of events was a conspiracy or simply the a natural course of NASA events. No matter what metric was used - NASA was pitiful at that time. The FBI was angry when I told them and they thought it was NASA "Obstruction of Justice".

By NASA changing delivery schedule Morton Thiokol was possibly taken off of the hook for certain contract compliance requirements and deliverables. Virtually a hand full of 'get out of jail free cards 'so to speak'. Even though there was a scheduled delivery date MTI had a contractual responsibility to immediately notify NASA about the each and every one of the catastrophic Hazards immediately as they surfaced. This responsibility was demanded via NASA documents. Still no NASA / MTI contract had emerged. Clearly we couldn't give anything to NASA now. It was obvious to System Safety personnel that almost everything on the SRM was in profound technical trouble. Most of us wondered if anyone at NASA even cared? We were all certain what MTI thought – *There are no problems of any kind anywhere - ever!!!!*

About this time the Rodney Commission descended upon the MTI Wasach plant. Since I was in my "low-key hiding mode", I did not get to meet these folks from Wash. D.C., Kent McKinnon, Don Thompson and numerous other MTI personnel had many orchestrated meetings with the Rodney Committee members. No information really ever came back to us from their meetings. Even Kent was unwelcome and in the dark. One day Malcolm and I were sitting in our back room offices when Don Thompson did a unique and strange thing. Don came down to our area and he was literally ashen colored.

Don looked as though he had just had the hell scared out of him. I was thinking to myself, it couldn't have happened to a nicer fellow, whatever scared the hell out of him – more power to it! I personally never had a harsh word with Don Thompson but I had certainly identified him as a man who would not work on my team anywhere under any conditions. Don was a consummate loser in my book. Don Thompson was the enemy of the free world, an enemy of morality and ethics, and possibly the worst example of a Mormon Bishop leader on two legs! I had always made an extra effort to be courteous to Don Thompson, not because I was afraid of him, not because he disgusted me. I was courteous because I needed to be at Thiokol for the FBI. I had sinned by refusing to become a Thiokol castrated corporate sheep. I could not continue my job, or to help the FBI if I was fired. The few times we spoke in the hallways, or at a meeting, or here and there was always cursory, polite. Both knew neither had any use for the other. Don Thompson knew I wasn't afraid of him and I saw through him like he was a clear glass. Neither of us wanted to be hurt by the other, a primordial stand-off between natural enemies sharing a common environment. Both had power, both had vulnerability, I was stronger and more dangerous to him than him to me, however I carried the torch for the American Space program.

Don Thompson leaned back against the wall, Malcolm to his left, me to his right, and Don literally took in deep gulping breaths. I thought to myself, either Don just had the crap beaten out of him by his bosses, had a personal disaster, or the Rodney Commission had twisted his nuts!

Don Thompson leaned back against the wall again, took more long deep breaths, and tried to get control of himself. He asked us how we were doing and proceeded to shoot the bull with us in a most courteous manner for the next ten minutes or so. He told Malcolm and I he was sure glad to have high quality folks like us working of the System Safety team and giving MTI a hand. He sort of nodded his head as he said it in a manner I interpreted to really mean he appreciated Malcolm. It was known to nearly everyone that Don had no use for me since I blatantly refused to kiss his ass. I really didn't give a damn who Don Thompson preferred or liked. After about 15 minutes Don left our area. Malcolm and I looked at each other in astonishment and curiosity. No doubt Don was sent (ordered) by someone. Malcolm said I have never seen Don ever act like that before, me either! Malcolm and I talked for the next 20 minutes about what in hell was up? I told Malcolm that I suspected a Rodney Commission member had found out how many Hazard Analysis were in the typing backlog or trash cans by now – hundreds. Or had they just read a couple of Hazards by accident and found out how seriously errant SRM engineering was. We sort of concluded his scare probably came from something to do with the Rodney Commission. Who knows, maybe the Rodney Commission told him Steve Agee was with the FBI? Nothing would have surprised me! I know for a fact that the FBI and Mr. Rodney were communing by this time. I had comments from the FBI some time later that Mr. Rodney was one of the few NASA types that helped the FBI at all.

About an hour after Don Thompson visited us, someone told us that members of the Rodney Commission were prowling the hallways and had found a huge stack of my HA's on Erma's desk. By now my backlog of typing with Erma and Wendy was in the hundreds of Hazards, thousands of pages with no hope in sight of anything ever being completed or returned in usable condition. All of us observed Erma and Wendy could type acceptable work for higher Thiokol officials, but they consistently put out crap or nothing at all for us in System Safety. Also of interest is Erma and Wendy's immediate supervisor, Sylvia Rollins, was known to be seen crying at her desk several times a day. Witnesses said Sylvia was extremely troubled by Don Thompson's pressure and all of the demeaning people wanting their typing who were perpetually pissed off by 'lack of performance and delivery'. Sylvia was sort of having a mental breakdown - who wouldn't?

A member of the Rodney Commission grabbed up a handful of hand written papers from Erma's desk and ran off with them. Other members of the Committee kept the stack of HA's for approximately 48 hours we later found out. Don Thompson was plenty upset about the committee getting its hands on these HA. We believed it was more than just my hazards, we believed some were Malcolm's, some Gerald Woods, and no telling how many others. Perhaps now NASA will wake up and get moving on some SRM system enhancements! I should have known better. I had such unreasonable expectations,. Silly me! We were then in the depths of Winter and we were frustrated by Thiokol obstacles adding to the dreariness. As individuals we experienced satisfying successes finding many bonafide problem areas in the SRM's. Each and every one of us knew that the two SRM's per launch were totally necessary to lift the orbiter and ET to the necessary altitude. Many of us asked ourselves, "if the SRM's are in this bad of shape, what kind of shape are the SSME's (Space Shuttle Main Engines) in, the obiter, the software? What about the antique magnetic core computer memory units in the obiter? Well, if it was not SRM's it was not our project.

Most of the System Safety people had by this time reached the uniform conclusion that they individually wouldn't fly in the Space shuttle under any conditions. In short, it looked like a sure way to lose your life. There were frequent discussions among the various technical experts in the System Safety Department about when we would launch the next Space Shuttle. At that time the target date was next June or July. None of us thought this was a realistic launch date at the current rate of MTI and NASA failures.

We all knew NASA was breaking their vendors bones to meet their published launch date for Discovery. We also knew there was no way in hell of meeting that date safely. Of course NASA and the White House breaking bones to launch STS 25 for the President's 'State of the Union Message' precipitated the Challenger Fatalities. Personally I predicted a date at least a year past June or July 1987. We also wondered if the Space Shuttle concept shouldn't be scrapped or phased out while moving from our present Space Vehicles to the projected National Aerospace Aircraft. Also there were frequent discussions about the looming probability of a big dumb booster, cheap, unmanned and robotic for the heavy jobs such as launching the KH-II Spy (Keyhole) Satellites, and the Space Station. Why send persons on every launch? In round numbers the cost of launching a human being into orbit then was approx. $14,000.00 per pound - vastly more now 30 years later. On any given manned mission the cost of keeping the crew alive is about 85% of mission cost. Do we have to send an astronaut on every mission? Definitely not. Wouldn't we be far ahead if we had an intelligent mixture of robotic heavy lift vehicles and one or more manned vehicles for use only when human activity was required. It's the nature of scientists and engineers to speculate and calculate and derive logical alternatives and conclusions while avoiding the ugly political games. Unfortunately political concerns ruled – putting men into space opened the Pork Barrels further. Men and women in space held citizens attention while NASA budgets were decimated in thousands of ways. Humans in space were partly public relations operations and such spectacles were controlled by liberal arts educated people. Much easier to obtain funds to keep Astronauts (national heroes) alive than keep robotic systems functioning doing important work safer.

I concluded that success really depends on what our national plans are in space. What were they? *No one really knew* then! Now 30 years later I am not convinced there is a cohesive plan for the Western World. Returning to the Moon is laughed at by many liberal arts educated Congressmen and Senators. They think welfare is money better spent. Going to Mars is laughed at by many. Some want to focus on an all purpose Space Station. Simultaneously, there were plans for a solely Military Space Station. NASA preached 'free enterprise missions' but it is in reality war operated in conjunction with the USAF and the Department of Defence, and the National Reconnaissance Organization, the U.S. Army (Star Wars), the National Security Agency (eavesdropping and code busting), and the CIA. Are U.S. Space elders and planners really looking out for the best near, middle, and long term interests in space? Few space insiders thought so. None of the experts I know think so and I don't think so. Currently there are approximately 169,000 human births a day, 71,000 deaths per day, and a net increase in human population of 98,718 per day. Earth is fast running out of space and resources.

When our space program collapses or is misguided will we end like Portugal when it fell out of the New World Land Grab and collapsed due to the treaty of Torricelli orchestrated by the Vatican. Will America be earth bound forever and dirt poor indefinitely? Damn good possibility! Absence of both Presidential and Congressional leadership allowed NASA and therefore aerospace contractors to fail. The DOJ Attorneys moved on this case as a tortoise crosses the Kalahari Desert – painfully slowly! The FBI refers to the DOJ Attorneys as 'limp dicks' due to slowness or no function at all. Low professional standards by all except for the odd flicker of hope in young person's eyes.

CHAPTER 21

DECEPTION

Time: February 1987

Most of us 'contract personnel' were generally gregarious people in a business setting. Gregariousness and communications are skills necessary to be successful acquiring data and moving in and out of these various client corporations deftly, and in acquiring the information and connections to "get the job done" fast and efficiently. All of us maintained our own circle of professional and personal affiliates within and outside of Thiokol. Contract and consulting engineers are frequently 'Project Storm Troopers'. They frequently work the toughest jobs, and the longest hours, the tightest schedules, and the most 'politically loaded' projects. They are classically treated as outsiders, but occasionally they are appointed such duties as, and it happened to me, 'Acting Vice President?' Contract and consulting engineers are generally a tougher lot than the 'in house' castrated 'cover your ass' employees. It takes confidence and gumption to be 'free and independent' and tell Boards of Directors, Generals, and Admirals, the harsh truths. I was concerned about who was regarded as a friend of mine because it might get them fired at MTI. After all I was 'persona non grata' to MTI Managers. I wanted to be a better friend to some of these experts but held most at arm's length for their own protection.

I desperately wanted to be a proper personal friend of Kent McKinnon's, but it was too dangerous and virtually impractical / impossible. Kent was a friend from a measured distance and in spirit. In addition to Space Shuttle problems, MTI had issues with the MX (or Peacekeeper) missile. [9]

Malcolm and I still car-pooled and I never thought that was necessarily good for Malcolm. Malcolm was at least someone I could talk to about a lot of things. He and I were working elbow to elbow doing damn good work. One of the Thiokol personnel, Mr. GS (Good Soul) was a wealth of historical and current information about MTI. GS had seen it all. GS was a high ranking Mormon Official at one time or another and had been at MTI for an astonishingly long period of time. This man had not played proper ball at Thiokol and was presently always used for one dirty job or another,

---

Appendix 9: Will Tragedy Delay MX ... By Guy Bolton, Tribune Staff Writer but clearly would never be moved into the upper echelon at MTI again due to the fact he was basically an honest man. *No Good Deed Went Unpunished!* GS inadvertently put up the red flag to me concerning a number of errant practices at MTI ranging from soup to nuts. Many of these errant practices were previously unknown to me. When I asked VR about GS's experiences and problems I was shocked at the information which precipitated. VR and GS went way back. GS was a fourth generation Utah descendent, he still held his original old family 'Homestead', and was trapped by the circumstances of his Mormon religion, family, Utah, and MTI.

Via Kent McKinnon's direction, I made a concerted effort to meet the Manager of Thiokol's Plant Safety, Mr. John Pontius. John and I hit it off very well. I orchestrated for him to help and lend his backing to Kent McKinnon and Kent's efforts to expose all of the SRM Hazards and do the System Safety job properly. Kent needed support at the highest level of Thiokol Management and John Pontius said he would support Kent in any way he could. He seemed genuine and I had no reason not to believe him. I trusted my instincts and judgement. John Pontius reportedly had daily contact with top Thiokol Managers.

During my discussions with John Pontius I found out that when the twenty two million dollar Solid Rocket Propellant Casting Building blew up at MTI a few years ago that the cause of the explosion was MTI employees grab-assing around the work area. The 'Casting' building which blew up, could have killed a number of people, but only injured a few. The explosion destroyed the 22 Million dollar building in a flash! It was the only propellant casting building in the world large enough to cast Solid Rocket Propellant into the SRM sized cases (diameter of 13.4 feet). The batches of solid rocket propellant were huge and dangerous.

John Pontius told me that the Casting building explosion occurred as a result of employees placing small amounts of rocket propellant on a small trolley track travelled by wheeled containers rolling on the upper floor tracks. When the propellant was pinched between the vehicle rollers and the track small amounts of propellant blew up, made a scary sound, and launched sparks. This had been great fun for years. The "snap, crackle, pop" classically scared the shit out of fellow workers. What fun! That time some of the sparks had escalated over the side of the high tracks and down into the main mixture bowl of rocket propellant. Fortunately, as someone ran to the upper story escape slide, and jumped he punched the emergency evacuate button that warned the rest of the people in the building – allowing a number of them to escape the building barely saving their lives. They cleared the building with *no seconds to spare!*

This was one of the most incredible explosions in Utah history. I asked John Pontius how it had all shaken out and he told me NASA had investigated Thiokol and they had all come to an amiable agreement and NASA funded MTI another twenty two million dollar building. I never investigated this very much further because I didn't have a great deal of interest in this recent historical event. This became a separate FBI Case. I do remember having the thought and wondering "are NASA and Thiokol so well in bed with each other that they can just blow up a 22 million dollar building and get another one for free? The answer is 'of course', after all it was GFE, which is Government Furnished Equipment. There were physical time constraints governing how far and how deep I could research additional problems. The FBI chronicled what I found out about the lost Thiokol Casting Building. Shortly, a few months later Thiokol blew up an MX Missile (also known as Peacekeeper Missile) Casting Building killing five workers outright. MX Missile propellant was much more explosive than Shuttle propellant by design.

Malcolm and I and others scoured over the MTI Plant looking for anybody who could do any intelligent typing. We found that in addition to generally inept and incompetent typists, completed copies for a number of Thiokol documents would come out of only one particular computer printer. This printer existed in a state of "broken down" 90% of the time. I expect that this machine literally held up a thousand pages of typing / printing a day, seven days a week. I thought to myself, "one more way of suppressing hazards"! Since there is not much time for the printer to print, when the machine was up and running, MTI managers such as Don Thompson would 'cherry-pick' what would be printed. [10]

Two of the System Safety Departments contractors (Arun and Rakesh) were superb computer experts. Both asked Kent if they could look into the printer problem. Kent told them to go for it. These two Indian Engineers spent a few days looking into our supporting computer and printer issues and then their verdict came in. The Thiokol person in charge of Thiokol central computer and printer was an idiot! Our experts said it would be impossible to do the system set-up any worse. Total debauchery. Our two engineers worked this problem intermittently for weeks. In the end, they turned their backs and walked away. They could fix technical problems in minutes. They could not overcome MTI management resistance. Was this MTI management stupidity, ignorance, suppression or a cost savings act? I

Appendix 10: List of Contract Personnel in System Safety and related departments at that time. FOIA Request

didn't know for certain at that time but suspected the sum of it all. It seemed to me that thousands of pages per week of backed up typing had to be more costly to MTI production than the cost of a printer or two that worked. In the end the two Indian engineers used the MTI mainframe computer to type their memos only. They had to battle Thiokol to obtain a computer access number on Thiokol's VAX Computer. All of the rest of us never tried. Thiokol's unwillingness and objections weren't worth the trouble. Later the FBI asked me why I hadn't typed my memos, etc.. I thought my request for a computer access number at MTI would be my final request there. After all, I and the others had offered to rent our own personal computers and were not allowed! I should note here that I was and am a very fast, and accurate typist since the 8th grade, was quite computer literate as most physicists and engineers are. Also I was massively busy.

Frequently VR related the Thiokol insider wars, the secretarial wars the Thiokol failures on the 'Peacekeeper (MX) Missile', and the MX LEGG failures. The LEGG is the 'Launch Egress Gas Generator' to blow an MX Missile out of its silo. An acronym for the bomb which blows the two hundred and fifty some ton missile out of its silo. So, not only was MTI castrating America's space program, but also America's Strategic Ballistic Missile Systems. Then one day VR laid out the story of how an MTI loser-type manager had wasted $10 million right off the bat when Thiokol got the SICBM contract. The SICBM was America's new small intercontinental ballistic missile. I thought MTI was probably the most dangerous company in the history of the world! Later I would learn of multiple failures of the ejection seat systems for Americas new Tactical Fighter on the *'Crest Program'*.

The U.S. Air Force fined MTI for more MX Missile Rocket Motor failings. Yet later, about March 20, 1989, a Trident II Missile launched from a submerged submarine flew for four seconds before its first stage Thiokol Booster blew up. The Trident II explosion was the culmination of a 38 billion dollar development program. Their list of failures went on and on...

CHAPTER 22

FBI IN HIGH GEAR

Time January 1987 – March 1987

I went home one evening to Brigham City tired as usual. I was driving down the street to the grocery store and a huge nondescript car loomed up behind me, passed me and someone made motions at me. "This guy drives like a maniac", I thought. Oh! It was Special Agent Tim Healy and Special Agent Don Kirby. Tim's Senior partner. I would learn later he was a Mormon and an FBI FCI Officer (Federal Counter Intelligence Officer). We had a brief meeting in the store parking lot. They needed directions to Malcolm Phillips place. I gave directions and off they went. Months later it would come out that Malcolm had played devious games and in fact lied to them both. The FBI is used to deceit and lies and has expertise dealing with such matters. 'TRUTH' has been an endangered species. Tim in particular knew Malcolm had lied. I never found out completely what Don Kirby thought. I can tell that while I felt Kirby was a first class citizen, he still struggled against the possibility of crimes at Thiokol would have massive impact and consequences to the Mormon Church . Kirby had been a Mormon Missionary in Taiwan during his youth. He still remembered hitting a bungee (sewage) wagon' with his bicycle! The smell and mess was impressed on his mind forever.

I had compiled a huge pile of Hazard Analyses (mine and others) for the FBI which I had either completed or was in the process of completing. I had the strong sense of escalating acrimony between the FBI and NASA. From the bits and pieces I had, NASA had a similar attitude to MTI – "NASA knew best"! "We're doing our job, leave us alone!" During this period the FBI was assimilating a lot of data and checking on everything they and I could think of. Some of the things they were checking on were as follows:

1. Did Thiokol hide their known safety problems so they would not incur high reliability insurance rates, or even worse not be able to obtain any launch liability insurance? { I hadn't thought of that at the time }
2. Had MTI lied to the launch insuring agencies?
3. Had MTI benefited from the crash of Challenger via insurance pay-outs?

4. Was launch insurance, plant insurance, shipping insurance, or any other type of insurance an issue in this case? { I had no idea? }
5. The FBI did sophisticated background checks on key MTI personnel and others.
6. The FBI did background checks on all of the corporations involved.
7. The FBI correlated this case to other FBI cases pending against MTI (there were several), but I didn't know what any of them were at that time.
8. The FBI was correlating my evidence with Roger Boisjoly's Qui Tam Cases. (Roger Boisjoy had filed a 4.5 Billion Dollar suit against MTI 'on behalf of the U.S. Government' as a result of the Challenger launch and accident which MTI violently opposed. He also filed a negligent homicide suit against MTI. He was proper to do so. My experience and my documents proved his case beyond any question.
9. The FBI was constantly interfacing with the U.S. Department of Justice on Civil and Criminal issues related to NASA and its subcontractors.
10. Research on the type and nature of other cases pending against MTI could not be discussed with me by the FBI at that time but they told me there were numerous cooking.

I asked the FBI if their other cases would help what we were working on, or if mine helped theirs? I was assured there were definitive mutually beneficial symbiotic relationships. My interest peaked and I was assured that I wasn't the only one who had the brass to throw a rock here and there at Thiokol and NASA. NASA had been America's privileged child agency and appeared "exempt". NASA was overdue for a trip to the 'wood shed'.

About this time NASA came back to the FBI with written rebuttal to FBI questions. Their rebuttal essentially said "we have looked at the problems and questions you (the FBI) sent us and *"there are no problems"*! *{Where have we heard this before?}*

Their rebuttal was of course ludicrous. I knew it and the FBI knew it. NASA's rebuttal was somewhat technical to include in the text of this book. When I studied it for the FBI I pointed out point after point that there were clear and obvious deceptions 'crafted' by NASA.

Simultaneously, Tim Healy and FBI No. 1 (Bob Bryant) requested one of the Special Agents in Washington, D.C. to request from NASA a complete copy of the NASA / MTI Contract for the Solid Rocket Motors. This Special Agent, Mr. Simpson, had been in meetings with high ranking NASA officials when he asked point blank for NASA to forward a copy of the NASA / MTI contract to the FBI. NASA, not wasting a heartbeat, looked him right in the eye and said "NO"! It was related to me by a D.O.J. Attorney that Special Agent Simpson was 'shocked', and pissed off?

NASA told the FBI that the contract was long, tedious, complex, technical, and that the FBI wouldn't be able to understand it! Tim asked me if it could be that complex and I told him I had seen hundreds and had never seen one yet that an ordinary human being couldn't read. They are written in relatively plain English with only a few items which are technical to the typical reader off of the street. Contract simplicity is standard so U.S. Congressmen, U.S. Senators, and accountants, and masses of U.S. Civil Servants can read and administer them plainly.

Tim and No. 1 were really pissed at being told the FBI couldn't have a copy of the contract. No. 1 called up FBI headquarters in Washington, D.C. and I believe it was through S.A. Simpson that he made arrangements to acquire the contract through other means than NASA. The FBI anticipated arrival of the contract so they (we) could better understand the scope and depth of any contract violations. No one can fault the FBI for being a bit in the dark without a contract in hand to bounce known events against. I was in the dark also. Perhaps this was a hermaphrodite contract that allowed all sorts of irregular contractor behaviour? Ummm - doubtful. By this time I had written about 180 Hazard Analyses regarding electrostatic discharge and improper electrical grounding of the SRM's. I was into a dozen aspects of the SRM such as:

1. The 'Safe and Arm mechanism' which starts the solid rocket motor off.
2. The 'Safe and Arm' electrical grounding
3. The 'Igniter' functions
4. Electrical signals to the 'NASA Standard Initiators'. {Nearly simultaneously at this time NASA sent MTI a bunch of 300 igniters (Standard Initiators') which *all failed* in firing tests! Would have been fatal to all on board a launch - incinerated at the pad!}
5. Numerous others ...

My personal life was not especially well. Because there simply wasn't enough time to devote to myself. I had some very fine lady friends scattered around the country and beyond, but no time to see or talk to them. This was long before the internet took over the planet. No time to visit or telephone, etc... I had met some nice women in Brigham City at the Thiokol Rocket Plant, and didn't have proper time to spare for them either. I knew that my strength was winding down like a spring and the situation at Thiokol was worsening by the moment. I knew I couldn't keep going like this much longer. There would come the inevitable physical crash. Of course there was the continuing possibility that I would be fired by Don Thompson in a heartbeat, or wasted by a well meaning Thiokol or Mormon mercenary! Perhaps a NASA public relations representative? Then one day a thought-provoking ESD passed through my mind. Could the Soviet bloc want to waste me? Why you ask? Simple! No doubt the Russian block knows the scope and impact of the damage Thiokol inflects on America. Russian block interests were greatly benefitting from MTI at the moment! I was therefore a serious threat to America's Cold War enemies. A provoking and frightening thought. I recommend to a FBI Federal Counter Intelligence Officer that the U.S. 'leak' all of MTI"s secrets and technology to the Eastern Block. MTI's secrets and technology would then be an effective mover towards Russian unilateral disarmament or collapse. 'Both nations' space programs and military programs foiled! The long sought military and space race parity! Both collapsed by MTI and NASA failures!

CHAPTER 23

'AGENT 99'

Time: Late February 1987

For ten years I had a long distance African (Rhodesian) girlfriend whom I had met in Pago Pago, American Samoa while doing business around the South Pacific. We had become true and fast friends there on the island of Tutuila and had spent more time in Honolulu on the way to the United States and my return to the U.S. mainland after living in Honolulu for a year. When we met I was living in Honolulu – Waikiki. After returning to the U.S. mainland, Fay and I stayed in touch over the years and occasionally visited each other on sorties around the world. Six months prior to my going on contract to MTI, Fay had flown through Seattle on a round-the-world trip from Sydney to Durban South Africa, on to Rome, and U.S., and back to Australia. On her trip through we visited Victoria, Canada. In Victoria, Fay and I talked about the unlikely possibility of getting together for a permanent relationship – i.e. MMmmmmamaaarrirriiaaagggee??? We were both overly cautious and scared. No rash action at that time. The mere thought of marriage made both of us tremble. Fay had never been married, I had divorced two women. One American and one Chilean. I viewed marriage as a Class A Felony with a long term sentence.

I had recently received a letter from Fay almost the same day I sent a letter to her. Her letter and mine were very similar. We ought to cut the bullshit and do something rash. We both thought we ought to spend some time together prior to possible marriage. Fay was scheduled to fly in from Sydney, Australia to Salt Lake, City. Fay worked for Ansett Airlines in Sydney then. The plan was for her to meet me and then we were both to fly off for an 11 day trip to Belize. Sociological Systems Safety Hazard Analysis for us both!

My life was confused. I hadn't adequate time to tend to it properly or at all for that matter for a long time. I knew I was going to die if I didn't 'throw in the towel' and get married. I was lonely. I also knew my life was bizarre to many American ladies' standards. I worked too hard, my consulting uprooted me frequently, and I was an incessant world traveller. Rather than socialise, I would rather sail or carve wood. I was and am arguably a recluse. Also I had previously divorced two women and had high standards and anxiety towards marriage. In short I lead a dynamite life. I honestly had a number of lovely marriage prospects. What to do? Knowing myself and them I was obligated to do no harm to any of us. No small task! I knew I could not do right by all of them. Could I do right by any of them? What to do?? No matter what I did would precipitate a portion of hurt and rejection. An ageless enigma. I actually loved several women, each in different ways and for different reasons. Do no harm is a tough standard!

At any rate, I was glad Thiokol might let me go because I was physically wasted far more than Kent or anybody at the FBI or Thiokol realized. I was going to Belize no matter what was precipitated from Thiokol. I was so exhausted I had a lot of internal twitching at night and in fact was quite ill and didn't know it yet. When I asked Kent if I could take eleven days off unpaid he said yes but wanted to verify it with Don Thompson. Don waved holy water on my absence. Wonder if he celebrated?

So, I left Thiokol for vacation later than I should have that day. I had to rush to pick up Fay arriving at the Salt Lake City Airport. It's a long drive from the MTI rocket plant to the Salt Lake City Airport. I drove like a bat out of hell through the ice, snow, and sleet. En route I witnessed one of the most incredibly beautiful storms I had ever seen in my life. A fast violent electrical storm. As always happens, I was low on gas, so had to find a gas station and fill up the car which made me that bit later yet.

I zoomed from the gas station toward Salt Lake City and finally arrived at the airport, parked the car in the snow, ran in to meet Fay and was only 39 minutes late. Damn! Fay is like me, she didn't like being left standing. I couldn't find her so I had her paged and a few minutes later we ran into each other in the middle of the main airport entrance. She was glad to see me and I was glad to see her. We grabbed her bag and ran for the car. – I told her I was in a hurry. I had an important meeting to go to and as we rushed from the airport toward downtown Salt Lake City, I told her I hadn't had time for breakfast or lunch either one – how was she doing? She was starving to death. I told her we'll go meet these people I have a meeting with and then we'll get something to eat.

She said fine. Who are you meeting with? I told her "you ain't gonna believe this! I've got to meet with the FBI and NASA right now!" "Oh, something to do with your work?" So I told her "Yes, something to do with being involved in some heavy duty stuff!" She says "oh playing James Bond" and I said "yes, quite literally". There is a stack of documents in the back seat and I need to deliver them to the FBI this afternoon and face a question and answer session with NASA managers and the NASA Inspector General's Office. Fay was shocked.

So we zoomed to the Salt Lake City Federal Building. Parked alongside the Federal building and went upstairs to the main FBI offices in a casual way.

This was a new one time only visit for me to show up at the offices overtly rather than covertly, and this was where all the officials were to be gathered. I told the receptionist I was there for Tim Healy – she said have a chair and wait. I anticipated to be rushed in and kept out of sight. After all, Thiokol Security Staff frequently came up here for meetings on rocket espionage and such. I was nervous being so in the open.

We waited. We were hungry and we waited. Finally I was pissed off, told the receptionist we were going to lunch at the ZCMI restaurant a block away. I told Fay I couldn't believe they would keep me waiting like this! I was far more pissed off than I let on. I was and am not spoiled but I felt "exposed to danger", hungry, rushed, and tired. We got as far as the Federal Building elevator when Tim came running down the corridor – "Wait for me, wait for me! I'm coming with you!"

I told Tim we were starving to death and introduced him to Fay. Tim looked in a happy but harried mood. Tim told us "You ain't going to believe what just happened!". We got in the elevator and went down and Tim was making friends with Fay and Fay was making friends with Tim. Fay was in a mild state of shock over the 'goings on'. We walked a few blocks up the ZCMI store across from Brigham Young's old mansion and went to their restaurant to have lunch. Tim asked me if Fay knew what was going on and I told him no I hadn't told her any of the particulars. I told her 45 minutes ago I was having meetings with FBI and NASA this afternoon and I thought he could tell her what he thought was all right to let out. He laughed and while I visited the men's room Tim clued Fay in that I was into some heavy duty stuff. An honest to God FBI spy! Subsequently known to her as "Agent 99". From then on she often referred to me as '99' (from the 'Get Smart' TV series.

Fay was surprised and shocked: She flew in from Australia and here she is in the hands of the FBI so to speak! When I got back from "the room" and sat down Tim started telling me what had been going on up in the Salt Lake City FBI Office.

I told him I couldn't believe they would leave me sitting out in the lobby like they did, and he said, well, they couldn't bring me in the back because FBI No. 1 was literally busy kicking ass on NASA Officials. There was blood and shit everywhere so to speak. I knew Bob {Bear} Bryant was a good man. Glad to hear Bob was having a good day!

Tim said that NASA officials, led by Mr. Jim Thomas, had sat down with Tim and other FBI Agents and were having a major meeting over the numerous identified problems and crimes at MTI. After they came to the end of their basic discussion – FBI No. 1 told Jim Thomas "now we'll bring in Steven Agee and you fellows can have a face to face discussion and ask any questions you want." Tim said that Jim Thomas leaned forward and said "we can't meet with Steven Agee!" "What would MTI officials think if they thought we were dealing with an FBI Source? NASA's relationship with Thiokol would be endangered!"

Tim alluded to the fact that No. 1 jumped out of his chair and kicked their asses right around the room and came down like two tons of bricks on the NASA officials in the room. I had met No. 1! and I knew he was a brilliant capable man and when push came to shove he'd be doing the pushing and the shoving. I had enormous respect for No. 1 and in fact I had enormous respect for all of the FBI personnel I worked with. There are FBI retards and losers, but I never worked with any. Tim said he was literally shaken by the fallout of it all. As a result of Jim Thomas's uncooperative and sold out attitude and unwillingness to meet with Steven Agee, No. 1 had essentially thrown them out of his office and sent them back to do whatever they do. Then I was given the impression FBI No. 1 opened another FBI case against NASA right then and there. Later I came to understand the NASA Case venue was moved and that there was probably a number of DOJ Cases against NASA. Probably it moved from Utah to D.C. but I didn't have complete overview and it wasn't really my business. Tim was very angry at NASA. It was clear to me I was cannon fodder to NASA! *Oh Shit, Why Me God???* Do I need this? Not only could MTI and the Mormon Hierarchy waste me, now NASA certainly hated me and feared me. My personal observations were that NASA was an arrogant and ungrateful pack of losers at that time.

Tim, Fay and I had a wonderful lunch. We talked about everything. It was agreed that Fay and I would meet with Tim and his wife later that night for dinner and then at 4 am in the morning Fay and I were flying off to Belize via Houston. After we ate a late lunch all three of us went back to the Federal Building and I gave Tim another stack of red hot documents. Tim took me on a tour of more Salt Lake FBI offices and we stopped in to visit Bob Bryant (No. 1) for a bit. I believe it was this bunch of documents that included a NASA document giving the blow by blow history of the many failures of the SRM field join o-rings, the putty and the joints themselves. This document was an extraordinary indictment of both NASA and Thiokol. Tim and Bob Bryant then interrogated me on the 'documents of the day' and did their usual write up to go with his 302 'case jackets'. Fay waited in the lobby and had some excitement. The Salt Lake FBI had a tip off the 'Unabomber' was in their area. They were in 'hot pursuit of a mad bomber' who had already killed several people. While Fay was in the FBI lobby a hotel employee came in with a waste basket full of bomb materials found in a local hotel room. The FBI guys and gals stampeded through the lobby armed to the teeth in hot pursuit! Exciting stuff! They missed the Una-bomber that day but years later the Una-bomber was finally caught and 'sent away' to permanent prison storage in Colorado.

Fay and I went to my apartment in Brigham City and I packed for the trip. Fay did a few odds and ends and then we were off to dinner in Bountiful, Utah. Tim and I had become very close friends. We were two men dedicated to the same things and we were working a problem we knew to be enormously important.

We arrived at the restaurant and Tim's wife Cher hit if off very well and we just had a splendid evening. This was the first time I met Cher. After the dinner, Fay and I returned to Brigham City, got a few 'winks' and headed for the airport at Salt Lake City. I was so exhausted I could hardly crawl! I had been too tired to dance with Fay or Cher.

CHAPTER 24

ESCAPE TO BELIZE

Time: Late February 1987

We jetted from the ice and snow of Utah on to Houston where the weather was mild. After a 1 hr. 45 minute layover in Houston, we boarded a TACA (San Salvadorian Airline) to Belize City. We both figured TACA Airline would be like 'bad tacos' but were surprised to find it was cleaner, neater, and more comfortable than most U.S. domestic airlines. I slept most of the way across Mexico to Belize because I had a month of sleeping to catch up on. Belize is south of the Mexican Yucatan and has the second largest reef in the world. At that time very few people knew anything about Belize. Belize is quite a new nation. It was formerly known as British Honduras., I felt physically out of shape, not particularly well, and rotten in general.

We arrived in Belize City which is an old world colonial sewer swamp. We caught a ride with a local missionary we met at the airport to the hotel area and proceeded to get beds for the evening. The St. George Hotel was an old colonial and was probably the number two hotel in Belize City and that wasn't saying much. The tropical climate and hurricanes had done their worst deeds to the buildings. We found out where to catch a boat to the outer islands and what things were going on in Belize at that time. This was not a well planned trip. However both Fay and I were very seasoned global travellers and capable of working through any problems. Of course *There Were No Problems!*

The next morning Fay and I literally caught a banana boat between the mainland and Belizean Caribbean islands. We took a 4 ½ hour beautiful cruise to the village of San Pedro on Ambergris Quay. A number of Belizeans on the boat, Fay and I being the only foreigners. The boat was about 60 ft. long, beat to pieces, no visible safety equipment, and the steering cabin was open to the sun wind and sea spray. The captain of the boat was a man in his early thirties and clearly a competent seaman. His leadership was reinforced by the rum bottle in his left hand and the marijuana cigar held in his mouth! Half of the people on the boat were smoking 'joints' or 'alfalfa' cigars! These were happy people. The *Caribbean* trade winds are wonderful. Warmth of the sun was wonderful! No thoughts of Utah. This boat transportation was about

USD$3.00 a head. Very reasonable. Mostly we slid between cays and over 60 ft. deep crystal clear water with dolphins, turtles, and millions of fish. The bilge pumps worked continually, what more could we expect? We communicated in Spanish and English.

I thought I had died and gone to heaven, Why in the hell would I ever go back to Utah? Who in the hell was the FBI? We arrived at San Pedro just at sun down. Grabbed our bags and walked up the sand street just as the sun set. San Pedro was a small quiet village. No paved roads. Simply sand roads and the people here were true beach people, the best kind. A few Americans and other foreigners, but mostly simple, quiet, lovely Belizeans of mixed colors and languages.

We asked for a place to stay and were told "Martha's Place" – someone made a phone call, arrangements were made and somebody in a pick-up drove us three blocks down the beach sand main street. We found what we were looking for. The facility was a $7.00 per night large nice, hotel room. Very clean and on the beach. Quite comfortable. After we got into our room we asked if there was a good place to eat and they all said "Elvies". Just across the street was Elvies – we looked out and it looked like a screened in framework with a huge Banyan tree growing out of it. We wondered across to Elvies and went in. There were crude 'raw wood' picnic tables built into the side of the framework walls which were screened. The restaurant floor was of course 'low maintenance' beach sand - very practical.

Sand floors and huge tree growing out of the floor which went right on out through the corrugated metal roof. The food smelled wonderful, we had a glass of rum and we ordered lobsters at about US$3.00 each and I knew I had died and gone to heaven. Smooth Brazilian rum, wonderful Belizean lobster, warm weather and no suppression of my work! I was not completely at ease however. In the far back of my mind was the thought that if MTI 'wasted' me in Belize they wouldn't have to answer to U.S. officials. *Just another ordinary dead tourist – for his own good!* While in San Pedro we took a glass bottom boat ride out around the reefs, we swam in, on, and around the magnificent reefs. We snorkelled on several sunken boats. We hiked in jungles and beaches. We ate a lot of lobster and drank our fair share of rum.

One afternoon during a tropical thunder storm we went into a thatched beach bar and had some anti-malarial rum. An American archaeologist was also there for ani-malarial rum. He was Dr. Herman Smith and his lady affiliate (camp follower as she called it). They had actually just that day discovered a previously unknown Mayan City further to the North along Ambergris Cay. They were ecstatic.

They could still walk, but the celebration was flowing. We had a great hour before the rain stopped and we went snorkelling some more. During that previous hour Dr. Smith told us several Mayan ruins we should visit. He was adamant about the Xunantunich (Maiden of the Rock) Pyramid complex which was inland near the Guatemalan Border. We took notes.

After about four or five days in absolute paradise on medicinal rum, we took an inter-island charter plane back to Belize City. We stayed at the Chateau Caribbean which probably was the best hotel in Belize City. As directed by Dr. Smith in San Pedro we caught the bus the next morning early and rode far up into the mountains near the Guatemalan border. The bus was an old school bus which had been in 'the wars'. San Ignacio was the last city and from there we caught a cab which would take us to the Mayan ruins . The cab followed a river which looked like Indiana Jones could have been using it. Finally there was a 'one car' ferry across the river, we drove onto the ferry and the ferry operator 'hand cranked' the ferry across the river (approx. 100 yards). It was simple and wonderful. Once we crossed the river we then drove up the mountain side a short distance and parked. We spent about half of the day walking around this great Mayan Temple complex, Xunantunich. Again I had died and gone to heaven. I had always wanted to fondle a Mayan Temple. That day I fondled a big one!

When we returned to San Ignacio we had lunch at a local general store and restaurant. San Ignacio was a sleepy Central American border village. We then caught an earlier bus than the regular scheduled one to Belize City and took the archaeologists advice and directions and bussed to Belmopan the capitol of Belize.

From the Belmopan marketplace where the buses stopped we walked a short distance to the central Belizean Government Offices and found the Archaeology Department in the semi basement floor. There we told the duty person Dr. Smith had instructed us to ask if we could see the National treasures. She said "yes, come, of course." When I saw the goodies, I was quite literally shocked, they were National and Planetary Treasures! Mayan Jewellery, long slopped skulls of Mayan aristocrats, jade trinkets, etc.. Again I had died and gone to heaven. Old bottles from all over the world from sailors, gold, and we could fondle all of it and we did. I was thinking my new career could be as a 'Contraband Artefacts Broker'!

When through visiting the treasures we ran back to the bus station at the city market place in heavy tropical rain. We caught the regular bus and were soon en route to Belize City. We stayed the night in Belize at Chateau Caribbean. We had arranged a charter flight to Big Creek the next day. From there we would be picked up and taken to the village of Placencia by boat.

We caught the plane and flew south and landed at Big Creek. We flew in a Cessna 182 and I got to log a few minutes flying time in Belize because I held a U.S. Commercial Pilots Certificate and the pilot there was happy for me to take over some. At the dirt air field at Big Creek a man with an old rusted out pick-up called for us. We threw our bags in the back of his pick-up and he drove us a quarter of a mile to a beach. Our bags were then loaded into a 40 ft. hand hewn wooden canoe with a small outboard motor on its stern and a guide. In less than a minute we disappeared into a Caribbean salt marsh and mangrove swamp – most primordial. We loved it.

Fay and I looked at each other wondering if we had returned to the jungles of an Indiana Jones movie? Here were clear beautiful Caribbean waters the sun beating on us, mangroves in all directions and I decided it was a fantastic trip we were having. After about 40 minutes into this canoe wondering through mangrove swamps and beautiful backwaters where no man with good sense had been before, we literally entered into a mangrove tunnel about 8 ft. high and 15 ft. wide, and a quarter of a mile long. I was just sitting there in disbelief and amazement as we rode into this incredible tunnel of vegetation. Extraordinary!

At the end of the tunnel was a small dock and a few scruffy looking boats. As we tied the boat a few small hogs were running and scurrying about, others were wallowing in the local mud Even the hogs looked happy! Our guide unloaded our bags, placed them in a wheelbarrow and said follow him.

From the dock area, we went to what was downtown Placencia. There were no roads, in Placencia, just a lot of beach, a few huts, a few minimal homes, and one narrow concrete walkway through part of the village. Our guide wheeled the barrow down to "Marie's House" which was simple indeed. Marie was a pleasant, plump caucasian, older Belizean lady. We paid our $11.00 a night for the use of a bungalow on stilts on the beach. We were then escorted over to our bungalow. Everything was set up. Rain water catchment supplied water, regular flush bathroom, kitchen, plates, cooking and linens – what more could we want. There was a swinging hammock on the front porch. Three steps from the front porch was the Caribbean. "Life's a beach!" Pure sand, coconut palm shade, no Thiokol or NASA or F.B.I..

There was clearly no reason to return to 'advanced western civilization'. Later every morning after returning to Utah I would get up in the morning and think what a dumb bastard I was for not being back in Belize. That night we went to the local restaurant which was basically a hooch out on the beach and had some disgusting concoctions of lobsters. We must have eaten a thousand lobsters on that trip. When we got tired of lobsters, we switched to conch which tastes similar to abalone. Someone had to do it!

One day, we chartered the same huge canoe and the same guide who took us there from Big Creek. We motored 14 miles straight out into the Caribbean to a small cay with some coconut trees on it. The trip was gorgeous beyond description. On route we could nearly touch dolphins, while we watched the shallow coral pass by. There was no change here since the first Mayan's came. Civilization had not defiled this place.

This cay was approximately ¼ mile long, 100 yards wide, and was beach, conch shells, and coconut trees Surrounding this cay was the most beautiful coral imaginable. Very few people came here and apparently no one had ever lived here in a dwelling. For an eighth of a mile there was a stack of conch shells 20 ft. wide, stacked 8 to 10 feet deep. What archaeologists call a 'Conch Shell Causeway'. Purpose unknown? I will bet the first conch eaten by the first Mayan there is in this stack somewhere. Millions of conch's had been killed and eaten in these waters.

Fay, the young guide and I slipped over that side of the canoe into the water and snorkelled the entire perimeter of the Cay. We saw millions of fish and every kind of sea life. Towards the end of the swim twice I was surrounded by schools of juvenile barracuda. Unfortunately for Thiokol and NASA, and the Mormon hierarchy those barracudas were not hungry. After we swam around the cay, we walked up on it. We spent a few hours there and then we returned to Placencia sunburned and happy.

A few days later we chartered a plane and flew from Placencia back to Belize City and then back to Houston. Then on to Salt Lake City. Then drove on to Brigham City. It was a wonderful break in terms of weather, mental health and all other things. Fay stayed with me in Brigham City about a week before she returned to Sydney, Australia. Initially after 10 years of knowing each other we spent some serious quality time together. I was not at my best, in fact I was at my worst. I did not feel well and couldn't figure out what was wrong. I was slow in coming to a conclusion. I was alarmed by my inability to handle my personal life at that time. A consultant and a transfusion were required to manage my personal life since I had put my entire being into getting America back into space.

CHAPTER 25

RETURN TO HELL

Time: Early March 1987

Flying back to Salt Lake City was dreadful to me. The weather was bad, my mood was bad! I did not want to return to MTI or Utah. My delinquent IRS bill was caught up and placated, so what did I need money for? There were other contracts in the U.S. now and we were past the first of the year and I was thinking how good it would be to have a new contract. If you play in dog shit you will eventually get some on yourself. We had a brief layover at Houston Airport and we scurried around doing a few things. From Houston Airport we flew to Salt Lake City and the damn snow was knee deep.

At the Salt Lake City Airport we rounded up our luggage and headed out into the snow to find a shuttle bus to haul us to one of the outer parking lots. My car was snowed in. After scraping the car off and thawing it out enough to drive we loaded up and drove to Brigham City. We had a wonderful trip to Belize and things were going well for us on a personal basis.

The drive back to Brigham City was cold and gloomy. Once back we turned on the heaters, thawed out the apartment and I started making ready for my return to the MTI grind next morning. I called Malcolm and made arrangements for carpooling and was set to go.

Next morning Malcolm picked me up. We went to work as usual - 'oh dark hundred' in freezing morning, the snow and cold. En route to work I asked Malcolm what had gone on while I was gone? Malcolm told me that work was absolutely business as usual! Still chaos and bedlam with both MTI and NASA. I hadn't missed a thing during eleven days away. He told me that everybody had tons of typing submitted with almost nothing typed or returned of course!

He also told me that Kent McKinnon was under more pressure than ever. Malcolm knew I had a relationship with the FBI. In fact he had been approached by the FBI,. Malcolm's FBI meeting was less than satisfactory. Malcolm apparently felt his income was more important than FBI and National needs. Malcolm had told me, "MTI problems are none of our business, Steve." In my opinion they were 'everyone's business.

We should just work and do our jobs to the extent we are allowed. Malcolm's resistance to FBI cooperation deeply offended me. The U.S.A. had accepted him as a U.S. Citizen and now he even carried an American 'Security Clearance'. Personally I felt he had a duty to help in return for what America had delivered to him.

I was nervous returning to the rocket plant, wondering what troubles would befall me. Did Thiokol know I was helping the FBI at this time? I walked through the plant, up through some corridors, back into the trailers and set up my desk and went back to work. Jumped right in where I had left off. Started attending the meetings on the 'Safe and Arm Device', and the 'Igniter Train' {sequence lighting off the Solid Rocket Boosters}. Found out there were no changes in attitude, disposition, or intent.

Quite a few NASA Personnel were meeting with the reliability side of Flight Safety and it was clear the Reliability Department was in serious strife with NASA. NASA rejected their *"Critical Items List"* (CIL) document and required Thiokol to do it over again. I never did find out the exact reasons NASA had for rejecting and yelling in various corners of the area that NASA was displeased. Personally I was tickled to death to find out that NASA was displeased. About time! There were a lot of brilliant contractors working Reliability issues, but I did not see exactly what was shown to NASA. I wondered if MTI was caught feeding NASA old crap? There had been times when I figured that NASA didn't know if it was day or night. At last there must be some issues NASA know something about. Maybe?

So one evening, I took Fay to the Airport and saw her off back to Australia. Usually when you solve one personal problem you create a bunch of others. In three million years people have probably never solved any basic human relationship problems. We are all in the dance together, going in circles aren't we??

CHAPTER 26

DID ANYONE UNDERSTAND OR CARE?

Time: Late March 1987

With a smile on his face Kent went down the hall, headed into Don Thompson's office to drop his resignation on his criminal boss. Kent later related to me that Don Thompson had stated that he wished he could find somewhere else to work as Kent had. Don Thompson felt he'd been stressed enough with the SRM Program. Who knows, that was probably an honest comment? Even though the FBI and I knew Don Thompson was guilty of a long line of false statements to the Government and safety suppression we also knew his superiors had placed a lot of pressure on Don. Don Thompson's "nads" were in tow and his heart and mind were following.

The next morning Kent McKinnon called an "all hands" meeting at one of the main conference rooms and we were all there and curious. Kent sat down in his most dignified way and said, "Ladies and Gentlemen of the System Safety Department, I have tendered my resignation. I'm going to another corporation and I have been directed to tell you that the SRM is a great project and will be going forward. System Safety will be going forward and will have full MTI support." Then he kind of rolled his eyes around and we all knew what he was 'not saying', i.e. tough shit folks! He said "I would like to leave you with quiet dignity and would like for no one to ask me where I'm going or what I'm going to be doing." I thought for a few seconds, 'that's strange'. Then I realized it was not. It was the only way for him to go. If he told everyone where he was going, what he was going to be doing, then his entire staff would want to be right there with him and he obviously couldn't take his whole staff with him. Also technically he could in no way compete with many of his staff. It would take him awhile before he could help anybody from where ever he was headed. I felt immense pride at having helped Kent score this new and no doubt better job. Kent then said it was not decided who would succeed him. Kent thought it would probably be Kerry Sanofsky who had been acting as Kent's right arm. Kerry was heir apparent. He had Safety Credentials (but no technical credentials) and was a local Mormon in good standing with his Tithe. The fact he was technically Inadequate to identify hazards or understand the technical complexities of the SRM Systems were not factors apparent to MTI or NASA. His Safety Credentials were the type that are applicable to insurance loss control work, a far cry different than hardnosed engineering or science and systems integration necessary to understand complex devices and their environments, processes, and consequences.

Likewise nearly all of the Thiokol Systems Safely Personnel, permanent and otherwise, didn't know much about insurance loss control - we were scientists and engineers, not paper pushers. I sensed that Thiokol placed great stock on Kerry's Utah Safety Certificates and in fact didn't even know they were the wrong stuff for the critical tasks at hand. Mahalon was on a power roll at this particular moment. He was running in and out of the meeting door as though he was so God Damned important he was just above Kent's meeting or anything to do with it. Mahalon, the contractor, clearly felt he had been elevated to the status of what he viewed as near God by Don Thompson. I sat there looking at Mahalon running around making a fool of himself and felt ashamed. I was guilty of bringing Mahalon to Thiokol. This stands as one of the worst mistakes of my life. I would have to answer to God for it. Thiokol in my opinion had done enough damage without adding Mahalon to the pile. *On second thoughts - maybe they deserved each other? Maybe Mahlon should become an MTI Director* - if the shoe fits, let him wear it.

Following Kent's last meeting Mahalon was making gestures and glad handing people in such a manner that everybody saw that Mahalon was campaigning to be the new SR&QA manager and or System Safety Manager. It was a disgusting and shallow display of egomania. It was also a great introspective lesson to me. We all needed to stand back and look at ourselves as others might see us.

Almost immediately Thiokol wanted to punish us consultants and get us "under control!" Also our hours, overtime were to be vacated. The method used to cut our overtime was that we were not to work without a MTI supervisor there. So, on Saturdays and Sundays there was often a shortage of Thiokol supervisors. They were probably elk hunting.

After we had all been punished / restricted for a while Malcolm, the Indians, and I had a talk. Malcolm recommended that our overtime was so valuable for accomplishing our mission on time "maybe we should buy it!" I told the FBI about this concept and they were interested in where it would lead? Would it connect us with Special Agent Kevin Kelly's separate huge procurement case against Thiokol? The work studying the Hazards was demanding more of our time - it was a bottomless pit. I was out of town the first time Malcolm and the Indians paid Kerry Sanofsky off to come supervise their Sunday overtime work. On the next weekend I was there. I agreed to "play ball" under FBI instructions, but rather than cash I wrote a check and a note to Kerry. Of course the FBI got a copy of them. They were fascinated and monitoring.

There were more people in the MTI hallways day by day. More NASA, more DCAS, more Air Force, more Thiokol, more contract personnel. The work was escalating and we were all optimistic that we would get a few licks in before it was too late and subsequent personnel were killed and the space program set back further. At this time I cannot remember how many Hazard Analysis I had written. It was well over the 180 mark and heading for 200 by then. I felt like I was ready to explode with Catastrophic Hazards. Each hazard times various flight phases the hazard lurked in. Sometimes only one flight phase, other times all phases of flight. Frequently MTI changed the definition of flight phases which really messed our HA's up. They had to be done over, sliced the technical pie differently.

Malcolm, the Indians, and myself were now looking at NASA and Thiokol schedules. We saw and realized there was no way our findings would be included into the next flight (Discovery flight 26). NASA had firmed its plans. Their minds were set! *There Were No Problems,* and there was nearly zero improvements from the Challenger launch. Launching at a warmer temperature would be the only difference.

There was not one permanent Thiokol employee other than Kent McKinnon who had or would sit down and talk to us about our Hazard Analyses or our findings. Kent had been willing, but he was kept very busy. It's easy for Thiokol personnel to hide behind the screen "well they were not completely prepared in the approved System Safety manner and ready for delivery to other departments." Damn poor excuse and bullshit! The fact is we should have been able to sit down on a daily basis and thrash them up one end and down the other to eliminate anything we might have overlooked and to add in anything. Engineers from other departments might contribute. This was definitely not the case. What was really happening was that nobody in MTI thought that anything they engineered could possibly be less than perfect. Thiokol's management position was generally that Thiokol does not give a damn (plenty of witnesses). Also, there was their thoughts that God would take care of it and them.

I was at the vortex of the most incredible conspiracy imaginable! *More people are going to be killed!* I was very unhappy and literally depressed over the implications to National Security and the future of those in space. From my point of view the United States might not have a future in space and therefore no meaningful future at all! I could see the collapse quite easily. From that moment to now the USA Space Program collapse has been accelerating.

The Russians had flown their Space Shuttle, the U.S. had returned to space in a perfunctory way and our American launches are still inconceivably unsafe and only half payloads were being launched until the Space Shuttles were grounded. Did the US Air Force understand the scope of the problem? Now 30 years later the space shuttles are retired and a new generation of rockets are under development. Isn't it interesting that the Shuttle was retired without a replacement system in place and with a U.S. Space Station in orbit!

Our U.S. International Space Station hangs unsafely by threads and is inhabited by Russians and other foreign astronauts. Access to the Space Station is now via Russian Rockets and commercially launched rockets 30 years later. Did the C.I.A. , the National Security Agency (NSA), or the National Reconnaissance Office, or the U.S. Navy or even NASA understand? I know for a fact that the Utah FBI understood. How many hundreds or thousands of aerospace vendors would be on hold while Thiokol diddled along with their *Crimes And Treason?*

Now RT Television (Russian) points out daily that Russia was:

1st into space

1st Man in space

1st Space station

Supplying crew and supplies to the space station

On 25th of May 2017 even New Zealand launched its first substantial rocket from a private launch site. Rocket labs 17m long (56ft). Electron nearly made orbit on its first attempts. This is a credible first attempt.

Chapter 27

RECRUITING MORE FBI SOURCES

Time: Early February 1987 - June 1987

I attended many more meetings with the FBI, and by now I was inviting numerous persons to dinners to determine if the FBI could trust and use their help. I did not see and hear everything at MTI, only a small fraction of what went on. MTI had almost 8000 personnel and thousands of acres and many buildings. I did not have the only point of view. Careful and controlled acquisition of additional 'personnel' resources was called for. Not being trained in covert activities I vastly underestimated the difficulties. Also I was experiencing an aspect of human behaviour which was foreign to all of my previous life's experiences. Some persons were deceptive, some lied, some were clearly out for themselves and unpatriotic, and some were totally dedicated patriots, no matter their inconvenience. The losers far outnumbered the winners in Utah. Would it have been the same if they were Catholics, Baptists, or others? I don't know.

I would meet with these selected persons over dinner, or at other types of quiet social functions and get a feel for them and evaluate their patriotic nature in my limited and optimistic way. Persons on the borderline or undecided were the most dangerous. I was shocked at the non-patriotism of some. Some were so ignorant or stupid that it didn't matter what their patriotism was, they were simply unsafe.

There was every sort of scientist, engineer and Thiokol worker that I talked to over a short span of time. Some never knew what they were queried for, others guessed, yet others were eventually told by the FBI.

On Feb. 6, 1987 Dr. Tony Laine's efforts came to a head. He was fired for doing his job. The following is extricated from his (and my) massive lawsuit against Thiokol long ago.

33. On Monday, January 19, 1987, Plaintiff Dr. Tony Laine commenced his employment with Thiokol. His initial day was spent with a Ms. Sharla Rose, who took Plaintiff, Dr. Laine through his orientation and provided compensation for his travel expenditures in the amount of $250.00 drafted on Thiokol's bank account.

34. Thereafter, Mr. McKinnon, who had previously informed Plaintiff Dr. Laine that he was being hired for his expertise in the field of electromagnetic interference ("EMI"), proceeded to orient Dr. Laine regarding the goals of the Safety Unit by having him read the Rogers Commission Report. She further had Dr. Laine review old FMEA's and HA's that had been generated prior to the Challenger Disaster. Plaintiff Dr. Laine noticed that of approximately 500 HA's referenced in these old reports, not one had ever been closed out (no remedial effort or waivers obtained.) {Note: Hardly any went to NASA and hardly any were catastrophic before the Challenger Accident}

35. For his first assignment, Plaintiff Dr. Laine was assigned to investigate the"O-Rings" of the SRM.

36. After one and one-half days of working on the O-Rings, Plaintiff Dr. Line was informed by Mr. Kent McKinnon that he would be working on the nozzles of the SRM as this was the area of greatest danger from EMI. Mr. McKinnon explained to Plaintiff Dr. Laine that he (Laine) had been retained specifically because McKinnon had a concern about "plasma effect" from rocket plume, and he was aware that Plaintiff "Dr. Laine was an expert in this field.

37. 'Plasma Effect' is a phenomenon where a surrounding electromagnetic field results from any high energy explosion, such as a nuclear explosion or plume from a rocket motor. Mr. McKinnon stated to Dr. Laine that he wanted him to study the SRM for potential safety hazards associated with plasma effect, and in particular, the possibility that electro- magnetic effect would penetrate the skin of the rocket and detonate the Range Safety Devices, explosive charges on the SRM that would permit operators to destroy the rocket shouldit pose a threat outside of the operations area.

38. Dr. Tony Laine immediately commenced his investigation into the Safety Hazard associated with the plasma effect resulting from the plume of the SRM. His initial effort was to pull together all research materials on the subject that he could obtain. Plaintiff Dr. Laine first obtained and referenced all of his own copyrighted research articles in the field. {His mind had obviously been in this space before he came to MTI}

39. Plaintiff Dr. Laine then went to the library of Thiokol's Wasach Plant and found that the latest volumes that it contained on the subject were published in approximately 1952. {30+ years old, Dr. Laine deemed them useless.}

40. Dr. Laine then contacted all of his associates in the field of nuclear physics and EMI and all unclassified information that any of them might have regarding plasma effects. In response Thiokol began to receive at the Wasach site many research reports on plasma effects, a few of which had to do with detonation of nuclear devices, however all of which were unclassified materials.

41. Within a week after he commenced working on the plasma effects of the SRM, Plaintiff Dr. Laine was approached by defendant Donald Thompson, Manager of Thiokol's Quality Assurance and Quality Control programs. Thompson informed Plaintiff Dr. Laine that he had heard of his research into plasma effects and he told Dr. Laine to cease researching this Safety Hazard immediately. Defendant Thompson informed Dr. Laine that the role of the engineers in the Safety Unit was only to 'cut and paste old reports" and not to raise or investigate any new safety concerns or problems with the SRM.

42. In response to this confrontation, Plaintiff Dr. Laine approached his immediate supervisor, Kent McKinnon. He explained what Thompson had told him and McKinon stated that he should not concern himself with Thompson but should continue to work on plasma effects.

43. Thereafter, Dr. Laine attempted to use the library facilities at Thiokol and found that an order had been issued by Thompson denying him access to the library. When Dr. Laine contacted Thompson, the latter again instructed him that he was not to carry out any original research but was only to "cut and paste old reports". Plaintiff / Dr. Laine again approached his supervisor, Mr. McKinnon, and Mr. McKinnon directed Plaintiff /Dr. Laine to continue working on plasma effects.

44. On three occasions thereafter, Plaintiff Dr. Laine was approached by Defendant Thompson, and was instructed to cease working on plasma effects. On each such occasion Dr. Laine approached his supervisor, Mr. McKinnon, and Mr. McKinnon directed Plaintiff Dr. Laine to continue working on plasma effects.

45. On Feb. 6, 1987, Plaintiff Dr. Laine was called to a meeting in the office of Defendant John Thirkil, Vice President of the Space Division of MTI. There he was told that he would be terminated for his work investigating plasma effects of the SRM. When Dr. Laine pointed out that he was instructed to do so by Mr. McKinnon Mr. Thirkill became enraged. Mr. Kent McKinnon was also in attendance at the meeting and verified the Plaintiffs story. Nevertheless, Mr. Thirkill terminated Dr. Laine on the spot, claiming that Plaintiff Laine had been insubordinate. {Note: MTI 'black balled' Tony in the Aerospace industry}

46. Thereafter, from Feb. 1987, to approximately September, 1987, Dr. Laine was unable to obtain employment, despite frequent and active efforts.

47. On or about approximately Sept. 28, 1987, Plaintiff Dr. Laine was contacted by a news correspondent who was conducting an investigation into Laine's termination by Thiokol. The news correspondent informed Dr. Laine that they had made indirect contact with Congressman Robert S. Walker, a member of Congress's Science, Space and Technology Committee. Through this channel, the news correspondent stated he had learned that Dr. Laine was terminated for violating National Security, for reasons of delving into "classified information" and that his educational qualifications and reliability were questionable. The correspondent further went on to say that Congressman Walker had transmitted to him that such information was the result of both NASA and Morton Thiokol "(undertaking an investigation)" Such allegations could only have come from Defendant Don Thompson and Defendant John Thirkill acting on behalf of Thiokol. {Dr. Laine did his graduate work next to USAF Colonels and Generals.}

48. Since this time Plaintiff Dr. Laine has been unable to obtain any gainful employment in the aerospace or defence industries despite frequent and active efforts. "

Thiokol wrongfully 'did Tony in!' Tony entered into a period of emotional disorder and financial ruin which would not be resolved ever and would in fact be contributory to his premature death several years later. He was attacked from many directions. Previously he had outstanding results for all of his clients inclusive of the CIA which had paid him $13,000.00 a week or $56,000.00 per month, or $672,000.00 per year which was phenomenal at that time in history. He was a brilliant hard working patriot to the point of psychosis! He was absolutely correct in his science. The issuance of his very high Security Clearances at the CIA verified his credentials and reliability.

I was aware of Tony's suppressions by MTI - blow by blow, and I documented them with the FBI. Many at MTI, and in particular System Safety were witnesses. Many others allowed themselves to be suppressed or in some cases used the suppression as a safety blanket. "They don't want anything so they relaxed and gave them nothing - easy!"

I learned a tough lesson on one individual. I worked with him on a regular basis and had high regard for his work. After feeling this person out on numerous occasions I asked if he would help the FBI. All of the colour drained from the face, a solemn demeanour over came the individual. "No, I do not want to get involved!" I was shocked. I elicited an 'oath of silence'. This individual knew plenty was wrong with Thiokol and NASA since he talked about a lot of clear cut NASA and Thiokol problems and crimes. One part of me has to respect this individual's right to his own decision. Another part of me downgraded this person to either a castrated sheep or 'mice balls'. If I could risk it all why couldn't this person risk just a little? Perhaps I was and am naive? Maybe the risk is all or nothing? Perhaps this individual and / or others had things to hide unknown to me or perhaps the FBI? Who knows what secrets persons or corporations have?

About this time I found my desk had been searched. This happened occasionally at Thiokol. Thiokol Security dug into everybody's business. This is also a common local Mormon trait. The Mormon Community is always checking up on individuals and even holds special church 'Courts' as a segment of their church functions.

After I identified people that I thought would help the FBI, I would set up a dinners and bring Tim and possibly his immediate supervisor along. Tim was of course introduced as a friend of mine who'd just escaped the U.S. Marine Corps and had flown Harrier Jets. After leaving the USMC he worked for some unheard of company somewhere training Harrier Pilots. We would all have a nice dinner (or two or three) and it would shake down that somewhere along the mid-dinner, we'd probe problems at Thiokol and see how the wind blew with these people (patriotic or unpatriotic). We discussed Waivers and Deviations, MTI or NASA irregularities, what cover-ups were affected etc. Some of these people were immediately transformed into ' FBI Sources'. Sometimes, Tim would just quietly look at me and kind of roll his eyes in a manner that said "no, no!" Other times it took several dinners to determine if Tim would or should recruit these people. In order to protect the innocent, and put the fear of God into the guilty, Tim had his 'ways and means'. I will not go into any more detail about who these people are. They may be with the National Security Agency, CIA, the U.S. Air Force, the Defence Investigative Agency, or yet others? Far be it from me to disclose how the FBI knew Thiokol was destroying documents, and other juicy morsels of interest. The FBI even know how many and which documents were destroyed I heard through the FBI grapevine. I leave it to you to figure out if the FBI had copies.

Now I was on a tremendous roll, acquiring documents out of Thiokol to make my points and claims loud and clear to the FBI. There were technical documents, management documents, Hazard Analyses, Reliability documents, historical documents, training documents, flight data documents and anything else that looked contributory. One of the favourite documents I acquired from source VR (Very Reliable) inside MTI was the following letter known as the, "Grimm's Fairy Tale Letter". It had been received by most MTI personnel. This was widely dispersed document which many MTI personnel respected and cherished ownership of a copy. Nearly everyone had seen the letter and agreed with it. I have often wondered if this letter did any good? The Grimm's Fairy Tale Letter by P. R. Dykstra is as follows:

MORTON THIOKOL, INC.
Wasatch Operations

Interoffice Memo

9 February 1987
A500-FY87-M201

TO: G. A. Bailey, W. W. Brant, G. W. Broman, W. T. Dolling, J. W. Edwards, D. J. Hammon, J. G. Hilden, R. J. Taylor G. T. Thies, J. R. Thurston, C. G. Wiggins, J. D. Wight

FROM: P. R. Dykstra, General Manager Strategic Division

SUBJECT: The Magic Wand Theory of Management

Up until the age of nine, I was an avid reader of Grimm's fairy tales. I was comforted to learn that no matter how difficult and complex problems might appear, they could always be easily, simply and effortlessly solved by a magic wand, Fairy Godmother or carefully phased incantation.

As I grew older, I switched my attention to adult fairy tales. I discovered that a majority of adults still believe in magical solutions to problems, such as -

A. Our basic conflicts with the Russians can be rendered potentially harmless if only we both sign an arms control agreement.

B. If we lower our educational standards so that everyone can breeze through school and be granted a degree we will have an educated population.

C. If we call discrimination based on race affirmative action it will wash away all the attendant intrinsic evils.

D. If we stop doing business with South Africa this will lead to racial harmony between the races.

The list could go on endlessly. I despair for the future of a country which elects its government based on how artful the candidates can be in generating novel fairy tales. Needless to say, in the real world fairy tales seldom, if ever, come true.

The reason for these opening remarks is that I see certain parallels in attitudes within the Strategic Division which should cause us to re-examine if we are unwittingly placing an unwholesome dependence upon fairy tales. By fairy tales I mean superficial, relatively painless solutions to complex problems that are not likely to resolve the problem.

Staff -2- 9 February 1987

We are all guilty of this behavior to some extent. I have spent thirty years here and have lamented our submission to fairy tale psychology over that period, but was comforted because I believed that I had eradicated it in my program office and it was the other organizations that were the perpetrators of the folly. From my current perspective I can see that the program offices are not basically different. "We have met the enemy, and they is us."

I detest writing proposals primarily because one cannot prepare a winning proposal without forcing creative personal involvement on all people involved, which is a frustrating, agonizing and debilitating effort. It always takes three to four iterations before everyone comes to the reluctant conclusion that they cannot slough through, but that they actually have to do some thinking and be creative and make the personal commitment to turning out a useful result. Since an acceptable proposal must be delivered on a given date, there is no alternative but to face the problem and solve it.

The same sort of commitment is necessary to solve our management problems, but is not as easy to enforce because evasion is easier. I wish all of you would re-examine your conscience and see if you have been evading your responsibilities by trying to promote fairy tales. If so, please mend your ways.

P. R. Dykstra

As you can see I was not the first one at MTI to question Thiokol's peculiar ways. The man who wrote the letter was an older fellow, in a different MTI division (Strategies), and a man who was close to retirement and had nothing to lose by speaking the truth and stirring the pot. I can only speculate from his letter, but it appears to me that he was surrounded by castrated MTI corporate sheep. He must have had a whole flock mulling around him.

Then another compelling letter emerged and I procured it. An MTI lady engineer wrote the following:

MORTON THIOKOL, INC.
Wasatch Operations

Interoffice Memo

19 September, 1986
L221:FY87:119

TO: A. R. Thompson

CC: G. P. Anderson, J. R. Kapp, E. G. Dorsey, D. L. Ewing

FROM: C. S. Ferrari
Structures Design - Ext. 4087

SUBJECT: Resignation

This memo is to notify you of my resignation. My last day will be October 3, 1986.

When my husband and I first came here three years ago, it was with the intent of making it a permanent career step. Unfortunately, my husband worked for a department manager who was totally incompetent and unprofessional, and was unconcerned with his employee's health and welfare. My husband was faced with the choice of sacrificing his professionalism and self-respect, or resigning. In our opinion, there was no choice, and he left the company to obtain a MBA degree. He has now completed his education and is ready to re-enter the workforce.

Since my husband quit, I have discovered that the attitude held by his former manager is a perfectly acceptable attitude here. People such as this manager are allowed to be abusive, unprofessional and incompetent, resulting in high employee turnover, low employee morale, and totally unacceptable working conditions. I have further noted that the company as a whole seems to have this same attitude. They have no real concern for their employees, and their response to complaints is basically, "If you don't like it, quit". They make no attempt to correct any of the inequities to retain professional, experienced people.

Now, at a time which is so critical in the space program, the company has succeeded in reducing morale to an all time low. The entire engineering workforce is upset, demoralized, and unconcerned with retaining their jobs. Their work output level is falling despite, and probably because of, mandatory overtime. Employees are frustrated since their recommendations are so often ignored. They are faced with unreasonable work loads and totally unrealistic schedules developed by management personnel

A. R. Thompson
Page 2
19 September 1986

with no technical training. It is only a matter of time before more of the engineering personnel get so fed up with it that they leave as well.

Although I have enjoyed the people I have worked with directly in engineering, who are extremely competent, I feel the company itself is not one I wish to invest my future in. To summarize, life is too short to spend it working for a company which does not know what it is doing, and far too short to waste it on a firm that doesn't care.

C. S. Ferrari

The above letter clearly speaks for itself. Mr. and Mrs. Ferrari were both engineers who worked at MTI and I think that her letter is apropos to much of what I experienced at Thiokol.

All of the documents we acquired for the FBI will not be shown or listed in this book. The total acquisition was thousands of pages. Remember, if you are interested in the entire set of documents you can request them from the FBI in Salt Lake City via a standard Freedom of Information Act request. Note that FOIA's are now manipulated by U.S. Government Agencies and they only give out what they 'orchestrate' unless individuals file law suits to get the documents which costs millions. Some of the acquired documents pleased me as I could see that I was not the only one who determined that Thiokol was into crime, shoddy workmanship, and low professional standards, and of course - dangerous and unreliable rockets.

Let's discuss 'low professional standards' now. The final deciding factor used to award the NASA SRM contract to MTI was that MTI had 'superior management' over the other SRM contractors perusing the NASA contract. It was a hotly pursued contract since it was very long term (at least 15 years) and could reach well over 20 billion dollars.

Thiokol did not win the bidding. The bids were based on 1,000 points possible and were as follows:

Lockheed	714
Thiokol	710
United Technologies	710
Aerojet	655

Most NASA managers did not want Thiokol for a lot of reasons. Against all best NASA judgement Dr. James Fletcher overrode everyone and of his own volition awarded the contract to Thiokol when other key NASA managers were sent on vacations. Other NASA officials were angry, and NASA internal wars were fought after MTI was awarded the contract. The history of this is well documented in the books *'Prescription For Disaster' And 'Challenger - A Major Malfuncition.'*

Later the Government Accounting Office launched an investigation into the NASA contract awarded to MTI, but later dropped it. Why did they drop it? After Dr. Fletcher awarded the SRM contract to MTI the GAO (Government Accounting Office) raised unholy hell with Dr. Fletcher and NASA. The 'wronged' vendors raised unholy hell with NASA, the GAO, and others. There were allegations of anti-trust, incestuous collusion, and abuse of power by Dr. Fletcher, the Mormon Mafia, and others. NASA observers have precisely chronicled Dr. James Fletcher's unilateral decisions and prejudices.

Six Astronauts and one Teacher paid for it with their lives! There was nothing superior about MTI management in my opinion. I felt they were the bottom of the bucket against all corporations I had served, and that was a bunch. Dr. James Fletcher, a high ranking and well connected Mormon was accused of Pork Barrelling for his Mormon friends at MTI in Utah. You can derive your own conclusions. When Dr. James Fletcher awarded Thiokol the Contract for superior management I wonder if he was aware that a bunch of managers and supervisors at MTI had school teaching degrees rather that technical or management degrees? If he knew, did he care? Thiokol attained the lucrative Solid Rocket Motor contract for the wrong reasons and in the wrong ways. NASA itself appeared to be sold out by Dr. Fletcher according to the books 'Prescription for Disaster' and 'Challenger - A Major Malfunction' NASA generally knew it had a "turkey" on its hands from day one when solid rocket motors were selected over liquid rocket boosters. Solid rocket technology is not considered reliable since Solid Rocket Motor propellant ages in an unpredictable way and is non-homogeneous. Solid Rocket Motors, once ignited cannot be turned off and cannot be throttled up or down. {Light them and ride them.}

Liquid engines cost more, but overcome all of the other problems. Additionally, liquid engines deliver substantially more energy per pound than solid propellant motors. Werner Von Braun, Germany's and America's greatest rocket designer, was dead against SRM motors being used for human flight. He felt they were simply too unreliable to risk human lives on.

Even if NASA cared, it was stuck with a problematic loser - Thiokol's SRM's. Most of the FBI cared, but had to perform their investigation in our American System's slow meticulous, 'politically correct', 'legally correct', 'freedom of religion' manner. After all, the Mormons were not the only religious group to be embarrassed by criminal individuals or subgroups, or as a key organization for that matter.

Did the D.O.J. care? I can't say. David Long a D.O.J. attorney I met with on a regular basis was a bright fellow, knew there was crime at Thiokol top to bottom, but did he care enough to push the cases forward? I did know he thought Thiokol should be prosecuted on both criminal and civil sides. Did the D.O.J. and the FBI have a free hand to prosecute? Doubtful as will be revealed later. Did the U.S. Congress and the U.S. Senate care? Clearly a mixed bag there. Did NASA care - *No! They were out of control and without any clear vision.*

Did the American public or voter care? For prestige, yes! *"For The Right Reasons"* - I couldn't tell at that time. I perceived the American public was oblivious to America's absence in the space race. How can you notice our absence from space when you are watching football, M-TV, Happy Days, or welfare handouts. Americans had their priorities.

While none of us realized it at the time this was the beginning of a massive decline in American high school and university standards. Also a dramatic decline of business ethical standards. The 'infection' was globally pandemic at all levels.

What did President Reagan think? I have no idea. I do know he talked a lot about America in space, but absolutely nothing of substance came from the White House during his term in office. Star Wars was much talk and very little to do with any reality. America's Space Station was much talk, great confusion and no real plan or hardware in place. The President's new Space Policy was considered a poorly funded ' bad joke'. The bottom line at the time was - damn few were informed, damned few were concerned! If more were concerned there would have been even more FBI sources. If more were concerned perhaps the Space Program wouldn't have been failing in the first place. Perhaps America would currently be resident on both the Moon and Mars?

CHAPTER 28

NASA COVER UP'S AND DECEIT

TIME: Mid-March 1987 - 1 April 1987

Kent McKinnon was now a short- timer and keeping an extremely low profile, going to meetings, being courteous to everybody, and in high anticipation of his new job and relieved to be leaving Thiokol. Even though he was still Thiokol System Safety Manager on the SRM he was in fact "stripped" of all powers. Mahalon Hollway, Lance Heller, Al McKinsey (MTI Plant and Facility System Safety), and Kerry Sanofsky were now carrying various components of the department in diffuse directions of course - adding to confusions!

Al McKinsey was now emerging as a power player. Al had a group of engineers and scientists in Plant System Safety (very different from SRM System Safety) who were doing facilities equipment and process Hazarded Analysis. These were not solid rocket HA's. His group had undergone substantial lay-offs and he was actively moving and shaking to take over SRM System Safety to conserve some of his current personnel. One of his gambits was now to mail some of his group HA's on the SRM's to one of the NASA groups mentioned earlier. Then Al started beating on Kent McKinnon and campaigning against Kent via Kent's bosses. Al was a much hated man. I personally viewed this as an additional delaying and stifling tactic to delay and or prevent forwarding of System Safety HA. Several of us wondered if Al McKinsey's efforts were to provide *"No Problems Hazards"* to NASA. I later found out NASA rebuffed Al's H.A.'s as *Unsolicited!* NASA to their limited credit "rejected" his HA's!

Kent experienced an emotional break with Thiokol. He'd fought the good war and was mortally wounded in a general sense. There was a lot of pushing and shoving between Thiokol employees and just one contract engineer as to who would finally replace Kent. Kent told me that Kerry Sanofsky would probably be his replacement, but he wasn't sure. Thiokol upper management was vacillating. Kerry Sanofsky was then polishing his act more regularly and trying to appear to be the heir apparent and was definitely making his move on Thompson to gain esteem.

Would Kerry sell his soul? Phil Greenhalgh was kind of watching it all in dismay, and was quite unhappy because he had not received promised pay raises. Phil was recipient of famous "put-offs". Phil desperately wanted to stay in Utah since he was a dedicated Mormon, but he put his resume out anyway. Kent had been pretty busy with the FBI lately. The FBI was getting what they wanted from him before and after he left. The FBI was glad Kent found a good job with a good company.

The loudest rumbling and shaking was Mahalon Hollway who was emotionally beating up on anyone who appeared in any position to replace Kent McKinnon. It was ugly and painful to watch. The spectre was reptilian! He would also beat up on any one who wasn't free to strike back. He could emulate MTI managers and he was indeed a cowardly fighter. He stayed clear of me, but everyone could see he hungered for the power to attack me and several others.

Mahalon now treated Kent McKinnon so overtly rude that quite a number of people in System Safety wanted to physically attack him. Even Kent's enemies were disgusted. Furthermore Mahalon was playing vicious hardball game with Kerry Sanofsky trying to show MTI how manipulating he could be with 'these children' at MTI. There were numerous clashes, and if it wasn't for the ugly spectacle of it all, in the human sense, it could have been theatrical comedy. If the same thing was brought to you on television it would be a grand farce.

Unfortunately the nation's future in space was severely damaged by all of this small minded bullshit! Kent's days were numbered and the number was less than thirty. I wouldn't be far behind him, I thought maybe just maybe I might beat him out the door! If it wasn't for my FBI affiliation I would have been long gone of my own volition. Under normal conditions I would have quit and withdrawn early on. There was a great deal of tension in the air. There were now, maybe a many as 2,000 HA's either complete or in process. Many were in sight of completion. Thiokol's System Safety department had grown by considerable numbers and we now had every sort of technical person. Nearly all technical disciplines were resident in our department. There was the 'Take the money and run' type, the 'dedicated' type, the 'go with the flow type', the 'not my responsibility' type and some other types. Other than Mahalon Hollway everyone in the department got along well and worked shoulder to shoulder as professionals regardless of their expertise. If managed at all properly this was one of the strongest technical teams on earth at that time - wasted and expensive though it was.

Simultaneously, there was one NASA Engineer at a 'Safe And Arm' meeting. He seemed bright enough and had reasonable interests in getting things done correctly. Then he missed the next several weekly meetings. He disappeared for weeks of training in various courses somewhere.

One day Don Thompson called a special meeting. Don Thompson looked "glazed" and he was uptight. Both System Safety and Reliability were there. We all went into a room. I sat by the Indians (Rakish and Arum). Don stood up with a strained face and gave a five minute soliloquy, on 'nothing about System Safety is being covered up and nothing is going to be!' I felt a jolt that that Don knew the FBI was upon him. I had my suspicions that Malcolm or either of the two particular other sources had sold their souls to Thiokol for their own 'contract longevity', cash, or whatever one sells his soul for? It was a year and a half later that confirmation would come in that it was Malcolm Phillips.

Malcolm reportedly got a sweet deal. He stayed on his approximate $17,000.00 per month contract when most others went out the door. Malcolm was to became 'Thiokol's man' at the Kennedy Space Centre (KSC) which was convenient for Malcolm since his house was just off base there. Malcolm loved all space vehicles, but he loved money more. The Indians and I frequently discussed how Malcolm never made any move unless it was to Malcolm's benefit. A week later Ross Bowman held another meeting and told us his door was open. However if he didn't like what he heard he could not be trusted. Ross Bowman's meeting was pitiful and he got a lot of laughs from all sides. Essentially Ross Bowman said they wanted to know of any and all problems, but he would decide what was really a problem. The time was mid-March 1987.

Ross Bowman and Don Thompson were already trying to cover their tracks from NASA, the FBI and others. How well informed was Thiokol? What top management spies and counterspies lurked about? I kept my cool and kept on trucking, but I was hurting from fear, dread, and stress.

Sometime in late March 1987 I went to another electrostatic discharge meeting. The meeting concerned minute critical details of electrical grounding, lightening and electric static discharges, accumulation of discharges and such. Electrical grounding and ESD had consequents on which could have massive and destructive effects on the SRM - killing people. I was astounded to find 42 people at this meeting - a world record!

For months prior we weren't able to round up 5 attendees. Frequently, only Joe and I showed up. Not only were there 42 people now, but they were all paying attention! They were actually talking about realistic concerns such as how NASA just came down on them like a ton of bricks to properly electrically ground the rail cars which move the loaded rocket propellant sections from MTI to California and then on to Kennedy Space Centre (KSC in Florida).

OH SHIT!! I've been had! My Hazard Analysis on unsafe transport of

SRM segments loaded with rocket propellant had gone to the FBI, then to NASA, and then to Thiokol management. I felt my life expectancy was now near ZERO!

When I originally met with the FBI I told them some town was going to disappear off the face of the earth one night when one of MTI"s rail cars exploded from improper electrical grounding, transient ESD (atmospheric or otherwise) or lightening strike! I told the FBI that Corinne, Utah could be the first to go since these special railroad cars were parked en mass at Corine, Furthermore, I knew rifles had been shot at the SRM cases and had not penetrated, but so what? The loaded SRM cases could be shot in the ends and this could ignite them. Perhaps armour piercing rounds - easily obtained. Utah is a 'shoot everything self - sufficient State'. One of the plant guards, and an MTI shipping employee informed me (when questioned) that as a cost savings effort the railroad cars were now unguarded. They had been guarded for years, but were no longer. Knowing boys will be boys (like I used to be), I was horrified. I was also informed that there were no security persons escorting the railroad cars in transit. I envisioned the United States to be "first to place a train and some tracks, and possibly a small unpopulated town on the moon: An unscheduled American Space project of the worst kind - a first!

I pondered if the FBI and I had stirred NASA awake and opened their eyes? I also wondered (feared) if NASA would "waste me". NASA was clearly negligent and had plenty to hide and a lot of their own 'management sheep' to protect. NASA and the meeting attendees were now interested in different types of electrical grounding problems. Dr. Leland Davis, who was the MTI resident electrostatic discharge expert was now attending, and was actually working on problems! Astounding! Dr. Leland Davis, a physicist, had just published a report on SRM propellant electrical ignition thresholds. Dr. Davis's new study (about 14 years late) received a lot of notoriety because his study concluded the SRM propellant ESD ignition thresholds and temperatures to be far more sensitive and dangerous than were previously known. Nearly any, 'small' current or spark could set the propellant off during the manufacturing process, the casting process, storage, transportation, SRM assembly, transportation to the launch pad, on the launch pad, or secondary improper ignition at the wrong point in time after primary propellant ignition and launch. So, for 25 flights, thousands of workers had been endangered every time they were in the vicinity of propellant loaded SRM segments because they thought the danger of errant ignition (and thus explosion) was minimal. I had shown The FBI and NASA the dangers were far from minimal. Thiokol had published minimized risks for 13 1/2 years into the SRM project and MTI needed more studies! *Fucking Insanity!* American Tax Payers had been RAPED and thousands of lives endangered. All astronauts had been endangered, seven were killed by errant SRM's!

Why didn't NASA jump on this problem first and foremost over other problems, I wondered. I would later find in the NASA / MTI contract that NASA was the liable party for transportation accidents or explosions of SRM segments, not MTI! Now that NASA knew it was liable and the rail road cars were astoundingly dangerous they immediately covered their back sides. NASA SHEEP came to understand NASA liabilities!

During the chain of events MTI determined Special Agent Tim Healy was on to them. NASA was there pounding on Bob Bryant (FBI No. One) while he was in Wash., D.C. to 'throttle down Special Agent Tim Healy'. After all, NASA now complained to FBI hierarchy in Wash., D.C. that Tim was young, inexperienced, and out of his scope. No one was surprised. The NASA drones had to fight for their positions, retirements and avoid incarceration. Couldn't let a little reality destroy their near useless existences. 'Bear Bryant' backed his Agents.

I was now just one of many activated sources. My other well placed sources were also on an acquisition rampage. The FBI was at maximum effort. The D.O.J. had been contacted, but the FBI / DOJ relationship was on a questionable threshold. The head U.S. Attorney in Utah was Mr. Brent Ward and he was a high ranking Mormon. Brent Ward wanted to be governor or Senator of Utah! Utah Senator Garns had displaced a Department of Defence pay load so he could ride the Space Shuttle for his own political purposes (well documented). Senator Garns was a high ranking Mormon. The FBI is owned and overseen by the Department of Justice (DOJ). The Attorney General who was head of the DOJ was a Presidential Cabinet Member. The Director of the FBI is not a Presidential Cabinet Member. The FBI was headed into tough times from a weaker than desirable political position now that the FBI Director, Judge Webster, was leaving the FBI to take over the C.I.A. The head of the Dept. of Justice, Attorney General Ed Meese, was under scrutiny and severe disfavour for suspected massive aerospace collusion and fraud. Why would I conceivably feel insecure? Why did the FBI feel insecure? Adding it all up you can only conclude Special Agent Tim Healy and FBI #1 Bob Bryant had to finger a lot of holes in the dyke! Forces against the FBI were overwhelming. I had never seen such massive crimes, let alone experienced them. Had NASA faced their problems they would have had the FBI as a heavy team partner. Once the FBI knew NASA was operating under the CYA (cover your ass) covenants and that NASA had lied verbally and in writing to the FBI, NASA was poisoned and was under investigation instead. And, so, the NASA / FBI relationship cycled from very close to overtly antagonistic. I can't tell all I know about the NASA / FBI relationship and I only had limited visibility on some issues and 100% visibility on others. What I can say for certain is the FBI scared the hell out of NASA and forced NASA to clean up its act. To what extent did NASA clean up on and to what extent did NASA cover up? At least NASA was now put on notice by the FBI - CLEAN UP YOUR ACT! Was there a caveat of 'clean up or else?' What would / could ' or else' be?

CHAPTER 29

"AXED"

Time: 3rd week 1987

After the first well attended ESD Meeting transpired the pace at Thiokol was accelerating in many ways. Thiokol was under duress over their railroad cars not being electrically grounded and for diverse other reasons. System Safety work by our group engineers and scientists was in fever pitch confusion and chaos was our environment. 'No System Safety Department product' delivered to NASA yet was our immediate concern.

When I visited with VR after work she verified I wasn't going crazy. Thiokol pressure was building. On Kent's last day, I was called in to a meeting with Kerry Sanofsky just before lunch. Thiokol had announced that Kerry would be the new and 'acting SRM Systems Safety Manager' for a while. He would hold that position until Thiokol made a final appointment. Tension was accelerating between the shakers and movers trying to acquire the coveted position of Manager of SRM System Safety. The shakers and movers all hated each other and were lined up to kiss Don Thompson's ass. Don Thompson was bent over to have it kissed.

Kerry and I sat down just before lunch in his office cubicle and he was clearly nervous and upset. Kerry even said, "my heart's beating very fast and I'm very nervous and upset". I said, "It can't be that bad". What's up? Kerry picked his words very carefully and I suspected he wasn't sure of what to expect from me in reaction to whatever was coming. Therefore what ever was coming probably wasn't good! My personal estimates of what was coming were forming fast and I thought I was about to be canned, but I was only speculating (odds: 10,000 to 1 against me).

There are forces at MTI which really didn't want you around Steve. "No Shit", I thought. Today is to be your last day Steve." A few seconds before he got it out of his mouth, I quickly computed, controlled, and measured my impending anger and response. Being under emotional stress was less common to Kerry than to me. I told Kerry, "well, I have done very good work here haven't I?" Kerry said "there are some persons at Thiokol who aren't satisfied with your work!" These (MTI) persons are afraid you are going to hurt MTI. Obviously those who were up the totem pole from Kerry wanted me out. I said "Well Kerry, Is this how you treat people who do good work? Absolutely no one in the department has done more research on your

SRM's, pushed harder for meetings and departmental communications, or cooperated better. I have given the new employees and contractors the grand tour and indoctrination for you!

"Do you just throw them out the door without any warning? Kent McKinnon has been my boss up until today and he has frequently expressed his appreciation and support for my excellent and insightful work. You have held Kent's office a few hours and now I'm being fired? Kerry was very nervous now, fidgeting and totally uncomfortable. I frankly didn't give a damn whether he was uncomfortable or not. Now it was clear to me Kerry was simply a Thiokol tool who was directed by Don Thompson or others to get rid of me on cue! And the cue had come! In a few hours, Kent would be gone from Thiokol forever and I had no one willing or capable of protecting me from Thiokol management. Kerry was enthralled with the prospects of being a young manager of System Safety on such an important project as the SRM's. Potentially, more prestigious for a non-scientist and non-engineer! Also if Kerry was to become the permanent manager rather than the 'Acting Thiokol Manager' he would be controlled by Thiokol. After all, Thiokol higher management knew best! So I slowly twisted the knife in Kerry - for the FBI's own good.

Kerry had never grasped the reality that he was the new Thiokol 'SACRAFICIAL MANAGER' if another SRM exploded. Kerry explained to me that for "reasons which he could not divulge that I was simply to be dispatched". I wanted to know exactly who wanted me out and exactly why! I said, "This is very curious since today (the same day) you are throwing me out, and you have also brought in new men. Do you think they are better qualified than me? I have worked with one of them at Boeing Aerospace and even though he is very nice, he isn't a down payment on my skill set to push projects or to look over systems. Kerry was very embarrassed. He is coming in because he was hired some time ago and there won't be any more persons coming in of course". I knew this to be a lie. Many others were coming in. I asked Kerry what I was supposed to tell future companies about why I was thrown out of MTI? He shrugged his shoulders. He didn't know and he didn't care. a willing MTI sacrificial victim to the next disaster.

Sharla Rose and others had kept a lot of us informed about new contractors coming in soon. I thought to myself, "The usual, bullshit!" MTI will keep hiring and hiring until they have acquired the appropriate number of corporate castrated contract sheep, and will cull out the winners as they have done for decades. I figured not only me, but several others were headed out the door for similar reasons - they have been doing good work. It was hard not to laugh when I knew a few months ago MTI had gone to a Salt Lake City shelter for losers and rounded up a bunch of 'alcoholic bums', put shirts and ties, and sports coats on them and bussed them up to be seen at work by NASA. NASA never knew if they could read or write. They were simply a bus load of alcoholics to be seen.

I and the FBI rolled on the floor laughing about it. I made a direct effort to choose my words carefully for Kerry, and in no way have harsh words with him. I was extremely courteous, and polite. I kept my anger and disgust well hidden. Every day I could buy at MTI was another day for the FBI and we had planned for this eventuality. Over the last few months I had helped Kerry with a lot of his problems and always made serious effort to guide him through the Thiokol *"Rat Maze"*. He asked for the guidance and he appreciated it. He was politically, managerially, and technically inexperienced. He had come to me and Malcolm Philips frequently for wise counsel on both technical and political issues. Kerry Sanofsky and Phil Greenhalgh both had acknowledged there was plenty wrong with Thiokol. They both knew Thiokol was suppressing reliability, and both recognized Don Thompson, John Thirkill, Dorsey and Russ Rasmussen as dangerous conspirators apparent. The FBI and I might have tape recorded Kerry or Phil (I neither confirm or deny). The FBI and I gave serious consideration to recruiting these two. In the end we decided against it for one "reason". We didn't think they were made of the right stuff.

I never thought for a minute that Kerry was getting rid of me of his own volition. He really wasn't comfortable with his new limited power, he did nothing of his own volition and he did not have the unilateral authority to let me go. Kerry hated Mahalon Hollway's guts, but could not make the simple call to ARC to terminate Mahalon. Kerry had told me Don Thompson wouldn't let him. But, now Kerry could fire me, not by calling ARC and having them do it, but directly himself. Kerry was directed and empowered. I assumed Don Thompson had pressured Kerry into "despatching me" or more likely ordered it. I will probably never know the complete details of events and it doesn't matter.

Was NASA involved? It wouldn't surprise me! A possibility to not be over looked as - would Kerry gain high Thiokol and Mormon and NASA brownie points if he dispatched me where Don Thompson hadn't? Was I cannon fodder for Kerry's career? A year and a half later witnesses would emerge who claimed to have both seen a memo which had reportedly heard a conversation between Don Thompson and Royce Mitchell (NASA SRM Office) which indicated that NASA's Royce Mitchell was a seriously interested party in removing both Kent McKinnon and myself! What was the whole truth? NASA was complicit.

Politely and courteously I told Kerry that I thought this was morally and ethically wrong and I really demanded (politely) that he tell me what this was all about. I was wondering in my mind whether any of my coded FBI conversations over the telephone might have been recorded. Except for one time, I used a bunch of different telephones.

Occasionally, I talked in one crazy code or another over the telephone to the FBI. "Hey Tim, are the skiers on the slopes?" "Yes Steve, better get your skis and join us around 8:30 PM!" "I'll wax my skis so I can move faster!" "Alright!" "I'll see you at 8:30 PM!" "yea!". There were many flavours of codes we used.

I was curious about thousands of things. Kerry said, "Well Steve I don't approve of us putting you out the door with no notice what so ever, but " I have to do it" or "I am authorized to give you one more week of work under special conditions". I asked who authorized him. Kerry declined to tell, but he said it came from above and from those who wanted me out. For this extra week of work you must promise to do MTI no harm. I then asked Kerry what harm could I do? The only thing I have done is write 245 Catastrophic Hazard Analyses, i.e., do my job faster and better than anybody else in our department. I personally gave good orientations to Thiokol's System Safety department, new contract personnel, and I had brought a lot of technical questions foreword that no one at MTI could answer. My 245 hand written Hazard Analyses (at that time) took five months and cost the taxpayers a lot of money. No one at MTI for that four and a half month period did more.

No one at MTI had ever officially looked at my HA's in any meaningful way or was interested in fixing the problems to date, err, my mistake *"There Were No Problems!"* What was I thinking? I told Kerry that he and I both knew I had pushed for their review. Kerry said, if I would promise him that I would not do any harm to him personally or that I would not hurt MTI in ways like Dr Tony Laine, that I could stay at work at Thiokol for one more week. "What a generous bastard he was!"

I carefully thought about it for 20 seconds, and told him there should be no problem since I really didn't mean him or Thiokol any harm! As you have now guessed, I lied! "Well gee Kerry, you're a 'real pal!" That's really nice of you! Whew! Thanks! How naive and insulting for anyone (Kerry or Thiokol or others) should think I would sell my soul for only one week's pay!

These Thiokol managers were so used to the castrated sheep types that they seemed to be totally at a loss about me or reality. I leaned on Kerry again and he reaffirmed to me that persons above him wanted me out, they were insecure and frustrated by me. "They don't want to acknowledge my HA's do they Kerry?" They don't feel there are hardly any hazards Steve.

They feel you are over-exuberant, but they will evaluate your work. "Kerry, our marching orders from both Kent McKinnon and NASA and the Rogers Commission are to identify and report all HA on the SRM to NASA no matter how innocuous they appear. Kerry, nearly all of my identified HA's are catastrophic, not by guess or imagination but by physics, chemistry, engineering, and mathematics. My findings are in fact irrefutable!" Kerry and I finished our conversation and I went on my way. A few minutes later, I grabbed Malcolm and took him out behind the trailers where we had our "serious fresh air meetings". I told him about the course of events. Malcolm didn't seem surprised, and we talked about how many others would be going out the door, or was I the only one? I don't believe either of us knew the answer at that time. I was stressed and becoming paranoid. Had Malcolm joined the Thiokol conspiracy? It would be over a year later before I found out Malcolm Phillips was well compensated by Thiokol for reporting what he knew of my FBI activities to Thiokol.

Remember Malcolm carried documents for me and the FBI and had been visited by the FBI several times and he had lied to them. Malcolm must have felt triumphant at that moment. He made a quick bundle of money and bought in a very long term Thiokol contract in Florida near his home where the Space Shuttle would be launched from Cape Kennedy. Perfect for him.

Somewhere in the early afternoon I ran into Kent in a hallway and told him I had just been canned and then given a week's notice and he looked astounded. He asked me why and I told him I really didn't know, Kerry refused to tell me. I would certainly appreciate it if Kent could ascertain why. It would aid our clandestine activities. Later that afternoon Kent and I ran into each other and he told me he had directly asked Kerry. Kerry, the young buck that Kent had trained refused to tell Kent why I was fired. Both of us thought Kerry's actions to be amusing and peculiar. We assumed they were directed from Don Thompson or higher. Neither of us was certain. It was unlikely that Kerry acted alone as he had limited authority. I considered calling Gene Pusey at ARC and raising hell, but then I realized ARC's only role was payroll and crimes. ARC had no hire or fire influence and was 100% controlled by MTI. ARC had only pecuniary interest in my employment. What was their legitimate slice? Did ARC have an illegitimate slice? I had never met Gene Pusey or set foot in any of his offices. Gene and ARC were only names and phone numbers to me. The FBI however had specific interest in Gene Pusey.

So at the end of the regular working day or slightly before, Kent McKinnon disappeared from the MTI hallways forever. Kent' s last significant act was to issue the following memo.

Pay attention to the note on the bottom of the letter by Don Thompson. The customary MTI action, deny the problem, pass the buck, and then ignore it! It was a year and a half later when I received a copy of this memo with Don Thompson's note on it. This memo clearly demonstrates

MORTON THIOKOL, INC.

Wasatch Operations

Space Division

Interoffice Memo

18 March 1987
8512-FY87-M136

TO: K. G. Sanofsky
SRM System Safety

CC: D. E. Thompson

FROM: D. K. McKinnon, Supervisor
SRM System Safety

SUBJECT: Loads and Stresses

The following concern was expressed by Wayne Tilse prior to his departure from Thiokol. I believe his comments may have some merit and this needs to be confirmed or reduced to the actual figures involved.

It appears that a complete re-evaluation of loads and stresses on the SRM needs to be done to include the effects of dynamic load inputs into the SRB from the ET, the action of the entire Shuttle System under dynamic loads and all sources of static, assembly, and residual loads and stresses. The Space Shuttle SRB Design Loads, Book 1 (SE-019-057-2H) does not adequately address dynamic loading, and probably not even static loads. The loads are 1 dimensional, and dynamic loads are reduced to static equivalents. This cannot produce an adequate stress analysis, since this method does not consider the dynamic response of the SRB System, or the dynamic interaction of the SRB ET and orbiter. The safety factor of 1.4 as required by the specification is apparently supposed to account for the unknowns, but dynamic (resonant) response of the SRB could produce stresses 2 times those produced by a steady state static load. This exceeds the Safety factor of 1.4 by .6 by 4490.

[illegible signature]
D. K. McKinnon

Kerry, I don't believe this!
please check w/ Arnie Thompson

Kent McKinnon's integrity toward the SRM's and other Space Projects. It also demonstrated Don Thompson's lack of concern and actions.

I went into extremely high gear to make sure that I produced every legitimate hazard conceivable to get in what licks were proper and appropriate before I was out of time at Thiokol. At this time I had a full 245+ HA's prepared in whatever the "format of the hour" was accepted. When I next saw Mahalon Hollway in the corridor he gave me his "I - know-all-about-it-look'. Thank you Mahalon! I really appreciate all of the unsolicited harm. You were a real patriot Mahalon! Then Mahalon proceeded to be as rude as possible to many of our department personnel. Over the next week various FBI people (new Sources), other than myself, and System Safety personnel quietly probed to find out why I was officially being axed and rolled out the door. I could only speculate.

I was tired, pissed, paranoid and glad to be going. My genes spoke to me that NASA had a hand on it. When I was with the FBI we joked that I was canned because I was an Ass Hole! Of course that was correct and what could I expect unless I acted in the manner of "mice balls"!

A fellow System Safety worker, Arun Murthi, came to me and we had a quiet discussion. He explained that earlier this week one of Kent's old friends, Tom Bradley had been assigned to five special and secretive HA's on electrical grounding and ESD and to show NASA "Thiokol had no problems"! Was Tom Bradley not a qualified scientist or engineer, no, but he had been a System Safety Supervisor at Ford Aerospace - a paper pusher type. From talking with him previously I found him interested in electronics, but he didn't know much. Did Tom Bradleyley come to talk to me about my findings to date in electro static discharge and grounding? Hell No! Tom had seen which way the wind blew at Thiokol and he made his decision to take care of Tom Bradley! Tom wouldn't cut it as an electrical technician let alone as electrical engineer or physicist.

Tom was now clearly identified by MTI management to be counted on to come up with the answers MTI wanted to feed to NASA on electrical grounding and ESD. Tom Bradley had been ordered to keep his work "quiet' according to Arun. Arun also told me I had been 'had' from all directions. Arun elaborated to tell me that a high Thiokol manager knew I was planning on tape recording Kerry Sanofsky in a couple of days (again?) and that some of my friends had sold me out for their own longevity and their continuing employment. I was learning a lot. Thiokol knew I was about to record Kerry for the FBI, but it must be true because I don't know of anything Arun was ever wrong about. How? Did the FBI have a leak, did the DOJ have a leak, had a phone conversation been intercepted, was my apartment bugged - I'll never know. The FBI discounted that Thiokol knew. I wasn't so certain. I was insecure. Arun didn't know what I was up to, so he sure as hell got it from somewhere! The FBI dug into it. No one should ever think for a minute the FBI is always right. The FBI wasn't particularly upset about this information, but I was inflamed - was my apartment and life bugged? Stranger things have happened. A Mormon conspiracy, MTI, DOJ, internal groups? Tim and I both thought it was business as usual at MTI. Was Thiokol trying to bullshit NASA some more? Of course! I was very angry about this turn of events, but grateful to know why Tom Bradley was avoiding me. I had sensed the evasion and had been sensitized. I reiterate here that Tom Bradley had been brought to Thiokol by Kent McKinnon to write an improved System Safety Plan first and foremost, and then HA's as needed within his limited technical bounds.

A few days prior, Tom Bradley informed me that his new System Safety Plan had been stopped dead in its tracks over a month before. Now Kerry was simply making minor revisions to MTI's old System Safety Plan which Kent and Tom Bradley considered archaic. I always wondered if NASA was officially satisfied with the crap MTI furnished them? Tom told me Kerry Sanofsky wasn't interested in it. I was despondent about it. Was it Don Thompson or higher? All of us wondered how NASA could buy such a bull shit System Safety Plan as was in effect. I wonder what ever happened to Tom's improved SRM Safety Plan? Ultimately, speculation was that it went into a trash can.

Tom came to me the day after Kent departed and urgently requested a "quiet meeting". That day or the next we went to a secure place (an empty cafeteria) and sat down over a cup of coffee. I asked Tom, "What's up?". Tom said he was very insecure. I told him that I was also insecure. "What exactly is wrong Tom?" Tom related he was under serious fire from Kerry Sanofsky and others at MTI. Kerry's new right arm, Donald Janacek, had severely verbally abused Tom. "What are the issues Tom?" "Well, Kerry and Don Thompson raised hell with me because I have spent vast amounts of time on our department's new System Safety Plan".

I said - "isn't that what you were ordered to do?" Tom said yes. How is the plan going I asked? Tom said "not too well!" What's wrong Tom? "He said the existing plan is absolute crap and his new one is moving very slowly. For months the typing had been a major hold up. Obtaining approvals has been an obstacle". Also he said Don Thompson resists everything. "Kent and I have worked hard on it, but I can't make any headway." I told Tom this was MTI business as usual! Then Tom told me there were other problems. Kerry Sanofsky and others had raised hell with him about not having produced some HA's. Tom said he was not even supposed to be producing any HA's originally! Notice Tom didn't feel at liberty to tell me about the five special

ESD HA's. Tom said he felt he was about to be fired. I told him I was already fired - effective in a few days (Friday coming).

I told Tom I had never seen such corporate bull before. We discussed in some detail aspects of HA and other documents supervision. We also discussed NASA confusion and the incredible amount of money being wasted. Tom was frustrated by the low quality level of HA's generally being produced. Tom felt if we were to do a proper job we needed an army of experts, he saw lots of problems. He seemed genuinely cornered and frustrated. I Told Tom that in my opinion he would survive. I reckoned that he was in the vice temporarily because he was "Kent McKinnon's Man". Donald Janacek was Kerry Sanofsky's man. They had both worked at TRW (A CIA aerospace corporation) and had met there . I told Tom that the sooner he took the preferred "Thiokol low profile" and "made way" for Kerry's man, Donald Janacek, the he could pass into oblivion and the sooner he would be safe. I didn't tell him the FBI had him under a microscope.

Since I saw Tom as a patriot at that time we had a 'beat around the bush' discussion if Tom would help if "someone brought Thiokol to task?" Wouldn't everyone? Tom said it needed doing and he would support in any way he could, but didn't know what he could do. Since Kent had trusted, liked, and supported Tom I appeared to follow suit. Tom felt better after our talk. I suspected I might have another FBI recruit, but was uncomfortable. I knew he was used as a "tool" to deceive NASA on the subject of deadly ESD poor electrical grounding there was a twinge in me that did not trust Tom. My knowledge and feelings had not all sorted out on Tom then.

As I was counting down to my last days, I accelerated my HA preparations and hoped one way or another all of my HA' s would all get into the NASA system. I was willing to defend my HA's against any team of engineers in the world. When Kerry was firing me he assured me my HA's would be reviewed and forwarded to NASA. Pigs are also known to fly south in formations on their annual migrations! Actually the FBI had all of mine and most others now.

During this week one of the contract employees who was now on the reliability side of the house was getting ready to leave Thiokol. His name was Steve Simian. He was returning to Seattle. A party was scheduled for his last day. It was really his scheduled last day and they added me on to the party as a matter of courtesy as I was being "kicked out the door" When the big day came for the party, I had my desk already half cleaned out and certain documents I gave to certain people who were working on pertinent systems related to my documents. I had the only compilation of electrical / electronic specs on the SRM's in existence at MTI. It had taken me months to accumulate and compile. I also had the only aggregate collection of shipping procedures via rail road. I had also compiled an excellent set of MTI documents, directories and references. I had compiled all of the NASA documents on lightening characteristics and added others from dozens of sources.

Malcolm was running around doing his thing in our office cube. A new man who reported in the day I was fired was settling in and had been talking to us about how we should just roll with the punches! I could already see this contractor was a "take the money and run" type! I was not impressed or amused. He pissed me off so bad I couldn't see straight. The United States will not be conquered, it will collapse as it becomes a nation of castrated sheep. Probably a dismal future! "Roll with the punches! Go with the flow! Not my responsibility! Not my Job! I only work here! The end of the world as we have known it. We have seen the enemy and they are us!

One of three fellows in our cube, Dale Donalds was in poor health and was doing his usual six hours of diabetic sleep a day in his chair. It paid the same. Another fellow, Fred Broadhacker, was going crazy trying to accomplish anything in our Thiokol System Safety Department.

He gave up on our departmental secretaries and was working on his new 'one-and -only' personal computer at MTI. He typed his own HA's. Our newly arrived, one-and-only PC was to be shared with so many people it was overwhelmed. Come to find out we had so many format and numbering changes Fred couldn't even keep up with them using a computer! Furthermore, the changes in the HA numbering systems and format were still raging. Fred was crazed by it all. No one else even tried. Twenty people from Reliability were after this P.C.. Access was a joke and it was a weak overwhelmed machine.

The last couple of days before I left, Malcolm continued grumbling over the continuing changes in the HA numbering systems again. When you have several hundred HA's hand written it is enormously time consuming and time wasting. Around lunch time on my last day, we all went to the party and it was attended by Reliability and System Safety department personnel. From the number of attendees I detected both department's head counts were way up. Don Thompson appeared at the party and frankly looked ashen and uncomfortable.

I do not think these were happy times for him and I think he was trying to make every effort in the world to appear innocent as a lamb. I sensed that he knew he was in serious trouble and he wanted to defuse it. Don Thompson shook Steven Simian's hand and then he came over and shook mine and wished me well. The poor criminal, treasonous, Mormon Bishop, ignorant Son of a Bitch!

He had a lot of trouble looking me in the eye. I had trouble shaking his hand. I was more inclined to attack him. So we ate the buffet lunch everyone had brought and I went back to shutting down my desk. I had a corner spot which gave me more wall space to hang documents and graphs on. Also, I had an extra table, a rare commodity at MTI.

I asked Malcolm if he wanted to take over my spot and he said yes. Having a table and a corner were major assets. So I basically moved out early in the afternoon and Malcolm moved in. My heart interrogated my mind. Malcolm resisted the FBI, yet he helped me carry out documents on twenty plus occasions. I was a victim of serious paranoia. Had Malcolm been in bed with MTI management in order to buy long term contract protection or not? I just didn't know then and did not have complete visibility. About 3:00 PM that afternoon, I loaded up the last "procured" documents I would be taking out of Thiokol for the FBI and stuffed them here and there on my person. Kerry came along and said, "Well let's go on and sign you out of Security and Thiokol." On the way down to security, I "leaned" on Kerry some more to determine why I was going out the door. Kerry was more emphatic than ever that he couldn't discuss the subject. He said his lips were sealed. Then he said, "you sure are a cool operator!" I wondered to myself whether that meant that he knew I was helping the FBI or if he just liked my style of operating or he needed something to say? What did "cool" mean to him? At any rate, my exit down to Security was relatively uneventful.

I had high anxiety of what awaited me at MTI Security. Were they going to pounce me with legal ramifications for messing with them? Could they waste me? Were they going to raise unholy hell with me? Did they know that I was helping the FBI? I really didn't know if they knew anything. I went in and signed some standard Security Forms and that was essentially the end of my employment at MTI and I had to transfer my Security Clearance. I was then escorted out of the Wasach Plant back door by Kerry as he was required to do. As we walked from Security to the back door by reception we pleasantly agreed to have dinner "on me" in a week at a Brigham City restaurant. I shook hands with Kerry and said good-bye. I then exited the Thiokol Wasach Plant for the last time. What a relief!

What a breath of fresh air to leave the Thiokol Wasach Plant! I have never been so glad to leave anywhere in my life, not even Vietnam. It was like a relief of having just passed a kidney stone. It was about 3:30 PM, 27 March, 1987. A sad and traumatic chapter in my life had just closed - or had it??

Since my beginning at Thiokol on 9 Nov. 1986, I had been living in the midst of a horror story! I was now 'born again' into fresh air, sensible hours, and probably life would be much improved henceforth. I questioned myself, "how will the FBI proceed? Would other sources do as well or better than

I had for the FBI. No doubt better! I was a much more obvious irritant to MTI than more recently placed sources who were wolves in castrated sheep's attire!

I was preparing to get married. I was busy shedding myself of all ancillary marital candidates. No small task. A number of high ranking candidates had me in their sights and their grappling hooks were on me. All were great gals. Those that didn't get me should thank God. My life is too dynamic (erratic) for most. Soon I would be off to Hawaii, Soon I would be out of Utah. I could go on out of the country. I didn't know what I would do next, but I was certain to enter a new phase of my life.

CHAPTER 30

DEPARTURE TO KAUAI

TIME: 1 May to 12 May 1987

Just after I was thrown out of my Thiokol contract Special Agent Tim Healy called me. He said, "Be calm! There has been a serious information leak at the U.S. Department of Justice. My name would not be coming public, but there would probably be a newspaper article out tomorrow that would tell of an MTI investigation by the FBI at Thiokol! Be cool!" "No sweat!" "AAaRrrrggghhhh!"

The next AM I stood weak kneed in front of J.B.'s restaurant in Brigham City. Headlines - "FBI Investigating Shuttle Contractor". *Oh Shit!* I'm going to be murdered in Brigham City today and I won't even have benefits of burial by the accepted local theocracy!

"FBI INVESTIGATING SHUTTLE CONTRACTOR"

"Washington (AP_ The FBI is conducting criminal investigation of Morton Thiokol Inc., the company that made the faulty rocket seals blamed for the 1986 explosion of the Space Shuttle Challenger. According to a document on file in U.S. District Court.

The FBI has divulged informants among the employees of the aerospace contractor as part of its criminal investigation, said the document, a private communication to a Federal Judge by the Justice Department.Details of criminal allegations under investigation could not be obtained. It is not known whether the FBI's investigation concerns the companies conduct before the ill-fated Jan. 28, 1986, launch that killed seven shuttle crew members or during subsequent investigations into the disaster by Congress and a Presidential Commission.

Phillips Wendy, the companies director of corporate development, declined to comment when asked about the FBI probe. The Chicago-

based company makes rocket-booster components at a plant in Wasach, Utah.

Susan Schnitzler, an FBI spokeswoman, refused to confirm or deny the investigation, saying only that the court document revealing the probe had been inadvertently released and resealed.

The Presidential Commission blamed the explosion on the failure of a joint on the right rocket booster. The joint failed because cold weather prevented rubber pressure seals from containing gases, the Commission concluded. The FBI investigation was revealed in court papers on file in connection with a lawsuit against Thiokol by Roger Boisjoly, a former engineer for the company who worked on the Space Shuttle project.

Boisjoly, who pleaded with his superiors and the National Aeronautics and Space Administration not to allow the launch in 26 degree Farenheight weather, has accused his former employer of lying to congressional committees and the Presidential Commission headed by former Secretary of State William Rogers."

The FBI now had a lot of sources and visibility inside and outside of MTI. They managed and used their resources as they needed. NASA was apparently shuffling to save itself from appearing totally incompetent to the FBI - too late now. NASA failed to convince the FBI of their competence. Fear ran rampant through MTI and NASA as I left. Political wars, real and imagined were being fought on all levels. Clearly MTI and NASA had enormous fear of another SRM failure, but not enough fear of failure to do their $500,000,000.00 SRM Redesign contract properly. All I could observe was MTI and NASA showmanship. Few real fixes. MTI would rather cover up the problems than repair the damage and move forward. Worse than childish. Ultimately dangerous and irresponsible to the entire Free World.

Let's talk about repairing the poor design of the SRM's. Could the SRM's be realistically be overhauled? No simple comprehensive fix, no single answer. Some of the problems were modest - add 3 inches of electrical grounding wire - Thiokol wouldn't and they could have easily. Many of the System Safety engineers and scientists were convinced the SRM Cases needed to be of a better alloy, better design, etc.. None of the Engineers liked the O-Ring Field Joint design which failed and caused the Challenger Disaster. Simultaneously many of the same group identified massive and avoidable problems in the SRM Rocket Nozzles. The "fix" on the SRM Cases and Nozzles was essentially to start from scratch - massive overhauls. To start over on the cases and the Nozzles could cost billions of dollars each and take years.

When you looked at the cost of a total redesign of the SRM's you would have to ask why not switch to liquid fuelled boosters? Werner Von Braun didn't like SRM's. He had good reason. History proved him wise and correct.

I was now getting ready to go to Hawaii and get married and my parting cover story to Thiokol people was that I had a new consulting contract sealed with the Department of Defence in Ogden. I wondered if I might get into trouble someday for using the Department of Defence as a cover? However, this cover would keep people from asking unwanted questions and would keep me from having to tell a lot of lies about what I was doing.

Everybody knew that the Defence Department had a number of hush, hush operations in the Ogden area. This story would be reasonably believable. As I left Thiokol I had set aside essentially two weeks time which I had planned to make available to the FBI 'free gratis' for whatever they needed me to help them with. After that I needed to crawl into a contract somewhere. There were a few emerging, but far fewer in the industry than anticipated. Two years of Gramm-Rudman and about 15 months of no Space Shuttles. The direct and indirect costs of un-launched shuttles and payloads is phenomenal. Many payloads have to be stored in dust free clean rooms, special temperatures, specific humidity's, and in some cases partial vacuums. Millions per month storeage in storeage fees per device or payload. Some required low pressure, some low light, others a 'clean room', others a 'Faraday Cage.'

From April 1st through the 7th, I had a number of phone calls and discussions with the FBI and was becoming impatient to get on with my own business and life. I discussed with the FBI what I should tell potential employers about my being fired at MTI. I hated Brigham City. In the early part of the week of the 6th of April, Tim told me he had cleared with the FBI No. 1, Bob Bryant, and higher for me to simply come to work for ether FBI at $50.00 per hour (that was very high then and would be equivalent of $300.00 hour in today's world - 2016) just as though it was a normal consulting contract. No overtime of course. There had been previous mention of me being paid to assist the FBI. I never believed it would come to pass. So instead of giving them two to three weeks for free, and then going on my way into another contract (if I could find one) I reported to work at the Ogden FBI office on the 8th of April, 1987. I wasn't an FBI employee, I wasn't undercover. I was just a "No Name Person" who was to be paid in parking lots with money in brown paper bags for services rendered to the U.S. Government. I became a professional full time FBI 'Source'! I was nervous and really didn't know how to act. I was an outsider. Besides Tim, there were two other FBI Agents in the Ogden FBI Office, Kevin Kelly and Don Kirby. Don did some very special things and also all the regular FBI stuff: murder, bank robberies, etc. Kevin was a relatively new agent and he had an array of cases under investigation. Don was technically in charge of the Ogden Office and was a distinguished, learned and smooth operator. He was also a practicing Mormon and had been a Mormon Missionary to Taiwan in his youth. I felt that every word Don said or heard to or from me had been measured to the closest thousandth of an inch, and I always felt under 'infidel scrutiny' from him. I am certain he knew my disdain for the Mormon criminals at Thiokol. Kevin had a number of cases running, and was seemingly disinterested in my activities. I believe "seemingly disinterested" was more of his facade than reality. As it turned out later, Kevin himself had a case going against Thiokol which will unfold later.

Kevin was astounded at the facts and figures, names and things I knew out of Thiokol and realized in certain ways they were meshed with his activities. I didn't know they meshed then, I found out later. Sources such as myself, were expected to answer to their FBI Case Agent only. If Tim, who was my Case Agent wanted other FBI Agents to know things that I had found for him, he would tell them. I realized that Agents in the FBI have a proven track record of keeping sources protected and under cover and to a reasonable extent controlled'. Therefore they try to keep their sources from talking to anybody other than themselves even within the FBI. For an FBI Agent to "make a case" he or she must carefully orchestrate everything, "making a case" almost takes an act of God! Also sadly, at some places and some times, there can be FBI turf wars and internal political battles.

Thousands of negative and complex forces and factors must be overcome. Most people can not be trusted. Frustrations of the process have to be experienced to be comprehended. Nothing seems to be good enough. Our revered America legal system and its inclusive safeguards are extraordinarily tedious and equally flawed in my view. Once a case is established is takes on an awesome array of constitutional and other legal protections. Brigham City became an isolated burden to me, I was exhausted from long hours at MTI and 10 to 20 hours a week additionally for the FBI. No human being in the world who has not worked like that can comprehend it. Even the battles I experienced in Vietnam were not so demanding. I felt good about what I had done for the FBI up to now and much more was to come. I still felt responsible to see the cases through, but now the FBI carried most of the burden and the initiatives.

I served the FBI *"For The Right Reasons"*. I looked around Brigham City and felt totally alienated. I was not a Mormon and the majority of the people there in Brigham City were not friendly to me. They were not rude, they were simply Mormon Xenophobes. After all, I was a foreigner to them, an infidel. There were a few nice people that I have met such as my land lady, Mrs. Dolores Hill and one of the other consultants' landlady's, Mrs. Patty Vanderstein. I had not had any time to really develop any deep and lasting relationships in Brigham City except with two couples and sources. I didn't think it would be good for anybody who wanted to live there to even be seen with me. Would Thiokol or Thiokol Mormons pressure them or treat them right if they were seen with me? Could they end up in a Mormon Kangaroo Court, or .... At the Cemetery?

I also feared for my life. My apartment had a lot of glass windows in front and back and I tended to keep my curtains closed for fear of snagging a bullet passing through. A Thiokol elk hunter's error! I really did not know what to think of MTI's reactions and how far they would go to protect their highly profitable 2.5 billion a year slush fund, errr, contract.

At this time I was making phone calls stirring the pot all over the U.S. for subsequent consulting contracts. There were some contracts floating around and maybe I could snag one. Normal thinking and anticipation for a consultant. This period in history changed the meaning of words and terms. Up to this time a 'consultant' was a very respected and lauded person. Because of the Gramm Rudman law consultants became 'unemployed'. So the meaning of 'consultant' became 'unemployed' for some years into the future. The new term for a working expert became 'Advisor' which roughly translated to 'working consultant' versus 'consultant' which meant 'unemployed'. Evolution and spin rule our modern world.

I would be developing future contracts and trying to consummate them. I stirred up a probable contract in Cleveland, no, it was three probable's in Cleveland. Also there were two groups of contracts cooking with my name on them in Atlanta, Georgia. This was the first time in my career that there were no viable contracts west of the Mississippi. What was I to tell contract houses and future potential employers about my time at Thiokol?

Even more worrisome is what would MTI say about me to inquiring employers? I was fired and thrown out of the MTI door. Being fired doesn't look professional, especially when it is from a bona fide industry loser like MTI. Perhaps I would tell them I was "laid off", but every contract house in

America knew that ARC was currently placing an army of technical persons into Thiokol. As a result of Sharla Rose's efforts ARC was now advertising in national publications for System Safety and Reliability personnel. I was in a bind and I was damaged.

My old standard contract houses and employers knew me and respected me, however any of the others yet unused by me (about 1,200 of them) would ask why I left MTI. One thing on my side was MTI's terrible industry-wide bad reputation. So, as I plodded along looking for future contracts I did as best I could in explaining why I was no longer at MTI. My options were truth or lies.

The lies were:	I was laid off End of job / contract I needed a change
The truth was:	I was wrongfully fired in the middle of the SRM Redesign Project - fired for doing my job!

On various occasions I did all of the above. On the superficial contract interviews I usually lied. On the serious contract interviews I was usually up front that had been thrown out of MTI for doing my job without mentioning the FBI.

A dinner with one of my well-placed FBI sources revealed some juicy and curious titbits of gossip. This source had a long talk with Tom Bradley a few days prior. The up thrust of the discussions were that Phil Greenhalgh had told Tom Bradleyley how Steven Agee came to be fired. Phil told Bradley, and Bradley told my FBI Source that Kerry Sanofsky and Phil Greenhalgh had decided while at a local Mormon Church meeting on the Saturday before I was fired that they would fire me. This was both fascinating and curious. Was this the truth or disinformation? Were they pressured? Was it a unilateral or bilateral decision? What was the truth? Neither the FBI or myself envisioned Kerry Sanofsky taking on such a decision on his own volition or limited authority. Kerry was well and truly 'on the leash.'

CHAPTER 31

CODE NAME "BRAINS"

TIME: 15 Jan. 1987 - Present

My work for the FBI was not as an FBI Agent. I was not an 'Undercover Agent'. Being an 'Undercover Agent' means just that. A full on FBI sworn law officer who is undercover. My relationship was as a concerned citizen and FBI source. Therefore, I was assigned a Code Name in the usual FBI manner when the MTI case was opened - "BRAINS" was my Code Name. Now that I was full time with the FBI I would be paid in cash and I signed for it using my "code name", not my real name.

I had to pay the IRS on the FBI funds paid to me, but there were no deductions up front as a regular pay check. So the IRS even invades FBI secrets! Could the Mormon Staffed IRS in Bountiful, Utah be trusted? We all knew the answer to that! Tim had a way of awarding code names. One of the code names he had assigned to some drug case hooker got him into a social bind. His little joke bit him when "hook" had to sign for the money. We all chuckled about Tim's indiscretion! My code name was "BRAINS" and was assigned very early in the case. That flattered (manipulated) me. BRAINS made me all the more willing to work from day one. After all - isn't there some of 'Walter Mitty' in all of us?

The early plans for my work for the FBI under code name "BRAINS" after I left MTI was as follows: I was to read the entire NASA/MTI Contract and point out to the FBI anything I could identify in the contract which was not being done by MTI or was being done incorrectly or apparently criminally. I was of course to emphasize pertinent safety aspects of the SRM engineering. Also I was to evaluate a huge mass of MTI documents which I and other sources and the FBI had acquired and 'shed all of the light on them that I could'. It should be noted here that documents and papers were now coming in from numerous other FBI sources. The FBI's and other government agencies tentacles had spread far and wide. I was at the centre of the spiders web so to speak. It was very interesting.

The Utah FBI office tried to acquire the NASA/MTI contract at a much earlier date through Mr. Simpson (high ranking FBI Agent at Washington, D.C. Headquarters). Remember, a NASA attorney told Simpson "No, the contract is long, tedious, and the FBI (most of whom are attorneys and accountants) wouldn't be able to understand it"!

The contract was finally acquired and arrived around the eighth of April at the Ogden FBI Office. The Salt Lake City office also ordered a copy directly from NASA against NASA's will. God knows, the FBI had never seen a legal document before! No doubt the FBI had not encountered complex documents before either! The FBI also ordered the acquisition of documents from other FBI resources in Washington, D.C. and directly from niche sources inside NASA. The FBI incurred the same resistance again from both NASA in DC and from the NASA Inspector General's Offices. The FBI couldn't share the entire story with me, but they were plainly 'jumping mad' at NASA. The FBI and I both thought it strange that a good solid, government agency such as NASA would be ashamed or bashful about their public domain contract. Very irregular we all thought! Did NASA have some gargantuan stupidity or crime to hide? Perhaps all of the above! Asleep at the helm?

While waiting for the NASA contracts and other NASA documents to arrive from FBI headquarters and NASA sources I was requested to make a list of all of the questionable activities and crimes I had observed while at MTI. The FBI and I both knew a lot more than we had previously had time to discuss. Almost by accident I was a gold mine to them. I really didn't appreciate how much I knew. My original list of irregularities and crimes was in two parts. What I saw to be NASA irregularities and what I saw to be Thiokol irregularities. The FBI Agents and I all understood that I was not the final judge of what a crime was and if crimes had been committed. I was not a judge and not a jury, nor were the FBI Agents. My list was simply to be used as jumping-off point to zero in on a number of various and sundry things and probably to tie the loose ends together - categorize them. Several final issues would boil down - unquestionably to prosecutable crimes. There are essentially two divisions of the legal arena, Civil Crimes, and Governmental Crimes. The FBI could pursue both, but specialized in and preferred federal crimes.

Tim and I often discussed different types of crimes. The FBI primarily investigates criminal activities and the Department of Justice classically prosecutes both civil and criminal cases. I had studied some contract law both in universities and out, but I had no training whatsoever in criminal law and criminal prosecution. Most of my knowledge centred around business contract law and the Uniform Commercial Code, (USA).

The FBI was profoundly concerned by a recent landmark aerospace case where the FBI and D.O.J. went against General Dynamics. The U.S. Dept. of Justice was defeated when General Dynamics "got off" by simply saying 'We really didn't mean what we said, or wrote'. The FBI was in a quandary about what would be required to prevent a repeat on each and every aerospace white collar crime in the future. This landmark case had set a recent irrational precedent which was a major obstacle to the DOD and the U.S. Supreme Court.

An amusing but devastatingly true FBI scenario happened about six times. Not Special Agent Tim Healy, but another Agent, kept 'acting out' scenes which the FBI feared. This Agent would pace around the office over and over and act out the Thiokol Cases as he saw them in the current Political Bullshit context. He was both great theatre and harsh reality. He was seeking a win.

"Here, Mormon ladies and gentlemen of the Mormon Jury, is Mr., excuse me, Bishop so and so. Bishop so and so is a manager at Thiokol's Rocket Plant.

He has a high school education.

He is accused of 'False Statements to the Government; and the fact is he is fucking guilty as sin!

Now, there were many witnesses to him and MTI RAPING NASA ladies and gentlemen, er Mormons of the Jury.

The specific victim of this fraud and rape was NASA and the American public.

How say you NASA? Did Thiokol and this Mormon, er Moron Bishop rape you repeatedly NASA?

"Yes, but we (NASA) loved it". In fact it was so good we recently awarded Thiokol $70 Million Dollars for good performance!

MTI's rape of us was wonderful! NASA loved it and wants more! Mormons of the ignorant Jury, do you think this Bishop here was an asshole or not?

What say you judge, er Bishop, guilty or not?

"INNOCENT"! "

Then he would, in a scholarly way, probe and assess how to "make" the Cases when NASA, the rape victim was willing, even positioned for 'best penetration'! 'Furthermore, the NASA / MTI Contract prepared by and issued by NASA was so poor the FBI and DOJ were concerned that it was possibly unenforceable since it was in fact "legal crap" in their words! In the end this Agent made substantial inroads to the cases. He was not being a bigot, he was dealing with very serious and complex issues of what we now call 'Political Correctness' and U.S. laws against violation of religious beliefs which was and is a very 'sticky' point in American Constitutional Law. The Mormon Church either directly or indirectly owns and controls the State of Utah and anybody who doesn't believe it is either naive or a moron.

For the readers' education I provide that civil wrongs are generally wrongful behaviour such as: "there is a contract, a participant in the contract fails to perform to the specifications of the contract and is therefore, guilty of non-performance". This is a civil wrong, unlawful and illegal: Such non-performance may be accomplished by intention, accident, stupidity, circumstance, bankruptcy or any of a number of other factors. Criminal Law is another matter and far more difficult to pursue and effectively prosecute. Under criminal law the prosecutor must show that the person had the intent and deliberately planned to break the law and then executed the planned crime. Criminal cases require incredibly more rigorous proof, investigations, and must survive stringent court procedural restrictions. If this sounds simple to you then I haven't explained it adequately. Civil and Criminal law are profoundly different and complex. Proof must be by a preponderance of the evidence thorough and must be clear and convincing. Criminal cases have to be proven 'beyond any reasonable doubt'. I am certain that I do not understand many of the intricacies. But then, most attorneys don't understand them either!

So I started creating my list of SRM Project irregularities. In my opinion the items on this list were not Kosher one way or another. I clearly realized that as I wrote them down many of these claims were not provable or they did not have to do with Solid Rocket Motor Safety. MTI lies were diverse. However, items on these lists were in fact part of the overall matrix of irregularities and crimes. Later on I learned that this list would help demonstrate trends to other cases. One of the ways of proving criminal activity and conspiracy is by demonstrating long standing trends / activities. The list I produced for the FBI on NASA irregularities was as follows:

1 Apparent procurement crimes - to the most mundane observer NASA was in bed with MTI in an abnormal manner in their procurement process.

2 'New SR & QA Dept. crimes' How could NASA not know Thiokol was suppressing safety information? Did NASA condone the suppressions?

3 Obstruction of Justice - NASA was actively working against the FBI and covering its tracks. The FBI documented NASA lying many times.

4 Hazard Analysis performance (past and present).
Past performance at MTI was terrible. Current HA's were apparently suppressed - to date.

5 Drawing tree and other deliverable document performance was terrible. (NASA would pay for used toilet paper)

6 Misuse and misappropriation of public funds - Paying MTI for trash SRM's, even condoning it

7 Failure Mode Engineering Analysis - poor performance was accepted.

8 Critical Items List (CIO:) - no acceptable performance or reality

9 NASA funds given to MTI to replace Casting Building (approximately 22 million). Why should the U.S. Government pay for MTI employees grab-assing around and exploding a major unique national space asset?

10 Contract administration of the 3300 and 3600 Documents. The 3600 document appeared to be MTI's escape instrument.

11 Below specifications on SRM nozzle materials and fabrication.

12 Below acceptable specifications on SRM nozzle performance: Degrees of motion, speed of motion, articulation, ablation rates, and other issues

13 Below safety factors on the SRM nozzle design & fabrication.

14 Below specification SRM Case design.

15 Improper nozzle qualifying tests.

16 NASA granted excessive deviations and waivers to Thiokol

17 NASA Officials possibly corrupted by MTI via "pay offs" and /or jobs at time of retirement from NASA. Revolving door policy.

18 Improper NASA STS and SRM Safety Support - JSC, MSF KSC cutting safety budget reciprocally as Shuttle launch rates were accelerated, (Even though they were substantially behind schedule).

19 Improper NASA QA on Thiokol Safety operations (HA's and FMEA's).

20 Long standing O-Ring / putty problem material fraud (14 years). NASA and Thiokol knew there were serious inadequacies and undependable properties of the putty used to seal the case joints, and these difficulties were well documented.

21 Difficult O-Ring / putty assembly problem not corrected - probably in-correctible.

22 Long standing, well known case joint rotation problem during assembly, ground transportation, during launch, and in flight

23 Long standing gas path leak check problems not corrected in the nozzle housing and the field joints.

24 Long standing case insulation problems not understood or corrected. They were generally ignored.

25 Case O-ring erosion problems not corrected and known and documented for a 14 year period.

26 NASA's award fees - extraordinarily high for a corporation which built poor quality equipment and killed 7 astronauts and damaged U.S. and free world's military security

27 NASA ghost effort on HA's and FMEA's - were they used to find problems or to cover NASA's liabilities?

28 Why is the 3600 Document not approved and is yet being used as a contract document (one year + on the redesign)? Probably to provide escape from responsibility potential for MTI

29 NASA Bullshitted the President's Commission on the Space Shuttle Challenger Accident (The Roger's Commission)

30 NASA had improper response to suggestions of the President's commission

31 NASA deception of the public, Congress, and the Department of Defence.

32 Improper testing of everything. NASA allowed Thiokol to do perfect case testing rather than worst case testing. Extreme negligence.

33 NASA may have lied to insurance underwriters to obtain a favourable rating for STS pay loads and SRM's.

34 NASA probably lied to the customers about the safety of STS pay loads.

35 Class "B" rocket propellant loaded on railroad cars and sitting all over the United States, improperly electrically grounded, close to the public, often unattended. {Explosion hazard, poison fumes hazards}

36 NASA paid for testing "by similarity" and no doubt about much of the same data over and over many times. {Sold to the Gov. over and over}

37 The NASA SRM concept and contract required 19 SRM CASE refurbishments and NASA well knew the Cases could never be reused that many times. Many engineers showed/calculated they were not safe for one single use.
SRM flight load books were disastrously poor. How could NASA allow Thiokol to use such ill-kept flight load books?

38 One dimensional finite analysis of components rather than three dimensional analysis

39 NASA allowed 1 or 2 Thiokol persons to do all Hazard Analyses for years (approx. 12 years). Several experts thought 35 were required while I was there . The 20 or so of us at that moment there were buried

40 NASA assisting Thiokol to downgrade specifications when Thiokol fell short. NASA shifting the dates when Thiokol did not produce results on schedule.

The next list is my Morton Thiokol irregularity list for the FBI:

1. Procurement Crimes:
   a. General purchasing crimes over a number of states involving a large number of persons.
   b. Purchasing crimes witnessed by Mr. George Schick and others at Thiokol.
   c. Purchasing crimes observed by Sharla Rose and others at Thiokol.
2. Systems Reliability and Quality Assurance crimes.
3. Probable auto fleet fraud in Thiokol history.
4. Hazard Analysis performance (past and present) - profoundly negligent orignored.
5. Drawing trees and other deliverables negligent performance.
6. Misuse / misappropriation of funds on SRM's.
7. FMEA - No performance.
8. HA's - No performance.
9. CIL's - No performance.
10. MTI Casting Building explosion - true cause probably covered up.
11. Fraudulent preparation of the 3600 document - intent of the document was to allow poor quality contract performance.
12. Below Specification nozzle materials
13. Below specification nozzle test motion
14. Below Specification nozzle safety factors
15. Below Specification on nozzle flight motion
16. Nozzle erosion problems apparently improperly reported to NASA
17. Improper nozzle qualifying tests

18. Requesting and obtaining excessive Deviations and Waivers. MTI survived on Deviations and Waivers rather than repairing problems. Improper reporting of problems
19. Probable Thiokol payoffs to NASA officials
20. MTI nepotism (Incredible)
21. Racial and religious prejudice (incredible)
22. Putty material for SRM Cases probable fraud
23. Long term difficult O-Ring assembly problems
24. Long term SRM field joint rotation problems
25. Long term gas leak check problems known and not adequately repaired
26. Long term Case Insulation problems not understood and not dealt with adequately
27. Long term O-Ring erosion problems documented for 13 1/2 years and not adequately corrected
28. NASA Award Fees to MTI sought by false representation of facts to NASA management
29. NASA choose to 'Ghost' Hazard Analysis due to lack of trust of MTI or other reasons. {or by FBI proddings}
30. ARC is essentially the sole source for contract engineers for 14 years.
31. Thiokol management threats against Kent McKinnon
32. Steven Agee (me) laid off / fired illegally
33. Dr. Tony Line was improperly fired
34. Approximately 1,100 Quality Assurance persons on staff and the training manager says 40% are untrained (440 untrained in QA) - very irregular. Was QA used to employ most of Northern Utah?
35. Improper testing on SRM components and systems

36. MTI probably lied to insurance carriers (and NASA) to get optimum insurance rates for shuttle flights.
37. MTI manager of Strategies Division waste of $10,000,000.00 on SICBM in first days of having the new U.S. Air Force Contract.
38. MTI apparently violates Federal law by requiring many permanent employees to work overtime and by not paying them properly for it.
39. Class "B" Explosives on rail cars all over the U.S. unescorted, improperly grounded, near highly populated areas.
40. Qualifying test articles by 'similarity' and selling the same data over and over to NASA.
41. Nineteen SRM refurbishments would be extremely dangerous.
42. Poorly kept SRM Flight Load Books.
43. MTI had a mass of unprocessed Flight Data laying around - should it have been processed, was NASA billed for its processing and the data falsified?
44. One dimensional finite analysis rather than three dimensional finite analysis
45. MTI crimes against Roger Boisjoly (who sued MTI using Qui Tam on behalf of the U.S. Government.)
46. MTI Crimes against Allen McDonald.
47. MTI has 2 persons do all FMEA's for years -ridiculous. {None were catastrophic}
48. General NASA / MTI Contract violations - possibly thousands.
49. Kerry Sanofsky's taking $ to supervise overtime so contract engineers could make overtime pay.
50. Thiokol blackmail employees by threatening to down grade their performance reports if employees wouldn't conform to irregular MTI demands.
51. Thiokol did not maintain adequate phone lines in and out of plant - improper project support.

52. Lance Heller and George McKage discussion of 'what's System Safety doing here at MTI?' Contract violations.
53. NASA / MTI Contract not available to MTI managers or workers.
54. MTI unable to provide NASA with a serial list of SRM Cases used on the Challenger Flight immediately after Challenger exploded and killed 7 Astronauts. {I am not sure they even did.}
55. Violation of standard U.S. Government Contract boiler plate 'rights of data' clauses, i.e. NASA asked for but did not receive data requested.

As you can see these irregularity lists were extensive. I have often wondered if 15 or 20 people from different MTI departments were honest and open about everything they knew, how long would these lists really extend? I assure you readers it would be massive. Also note that this was my first cut, later I extended the list considerably and each item had a huge amount of documentation. Almost any department in any corporation and almost every corporation one way or another will bend some rules for their own profits, but this was ridiculous. Furthermore, a lot of additional problems raised their ugly heads as time went on and I finally read the NASA / MTI contract. My original lists were only the tip of the iceberg before I read the NASA / MTI Contract. I wondered what lists the other sources and the FBI were accumulating? Not counting my lists, what else was really wrong with the solid rocket motors? In general, the following issues were definitely wrong with the SRM's -

1 Severe problems in SRM electrical grounding all through manufacturing, transporting, assembly, pre-flight, and in flight. Catastrophic explosion could occur at any time prior to or after launch.

2 The SRM Cases were severely structurally weaker than required for even one single flight. The concept of 19 reuses was a joke and NASA had know it for years. NASA lied about 'REUSABILITY" to the U.S. Congress and U.S. Senate to obtain adequate (profitable) funding!

3 The SRM nozzles were truly "Rembrandts" - all unique, experimental, and dangerous works of art not of engineering and exact duplication.

4 The field joints with O-Rings had a 13 year history of test and flight failure.

5 The SRM insulation (liner) was nearly black magic in concept.

6 The SRM Systems Tunnel was full of serious problems.

7 The SRM Ignition Train had numerous potentially disastrous faults.

8 The SRM propellant was not characterized by density and properties as well as it should have been.

9 Manufacture and assembly plans were faulty, not updated, and in some cases missing.

10 Much of the ground support equipment was not electrically grounded.

This list is only cursory and an overview. The complete list would go for pages and bore you to death. It did not bore the FBI, however it overwhelmed both the FBI and the DOJ.

No doubt certain NASA and Thiokol officials would have been outraged when they look at these lists, threaten me with law suits, scream, yell, and even defecate in their pants! So what! The truth is often painful! These were legitimate concerns that I and many others observed. Every one of these items listed can be substantiated in various ways.

Substantiated by personal observations, other observations, FBI recordings and documents in FBI possession and other appropriate means. I would welcome a courtroom airing of the facts and truth - even their lies. These were planetary geopolitical crimes of inconceivable consequences and planetary importance.

After the FBI Agents studied these lists they felt that it would almost break the national budget to try to prosecute all of the identified crimes and criminals. Much wiser the FBI thought to boil down my list, others' lists, and the FBI's own list to the "pure stuff" and simply prosecute a few selected cases and bar MTI from future Defence Contracts. By now the FBI had confirmed from a substantial number of sources that Steven Agee's opinions were not insanity and were relevant. I am also certain they were satisfied I was not 100% correct. No one ever is. Some items on these lists were irregularities - not crimes. There were some FBI Agents who though MTI was a typical Aerospace Corporation - rip, gouge, steal! I assured these Agents, MTI was far worse than most. One agent couldn't believe MTI could do any wrong (he was of course a Mormon). Other Agents thought I have found only the tip of the iceberg- and I am certain that was the case.

The FBI knew without question there was great evil being done in the northern Utah desert. What would the FBI do? How best to do it? When I originally met FBI NO. 1 over dinner he thought the crime was *"Treason"* and punishment should be "dispatching by his Swat Team". Justice by *'Swat Team'* was comfortable to me.

The Thiokol and NASA Cases were not apolitical. After all NASA had its own political friends, enemies, and budget to procure and protect. Thiokol is a large influential corporation in Utah and the United States. Morton Thiokol Inc., as in 'Morton Salt' is an old Chicago company representing 'old money'. Code Name 'Brains' asked himself if the DOJ could handle the volume of problems, crimes, and forces involved? Was this case bigger than the FBI? Bigger than the Dept. of Justice? Too big to prosecute since it directly or indirectly affected millions of persons and all of the nations of the free world. It impacted subsequent world history.

One thing for sure, no problems which are hidden will ever be properly corrected. First bring the problems forward, then solve them. I, Code Name "Brains", had exposed the problems, now the FBI and others had to fix the problems. Could they? Would they? How much danger was I in?

CHAPTER 32

ROGER BOISJOLY - QUI TAM LAW SUIT

TIME: Filed 13 Sept. 1988

Early on the FBI had asked/told me they had been requested by the Department of Justice to look into Roger Boisjoly's Qui Tam Law Suit claims. Did I know Roger? What did I know about him? Mr. Roger Boisjoly was suing MTI for billions of dollars and negligent homicide under a Federal Qui Tam legal action. What Qui Tam means is that Mr. Roger Boisjoly was suing MTI on behalf of the United States Government under this special set of legal rules called the "False Claim Act". The rules of Qui Tam were not new and had been successfully tested in Federal courts, however they are unique in their own right. Once a person files a Qui Tam action all other persons associated with the same litigations are barred from filing additional Qui Tam actions. If it was not for this "one only" rule I would have filed a Qui Tam myself immediately. Under a Qui Tam ruling the U.S. Government could award Roger from 15% up to 30% of the total recovered amount. Mr. Roger Boisjoly was claiming that billions of dollars of funds had been used erroneously by MTI.

In my opinion the FBI did not feel Roger Boisjoly sued MTI for personal reasons, animosity, or for his own personal gain. For his reasons (many of which I never knew) it appeared to me that Roger Boisjoly had sued MTI *'For The Right Reasons'* - motherhood, apple pie, American exploration, and National Security. Accumulatively, proper reasons. Mr. David Long, an attorney from the Department of Justice (Wash., D.C.) was assigned Roger Boisjoly's case by the D.O.J. and the cases against MTI and NASA as I understood it.

The FBI had been called into Roger Boisjoly's case by the D.O.J. to investigate and determine if the U.S. Government wanted to or should join in Roger's Qui Tam case. Very soon Special Agent Tim Healy would be going out to see Mr. Roger Boisjoly. As you know the FBI is a branch of the DOJ and one FBI Agent had told me that Ed Meese (then Attorney General) knew me by name as a result of FBI investigations of Thiokol and NASA. That was not good in my opinion since he was about to be ousted from office for his many crimes well exposed in many news articles. He was the 'Attorney General of Ill Repute'!

Mr. Roger Boisjoly was a highly qualified and respected Mechanical Engineer. His speciality was mechanical design and structural analysis. He had been at MTI for six years. Roger had held the following positions: Senior Lead Engineer, Project Engineering Manager, and Senior Scientist on staff to MTI's Manager of Applied Mechanics. Roger was a key player in the critical prelaunch meetings between MTI management and NASA management to 'Certify the Challenger as ready to launch'. The following are letters from the President's Commission on the Challenger Accident report relating conversations between MTI management including Roger Boisjoly and NASA. One of Roger Boisjoly's weekly activity reports follows:

ACTIVITY REPORT

The team generally has been experiencing trouble from the business as usual attitude from supporting organizations. Part of this is due to lack of understanding of how important this task team activity is and the rest is due to pure operating procedure inertia which prevents timely results to a specific request.

The team met with Joe Kilminster on 10/3/85 to discuss this problem. He wanted specific examples which he was given and he simply concluded that it was every team members responsibility to flag problems that occurred to organizational supervision and work to remove the road block by getting the required support to solve the problem. The problem was further explained to require almost full time nursing of each task to insure it is taken to completion by a support group. Joe simply agreed and said we should then nurse every task we have.

He plain doesn't understand that there are not enough people to do that kind of nursing of each task, but he doesn't seem to mind directing that the task never-the-less gets done. For example, the team just found out that when we submit a request to purchase an item, that it goes through approximately 6 to 8 people <u>before</u> a purchase order is written and the item actually ordered.

The vendors we are working with on seals and spacer rings have responded to our requests in a timely manner yet we (MTI) cannot get a purchase order to them in a timely manner. Our lab has been waiting for a function generator since 9-25-85. The paperwork authorizing the purchase was finished by engineering on 9-24-85 and placed into the system. We have yet to receive the requested item. This type of example is typical and results in lost resources that had been planned to do test work for us in a timely manner.

I for one resent working at full capacity all week long and then being required to support activity on the weekend that could have been accomplished during the week. I might add that even NASA perceives that the team is being blocked in its engineering efforts to accomplish its tasks. NASA is sending an engineering representative to stay with us starting Oct 14th. We feel that this is the direct result of their feeling that we (MTI) are not responding quickly enough on the seal problem.

I should add that several of the team members requested that we be given a specific manufacturing engineer, quality engineer, safety engineer and 4 to 6 technicians to allow us to do our tests on a non-interference basis with the rest of the system. This request was deemed not necessary when Joe decided that the nursing of the task approach was directed.

Finally, the basic problem boils down to the fact that ALL MTI problems have #1 priority and that upper management apparently feels that the SRM program is ours for sure and the customer be damned.

Roger M Boisjoly

Roger Boisjoly 10/4/85

In this activity report, Roger Boisjoly expresses his frustration with the slow progress of and lack of management attention to the seal task force.

MORTON THIOKOL INC COMPANY PRIVATE

Wasatch Division

Interoffice Memo

31 July 1985
2870:FY86:073

TO: R. K. Lund
Vice President, Engineering

CC: B. C. Brinton, A. J. McDonald, L. H. Sayer, J. R. Kapp

FROM: R. M. Boisjoly
Applied Mechanics - Ext. 3525

SUBJECT: SRM O-Ring Erosion/Potential Failure Criticality

This letter is written to insure that management is fully aware of the seriousness of the current O-Ring erosion problem in the SRM joints from an engineering standpoint.

The mistakenly accepted position on the joint problem was to fly without fear of failure and to run a series of design evaluations which would ultimately lead to a solution or at least a significant reduction of the erosion problem. This position is now drastically changed as a result of the SRM 16A nozzle joint erosion which eroded a secondary O-Ring with the primary O-Ring never sealing.

If the same scenario should occur in a field joint (and it could), then it is a jump ball as to the success or failure of the joint because the secondary O-Ring cannot respond to the clevis opening rate and may not be capable of pressurization. The result would be a catastrophe of the highest order - loss of human life.

An unofficial team (a memo defining the team and its purpose was never published) with leader was formed on 19 July 1985 and was tasked with solving the problem for both the short and long term. This unofficial team is essentially nonexistent at this time. In my opinion, the team must be officially given the responsibility and the authority to execute the work that needs to be done on a non-interference basis (full time assignment until completed).

As you can readily see from these MTI and NASA documents, MR. Boisjoly was in the heat of life and death argument and he lost. Roger Boisjoly, a highly qualified and experienced engineer, was overruled by a man with a degree in education, not engineering. Basically top mechanical Engineer vs. auto-body shop repair school teacher! An auto body shop school teacher made the final decision, seven people died, the Space Shuttle Challenger at 4 1/2 Billion Dollars (then) was lost, important mission equipment in the order of 300 million was lost, and the U.S. was put out of space operations for years into the future. Awesome! [12]

Later Roger Boisjoly would write a paper for the American Society of Mechanical Engineers titled "Ethical Decisions - *Morton Thiokol And The Space Shuttle Challenger Disaster.*

Roger Boisjoly's article abstract is as follows:

ABSTRACT:

A background summary of important events leading to the Challenger Disaster will be presented starting with January, 1985, plus the specifics of the telecom meeting held the night prior to the launch at which the attempt was made to stop the launch by the Morton Thiokol Engineers. A detailed account will show why the off line telecom caucus by Morton Thiokol Management constituted the unethical decision making forum which ultimately produced the management decision to launch Challenger with out any restrictions.

The paper will continue with the post-disaster chronology of my working relationship with Morton Thiokol Management and conclude with a discussion on accountability, professional responsibility and ethical conduct which should be practiced in the work place, plus statements from the academic community about the plight of whistleblowers and my closing remarks.

See Appendix 12a and 12b and 12c for articles:
Engineer Sues Thiokol for 'Criminal Homicides'.
By Nancy Lewis, Washington Post Staff Writer
Engineer Who Opposed Launching
By Phillip. M. Bosfey, NY Times, 1/29/1987

Roger's article goes into succinct details of the pre-Challenger events which ultimately ended in the enormous catastrophe. A the end of the article are the two following sections:

ACCOUNTABILITY:

I have filed a billion dollar law suit against MTI and a ten million dollar law suit against NASA plus a False Claims Suit against MTI on behalf of the U.S. Government. First, it is my intention to secure compensatory damages for my lost salary and ruined career and second, I hope the suits send a serious and significant message to MTI in particular and to executives of other companies and government agencies that they cannot make arbitrary irresponsible decisions that kill people and ruin the lives and career of their employees without accountability. In my opinion, there has been zero accountability to date.

Everyone involved with the decision to launch Challenger at MTI and MSFC has been either transferred or taken normal or early retirement without any penalty for their actions. Further MTI has not paid the $10 million penalty in their contract for their hardware being responsible of the disaster and loss of seven lives and the destruction of over $2 billion in hardware. An arrangement between MTI and NASA has been reported in the media that MTI will take $10 billion less in profits when the production contract resumes. I would like to raise several very basic questions about these conditions.

First, who receives the interest / investment benefit (millions of dollars) by not paying the contract penalty? Second, why did the cost of the boosters increase above the original contract price when it was common internal company knowledge that MTI was in the process of negotiating a lower price prior to the Challenger disaster to protect their single source contract status Now, my colleagues, that is what I mean by zero accountability.

CLOSING REMARKS:

All of you must now evaluate your careers and emerge with the knowledge and conviction that you have a professional and moral responsibility to your selves and to your fellow man to defend the truth and expose any questionable practices that will lead to an unsafe product. Don't just sit passively in meetings when you know in your heart that you can make a constructive contribution and also be prepared to share your ideas with others and to compliment others for their ideas, especially when their idea is better and may even replace yours. After all, you have a responsibility to promote the best product for a company and put personal pride aside, this is the best way to cultivate colleague respect and friendship which in industry always results in a positive long-term benefit for you, the company and its prod*uct line.*

I wish that the Challenger disaster had never happened. Since I cannot turn the clock back, I hope that if anything good can result from this tragedy, then I desire that all academic institutions and professional societies will recognize the importance of teaching ethical behaviour in decision-making situations by using case histories like this one to demonstrate what was wrong so everyone is aware and prepared for want to expect when confronting a similar situation requiring an ethical decision.

I have been asked by some if I would testify again if I knew in advance of the potential consequences to me, my family and my career. My answer is always an immediate yes. I couldn't live with any self-respect or expect any from others if I tailored my actions based upon potential personal consequences resulting from my honourable actions. As a result of this paper and other exposures to real case histories, I hope that your answer will also be yes.

I hope and expect a drastic improvement in ethical decision-making practices and employee treatment for promoting ethical conduct as a result of my law suits and eliminate or at least significantly reduce unethical decision-making practices within our industrial and government communities.

I will never forget and I hope this nation will never forget, especially the engineering community, the supreme sacrifice that the seven Challenger astronauts paid by forfeiting their lives for such an irresponsible launch decision. May we always remember Astronauts Jarvis, McAuliffe, McNairy, Onizuka, Resink, Scobee, and Smith for their courage and dedication to this nation's space program.

As you can see from Roger Boisjoly's article he was a sincere and concerned man. He had done what he did *For The Right Reasons*, he was a substantial drip on earth's moral drip stone. In the end I feel Roger Boisjoly was "done wrong" by our American Legal system and the DOJ. In my well informed opinion the DOJ should have joined into his Qui Tam law suit and tried to recover some of American tax- payer's hard-earned , ill-spent money. When Roger's cases were thrown from Washington D.C. to Utah (for whatever reason) the cases then fell into a highly Mormon-controlled political court arena. In the Utah Federal Courts there was every expectation of MTI pressure, Mormon pressure, Gubernatorial pressure (Mormon), Senatorial pressure (Senator Gars and Orin Hatch - Mormons.) and who knows what else. Utah was quite simply the last state Roger's cases could find a fair trial. Roger Boisjoly was also a Mormon - but he stood his ground. In the end a Utah Judge chopped up Roger's cases like French fried potatoes and threw them out of court. It was a sad day for our Constitution and laws, America's Space Programs, American and Free World National Defence, and hard working American taxpayers. The U. S. Dept. of Justice was a failure!

Parallel to my activities at that time, around the 8th of April 1987 I began to read and study MTI documents I had procured but had not necessarily studied while working at MTI. I should note here that all of the documents I procured for the FBI were not necessarily important to the FBI. Some could have been obtained from NASA if NASA had cooperated. Some of the documents couldn't have come from other sources. On the 8th I worked 8 hours. That afternoon, Tim took off to interview another MTI source, while I went to take Kerry Sanofsky to dinner at Maddox Restaurant in Brigham City as agreed to after I was fired and before I left MTI. I had set up this meeting with Kerry just as I was pushed out the MTI door. Kerry and I were quite civil. I had liked Kerry very much up to the point where he had been 'turned.' Tim and I had anticipated me wearing recording equipment and trying to find out why I was fired and whatever else we could. We were hoping to get more information on Don Thompson, Russ Rasmussen, John Ewell, Ed Dorsey, Al McKinsey, John Pontius, and a number of others. You never knew what might fall out on the table over dinner!

Kerry and I had always been amicable and I made every effort to keep things friendly. Kerry needed assurances from me that I would not hurt him politically. I assured Kerry I meant him no harm. I wanted answers. Kerry's current attitudes incited outrage in me, but I successfully kept a lid on my feelings. Such a fine young man basically - why would he join a Thiokol conspiracy? Financial forces, religious forces, political forces, ego - all of the above? Clearly he had joined the conspiracy either actively or passively. What matter if active or passive? Kerry now had to demonstrate his disdain for Kent McKinnon (his mentor) to MTI and he did. Kerry had taken up the MTI sword and he carried it high for Don Thompson and others. What really drove him? On April 8th, Tim and I decided we would not record anything from Sanofsky. That particular evening Kerry probably didn't know very much and he was now enthralled with being "manager of SRM System Safety at MTI" at such a young age and unqualified for the job. Also Kerry was right in line at Thiokol and was "under control" by Thiokol management.

Kerry was prompt. We politely visited in Maddox's pleasant, large waiting lounge. As we went to sit down and eat, I cautiously took off my coat so Kerry would not be suspicious that I had recording equipment in my coat. He will probably never know whether I really did or did not record the conversation for certain? After all, some of the FBI recording equipment can be "dangled" down your pants. Some could be mounted under the table, and there were yet other alternatives.

We sat down and had a pleasant dinner. Kerry was clearly cautious and cagey and I politely tried to extract from him the reasons I was officially fired, but he resisted meaningful responses. I chuckled over a probable $30,000 a year (at that time) liberal arts person having the audacity to fire a $300,000.00 a year consultant virtually approved by the U.S. Congress. However such is the reality of Utah - Mormon politics, low end aerospace politics, and fear in executives trying to stay out of prison.

Over dinner I asked Kerry what had happened to my 245 Hazard Analysis? Had Erma or Wendy found the other 20+ they had supposedly lost? Kerry said nothing had happened to them yet and they had been saved. I told him it was a shame their author (me) wouldn't be there to defend and explain them. Why did they put me out the door Kerry? Kerry said he couldn't discuss it. During the course of our dinner I mentioned to Kerry that I didn't think it was right to be put out the door when my contract and the Solid Rocket Motor redesign System Safety tasks were less than one half complete. I identified a lot of ways the Space Shuttle and crew could be wasted to Kerry. I told Kerry I had come to Thiokol from Seattle to work until the end of the System Safety project. I told him I was frustrated by not being allowed to complete my work. Kerry assured me the work would be diligently continued. I laughed to myself and then told him I was very happy with my new contract at the Department of Defence. I didn't tell him how pleased I was to be out of MTI.

As I left MTI'S System Safety, scientists, engineers, and others were going crazy over their work being deliberately lost, scattered, and ignored. Pushers and movers were being controlled. Losers and conformers were being rewarded. A 'losers and conformers' reward was to keep your nice, undemanding job and vegetate at your desk - pay day once a week and talk about elk hunting anytime.

Kerry's heart and soul were now owned by others. In many ways he acted much the same as Jerry Fonnesbeck had been controlled. Since Kent McKinnon's resignation had been tendered Kerry was not the same person. Kerry weighed and measured every word. He was obviously under stress. Who knows, maybe he wore recording equipment? I sensed Kerry had been prepared and prompted for our meeting. I also figured there were MTI personnel observing our meeting. The meeting with Kerry, as Tim had wisely predicted, was relatively innocuous and uninformative. I was wasting my time. Why did Kerry even come to our dinner meeting? Was he seeking information for Thiokol? Could it be there really was no other reason for him to show up others than a free dinner? The food was good there as always. Did he come for Don Thompson or others? I was wasting my time with Kerry and he was the smallest fish anyway. Kerry and I parted in the lounge on a friendly word and handshake. I wished him no harm. Also I no longer respected him. I now computed him as a 'Thiokol Criminal', and a 'traitor to the United States of America'. Tim and I were after large sharks. Kerry was a misguided minnow.

Later on Thiokol actively *Black Balled* me from at least one consulting contract on the premise that I had a fight with Kerry. As I look back now I find it hard to fathom MTI saying anything like that. Kerry came to dinner with me, and we parted on a friendly basis in the huge waiting room of Maddox Restaurant in front of over thirty people. Never in my life did I ever have a harsh or cross word with Kerry Sanofsky or any other Morton Thiokol employee. More of their games and B.S..

On 10 April, Tim had a lot going on with some of his other cases - cops and robbers stuff - and I only spent three hours reading documents from Thiokol that day. Tim's cops and robbers cases were exciting and pleasing to him - instant satisfaction. An outlet for long term frustration. They satisfied his adrenalin demands.

Tim was so frustrated over Thiokol and NASA he needed to do a little head banging. Tim's cops and robbers, peddlers and hookers, and other delays frustrated me to my limit. I didn't get to knock heads, shoot at the range, arrest worthless bastards. While Tim played out his life's fantasies of 'cops and robbers' for real, I read thousands more pages of incredible documents. I sometimes felt left out and was perpetually pissed at our MTI Case. The weekend of the 11th, 12th, and 13th was spent hiking around Brigham City and taking care of personal business. I was under enormous emotional stress and wanted to feel better and have my life in better balance.

The only thing right in my life at this time was my new marriage, but I was so stressed I wasn't enjoying marriage perfectly but my Rhodesian wife backed me well at that time. She knew I was bashed in and she married 'the package'.

On Tuesday, 14 April, I started writing commentary on the state of knowledge of the critical players at Thiokol. These comments were difficult for me because again I really didn't know exactly what Tim and the FBI were looking for. I had an incredible mass of information in my mind, on paper, and tape recorded. I was not a trained legal type or FBI type so I was often confused as to what the FBI needed and wanted. As I drafted write-ups on various things I expanded them and as I found out later these were often not quite what Tim wanted or needed. There was so much crime I was overwhelmed. Tim wanted the "good stuff", he would know "it", but he and I had not yet formulated what "it" was and what form "it" should be in. All of this I considered to be the natural course of events but my mind was poisoned by "what in the hell is it"? Both then and now I felt there should have been a DOJ attorney with me from day one. The facts dictate what law applies. The facts would have been easier to point out if anyone had furnished me the elements of each crime to be proven. On 15 April, I worked 8 and 1/2 hours on write-ups on crucial players and was finding more and more historical and current points to ponder. April 16 I worked 8 hours and completed write-ups on a bunch of the basic safety and reliability personnel at Thiokol.

You may ask who in hell I was to write up information on these people? A reasonable question. Perhaps I was the only one willing to risk it all for America on this project. I was so compelled by the clear dangers to America I would have done anything. All I could see was, 'America the Lost!' A nation collapsed by corporate criminals and NASA drones. How our enemies must laugh at our sad lack of national diligence. Our space efforts had become deeper in disgrace.

April 17th, I worked another 8 hours examining a few contract documents which had arrived from other FBI and NASA sources. I began listing components of the NASA/MTI contract. NASA was not wrong when they told FBI that the contract was lengthy. Thousands and thousands of pages. NASA was not telling the FBI the truth about the contract vernacular - it was generally in plain simple English. In fact it was too plain loose and generally non-specific indeed. In the contract, millions of dollars were given to MTI with no apparent explanation other than a statement of the hardware item ordered. The 18th and 19th I took off and enjoyed the weekend. I had a creeping problem which was submerged in my consciousness. Due to stress and overwork my health was declining rapidly. I knew it, but was unable to overcome it. Something was very wrong but I couldn't quite put my finger on it.

On April 27th I worked 9 hours at the Ogden FBI office. I realized that talking to people about the various aspects of Thiokol and NASA was so complex that I needed a time line chart for "show and tell" of how all of the facts fit together. So at lunch time, I stopped at the local stationery and engineering store across from Nordstrom's in Ogden and bought graph paper and other necessary materials. I took them upstairs to my office in the Federal building on the 4th floor and made a chart about 17 feet long and 3 feet wide. This chart covered 13 years of the contract and I left enough room to add two more years, assuming the FBI would take a while to finish the case up and NASA would subsequently regroup. The chart was similar to the following chart. This chart is a clear indictment that NASA was in no way sensitive to the O-Ring failures over the years. How could NASA know this and put astronauts or valuable defence cargos aboard the Space Shuttle? Originally the Space Shuttle 'Atlantis' was built to be the Dept. of Defence vehicle for 'Classified' pay loads. Over time the 'Classified' loads were spread to some of the other shuttles. [13]

At the same time I started cataloguing the 735 contract documents we had received and I briefed Tim on the status of my efforts. I also added to my list of allegations, and did some more Hazard Analysis related to others' HA's. I phoned several people at MTI and had seemingly innocuous discussions for the sake of probable MTI telephone "taps". we used various codes between us and I ascertained further Thiokol information via some of these phone calls. On the 27th I finally felt I had my teeth into what the FBI could really use from me and I was on the move for them. The FBI never pressured me in any way but I felt internal pressure from myself. The Utah evil had spread too far and had to be remedied.

In non-technocratic language America being out of the space business was analogous to America without the Stars and Stripes.

Appendix 13: Time Line on MTI and NASA contract.

On the 24th I briefed Tim on the charts I had completed and began work on 900 plus CO's (Change Orders) and SA's (Supplemental Amendments) which were part of it. I read and I read! [7]

On the 29th NASA 'unwillingly' supplied a partial MTI / NASA contract which arrived in the Salt Lake City FBI offices. The next day Tim brought the contract to our Ogden office. I did a gross examination of the NASA / MTI Contract for the first time and began an inventory of it. I reviewed the accusations chart I prepared with Tim and we began to zero in on special aspects of clearly visible and some nearly invisible crimes.

Tim was under more pressures of his own now. His family life was now consumed by the NASA and MTI Cases. This brilliant and capable man was being kicked around. I knew FBI #1, #2, and #3 were pressuring him to do more, better, faster, ... They were great people and wanted it "all". Tim was trying to give it all. He was blocked from his MTI/NASA cases by the pressure of less important work. In the FBI scheme of things they were all of equal importance.

On 30th April, I was organizing the NASA supplied contracts by different numbers and types and inventorying the remainder of the contract documents. The ultimate clerical job. I prepared a missing items list for Tim on what I found to be missing from the contract obtained by FBI sources and furnished from NASA. There were differences. Between the partial contracts we had approximately 85% of the total contract at that time. Spook work isn't all "Sex and Rock and Roll"! Somewhere someone has to do the work. The work part is difficult, tedious, and frequently boring, however the scrap you overlook or miss is probably the 'missing link'.

Early in my relationship with Tim Healy, while I was still at MTI, he had asked me to write up a list of my recommendations to FBI for their own use and so the FBI could eventually pass them on to NASA and other U.S. Government Agencies. Soon after that, NASA and FBI relationships deteriorated and the FBI was very wary of what they could or would give to NASA. Possibly a kick in the head. So my comprehensive recommendations were temporarily suspended. My recommendations would probably only aid NASA to cover its tracks. It is not a case that the FBI did not want to pass recommendations on to NASA but it seemed to me that the FBI was probably going to have to prosecute NASA in the end and therefore the FBI shouldn't give NASA any additional heads-up warning. Tim wanted my recommendation document seemingly with urgency when it was originally requested and I gave up my last few hours of sleep in that early period of the investigation to prepare

---

*** See Appendix 7: "Solid Rocket Motor (SRM) Chronology"**

my recommendations. I was so exhausted and overloaded I couldn't think clearly. Then months later the final recommendations were ready to go to the typist again. I had completed it months before and Tim had the original on file. It would ultimately be typed and reviewed two or three times and then put with permanent FBI records. My recommendations and letter to the FBI concerning my recommendations were in my April 17, 1987 letter to Tim Healy.

The total number of 302's on the MTI and NASA cases and their contents were FBI secrets. All of this information would eventually come out in pending FBI cases unless censored for ever in order to protect the high level government 'guilty'. Tim somehow included my personal recommendations to the FBI with some of his documents. He did it in such a way as to protect me. Tim explained that my recommendations on specific SRM parts was fine, but I wasn't authorized to comment on what should be done with NASA and such huge sweeping statements. I thought *"Bullshit"!* I was enormously well informed. I was apolitical and I was honest. Any way you sliced it my opinions were worth a thousand D.O.J. opinions or Congressmen's uninformed opinions.

My recommendations to the FBI were as follow:

Brigham City, Utah
April 17 1987

Mr. Tim J. Healy Special Agent, F.B.I.
Room 3203 - FOB
125 South State Street
Salt Lake, Utah 84138

Dear Sirs,

Having recently prepared and completed two hundred and forty-five HAZARD ANALYSIS (HA's) on MORTON THIOKOL, INC.'s (MTI) Solid Rocket Motors (SRM's) for the Space Shuttle status followed by my personal recommendations for both the Space Shuttle Program and the SRM Program. The following synopsis and opinions are my own and are derived from my personal experience with the SRM Systems and my interaction with associate scientists, engineers and managerial experts from varied and multiple disciplines who also wrote hundreds of unique and separate HA's on MTI's SRM systems.

SYNOPTIC SRM STATUS:

Each Space Shuttle requires two solid rocket motors. By NASA original contracts requirements the SRM components are to have a five year shelf life and be capable of nineteen refurbishments (reuses) for most non-expendable components. The basic design of the SRM's is fourteen years old (circa 1973). Over the elapsed fourteen years scientific engineering knowledge has progressed dramatically. Simultaneously here have been twenty-four successful launches (48 SRM's) and one catastrophic launch (one failed SRM). The SRM failure rate to date is therefore four percent failure. The SRM system is composed of the case (skin) subsystem, nozzle subsystem, ignition subsystem, lightening and electrical grounding subsystem, instrumentation subsystem, interface subsystems, ground support equipment systems, SRM to external systems interfaces, operational and support systems, orbited and ET (external tank) interface systems, propellant system, recovery system, and component refurbishment processes.

ELECTRICAL, GROUNDING SYSTEM	Grossly Inadequate
LIGHTENING PROTECTION SYSTEM	Inadequate and therefore dangerous
INSTRUMENTATION SYSTEM	Too few instruments during both test and flights. Overcrowded wiring. Improper grounding and shielding of wiring harnesses
INTERFACES	All of them are structurally questionable. Most serious concern is the nozzle to case joint and the struts attaching the SRM's to the external tank.
GROUND SUPPORT SYSTEMS	Electrically ungrounded, much new design and testing to do. Procedures are in need of refinement. Even after 25 flights the ground support equipment are not refined.

SRM TO EXTERNAL SYSTEMS INTERFACES	Sparse interface technical exchange between NASA and Aerospace Contractors. Poor documentation, questionable reliability and understanding of interface connections, construction interacting forces and interface requirements, Uncertain communication routes.
OPERATION AND SUPPORT	Personnel and physical resources are currently at low ebb in training and enthusiasm due to the Challenger Disaster.

HAZARD ANALYSIS by MTI's System Safety Group to date have produced approximately 1,000 completed Hazard Analyses. The total number of HA's generated is anticipated to be 5,000 and could escalate to 8,000 identified hazards if the system's safety program is properly completed. The identified hazards to date derive the following synopsis -

HAZARD ANALYSIS:	Nearly all HA's to date are rated as catastrophic in nature.
CASES:	Never were fabricated to contract specifications. Very limited reusability (safety problem), composed of improper steel for use in sea water. Multiple joint design faults. Dome too weak to meet specifications. Seventy-four gas leak paths through dome / igniters system.
NOZZLE SYSTEM:	Experimental, limited testing due to expense, questionable reliability. Qual. testing with non- fidelity nozzles. Uncertain chemical curing process. Questionable raw materials. Numerous other problems.

IGNITION SYSTEM:	SRM ignition has been reliable. Very serious gas path leaks in violation of specifications. Initiator and igniters are structurally questionable when subjected to ignition pressure impulse. Inadequately electrically grounded ignition components (extremely dangerous).
ORBITER AND EXTERNAL	Limited and inadequate interface data TANK LOCK SYSTEM exchange, poor documentation, questionable reliability and understanding of the forces. Questionable metallurgical practices.
PROPELLANT:	State of the Art with attendant problems. 1987 electrical properties characterization studies of the SRM propellant proved that the propellant is much easier to ignite and is more dangerous than was previously known.
RECOVERY:	Ocean impact and recovery has been and can be more damaging than anticipated. Residual damage can not always be repaired by refurbishment.
REFURBISH:	MTI shop - excellent. Support Eng. / Safety / QA, etc. are questionable and variable. SRM design / functional faults found during refurbishment are not always reported to other MTI departments or NASA groups and therefore go unresolved or are improperly resolved.

RECYCLABILITY OF SRM COMPONENTS	Nowhere near specification of nineteen: refurbishing. In reality only "one shot" is a realistic target. No refurbishments unless better materials, design and processes are used.
DESIGN SAFETY FACTORS:	"Under" specification / requirements in most cases.

In my opinion the design of the SRM is seriously inadequate to meet the specifications or mission life cycles. In my opinion, the aerospace contractors and NASA have inadequate interface technical interchange activities. Inadequate interchange is continually diminishing system integrity and reliability. In my opinion, the MTI SRM's are currently in dangerous condition and will continue to be so unless extraordinary efforts are made to improve nearly everything about them. It is a virtual miracle that no SRM's blew up prior to the 25th flight.

Based on the above opinions and facts, I recommend the following actions:

1. Transfer STS operations from NASA to the USAF. NASA is not an "operations" oriented organization. NASA was and is best at research, development, and planning operations. To the best of my knowledge, the U.S.N. and the U.S.A.F. have been successful at directing MTI contracts to successful completion.
2. "Project Management" and "Project Integration" should be taken away from NASA and be assigned to an independent contractor, not controlled by NASA. In my opinion, NASA has failed miserably at project management, project integration, contract supervision and quality verification.
3. Assign tough, new personnel as resident NASA and DCAS Officers at aerospace contractor facilities. NASA residents are far too friendly with MTI management and personnel.

4. Resume launching STS's in next 100 days, launch next STS without a crew (this is possible and practical), and with a minimum payload weight during warm weather. Follow on with missions using only military personnel and light loads until all of the redesign / new design is incorporated. Accept the risk of SRM failure.

5. Develop strict guidelines on substituting 'modelling', 'similarity' and 'ATE' activities for testing. NASA, the USAF and other U.S. Government Agencies are frequently incapable of evaluating results or controlling costs. These activities are virtually a 'license to steal' by Aerospace Contractors. Rather than saving money they usually radically increase testing costs and lower reliability.

6. Do not allow NASA to call for unreasonable, beyond state of art, and inappropriate requirements and specifications in their contracts. Example: Nineteen Refurbishments on the SRM Cases is ludicrous. Reality is that very few if any refurbishments can be safely executed.

7. Demand adequate contracts be used and enforced at all times. Example: Morton Thiokol Inc. is on a questionable contract to NASA.

8. Demand adequate prosecution of aerospace contract abusers.

9. The USA is wasting enormous resources, enormous amounts of moneyand energy by not having a succinct national policy and 'National Plan' for America's exodus into space. A plan / program not sensitive to political views - a continuing reasonable national dedication.

10. A new 'can do' attitude in the USA Space Program is in order. Presently the worst contractors are always awarded the greatest rewards. Aerospace contractors who perform well are penalized for high profits. ASCONS who fail are paid to do the work over and over again. The entire industry therefore plans and thrives on failure.

11. All NASA and military projects should have an external safety, reliability and quality Assurance organization one hundred percent separate from both NASA and the Aerospace Contractors. In effect, an honest referee.

12. Develop a second source to MTI to build SRM's.

13. Drastically reduce the "paper system" to spend the money on hardware and design rather than vast amounts of superfluous paper work.

14. Restrict contractors to doing what they do best. Example: MTI is good at casting solid rocket propellant, MTI is poor at rocket hardware design and testing.

15. Prosecute Aerospace Contractors for TREASON who knowingly endanger the U.S.A..

16. NASA Engineers are very poorly paid, therefore, they are not a collection of the best talent available. The result is poorly conceived projects, poorly supervised projects, poorly executed projects, all of which is a National Security and a National Defence hazard.

I hope these recommendations assist you. We have placed the U.S.A.'s emancipation into space, its national defence systems, its international communications systems, and in fact, our National Security in far less than the best care with NASA and Morton Thiokol, Inc.. I am extremely concerned. If I can be of any assistance, feel free to contact me.

RESPECTFULLY YOURS,
(signed)
STEVEN W. AGEE
XXXXXXXXXXX
Brigham City, Utah 84302 "

As you can see from the above letter I felt free and confident to set the FBI on the track. I have often wondered if this letter did any good. It's entrails are included in one or another of my several 302's. I still feel these recommendations were valid and should have been instituted.

Next Friday , the 5th I spent 10 hours reading Contract Change Orders and SA's. I extracted a number of contract documents and shuffled them together with other FBI acquired contract documents finally accumulating a nearly complete NASA / MTI contract. There was a list of important documents missing from the composite. We worked with what we had in hand at the time. There were some gaping holes in the documents. Since the FBI did not acquire the contract in the usual 'up front' manner there was no telling where the odd bits were? I never did find out how the FBI acquired most of the NASA / MTI contract.

I believe it was by skulduggery of some sort since it was an established fact that NASA was not cooperating with the FBI yet. Also about this time I had a number of documents and contracts components to look at which had been prepared by David Long, the DOJ Attorney who was also looking into Boisjoly's matters and claims. We had some good discussions. David Long was a pleasant and intelligent man. A cool calculating legal mind. He was definitely of the 'right stuff'. He was contributory to Tim's efforts. Also David Long thought MTI should be charged with both criminal and civil charges. David was considering what to do about NASA's failings and crimes.

That pm, Tim and I had meetings with two MTI persons who were now FBI Sources. These sources were from different MTI departments. This was a hectic meeting since Tim had a whole bunch of drug related cases simultaneously and he was literally running his buns off - 'Cops and Robbers' stuff! One drug pusher was going to kill some other drug pusher! I was thinking good riddance! Why should such a mundane crime obstruct the U.S. Space Programs and America's future? Fuck the bastards!!

In no way do I demean the importance of suppressing drug peddlers and drug related crimes in the U.S., however, I did not at that time understand how something as important as what we were working on could be delayed over a few drug addicts or pushers being murdered. I suppose it all depends on your point of view; it was probably important to the pushers? I envisioned the U.S. being out of space activities and so did Tim. However, Tim's performance appraisals required him to be a good FBI Agent and that required a lot of his time spent rounding up drug addicts, drug peddlers, bank robbers, kidnappers, etc., etc.. Justice for all! After experiencing this dichotomy of trivial to astounding I criticize the FBI for demanding so much diversity of their agents. Most FBI Agents aren't up to the demands of mixing inconsequential and planetary issues. Tim was up to all of it, but it was in fact killing him. In the long run Tim worked 100 hrs. a week for his entire career.

Tim anticipated he would be late for our pm meeting with these two sources and I should go meet these excellent sources on schedule. So, I went over to the Hilton Hotel in Ogden where we were to meet and intercepted them in the hotel parking lot. They were very nervous, cagey, and glad to see me. I got in their car and we spluttered around the hotel area a little bit waiting on Tim who turned up late and 'crazed' almost an hour later.

When Tim showed up, we talked from one car window to the other and he said to follow him. He was going to clear the area and we would go to the Prairie Schooner Restaurant in Ogden. I rode with Tim and he wanted to know how our sources were doing. I told him they were incredibly nervous and cagey. They really appreciated Tim's efforts to protect them. They were afraid of MTI and the 'Mormon Mafia' even more than I. We literally drove about 10 miles up and down Ogden streets, alleys and parking lots until Tim felt there was no one following these sources and these persons fully understood Tim was protecting their interests. These two paranoid sources were grateful for Tim's consideration. Looking back now it is clear that Tim was truly concerned with them being followed. By now I had become so accustomed to these activities and really didn't treat the spook practices with adequate respect. I was tired and crazed and down deep I assumed Thiokol was in my back pocket watching every move I made. Familiarity and exhaustion breeds contempt. Looking back, now I can see that I was so physically and emotionally wasted that I couldn't keep my guard up. I was approaching the end of my abilities to cope and the end of my usefulness.

When we got to the restaurant, Tim had to make some phone calls. He was like a cat on a hot tin roof! Seems Tim was standing between several murders. My God, let them waste each other and then let's get on with our business! I might even go so far as to say I'm prejudiced.

Our source meeting was held and information exchanged, courtesies extended, 'thank you's given. These particular two sources had a ton of information to share. They were both top professionals. Their tales of suppression were staggering, their tales of Thiokol engineering incompetence were frightening. They told an interesting tale of "Sanofsky's man". They reported that the long arm of the law had arched out in Brigham City and nailed Don Janacek on a Driving Under the Influence (DUI). Didn't surprise me since he often smelled to me and others as though he had been into booze. Evidentially due to previous DUI"s Janacek was headed to jail in a few days, so rather than attend jail in Brigham city he quietly hooked up his '5th wheel' trailer and returned to Texas. He was no longer Kerry's man, he was gone! Numerous persons supplied by ARC had been victims of alcohol issues. Mahalon Hollway, and several others. These two sober sources were pissed at both Thiokol and NASA for suppression of their work and keeping them from important meetings and matters. Our meeting completed about 11:30 pm and Tim was driving me back to where I was parked for our meeting. Radio information came on indicating that Tim's source was about to get murdered! I was pleased! As we whizzed into the parking lot where my car was, I jumped out of the FBI car and Tim zoomed off into the night. I had wanted to talk more with Tim that night but a pending murder simply prevented it.

I had serious indigestion with an 'almost source' cooking and wanted Tim to deal with it. Time was up and this problem would pass to its eventual culmination. This is the last I would see of him or the FBI for 8 days since I was leaving the next morning to get married on the island of Kauai in Hawaii. Not only did I need a break from the pressures I had been under, but Tim also needed to get a bunch of FBI work on all of his cases under control. I sensed Tim's rubber band was stretched to the limit - too much 'cops and robbers'! Too much MTI and NASA! Not enough home and family. Also, Tim could get snappy if he didn't get to his American Doe Karate training in on schedule. Tim's Karate training allowed him to kick in a few heads. No doubt Tim would someday have his black belt. Also, at this time Tim was acquiring his FBI Pilots certification. He was of course a great Naval Pilot in Harrier Jets, but had to jump the hoops to fly slow civilian aircraft for the FBI tracking spies, etc.... Some years later Tim had three Black Belts in three different Marital Arts. He could kick the crap out of 'em, or shoot 'em. Both better than relying on the D.O.J. LD's (limp dicks).

Tim's Thiokol and NASA cases had placed a serious strain on Tim's love life. Tim had a lovely Italian wife and four small boys. His Thiokol and NASA cases were all consuming. He still carried more than his share of other case loads. There was the American indicted murderer, various drug cases, several bank robberies, a few telephone fraud 'boiler rooms,' dozens of 'hookers', and much more. During the next year, Utah had 19 bank robberies and Tim arrested all of them. Bank robbers talk; hookers listen; the FBI follows the hookers. Utah FBI #1 was tremendously supportive of all his agents. I sensed Bob Bryant especially supported Tim. Tim was a brilliant powerhouse of effective activity. A man among men - the kind who is always called on while the losers sit around and play with themselves. Something about the FBI which troubled me deeply was the Agents work load and case mix - outrageous!! I was not an FBI manager, I was not trained in law enforcement, but I was and am an observant citizen and an effective world class project manager. Watching the FBI close-up in action, I was amazed that they would always be working such a mix of cases. A telephone pest in the am, bank robber after lunch, espionage in the late afternoon! It seemed to me a lot of profoundly important issues were diluted by Agents required time at the shooting range, FBI training sessions, FBI meetings, incredible numbers of Federal holidays, and abundant Federal employees vacation time. One of the senior Agents said it was an FBI joke - "anything to keep FBI Agents off streets!" The actual 'case time' available to fight crime is a pitiful small percentage of an 8 hour day or 30 day month. Crime and criminals go around the clock 365 days a year. Legions of full time criminals against a hand full of regular work hour FBI Agents. Any FBI Agent will confirm to you 'Crime Pays Well'. Whatever we pay these men and women, it isn't enough. These Agents face every obstacle to obstruct their work, success, and effectiveness. They start off overloaded, they are encumbered by internal FBI / U.S. Government bureaucracy, they usually encounter resistance from everyone, they virtually have to do the prosecuting attorneys' jobs and prepare the prosecutors' cases for them, and in the end they take the accused into liberal courts hostile to law enforcement and sympathetic to criminals. I can't figure out why anybody would want to be an FBI Agent? Can the rewards possibly outweigh the obstacles and frustrations? After many cups of coffee with senior agents the answer is clear - most are sad, frustrated, disappointed - even unfulfilled. A sad state of affairs for such an excellent assemblage of personnel. Remember, FBI Agents had to meet much higher requirement than the vast majority of federal employees or other law enforcement agencies.

Now, for a second time I was temporarily away from the FBI. It was time to go to Kauai and get Marrrriiiiiedddddd?! *Was, Or Is Marriage A 'Class A Felony' With A Long Term Sentence?*
Frequently A Longer Term Sentence Than Murder!?

CHAPTER: 33

MARRIAGE

TIME Time: Early 1987

I won't bore you with much. Long story short, Fay and I flew to the Hawaiian Island of Oahu and then on to Kauai. We stayed a day or two at the old and famous Alamoana Hotel at Waikiki where we met her parents from Durban, South Africa and they met me for the first time. Fay's brother also came in from London. He was very nice. They all seemed nice enough. That pm I took them to a wonderful dinner at the Royal Hawaiian Hotel next door. The same hotel where Robert Louis Stevenson wrote many of his books. At the end of dinner was a Polynesian show. Native dancing, meat-cleaver throwing, good music, etc... Drinks of course.

Then the next day we flew off to Kauai. Without question Kauai is the most beautiful place on earth to me. Paradise and splendour. One of Fay's lady friends, Romalee, the one who went to Yellow Stone with us was an executive with Sheraton Hotels. She had acquired a very reasonable rate for a great executive room.

After settling in I was off for a long snorkel in the adjacent lagoon. Just me and sea life and the reef. Was wonderful. I pondered MMMmmmmaaaaaAARRriage??? Too late to quit now me thought??? Well it should work? It might work? It won't work? ERRrrrr!!?

We had our measles shots, and we had some difficulty running down a Magistrate to issue a license. In the end we were licensed. Then the next day I believe it was a very pleasant Hawaiian gentleman judge came to the hotel and we were married under a coconut tree by the lagoon. My new mother in law had beat on my new father in law and her son to get the cameras all batteried up and get good photos. Then both of them 100% blew it. It was a premonition of events over my horizon.

That evening I took everyone to a wonderful dinner. In a few days her brother 'Willey' was off back to London. Her parents off back to Durban, South Africa. Fay and I back to Brigham City, Utah.

We were married! What more is there to say, the die was cast so to speak? Well as a matter of fact there was far more to say and it will develop as we go along. A well meant start to a Shakespearian tragedy for all concerned.

CHAPTER 34:

FBI SHOPPING LIST

TIME: Early May 1987

On 11th of May I returned to the Ogden FBI Office and worked for 4 hours getting back up to speed. On the 12th I worked 8 hours and was in meetings with Tim, and U.S. Department of Justice Attorney, David Long. We discussed some DOJ letters, National Research Council letters, contracts, and Mr. Boisjoly's law suits against MTI. David Long was interested in the MTI and NASA cases and I believe quite helpful and contributory to the pursuance of the cases. Our conversations were to some degree limited. I could and did tell them all I knew. Both of them were restricted in what they could tell me. They told me plenty but not all. I perceived much more, but not all.

I was elated by these D.O.J. meetings. Finally, I felt that the case business was now under way. I was instantly tired again. I was probably depressed from continuing uncertainties and stress. Also there had been a disastrous development. Early that morning, the Ogden FBI received a call from a known and frequently used high level MTI FBI informer who was a member of MTI security offices. The call came in while I was sitting there. This FBI source (not one placed by me) told the FBI that an asshole, Steven Agee, evidently had been swapping sex for information! The conversation then went on to explain how a friend of a certain lady source, VR - (Very Reliable), who worked at MTI Security told MTI Security that Agee was an FBI undercover Agent! Agee and VR were evidently having sex and she was worried about Thiokol's welfare and therefore her job. So, Thiokol Security called in VR and put the thumb screws on her. The conversation was quite lengthy and finally came around to Tim asking this Thiokol source if the contract engineers were any good. The source told Tim none of them were any good. I personally took this as an indication he was a Thiokol disinformation operative for MTI. He was acting as a 'double agent'.

I will never know what was really said, or what was asked. VR and I were not now on speaking terms and had never been lovers. We had been the best of friends. This MTI security source was considered to be a reasonably reliable FBI source by his controlling agent (Tim was not his controlling Agent). I was sitting next to the FBI Agent (Tim) taking the call. I thought to myself - is this a disinformation phone call and is Thiokol so plugged into the FBI they can just call in and spread bullshit, hate and discontent or is this a legitimate source telling what he knows as best he knows?

I thought to myself - what a cesspool the Thiokol Wasach plant was! I could easily write another book on incest, theocracy, and other sewer swamp activities in the high Utah desert. Miss VR was a conservative and nice person. She was knowledgeable about a vast amount of sordid Thiokol history. She had been "felt out" in preparation for eminent recruitment to become an official FBI source. Tim was ready to set her up as a source on two occasions, but was interrupted by his 'cops and robbers' activities. Miss VR had supplied a ton of information to the FBI through me already. VR was a de facto FBI source without any protection or benefits whatsoever. VR was in deep shit thanks to her confiding in a loving fat Mormon girl friend. With friends like this she couldn't afford to have any enemies.

Tim and I were both shocked and temporarily devastated by these developments, we felt the thumb screws. How much more did she know? Did her girl friend really screw her this bad? Why? Tim and I knew why, the 'Mormon consortium'.

Did VR turn coat? Oh shit! What really happened? Was this Thiokol misinformation to discredit me? As I look back on it now, I suspect that phone call was contrived by Thiokol - a covert disinformation activity designed to throw the FBI and myself off balance or in their eyes 'redirect' the FBI. Thiokol knew they were in deep shit and they are historically serious players at these types of games. After all, the MTI Security Dept. was instrumental in *Black Balling* and essentially destroyed Dr. Tony Laine's and others careers. A year and a half later I talked to a reporter, Mr. Dick Powell who had records of over 30 persons suppressed, blackballed, and damaged by Thiokol. There was even a suicide reported over the improper dismissal of one individual. Strangely, on the same day, my recruited FBI source GMC was standing by Mr. Thirkill's (VP of SRM Project Office) door when this same Thiokol Security Officer came running in the door screaming "It was Agee!" to the V. P. Draw your own conclusions! I drew mine long ago.

MTI had double agents! My mind was on fire. Information coming at Tim and me from all sources was confusing and contradictory. I was bombarded with irrational data and I was becoming more confused and more paranoid. There was far more information available than could come from me or any one other FBI source. Some people knew things they shouldn't Others didn't know what they should. Did the FBI or DOJ have leaks?

Will I be shot at home tonight by a crazed Thiokol manager? I believed physical harm to have been a serious probability. The probability was increasing. MTI had a 1.3+ billion dollar a year contract to protect. MTI also had the money to substantially protect itself. Simultaneously NASA, in concerted efforts to "save itself" from the FBI and DOJ and negative public opinion was lying to the FBI and trying to "fix things" on a daily basis. Many of the perceived fixes by Thiokol were simply changing document delivery schedules. HA's and FMEA's had been due soon, but not now - so no problem! Thiokol and NASA were on schedule by virtue schedule was altered by NASA to eliminate MTI's delivery delinquencies. This is the simplest example of the easiest cover-ups. The lies and deception went on and on! On the 13th, I worked 8 hours and I reread in great detail the SR and QA 5300.4 document and I made detailed notes on it. The 5300.4 document was NASA's bible on System Safety reliability. My mind was not as clear as usual. I was troubled by uncertainties about Miss VR's dilemma. Later that day, Tim scanned what I had done, and Tim told me it wasn't quite what he was after. Sometimes I was very frustrated trying to determine what Tim was really after. In fact a couple of times I was actually pissed off! Soon I realized Tim really didn't know what he was after anymore than I did. Essentially, he 'would know it when he saw it'. Tim Healy was a relatively new Agent and had a lot to learn about how to make a complex case go together effectively. It's a tough complicated task. I knew nothing of how to make a case. He clearly admitted this and in no way should this inexperience be interpreted to mean that he was not an excellent man. He simply had an enormous task in front of him. He was being diligent and was trying to do the best job he could. He was trying to get all of his cases under control. By any measure the worst job Tim could do would be a damn good job. I would not want this brilliant and diligent man after me if he thought I was guilty Later I would look back and realize that if an older, more seasoned agent had been my first contact with the case then the MTI cases may never have ever been opened. Some FBI agents are lazy, some are political, some not so smart, others have been burned by the FBI bureaucracy. Most were burned by lazy and inept DOJ attorneys, others were essentially "destroyed" as they faced liberal pro-criminal judges. Political Correctness was becoming a pandemic disease. Special Agent Tim Healy understood all of these career hazards completely and rose above it all. Tim had the intelligence to rise above the negatives - the arrogance of a Harrier pilot to avoid the dangers, and the dedication to do the right thing *For The Right Reasons*.

This Thiokol case would probably launch or break Tim's career. This case could make or break the United States' future in space. A failure on this case could severely weaken the United States. Tim had an enormous responsibility on his hands. Tim had 1,000% support from Bob Bryant, Utah FBI #1. He had 1,000% support from Vernon Kohl, Utah FBI #2 and the same of Dave Latin - #3. Dave Latin was Tim's closest co-worker on the MTI case. #1, and #2 were there as needed. Eventually a lot of the case actions moved to Washington, D.C. Tim had to commute. The forces against Tim were incredible to such an extent they couldn't even be rationally evaluated. Tim was confronted abruptly by FBI internal jealousies of other agents. An army of Thiokol attorneys was trying to cover felony tracks, NASA belligerence and deceptions, difficult acquisition of documents, lack of interface and or cooperation from other associated U.S. Government agencies, an army of fearful intimidated witnesses, a state loaded with Mormons prejudiced to protect other Mormons, two local U.S. Senators who were both self-seeking, a DOJ Attorney General (Mr. Edwin Meese) whom was under investigation for aerospace contract fraud, a U.S. Congress and Senate who apparently weren't especially informed about space matters, dedicated well organized aerospace criminals, desert isolation of the suspect firm, and a hundred other negative factors. Tim and the FBI, the United States, and the free world were in deep shit and only a hand full of people in the world understand it or cared. Not for one minute did I ever think Tim couldn't or wouldn't bring the criminals to justice. I also knew there would be hell to pay before it was all over. Even if the appropriate criminals were brought to justice could our space program be salvaged? I had my doubts. The problems were deep rooted. The SR & QA 5300.4 document contained the guidelines NASA had ordered Thiokol and all other space equipment vendors to use as the base safety document in running Systems Safety Departments, Reliability Departments, Quality Assurance Departments, and many other related activities. The 5300.4 was a contract basic cornerstone document, written in simple plain English which didn't leave much to the imagination.

On the 14th of May I was reading the base contract and other documents and making detailed notes for Tim. We had several meetings at which Tim had more discussions with me about finding and defining how to prove "intent" of people who commit a crime. Since I saw many crimes committed, there was no question in my mind that crime had been committed. Tim and David Long (attorney) knew from multiple sources that Thiokol and NASA were into crime. However, I had to lead them to the point where they could demonstrate in a courtroom that there was corporate intent' beyond any reasonable doubt'. This basic conversation would be repeated at least a hundred times in the next few weeks. Also that evening, I was paid for the first time by the FBI. I followed Tim to a parking lot rendezvous where we were met by other FBI agents. We all piled into my car and they each counted the money and verified between themselves that it was the exact amount due me. The FBI doesn't want the money stolen! Then the Agents counted it for me and had me count and verify the amount. Cash in a brown paper bag of course! I then signed the appropriate documents with my code name 'BRAINS'. This was a good sized pay check and I was really nervous having this amount of cash on me. Handshakes and congratulations all around. I jokingly said now you probably want your 10%? They all screamed and groaned "NO! NO! Don't even joke about such a thing"! I had the distinct feeling these were reasonably honest men! My little joke was in bad taste and untimely. So I'm a peasant - a chastened one. I had only been robbed once in my life and that was in Olongapo, Philippines (Subic Bay) at the point of five butterfly knives. I didn't want to experience a second robbery. I had to run like hell to my bank to make the deposit that pm. The bank had just closed before I arrived. So I beat on the glass door. Fortunately a bank officer saw me, came to the door and asked me what I needed. I told him I had a large cash deposit and wanted to deposit it ASAP. He said "let's see it" , I showed him a huge wad of bills. He let me in and politely accepted the deposit. In the back of my mind was a paranoid thought that this bank officer could be wondering if I was a "Cash Only" drug dealer. I wondered if he would turn me in to the FBI or police the next morning? It wouldn't have surprised me.

On the 15th of May I was reading additional contract documents and making more notes for the FBI. I had a long meeting with Tim and discussed my findings so far. NASA had been feeding Tim a line of crap that all of the SRM problems were just "problems", not crimes. Well, it isn't that simple! Tim and Bob Bryant were "pissed" at NASA. We were stressed. The crimes were blatant and obvious, but matching documents was a monumental job. It was pretty much an ordinary day as my life went at that time. On the 16th and the 17th I worked 8 hours each and reread some of the base contract and made more detailed notes regarding "Thiokol intent" to commit crimes. I reread the SR & QA 5300.4 and made more notes on intent. All trails from System Safety, Reliability, and Quality Assurance came from and returned to this key document which was hardly to be found at the MTI Plant. I asked Tim "where are the LD's?" The FBI often refers to the DOJ attorneys as LD's or 'limp dicks' because many of them will do anything to avoid actually preparing or prosecuting a case! Tim said an attorney was being "worked". No one from the local Utah DOJ office was assigned yet! Tim wasn't clear about what would transpire with the DOJ. There had been some serious FBI dissatisfaction with one or more of the DOJ attorneys at the local Utah FBI offices. Also, the regional chief prosecutor was Mr. Brent Ward, a serious high ranking Mormon. Would a Mormon prosecute a bunch of fellow Mormons? Would a Utah Mormon put the largest employer in the state out of business or would he cut disgusting deals? On the other hand Brent Ward reportedly had high aspirations. It was known all over Utah that Brent Ward wanted to become Governor of Utah or a Federal Judge, a Congressman or Senator. Brent Ward was reported to be a dedicated and political civil servant. You can guess the rest! I never met Brent Ward and pass no personal judgment, however I speculate a lot! I am naturally cynical. The FBI were seriously considering bringing a special prosecutor or DOJ task force from Washington D.C. I believe they held that thought close for some time.

On the 19th I worked 8 hours and I read the 610 and the 735 contract documents and assorted NASA letters. I made very careful notes on 'intent' while relating to these. Also I read contract SA and CO's (Change Orders) from Number 1 to number 99 document. These were early alterations to the basic NASA/Thiokol document. On the 20th May I worked 8 hours and began reading SA's and CO's. CO's are change orders that change the contract. SA's are Supplemental Agreements which also change the contract. The modification documents numbered from 100 to 200 were digested by me. More careful notes on possible intent and their relationship to the pending cases.

On the 21st I worked 8 more hours and read more contract SA's and CO's (Change Orders) pertinent facts. This sort of thing was becoming quite routine now and probably only a handful of people could have any better insights into it than I did. I was the only one I knew who ever read the entire, though incomplete, document. Most corporation and consulting technocrats are in specific disciplines and niches. Few have the opportunity or scope in job description enabling them to move through all aspects of a company as we systems people could. Also, from my Project Management background I had a good idea what to look for. Thiokol wanted to restrict System Safety's movements and academic inquiry and to a large extent they succeeded. Even so, by comparison to other departments we had broad leeway. NASA didn't pay any attention and Thiokol didn't perform! Come to think of it why should Thiokol perform? NASA wanted to hear good news only. NASA paid bonuses for good news. By now the FBI and I felt that NASA's 'Ghost Efforts' paralleling MTI's work were not for actual quality assurance, but were NASA's insurance policy. MTI and Ghost Effort groups could be blamed, or dilute the blame if there was another SRM failure. NASA could be nearly blameless. From where I sat it appeared the vast amount of the $500,000,000.00 SRM Redesign money went to "White Wash", not into "real" hardware or repairing the real problems. Superficiality!

On the 22nd of May, I did more of the same. The 23rd and 24th were again weekend days and Fay and I found various things to do. I was not enjoying married life as much as I thought I should, simply because I felt under enormous stress in Brigham City to the point where I did not feel well. Our marriage was good, but I wasn't. Our phone rang twenty times a day with no answer on the other end. Was it Thiokol harassing me? Was it an FBI source who had gone wrong and turned to Thiokol? Was it a local teenager with nothing else to do? Was it a Thiokol employee who was pissed at me? Was it a Mormon mercenary? I'll never know for certain. A year later I would find out a number of others had been telephone harassed also. Some were threatened. Dr. Tony Laine was both frequently telephone harassed and he and his wife's lives were threatened. The ultimate sources have never been determined.

I was constantly concerned about being shot or otherwise dispatched. When I went out on my bicycle, I was always looking in the rear-view mirror of my safety helmet and over my shoulder. Who knew, I might even be run over? Fay and I had discussed my personal safety with my attorney at that time - Mr Dick Giauque in Salt Lake City. Dick thought I should be careful and exercise due caution. Of course he would say that. He couldn't get rich from deceased clients.

On the 25th of May, I worked 8 more hours all along the same lines. I felt at this time we were making substantial progress and I was frustrated by how long it was taking me to dig through the avalanche of documents, letters, etc. On the 26th I found some exciting and pertinent items in the contract SA and CO's. On the 27th I worked 8 more hours and found more critically pertinent items. My discoveries cheered me up. The cases against MTI and NASA strengthened by the minute. Now Tim and I both felt like we knew better what directions to go.

On the 29th it was pay day again and this time I was paid at the Ogden office going through the usual rigmarole the FBI uses to pay sources. After being paid, I "ran" the money to my usual local bank. Tim, another young FBI Agent, and myself were off to a nice lunch. Hilarious! On the 20th Fay and I toured more of Utah. Lots of beach, darn short on palm trees.

On Monday the first of June, 1987, I worked 9 hours and I made an exhaustive list of persons having notable knowledge or influence on the Thiokol case. I have an obligation to say here that the vast number of these people were to my knowledge not involved in any criminal activities, some were witnesses to activities, people who were knowledgeable of activities, and every other relationship to the case imaginable. I repeat, most of these people were not guilty of criminal activities or even suspects so far as I knew.

On the 2nd I expanded my list and I was working on FBI Green jacket items. FBI Green jackets are case files. They contain records, data, etc. Sometime in here Tim was off to interview Mr. Roger Boisjoly. While I worked on at the FBI offices Tim went enthusiastically on his mission. Later that day Tim called and said, "Steven, 2,000 people are going to go to prison!" Later Tim returned with eleven boxes of 'red hot documents'. These eleven boxes of documents smacked of crime upon crime. Chronicles, and memos of some of the worst engineering in the U.S. Space programs history. Specific, in depth details of well known problems not dealt with yet! Roger Boisjoly, Tim, the FBI, and DOJ had a ready-made case. How would they handle it? What would they do with it all ? Would the U.S government join Roger Boisjoly's Qui Tam lawsuit on the government's behalf? A high multi-billion dollar suit against Thiokol on behalf of the U.S. government. Tim knew beyond a doubt that Roger Boisjoly's claims were valid. They fit together perfectly back to back with mine and others findings. Tim and the FBI would do their job and pass their recommendations on to the DOJ attorneys. The U.S. government could join onto Roger's Qui Tam any time within 60 days of his filing and recover billions of wasted Government money. On the 3rd, 4th, and 5th of June I expanded on the same activities. On the 6th and 7th Fay and I toured some more. By now Fay was bored and started making an incredibly complex quilt. She was bored to death with the possibilities of Brigham City. Fay was used to the hustle and bustle of downtown London, Sydney, Australia, and many more. Furthermore, an Ansett Airlines office where business was hot, heavy, and international. My entire focus at this time was to support the FBI. I was borderline worthless as a husband. We did not feel we could safely meet new friends, talk with people in the community or be associated with them. It could be catastrophic for anyone to be associated with me. By this time it was rumoured to the FBI that Roger Boisjoly's friends at MTI, who came to visit him, were "let go" because of their association with him. The Seattle Times later found this not to be true. On the 8th I worked 5 hours. Tim and I had long discussions regarding various aspects of the MTI and NASA cases. I took some NASA/MTI reading home with me. We didn't feel it was good for me to be seen around the Federal Building all the time so I would start taking some work (reading) home. Of course taking work home mandated the slightly additional burden not to lose any documents or have them seen around my place. Can you imagine - car wreck, MTI, NASA, FBI documents scattered! On the 9th I worked at home all day, and one of Fay's girlfriends from Sydney, Australia flew into Salt Lake City. She'd be staying with us for a while. We picked her up about 8 pm and brought her home. Wow! No wonder my wife had never mentioned her girl friend to me before! This was the finest looking woman imaginable. Surprising Romalie had any girl friends at all! She was actually a model working in Australia. A FOX!!

On the 10th I finished a list of persons I knew were involved with the cases and I read more of the remaining documents. On the 11th, I sorted a stack of documents and began a detailed interview by Tim taking his first cut at preparing his FBI Green Case Jacket called the "302".

We had never visited Yellowstone before. Our trip to Yellowstone was beautiful and pleasant. We stayed at a cabin near the main geyser grounds for the evening of the 12th. We had to kick a munching buffalo out of the way to get into our log cabin close to Old Faithful. These were "lap" buffalo - pets! A lot of tourists were there, but that's to be expected since Yellowstone Park has millions of visitors a year. The next two days we toured all over the park and generally had a wonderful time looking at the geothermal sites and "wild animals". On the 15th June, I didn't work. Tim had a bank robbery trial and wanted me to bring Rosmalie and Fay to see him "send off a bank robber"! Since we didn't have anything else to do, that's what we did. We drove to Salt Lake, found the Federal Court House, went through security measures in the lobby and up to the appointed court room. We watched the attorneys argue and haggle over a bunch of erroneous crap for an hour or so. Tim caught up with us in a few minutes at the ZMI restaurant in downtown Salt Lake City and we enjoyed a light lunch. Tim was enthralled with Rosmalie. Could any virile man not be? Hell No! My God she was a good looking woman! Then Tim left and went back to his trial and we went about our own business. Ouch, I left my car keys in the car! Oh Shit! Japanese cars are really tough to "coat hanger". $25.00 later the tow truck driver burglarized the car and we were off. Oh yes, the bank robber finally "pleaded" and was sent off for fifteen years!

I did FBI work at home the next day and made my notes on my last stacks of documents. On the 17th of June, Tim began interviewing me in depth to formulate his 302 interview of my findings on the basic NASA/MTI contract. This was a gruelling and very important interview In order for Tim to consummate his case. On the 18th, my "interrogation" on Contract 302 was continued. Also I read a pile of Roger Boisjoly's documents. Roger Boisjoly had kept an accurate log book of all of the Thiokol madness he had witnessed - "who had screwed who, how badly, and why". Roger Boisjoly had 'red hot documents' there and Tim had told me that Boisjoly's documents intermeshed with mine perfectly. I could see it. This was a relief to me since I had felt like the "lonesome stranger" for a long time. As soon as I dug through this box of goodies (one of 13 boxes) I realized between Boisjoly and myself that Thiokol could not stand against us. Thiokol was doomed on Contract Violations, false statements to the government, RICO violation (racketeering), and non-performance. I wondered who would fall first -Thiokol, America's Space Program, NASA or the Free World? Then I was back at work on the 22nd. I worked with Tim reviewing more contract documents, SA's (supplemental amendments) and CO's (change orders) from 100 to 200. We also worked on a huge list of Hazard Analyses on the SRM's which had been prepared by a number of engineers. Not just Hazard Analysis prepared by me, but by many others at MTI System Safety. On the 23rd, I worked with Tim on the 300 series and more Hazard Analysis.

Fay and I felt at this time I would soon be called to the East Coast for any one of several substantial pending consulting contracts. I had a large number of contracts close to signature so we decided to make a quick trip over to Seattle to take care of some personal business, and then return to Brigham City.

On June 30, I worked on a synopsis of the acquired HA's from my own and others' work. By this time, I was weary of working for the FBI and the frustrations involved with it. This is not to say I did not respect them or their efforts. I was simply weary. I am a methodical plodder type person and was continually frustrated by the interruptions - bank robberies, drug cases, and various unsavoury other cases which continually interrupted our work.

Some of the FBI Agents were quite concerned this case was so large and politically hot that it might be "stopped" by high officials including the President. So, they planned ahead to prevent suppression. It's peculiar when you think about it - Agee's work and findings are suppressed and he goes to his boss, his boss's (Kent) work is suppressed, then Agee goes to the FBI who has to plan for and actively avoid suppression from high American government officials inclusive of the President of the United States. Have we or are we a suppressed society? I will not expand this scenario. It could cause some good men serious problems. I wasn't supposed to know how sick reality actually is. Our government appeared to be lost somewhere between imperialism and totalitarianism. The average bank robbery in the United States yields about $2,000.00 to the robber. I estimated the Thiokol accumulated negative impact on the U.S. to be in the hundreds of billions of dollars. It would take a lot of bank robberies to equal the financial assets and military assets lost to the U.S. by Thiokol and NASA's indiscretions not to mention the destiny of the human species.

In many ways, it was as though I was on the witness stand continually myself while I was under cover and when I worked with the FBI. Not for you to understand and think that I was being ill-treated by them, because

I wasn't. They were continually probing deeper and deeper to refine their case and make it stronger and clearer. I knew so much about the inside of Thiokol and their Solid Rocket Motors it was awesome! Almost any aspect the FBI could bring up on the Rocket Motors or Thiokol I could provide some pertinent information. Tim and I figured that we were probably about 30- days from finishing up my efforts at this time. How to best plan for separation from the FBI? I needed another consulting contract. I was calling contract houses. One of them had a firm contract, but I turned it down. It had an unacceptable low hourly rate. Many other contracts were cooking, and cooking, and cooking - adnauseam. The U.S. high tech economy was stagnant. Also I opened more questions than I answered. The FBI had a wild tiger by the tail.

Steven Agee (left), friend Carl Haller (right) in Wash., D.C. USA

CHAPTER 35

SOURCES AND LEAKS

Time: July 1987

On the 1st of July 1987, Tim and I continued work on another of his FBI 302 documents. It was a gruelling process. Tim was dedicated to writing and weary of always having to educate everyone around him (NASA, FBI, DOJ, others). There were certain people in certain government organizations that seemingly didn't understand anything. Ignorance was their security. Tim was determined to make these 302's successful and have everybody able to read and understand them. On the 2nd of July, we continued on the 302's. Tim knew these 302's would probably go up the chain to Ed Meese, America's faltering, Attorney General who was one of the criminals apparent. He had to do the 302's right. Tim's future was on the line in direct proportion to his visibility in the FBI and DOJ. *"No Good Deed Goes Unpunished".*

The 4th of July was the Independence Day holiday, and my wife and I enjoyed the three day weekend. On the 6th, 7th, 8th, 9th, and 10th Tim and I continued on his 302 document. It was impressive though still raw and hand written. On the 13th of July, Tim and I had a tremendous discussion in reviewing the 302 and its inclusions and attachments. They were broad, bold, to the point and decisive in my opinion. I thought any "fucking idiot" could understand them. They could not possibly be sidetracked as legally insufficient. It had now been some time since I had talked to any of the internal Thiokol sources. I believe Tim had recently talked to some of them. On this day, I called WI (Well Informed), in the evening (at his home and talked with him) to get a feel for how his activities were progressing at MTI. Tim had requested me to contact "Well Informed". WI seemed notably up tight. WI said there was a lot of "tension and bull shit" going on currently. The 15th, 17th, and 20th and 21st were busy days finishing up several other 302's on my other associated activities. There was a 302 covering my time inside Thiokol, and a 302 on what I found in the contract and other 302's - FBI cases to spare!

Tim also followed up with a call to WI a couple of days later. Tim found out that WI didn't want to meet with me again for some time or ever? Thiokol had really put pressure on WI for associating with me. In fact they kicked WI around "hard". They had evidently threatened him. Also WI had witnessed new and horrendous crimes in the Reliability Department and Don Thompson was involved as usual.

On 4 July 1987 Tim and I were in Tim's Ogden FBI Office working our guts out on a 302. His office was literally up to our asses in stacks of thousands of Thiokol and NASA documents. The Ogden FBI Office rarely had visitors. As we were working on a 302 document there was a knock on the door, Tim stepped to the outer office looked through the peephole and opened the door. The visitor introduced himself as Daryl Lee, an attorney representing Morton Thiokol Inc.. He wanted to see Tim. How did Thiokol know where to send him and who to see? Tim asked him in and came over and partially closed his inner office door on me without saying a word to me. He of course knew I could and would hear everything. I was in a state of shock and terror. As I looked for some "last detail" document for 302 regarding MTI I listened as best I could. I was trying to be "cool". Actually I was outraged that Thiokol would send an attorney "fix it" person right up to the FBI office. They were bold in their crimes! I knew if they wanted to intimidate Tim they had picked the wrong man. Left to my own solo devices I would have thrown Daryl Lee out the 4th floor window right after I beat the shit out of him. We knew Thiokol and the Mormons were into world class crimes. Therefore, I was not impartial about a Thiokol "fixer" from their legal department could not be anything other than a criminal, sanctioned and licensed by Thiokol or Mormons. Daryl Lee was of course a Mormon. I couldn't hear it all but what I heard inflamed me. Daryl Lee had come to tell Tim and the FBI that "Steven Agee was no damn good, and a troublemaker". Daryl Lee was a Mormon brought from Southern California to deal with Thiokol's numerous Challenger Disaster related problems. Daryl Lee probed Tim to see if Tim was a Mormon. Daryl struck out on that one! Tim was a pissed off Irish Catholic. Daryl asked a lot of questions and Tim stopped him in his tracks. Tim was more than up to the query by this slime bag attorney.

Daryl demanded to know if there was a case against Thiokol or not! Tim told him to go see FBI # 1 in Salt Lake. The FBI didn't give out such information! Tim and Daryl sparred around the outer office for 30 minutes. The hunted and the hunter. One a great American, the other a paid and willing representative of a great evil - 'attorney scum'.

On July 1st Tim and I had set up a dinner with Source GMC. GMC had literally replaced me in System Safety. He went to dinner at a well known restaurant in Ogden where privacy was reasonable - The Prairie Schooner. Source GMC gave us a clear cut and excellent update and status of what was currently going on inside Thiokol and the NASA / MTI mutual symbiosis. GMC was a professional and a gentleman. He had a good balance about him. GMC indicated to Tim and I much of what we already knew. One of my well placed sources, Miss. VR, had caused a great deal of trouble inside Thiokol either willingly or unwillingly. A fucking disaster actually! GMC said that mortal fear had shot through Thiokol over Miss. VR, leaking some information. VR knew about my being an FBI source. She was within two days of being a formal FBI source when I left to get married. She was already a defacto source. Had it not been for Tim's cops and robbers she would have been a formal FBI source. VR liked me a lot and she was crushed when I decided to marry African rather than American/Mormon. Later, I found out when another man had made a similar decision years previously she almost committed suicide. I now realized she was off centre a bit. She had been a good friend to me and the FBI. From what we found out she had spilled some beans to Thiokol. She had confided in her Mormon girl friend who turned VR into Thiokol's Security Offices as we had previously found out.

The FBI had a long documented history of problems with female sources. The FBI walks on eggs not to be prejudiced, but realities are that men and women are different. Women traditionally talk more than men and they are traditionally more emotional than men. Frequently they are more honest than men, especially with another woman.

When VR's Mormon girl friend turned her into the Thiokol Security Force, and no doubt the Mormon 'fruit inspectors', an avalanche of trouble for me and the FBI had been set in motion. Thiokol Security interrogated VR and no doubt used and abused her. She was alone. I suspect Thiokol turned her every way but loose. I don't know the realities and details. Even so - Thiokol was plenty scared and dangerous! If she told nothing they should be frightened. If she told anything they should be terrorized. To this day I do not believe VR revealed anything of importance, but she was brutally victimized by her good Mormon girlfriend, her church, and her employer. I have always felt bad about that, but I never claimed to be perfect. Also the FBI wanted any and all information and VR was a 'goldmine'. If there were faults they were with me and the FBI for not absorbing VR as an 'official FBI source' earlier.

FBI Source GMC was now becoming a wealth of information and related to us that NASA had also "gone crazy" in his view. GMC went in depth about NASA insanity and foolishness. GMC went on and on in depth and detail about NASA chasing its tail on all fronts. NASA wanted what it wanted "now!" NASA didn't want to hear the truth about anything. NASA wanted a clear bill of health on their SRM's. Didn't want to spend any money, and was crazy to keep launch schedule. Tim and I were both thinking that the FBI had scared the hell out of the NASA organization, but was it doing more good or harm? Was NASA fright going to kill more Astronauts or workers?

GMC said he knew of NASA's Ghost Efforts on Hazard Analysis and FEMA's by Martin Marietta and USBI. Tim and I had already validated that 'ghost efforts' meant that anything found wrong in System Safety at MTI was analyzed by others at another company with their anticipated findings produced by unqualified or qualified persons without documents to work from - *Spin And NASA Cover Up!*

At a later date GMC told us Boeing was reviewing MTI's efforts at NASA's direction. One other aspect of the NASA / MTI 'cover-your-tracks' routine was telling Tim and me how MTI had a System Safety Engineer from their Elkton Division reviewed all of my catalogued 245 Hazard Analyses and render the verdict that all but a few were of 'no value'. The chosen lady engineer was a lady I never met, Rita Kerrigan. Tim and I wondered if she was paid huge amounts for her huge expertise, if she had any idea what an SRM was, did she study thousands of pages of data and make calculations, and we knew - she was designated 'cannon fodder' just like Kent McKinnon had been. Tim and I felt sorry for her. It was disgraceful and dishonest. GMC explained that Rita Kerrigan was probably not well informed. After all the MTI Elkton Division's job on the SRM's was to clean the SRM's Solid Rocket fuel ashes out after they had been recovered from open ocean after each launch. Thiokol's Elkton Division had nothing to do with the redesign of the SRM's or SRM System Safety. The Elkton Division probably had no notes and documents of any consequence on the SRM's. Rita Kerrigan probably never understood Thiokol's contractual duty to NASA and the U.S. Government. She probably hadn't been told to forward any and all Catastrophic Hazard Analysis to NASA within 24 hours! Rita probably did just as she was told and never knew her opinions weren't worth the paper they were printed on in the criminal scheme of business as usual. Much was known to the FBI about MTI coercing employees to sign off on many such things against their will. Another thought to consider is that in the Mormon culture women maintain a well defined role and very subordinated positions. 'Your job or your signature!' Men know best, Mormons know best, and God knows best. GMC said that NASA was now attending a substantial number of MTI working meetings, but nobody was seeing any results from NASA's efforts yet. We felt NASA was now motivated and concerned, but was decisively unable to deal with their SRM problems. How could they? NASA needed a new SRM vendor and a new design but time and money was out to make a better machine. GMC was far less technically trained than me and many others there but he still knew enough to see NASA was incompetent to the point where NASA had no idea what was even going on. Other sources felt Thiokol to be inept in numerous and visible ways, but also felt Thiokol to be adrift in a sea or chaos and confusion caused by NASA inadequacies. None of this could overcome Thiokol's deliberate cover-ups and deceptions, and absolute ignorance.

GMC was an educated responsible man. He was not an engineer or scientist. GMC told us a set of information which would indicate a damaging "leak" in Tim's case! I can't remember the exact details of this conversation but it shocked Tim and me. I really felt extreme pressure and stress. I had lost control of my numerous well-placed sources in Utah. Tim was trying to be in control, and the uncertainties were vast and ominous. I was emotionally shaken. My mind was on fire. Miss VR absolutely did not know all that was leaked, but the indications were that she was more of a problem than previously thought. Logic was that she knew some of what was leaked, furthermore GMC was now talking more kindly of MTI and less kindly of NASA. What was going on? Had GMC rolled over? Was GMC a "leak" or was he now a Thiokol disinformation operative? No doubt in my mind there was profound disinformation all about. Tim and I were massively informed but not without threads of confusion and alarm. We basically liked and trusted GMC, but had he changed? GMC was a Mormon family man.

Tim and I were both aware of the earlier DOJ leak in Washington, D.C. and were aware of VR's girlfriend's accusations. There seemed to be others. I believe Tim thought so. Malcolm Phillips had carried papers for me. Had he sold out the U.S. and his soul? Yes, but he didn't know nearly enough to do all of this. Malcolm had limited, discrete knowledge.

I was tired, confused, frustrated. Was the local D.O.J. office compromised? I thought so then and more so now. One of the lady D.O.J. attorneys in Salt Lake was married to an attorney on staff at Thiokol. She had been prosecutor for some of Tim's cops and robbers cases. She was a Mormon. It is impossible to estimate how deep Thiokol and or Mormon influence penetrated everything in Utah. Actually there could have been more than one leak. Their influence akin to tyranny. This entire experience was frightening and enlightenment to me. Enlightenment into the dark world of theocracy paranoia, confusion, and real danger... Did the FBI and the DOJ have any friends? Damn few at best! As I was writing this book and looking back I have derived a certainty that both the FBI and the DOJ had their own individual leaks. Tim's excellence bred enemies. Thiokol courted the FBI, DOJ, and any other governmental influence. Morton Thiokol Inc. had the power of the Madam at a great brothel! Senators Orrin Hatch and Jake Garns were all powerful Mormons. Security is difficult. I could not control my sources to my satisfaction. I could not control other sources at all. I was not a perfect operator. FBI personnel really couldn't control sources completely. I had always known people were unreliable, but I had vastly underestimated the true extent of human unreliability and treachery. Possibly this is because I was an '"only child" and grew up with limited social intercourse? Beyond human undependability was a phenomenon strange and inconceivable to me! A significant number of relatively ordinary and generally innocuous persons will witness serious crimes, even be a second or third party to them, and act like proper castrated sheep. Then, when the FBI is sniffing around these persons do their unpatriotic best to blow the whistle on the FBI, obstruct justice, and specifically thwart legitimate legal efforts. My list of examples would be long. These persons really aren't criminals, they are just common, incapable persons I suppose. What was really going on? I really didn't feel like doing any more "spook stuff!" I really wanted to hold a press conference and blow all of the bastards out of the water. I felt America was losing 'the works' while their 'children played'. Tim felt the same.

Proper legal prosecution is 'the right way', but 'trial in the media' just might be faster, more effective, and would liberate me from an inflamed need to get the U.S. Space Program back on track. I was exasperated, pissed, confused! Did MTI have an army of disinformation officers in place? Absolutely! Were our well placed sources sound? God only knew? Were there forces afield none of us had identified yet? Probably several.

Now, years later I have much better 'hind sight'. Looking back I see a cauldron of conflicting aspirations.

Tim was like a crazed lion trying to plug the holes in his cases and cut his losses. I watched, and allowed Tim to manipulate me and others as he needed. It was part of his job, part of his mentality, and I considered his needs and America's before mine. This was enormously difficult for me at that time. I didn't always think Tim was correct, but he had the direct case responsibility and more expansive visibility. I knew in the end he would have all of his cases' "ducks in a row". Additionally, I watched Tim manipulate others as he needed and when he needed . Usually they had no idea they were being manipulated. Tim had qualities beyond intelligence. Tim had clear far reaching perceptions into situations and people. I was "spent". Tim was just coming up to speed. I returned to Brigham City exhausted while Tim and GMC continued to talk in a car at the restaurant parking lot. I know Tim had to weigh every morsel independently for truth, insight, understanding, innuendo, continuity and manipulability. I had every confidence in GMC up to that pm meeting. I then developed an unspoken sense of creeping mistrust. Then I wondered? GMC was the man I had used to contact the FBI. What if he sold out? The way the cases were going I was really going to be shot dead.

On the 22nd Tim and I had worked all day as usual. Tim said he thought it was time to "wrap it up". I was glad. I wanted to renew my life. I was profoundly weary of the entire mess to the point where I felt wasted and ill. So the 24th was my last full day of work for the FBI. I felt the GMC meeting was rushing me out of the FBI door. I felt the GMC meeting was now pushing the cart? Had Daryl Lee's first contact with the FBI proven I was too dangerously exposed? When my departure actually came I had mixed emotions. Yes I'm glad to be through (for now). Which of the nine pending contracts will I close first? Where will I go next? I must leave Brigham City immediately! Our crime project was in good hands with Tim Healy and the FBI team working the case. Even though I was leaving Utah I would continue contact with the FBI to aid them as needed when needed. I still had answers to more questions. Completion of my full time work for the FBI came a little quicker than I had expected. It seemed to me that the FBI was now very much in control of America's Space exploration destiny. Much more so than NASA. God knew they had a good start on it thanks to me and other sources. No one ever said, but I assumed I had become a liability to the cases to some degree now? No doubt MTI watched and pondered all of my movements, my efforts. I was exposed. One of Tim's many difficulties was judging into the future when various aspects of the case would start and finish. Too many variables and unknowns for a simple solution. There were so many interruptions that activity was more of a guess than an estimate.

It was with mixed emotions that I drove back to Brigham city after finishing up my last day of work for the FBI. There was more work to do, but on the other hand I felt it was time for the FBI to gather their full forces and move forward themselves. I was shattered. It was July 26th. The next day, my wife Fay and I were packing and preparing to leave Brigham City. We weren't sure where we were headed. I had an unusual number of consulting contracts about to close. Under normal conditions, I always had contracts brewing on the U.S. West Coast. However this year there were absolutely none in the West. All of my current contracts were brewing on the East Coast and were scattered from north to south.

On the night of the 27th, Fay and I had dinner with Tim and his wife at an Ogden restaurant. We had all become close friends or more like family. We had a lovely dinner and great discussions. It was the final passing of the baton of this case from me to the FBI. So now they had to 'solo'! A tremendous amount of information was now totally in FBI hands. I had already been through the most gruelling and hostile period of my life up to that point. My shootouts in the Vietnam jungles seemed tame by comparison! In Vietnam we simply killed our problems, much more effective and simpler. Political correctness of that time.

In Vietnam identification of the enemy was simple - if it had slant eyes we wasted it!" A place, time, and circumstance in history. Orders from headquarters! Corporate America enemies frequently looked like church officials, sounded like statesmen, acted like assholes, and appeared as pillars of communities and governments. Things are never what they seem. No way would Tim allow himself to sink. He would prevail in the end if he was allowed to. But not only was Tim submerged in an enormous volume of documents and data and a complex history of the case, he was also surrounded by numerous other obstacles previously mentioned and others yet to emerge. Agencies and groups within the U.S. Government were hostile by nature to outside examination. After NASA had sent America and the first persons to the moon, launched an untold number of satellites, built the Space Shuttle which had 24 successful flights! They were composed of professionals and also losers who radiated arrogance from others previous successes. Arrogance never usurps facts and professionalism. Perhaps NASA lost its visibility to manage operational projects, money, safety personnel, and develop effective and credible long range space exploration plans. Perhaps NASA lost sight of the fact everything they did literally cost 5 to 10 times what is should have! Perhaps they lost sight of the fact that the cream of NASA personnel had drifted away soon after the Apollo Moon Missions when NASA budgets were cut to the bone. NASA in 1987 was a shadow of its once potent self. NASA was then a bad and expensive joke and only foreigners laughed. Americans were oblivious to NASA's collapse and incredible failure for two reasons. Reason one is there NASA is a huge public relations machine and it can and does lie. The second reason is that Americans are not much interested in space activities and not currently well educated and informed. USA educational standards dropped dramatically since the late 1960s. Perhaps those NASA officials hadn't observed the gradual changes in NASA and were not aware of them. So, naturally NASA was defensive, if not uninformed. No one can deny that many NASA officials were civil servant drones and would do anything to hold their status. Not everybody surrounding this case had America's interest at heart like the FBI and myself. John Wayne would have been pissed at the state of affairs. Our national environment was obstructive and hostile.

On the 28th of July, Fay and I packed boxes, and were loading the car and our 12 ft, Ranger sailboat. On the 29th and 30th of July, Fay had some medical things taken care of. She was 'off' for a couple of days. On the 31st of July, we finished loading up and packing, hooked up the boat trailer, said goodbye to our landlady and GM's family, and headed out of the State of Utah with exhilaration. We stopped for a few minutes in Ogden to say good - bye to Tim and it was an emotional departure because so much water head gone under the bridge. Emotions had run high and *For The Right Reasons*.

Tim thanked me for himself and told me I had done "a good one for my country". Without working at it, we had become family. We were enmeshed with genetic 'oneness' to do the right things *For The Right Reasons* trite and stupid as it will sound to many people - Tim and I were and are 'shoot to the last bullet', and then use a club if necessary patriots. If you can't understand this - well, so be it! Our efforts were for the entire free world and the future of the human species. Our efforts caused us substantial anguish and suffering. There had been no 'strokes' or satisfaction so far other than Tim and the FBI - just work and anguish for me. Lost consulting, lost income, lost friends, damaged expectations, and eventually loss of comfortable U.S.A. residence and ultimately much more. Fay and I then got into our Toyota station wagon and swore to ourselves, we wouldn't stop driving till we were over the border and out of the state of Utah. It had not been a good place for either one of us.

This was the end of my full time FBI activities for some years to come, but not for ever. It was not the end of my concern for American Space Supremacy. Now I would only be a part time FBI helper when needed. This was only the first half of the story. Also this was just the first half of "source information". A lot of information would come in to me and the FBI for years to come. Just one example of the titbits that come in is that I received a phone call from a source who told me Mr. Ray Ave had overheard a conversation between Don Thompson and John Thirkell, V.P. trying to put their plan together to sabotage my and the FBI's activities. Ray Ave overheard these discussions prior to and after my leaving Thiokol. It was reported that Ray Ave flew back to Indianapolis and reported all of the Thiokol crime and B.S. to Mr. Dick Dice the manager of ASSETS (Action Systems). So one more contract engineer was horrified and had a pile of good information. Of course I passed the information immediately on to the FBI. Who was ahead? The FBI and the sources, or the criminals and the leaks? I couldn't tell?

Arizona Republic, Tuesday, Nov. 15, 1988

CHAPTER 36

FREE OF MTI AND THE FBI AT LAST

TIME: Late July 1987

Fay and I stopped driving well into Wyoming on the evening of 31 July 1987. We found a clean and neat motel and went to have a Mexican dinner. I celebrated vacating the nearby liquor dry state of Utah with two Pena Coladas. I was a physical and emotional wreck and my wife and I both knew it. In the next few days we crossed Wyoming, Nebraska, Iowa, and then Missouri. Our destination was Lees Summit, Missouri. There we visited with a couple I had attended the University of Texas at El Paso with. They were both senior geologists, Frank and Shirley Abshier. While there visiting, I settled into a routine of locating my next consulting contract and trying to physically rehabilitate myself. We arrived at Frank and Shirley Abshier's on the 2nd of August. I enjoyed seeing them again. Fay enjoyed meeting them. There was a vast amount of gossip to catch up on. These are the Frank and Shirley I visited with in Denver at Christmas. Every one enjoyed each other's company. Fay and I anticipated being there three to five days. When we left Brigham City three days earlier I had nine 'dead serious' consulting contracts brewing. Two of these nine were imminent contracts in Cleveland and Atlanta. When I 'settled in' at Missouri I broke my 18 speed bicycle out and started taking daily 15 mile rides. Tim had done right by me on funds payment. Evidently he and No. 1 had wisely anticipated I was at the end of my career. Since most of my Thiokol income went to support the US IRS and the State of Utah taxes, both Fay and I were glad for the FBI income. Both the exercise and change of environment were long overdue. Next to Frank and Shirley's place was a huge park with perfect rolling hills for a good bicycle work out. Fay was especially relieved to be 'liberated' from Utah. She hated being treated as an infidel there by the Mormons. Fay was horrified by the typical life invading practices of Utah Mormonism. Fay and I support the Mormons right to their own religious practices, but it was not our choice. Fay was more attuned to the Church of England.

After three or four days in Lee's Summit I was leaning on my most reliable 'contract houses' for a contract 'NOW'! I was pushing to close on any of these contracts. All nine of my pending contracts were sliding - not coming to consummation. The aerospace / high-tech companies wanted me, but there was no funding! "No funding" was heard from all over the USA. I was not alone! After checking with associates across the country, I found everyone complaining about how tight the U.S. economy was. No funding equalled no work. Curtailed projects abounded. No Space Shuttles flying meant a huge backlog of payloads. No problem, I was respected and well known . I would overcome the industry slump, I surmised. WRONG!! The structure of the U.S. high technology economy was similar to the following pyramid at that time.

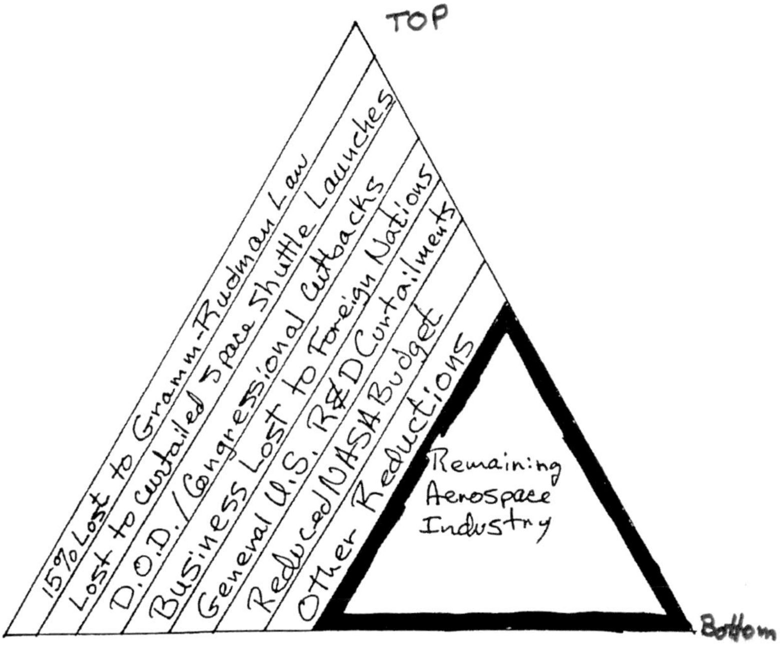

As you can imagine a 15% cut across the board hurts both the upper and lower level personnel first and foremost. Around the country they were blowing phones off the desks at the 'contract houses'. There was near total evaporation of upper level contracts. Virtually all of the employed upper level persons were on contracts substantially lower paid than normal.

My associates, myself, and Fay sat down and looked at the late 1987 forces against us..

1. Continuing 5 years evaporation and tightening of all upper level contracts looming
2. The Space Shuttle had not flown for 18 months (at that time) - stifling thousands of projects either directly or indirectly.
3. Unclear schedule for next Space Shuttle launch.
4. Continuing Gramm-Rudman pressures on the U.S. Military budgets for 15% annual cutbacks every year.
5. Definitive lack of U.S. National commitment to ongoing space programs - military or civilian. More funds for welfare was the cry!
6. Vicious U.S. Government Agency and Military strife over every Congressional budget dollar available - U.S. Government Budget Wars.
7. Improperly informed and relatively unconcerned American public.
8. A huge portion of America's high technology work had migrated to Southeast Asia, and Europe where their students have excelled while America's students have performed poorly with declining U.S.A. standards.
9. A Presidential election was looming.
10. Declining value of U.S. Currency.

Adding all of the above up, the total accumulated effect was frightening. Not only millions of individuals working directly in high technology industries, but to America's national status. The sum of the negatives in our high technology economic environment were allowing some high technology firms to severely depress wages of those few offered work. Supply and demand. American industry has never wavered at an opportunity to cut wages to the bone. National insanity, a drug abusing basketball player could make millions per year. Great neurosurgeons can hardly afford mal-practice insurance, the most productive engineers in the history of the world weren't working, and the most productive farmers in the world were mostly bankrupt by national banking design.

On one of my regular phone-in contacts to the FBI I caught Tim in a down mood. He and another Agent had interviewed certain persons who had left Thiokol out of conscience. However by the time the FBI got back to them, these persons had been put on consulting contracts to Thiokol at much higher than ever wages. In other words Morton Thiokol had bought them off.

Tim said they turned red in their faces, squirmed, and lied their asses off. Word by word, lie by lie Tim and one of his partners watched as they interviewed these witnesses. There were no problems! Frequently a string of lies leads to truths. I was incensed. So, I and others persevered on looking for "elusive contracts" all over America. What else could we do? When you are an ant on the hillside and are swept into the river you either float with it, swim for it, or drown.

The FBI still owed me a substantial amount of money and was way overdue to pay by this time. I suspect an entire book could be written about what it took to pay me that check. That check represented 1/3 of a senior FBI Agent's annual salary. I suspect a number of FBI payroll bureaucrats resented anyone getting that much money. Clearly Tim was being educated about the inner workings and complexities of the FBI - hell, nothing's perfect! Tim was relieved to finally pay me after months of problems and delays. I sensed there had been wars, maybe even assassinations over getting me paid. There was one more small check due after that big one. God only knew how long that would take. This check was about 90 days overdue and the clock was ticking. It frustrated me and made me financially nervous. During my life I had been financially screwed by almost everyone, now by the FBI! Why should the FBI be any different?

Around 15 August 1987 Tim told me my money was wired to the Kansas City FBI office. He was both exasperated and pleased. The next day Fay, Shirley Abshier and I loaded in the car and went to downtown Kansas City to the Federal Building. In the Federal Building foyer, we went through a serious high security entrance. Upstairs I asked the FBI secretary specifically for the Agent Tim had told me to ask for. A few minutes later an older Senior Agent emerged and asked for me. We shook hands, and I followed him into a room. He and another Agent gave me a check drawn on Bob Bryant's personal Trust Account. They asked if I had been signing with my name or my code name "BRAINS". I told them I signed "BRAINS". I appreciated Bob Bryant expediting the solutions to whatever bureaucratic SNAFU'S had occurred. The Agent who gave me the check had asked 'Dare I ask what you worked on" and I told him the 3 minute story overview. He screamed "Oh Shit!" and he was in a state of shock. He congratulated me and we shook hands. This was a welcome overdue payday! We three went to a steak lunch in downtown Kansas City. Afterward we went up to the war memorials in the park above the city and looked them over. I was still looking over my shoulder. One day I would think no one could possibly waste me, and then I would think a Thiokol 'contract person', a 'misguided Mormon' - who could guess? I had damaged them irreversibly to their core. Tim's father-in- law, Frank, was of Sicilian descent. He came from Chicago and had lost his father in a Chicago Mafia shoot out. Tim, the Case Agent, felt there was no concern for my life and that the Thiokol managers were scared wimps. I had no experience to judge from. Was I in danger? I was a central figure in an enormous crime. Persons get 'wasted' over a marijuana 'joint'! HUUummm?? I had tripped up billions!

My contract hunting went static! Fay and I were frustrated. I needed to get back to my normal consulting to divert my emotional frustrations and fears. After my AM bicycle ride one day I went to a local restaurant. Stopped off at the men's room and was shocked to find myself pissing bright red blood! Damn, just what I needed!

Special Agent Healy had his own problems back in Utah. Tim was literally under gun fire! Tim was driving the armoured vehicle in the following article. Tim had escalated from 'cops and robbers' to 'cops and robbers, bullets, and deaths'! [14]

Tim was a major player in the Utah Shoot outs. For six months I had been feeling extremely rough. I had a twitching in my side at night. Could this be another kidney stone? Oh No! Oh Damn! Oh Shit!! Better death than to pass another kidney stone! I had been the unwilling possessor of three kidney stones from my left kidney previously! Each of those hurt me so bad as they passed 'I was afraid I might not die'! Better dead than to piss rocks! I zipped into the local 'saw bones' ASAP. After looking me over he told me 'the worst' it wasn't cancer or kidney infection, it was a rock in his opinion. CRAP! Thiokol and Mormon revenge!

Next AM was X-rays. My rock was 6 mm and about to fall out of my kidney into my plumbing pipe. This was a good sized rock as they go and I sensed it was going to be a 'rocky journey' to pass it. Maybe I should waste myself? The radiologist said he didn't think I could pass it. Oh No! Surgery or liptotripsy - AaRrrghh!! For a few weeks it was tests and X - rays. By this time I really felt rough - desperate as a matter of fact. I tried to drink 1/2 of the world's water to help the rock along. Several days later I was out on my bicycle and almost didn't make it back. Business as usual for kidney stones.

Appendix 14 for article: "Utah Stand Off Ends With Officer Slain " Rocky Mtn. News, Jan. 29, 1988

On September 25 I went to Dr. Austin (a urologist) in Kansas City. I felt poor and he didn't offer much hope of me passing this one. After seeing Dr. Austin, Fay and I went to find Frank James's (the famous train robber) grave. She and I both enjoyed reading old epitaphs - visited a few shops and returned to Frank and Shirley's.

Soon, Fay and I re-evaluated our situation. We were marooned in a further collapsing high technology economy. As it stood the Gramm-Rudman law had just come active again as congress couldn't settle an appropriate budget. Disaster! Automatic 15% across-the-board curtailment of all U.S. Government agencies and their personnel and projects. The government drones would be shuffling for their lives now! We consultants would be the 'first screwed'. Oh Shit! We were industries 'cannon fodder' again and as usual. What in the hell was the point of being a great project manager if there were no funded projects? We decided the most likely contracts to gel would be in Georgia. We should go there and be on their door step. Two days after the Gramm-Rudman law went into effect the U.S. Navy stopped my pending Cleveland contracts dead in their tracks. There were now two contracts in Savannah, Georgia, and four in Atlanta. The contract house assured me I would be on one of these contracts any minute. NO question about it they said!

On Sept. 15, 1987 I received a call from Dawn Reynolds in Huntsville, Alabama. She worked for Technicor and she asked if I knew about the Space Shuttle's Solid Rocket Motors (SRM's). I told her I had just left Morton Thiokol inc. in Utah and knew their SRM systems inside out - in fact I had just completed 245 Catastrophic Hazard Analysis on them. Dawn was ecstatic and she was certain she could land me a contract with Boeing Aerospace regarding them and at a very good rate. I would be called the next day for an interview by Mr. Clint Rogers from Boeing.

When I told Fay I might be on contract to Boeing Aerospace to supervise Don Thompson at MTI she almost went crazy. She said 'it's a toss-up who kills you!" What a fucking predicament!" The only contract for me was in the lion's den! I needed the work and income. I didn't need Thiokol. I called the FBI and told them what was up and they were ecstatic about the possibilities! What a coup if it could be? Just think of it - to have Don Thompson and Kerry Sanofsky by the balls on behalf of Boeing Aerospace, and the FBI! I couldn't believe it could really happen. Besides my new wife was going to kill me first. Then, Morton Thiokol Inc. 'Black Balled' me!

The following is from my FBI Logbook -

9/15/87 at Kansas City Mo..

Technicor in Huntsville called me. Laura left her name for me to call back. I did and Laura passed me off to Dawn Reynolds. Dawn told me they had a contract with Boeing Aerospace which would be 7 to 8 months in length. It was to go until after the next obiter flew. She asked me if Technicor could submit me. I said 'yes'. They submitted my papers to Boeing Aerospace Corp..

9/17/87

Boeing Aerospace man named Mr. Clint Rogers (205)544-1788 called me and interviewed me for the above position. The interview was brief and cursory. I was immediately accepted by Mr. Clint Rogers. There were several discussions with Dawn prior to and after the telephone interview. After the interviews I was told by Dawn that Boeing told her it went very well. Boeing liked me, wanted me, and 99.8% of the contracts paper work was complete already. Technicor would notify me tomorrow of the final paper acceptance and they asked me how fast I could be available - I agreed on 24 hours to be at work from final approval.

9/18/87

AM call to Dawn confirmed that Boeing had completed the paperwork. I was offered the contract. My wife and I were in Kansas City, Mo. and we proceeded to load our Toyota Station wagon & boat for a 9/19/87 departure to Huntsville and an agreed start date of 9/21/87. The contract was very substantial money for that time in history and I was to be the Boeing Aerospace representative at Morton Thiokol Inc. at the Wasach, Plant. I was to supervise Mr. Don Thompson and Kerry Sanofsky of MTI and was to have frequent interface with Mr. Tom Lewis of MSFC a NASA representative. I would answer directly to Mr. Joe Helmsteder of Boeing at Huntsville. All of the work was to be on the SRM for the STS. PERFECT!

At 2:30 PM on 9 / 18 /87

Dawn called from Technical and told me 'don't leave', we were just notified by Boeing AC that the contract is on hold, we will get back to you Monday. A few minutes later I called Mr. Clint Rogers at

BAC and asked what was going on? He told me MTI was adamant they did not want me. They told Joe Helmsteder that I caused a lot of trouble there and had a serious fight with Kerry Sanofsky. MTI doubted they would let me on the MTI facility! So, I asked if the contract was 'dead' and Clint said yes. I thanked him and hung up.

9 / 19 / 87

Talked with Dawn at Technicore - she had not received any word from Boeing Aerospace. In other words she didn't know MTI forced BAC to dump me. She had faith in me and was then continuing to look for other contracts.

Don Thompson and Thiokol had of course lied to Boeing Aerospace about me. I was pissed off. Thiokol had laid fertile grounds for an enormous law suit. As usual Thiokol had shit in their own drinking water.

So, Fay and I loaded up and went to Atlanta instead. We took our time and took in fifty or sixty sights. Civil War battle fields, Ruby Cave, the Chattanooga Choo Choo, President Andrew Jackson's home, etc., etc.. We immensely enjoyed the tours. I felt under the guillotine with my kidney 'rocks at large' problem. I was out of work, blackballed, ill and insecure.

Orgonian, Jan. 6, 1988

CHAPTER 37

ATLANTA

Time: August 1987

We arrived in Atlanta in the afternoon after travelling through some beautiful Georgia countryside. We checked into a motel room and hoped for a fast resolution to our loose ends. The next day I met with a contract house executive face to face. They were very upbeat, and optimistic about their six projects in Savannah and Atlanta Georgia. Next I had to locate some 'on tap' medical help. After a dozen calls to locate an affordable liptotripsy machine I failed. I had another X-ray showing no improvement. I was miserable. Then I had great luck. For years I had been absolutely 'afraid' of going to a Veterans Administration Hospital. I feared them because a friend of mine quit his VA career sixteen years into it as a physician because he thought VA medical standards were dangerously poor. He scared the hell out of me about the VA's inadequate ways! Also I was keenly aware that the VA's funding has been cut to the bone and then some - Gramm Rudman. Frequently the VA is so poor they were months late on my then minor Vietnam Disability Check. If you can't get your small check how in the hell can you trust them with your life? The VA was broke and it really didn't care about wounded and ill war veterans!

Great luck as I said. In absolute and final desperation I called the VA Hospital and asked if they had a liptrotripsy machine. A pleasant woman went to some trouble to avoid the regular bureaucracy and connected me directly to a Senior Resident in Urology - Dr. Rolando Lopez. We talked and I told him I was a consultant 'on the move and in trouble'! He said the VA Hospital didn't have a machine, but they bought time on Emory Medical Schools', a few blocks away. Emory I trusted! He set an appointment next Wednesday at the VA Operating Room lobby.

Fay and I were there with a stack of X-rays. We just sat there, while he held them up to the ceiling light and Dr. Lopez said he could set up liptotripsy in Jan. 1988. We beat around the bush and explained 'better now than when I had a consulting contract' pending. He instantly agreed and shuffled his busy schedule and gave me admittance papers for the next week. He was great and I have been forever grateful.

Dr. Orland Lopez and his lovely side kick Dr. Kathy Easterling were bright rays of hope at a dark and troubled time for me. The next seven days dragged on and the contract houses were drowning in their own economic collapse problems. The Aerospace Contractors they supplied were stopped dead in their tracks for all of the reasons previously mentioned. Poor U.S. economy, the Gramm-Rudmann Law, and no space shuttle launches. No luck that week. Damn!! What in the hell does it take? Experience, education, brains, track record, were of no value. I was out of control of my destiny then. Reluctantly I checked into the Atlanta VA Hospital. Had to do it! The admittance lady was young and had a nice smile, and was the rudest and most impolite bitch I had met in twenty years. After passing her rudely poised questions I spent 2/3 of a day being admitted. Blood, X-rays, etc... I eventually arrived on the 8th floor ready to go off to liptotripsy.

Liptotripsy is where sound waves are focused on the kidney stone and the sound waves smash it. Then instead of pissing a large rock you can piss beach sand. Hell of an improvement - faint!

The next AM I was off by cab from the VA hospital to Emory Medical School, walked in, checked in, and was at the liptotripsy clinic. Dr. Lopez and a pleasant team of physicians and assistants went to work. Instead of being my usual bold self I was 'wimpy' that day. I had severely overdosed on too many heavy duty activities in 1986 and 1987 and they had solidified in my kidney. This medical team was confident, fast, and I cannot laud their professionalism enough. They immersed me in a stainless steel swimming pool. Then they blasted the 'rock' to pieces and had me out of there in well under an hour. It turns out the stone was only a hairs distance away from being unreachable by litptotripsy. An hour later, a day later, a week later I would have been in nasty shape and serious trouble - probably surgical. These wonderful people 'saved me' physically and financially. Six hours later I felt as good as new and was well. I won't go into the pain of the interim six hours passing 'sand'. The only thing wrong with Liptotripsy was the cost - staggering if you have to pay for it. A veritable fortune at that time.

A day and a half later I was out of the hospital. On the way to our motel Fay said "Steve, I don't think there will be any more contracts in the U.S. this year!" I couldn't admit it to myself, but I sensed she was right. We decided in three days we could go back to Seattle and set up housekeeping if the Atlanta bunch, or any bunch for that matter couldn't produce a signed contract. At that time I did not understand that world history was at a juncture and the financial and moral strength of America had just gone over a high cliff - never to return to its former power or glory.

What should I do? Would aerospace and high technology return to former strengths? How extensively was I 'black balled'? Should I change careers?

© 1988 MIKE PETERS—DAYTON DAILY NEWS

'Don't worry, Wendy . . . this pixie dust is guaranteed to work by an engineer from Morton Thiokol'

NEWSWEEK : JANUARY 25, 1988 **13**

CHAPTER 38

RETURN TO SEATTLE

Time: Sept. 1987

Fay was right about the collapse. We would return to Seattle. Driving back I was a touch sore from liptotripsy. How could I fight current changes in our American and world economics? I was one of the strongest Project Managers in the world and I couldn't even get arrested let alone get a job! The better I became, the less work there was. The broader my experience and responsibilities the fewer projects to work on. I considered my friends and associates. We were the tail end of the best generation of technocrats ever produced in America and the world. All of our lives were being torn apart. Our industry was dying at this point in history. America had turned a corner and was declining in every way. The free world was dying while Americans watched athletic contests and Soap Operas. I wished I could shake my ominous feelings by watching "Happy Days". This crap had been creeping up on us for the last five years. Now I sensed our National Defence programs, Space programs, and the U.S. Economy were irrevocably 'over the cliff'. I thought time would tell, and indeed time has told. While our national leaders told us we were better off than ever, we all know our financial strengths were less and less. Our currency and our global respect have devalued to near nothing from where it had been.

We had a fast trip across the southern states and superb weather back to Seattle. We dreaded even crossing Utah. I was uncomfortable and Fay drove all of that day. She crossed Utah like a bat out of hell. We crossed the Bonneville Salt Flats and arrived at Wendover, Nevada, a welcome sight. We arrived there after dark. Flashing neon signs, gambling, drinking, prostitution just over the state border from Utah, all serving the general population and Utah Mormons.

On route I considered and listed all of my obvious options. My options were few. I had been on contract at various Boeing Corporations and Divisions in the past. Maybe I could get in there again. I would try. I had powerful connections at Boeing, but also knew Boeing Aerospace and Boeing Ballistic Missile Division had been laying off thousands of people. Boeing was then forty four separate corporations, all competing against each other. I had consulted to seven of them.

I decided to write *"For the Right Reasons"*. I decided to explore bringing a lawsuit against Thiokol for 'black balling' me. I also decided to slowly and quietly start creating a global business that would shelter me from America's retreating power, position, and economy. Up to this moment all of my successes had been negated by evolving aspects of America's decline.

CHAPTER 39

THE GREATEST BANANA REPUBLIC

TIME Nov. 1987

Our travelling time was not wasted. I was considering new businesses to do. New thoughts for the FBI, how to get a technical/management consulting contract in the greatest Banana Republic of all - America! Yes, Banana Republic! I believe a Banana Republic is defined as a country which changes governments frequently, is bankrupt, borrows excessively, and is powerless in the international scene. You tell me - were we a great Banana Republic or not and that was 30 years ago. Much more devastating now in 2017.

My mind was on fire to get started on this book. A million thoughts on this book were circulating. This book was to have several purposes. If the FBI should fail by a fluke, or be turned off by politicians the fight to keep America in space had to proceed. "*For the Right Reasons*" would be published either piecemeal in a newspaper, or as a book. I suspected this book would become a historical landmark in the U.S. Space Program and new laws, policies, and practices would conceivably emanate from it. I knew my style as an author had questionable merits, but my story should be pungent. I felt physically, emotionally, and financially whipped and wondered if I was up to writing this book then. A huge job! Whether I wanted to write this book or not - I felt a moral obligation to do so. I had felt similarly about going to the FBI about Thiokol's obvious criminal activities. My book may never be sold, but my conscience would be clear that I had done all I could for our national welfare to date. Some of the person's the FBI and I approached said the problems at Thiokol and NASA were "none of their business" and "none of our business". I say "bull shit"! It is all of our business how our tax dollars are wasted, how our resources are wasted, how our nation and freedom are jeopardized and squandered. I cannot say loudly enough how closely our futures and our progenies' futures will be tied to space habitation and explorations! Listen before it is too late! Act before it is too late was and is my credo!

Time was very short for America to alter the evolution of space-related events to our national advantage. Otherwise, "space will belong to our economic and military enemies and America would be grounded on earth. Literally trapped - fewer resources, less space per capita, diminished economic possibilities, lower standards of living, an exploded population, and fewer sources of sustenance and happiness. A sad state of affairs for what was the greatest nation ever. I felt anger and violence toward those who think and say it isn't our business or responsibility". I feel literal vengeance toward those who encumbered the FBI's investigations. I feel disgust for all of those who lied to investigators, and there were many. Castrated sheep take many forms! Half eunuchs, full eunuchs, mice balls, non-patriots, treasonous bastards, and some corporate and government attorneys.

After arrival in Seattle I checked in with the Utah FBI office. There had been stirrings and meetings about what attorneys would prosecute MTI and NASA for. Attorney selection was still in progress. I was offended. The FBI needed an attorney or an army of them on the case on about 1 May 1987 in my opinion. Now in mid-November 1987 there still wasn't a prosecuting attorney assigned? What's going on? Did the FBI drop the case? Was the FBI bull shitting me along? Did the attorney general (head of the DOJ), Edwin Meese, stop the case because he was implicated? Did President Reagan stop the case? Are the DOJ attorneys incompetent? Did Utah Senator Garns squelch the case? Did the D.O.J. Federal Attorney, Brent Ward, squelch the case? All of the above and more? Did the Department of Defence or CIA, or NSA intervene? Did the FISA Court short-circuit normal prosecution? Did anyone give a damn?

Deep down I knew the FBI was performing, but have learned to trust no government agency completely. It's a shame we have to rely on government agencies at all. If a high ranking FBI manager told FBI Utah #1 and Tim to cool it what could they do? With or without the FBI I was going to take care of Thiokol and NASA. They had to be taken care of. I was one of the few persons in the world with the tools available to take care of them. Repairing Americas' Space Program was simply more important than me.

At this time I sounded out some of the FBI sources at Thiokol for Tim. When I talked to Tim I could tell he was back to heavy duty "cops & robbers" and this case was apparently on the FBI back burner. I was angry. Let the drug pushers kill each other! Let's fix the country! Tim did not have the same sense of urgency about this case as the million plus persons unemployed either directly or indirectly by the stoppage of the Space Shuttle. Tim Healy was now telling me "There is no rush", "we have plenty of time"! Who knows what all was going on - probably not even Tim, certainly not me. In retrospect I suspected a group of political wars were in progress and he had to step aside for a time while they resolved themselves.

By this time sources still in Thiokol had found and documented additional crimes in other Thiokol departments and continuing crimes in System Safety. I had the impression the FBI had more hooks deep into NASA, but I couldn't ascertain the full details and scope. Tim protected me from too much case knowledge through all of this. In this book I have to protect Tim, and other diligent participants to a reasonable extent. I do not have to be politically correct of course.

Concurrently, Dr, Tony Laine was pursuing his suit against MTI. Dr. Laine could have gone to the FBI, but instead he wrote to the President of the United States, the Vice President, numbers of Congressmen, Senators, and other government officials. They should have listened to him.

Dr. Laine dropped a $250,000,000.00 law suit on MTI for wrongful dismissal. Many of us who had been in Thiokol System Safety felt that Dr. Laine had acted strangely at Thiokol and there were a bunch of credible witnesses. Via Kent McKinnon, the FBI literally had a witness at the improper firing. Dr. Laine came to Thiokol after me and he verified my original findings on poor design and quality of the SRM nozzles. Dr. Laine wrote other Hazard Analyses on the electrical grounding and plume gas electrodynamic characteristics of the SRM. Before he came to Thiokol he had written a copy righted paper on missile gas plume electro - dynamics. He verified and expanded my findings. On both the SRM nozzle and electrical systems he identified original and unique categories of Catastrophic Hazards at MTI.

Thiokol did not like Dr. Laine. He could not be "controlled" and suppressed! That one blew up in their face. Thiokol failed to control him and they had no interest in his findings (or anyone else's) and he blew up in their faces. When I started writing this book, Dr. Laine's letters to U.S. Congressmen had filtered down through the FBI to Ogden and the FBI was frustrated by them. For example one of his letters states all of the real problems definitively and are corroborative verification of other sources allegations in the first 1/2 of his letter. Then, in the last half of his letter, goes into things which could possibly help Thiokol. Dr. Laine was reported to be seriously physically ill. Was he thinking straight? Did he have a mental breakdown of some sort? I felt Dr. Laine had. Tony had talked directly to Don Thompson several times and always came away dismayed at Don's ignorance and suppression. Tony tried to enable all of us to lease our own personal computers and Don Thompson refused to allow it. Tony wanted to work the problems and was even willing to invest in them. Tony was in serious trouble with MTI for trying to enhance their antiquated library. Tony was acutely stressed by Thiokol's lack of concern. I and others observed Dr. Tony Laine to have his spring wound tightly when he came to Thiokol. When he was fired by Thiokol his spring broke. There was clearly two different Dr. Tony Laine's - pre and post firing. Post fired Tony was a psychological and sociological tragedy. Tony's life crashed and burned which will be expanded on later. In my view and the view of others MTI and the Mormons literally killed him in the long run.

Also concurrently Mr. Roger Boisjoly was then preparing to go back into court on 19 Jan. 1988. Since he had become a prominent media spokesman on the Challenger Disaster and American Space Program he was most likely out of physical danger. No one knew what would happen on Boisjoly's case once he entered the Federal court system. Roger filed several suits. One was a Qui Tam suit - on behalf of the United States government. His total suits amounted to billions of dollars returned to the U.S. Government. My attorney, at that time was Mr. Dick Giauque of Salt Lake City. He was listed as one of the best attorneys in America. In the first moments of meeting him, any rational person would realize he is an extraordinary man. He is simply a "superior intellect". Nearly 20 years ago Dick Giauque's firm had taken Thiokol apart over wrongful deaths from an explosion at another facility. Dick proved that a metallic substance can never touch a rocket propellant chemical component called Ammonium Perchlorate. Also Dick had exposed some directors of Morton years ago for fraud. Dick had been deep into MTI's board room pockets. Dick had also whipped the Mormon Church numerous times. After I came to know him I realized he virtually didn't have much interest in cases under $3,000,000.00 or so. He was apparently a seven digit annual personal income person. He said it was the same trouble and expense to sue for one million as 20 million. He drew his line at three million. He would do no cases under that amount. I felt Dick was clearly a brilliant, alcohol damaged, worshiper of money, an attorney to be feared.

"The crash of 1987"! On 9 November 1987 the U.S. and in fact world stock markets crashed and then they went tripping along in a death spiral - Black Monday! Approximately 51% of the United States invested funds were wiped out in one bold stroke. What do you think that did to my pending consulting contracts? Shit! From Nov. 9th and on into 1988 there weren't contracts for consulting flea pluckers, let alone anything useful. I just worked on this book. No sense even thinking about a contract until sometime in 1988. Since I had been making a serious income for years, it was impossible to move backwards in the American system. 'Overqualified' was a terminal disease. It was very painful to be out of work. The I.R.S. was going to be pissed off, not being able to tap my income in the usual amounts!

What would the "crash of 1987" do to the aerospace corporate stocks? Was the United States now too broke to pursue expensive space projects? Will we be conquered because we can no longer afford to progress and produce? Who really had the most control in the balance (or imbalance) of world economic power in 1983. And now in 2017 after Brexit where the UK voted to leave the European Union the collapses continue. The U.S. dollar was and is substantially devalued all over the world, even more so now 30 years later. How would this affect our high technology efforts? Profoundly no doubt.

All of the high technology industries managers were terrorized. So were my associates and myself. You may be wondering what NASA was up to about now. The usual - confusion, chaos, myopia, waste and public relations campaigns! Sources inside and outside NASA, extremely close to the problem told me that NASA was still jumping through its ass. NASA was still changing Hazard Analysis formats frequently in November 1987 - 10 months after the FBI and I told NASA of the problem! FBI spies inside MTI were more pissed over NASA confusion than MTI crimes. I was alarmed that the FBI put NASA on notice and NASA still couldn't get its act together. NASA seemed to be beyond help. It was now crystal clear to me that NASA needed to be abolished and replaced as I had suggested to the FBI much earlier. The U.S. Air Force was a far better project manager.

On December 29, 1987 Thiokol experienced another tragedy. Five lives were lost in an explosion of an MX missile first stage propellant casting building. [11]

Note the attendant article on the Space Shuttle test failures.

The five persons killed were:

Michael H. Smith 24 years - married.
Laurance W. Rothenberger 47 years.
Jeffrey W. Ellis 28 years - married.
'Del M. Loosli 42 years - children.
Richard J. Goodsel 36 years ?

It was later published that the electrical grounding of the building was in question! When I was at Thiokol, there had been quite a stir when it was found that many rocket motor process buildings were not properly electrically grounded. It was well witnessed and documented that MTI didn't know how to ground a building. Fourteen years on this contract and NASA hadn't

---

Appendix 11: Tragic Day for Utah's Thiokol Eric McMullin, Tribune Staff Writer

figured anything out yet! The more the O-rings on the field joints leaked, the more convinced NASA was of the excellence of MTI's SRM's!! Retards!

NASA had issued a contract to Boeing Aerospace System Safety to look over Thiokol's shoulder, but NASA wanted "something for nothing" as usual. Several sources informed the FBI and me that NASA asked Boeing Aerospace System Safety to do a "Fault Tree Analysis" on the SRM's, but to do it for nearly nothing. Boeing originally thought they could do it "quick and dirty" but it didn't quite work out that way. Everywhere competent people looked, NASA was being inept and poorly directed. NASA wasted enormous amounts of taxpayer money. NASA was then currently only an ethereal shadow of its former self! The great NASA of the Apollo Moon Missions no longer existed. In the opinions of many, after the elegant and effective Apollo Moon Missions NASA disintegrated and became totally incompetent and incapable of managing complex and ongoing operations. Operations and mission management currently would be in far better hands if run by the U.S. Air force. The U.S. Air Force isn't perfect, but it is a long way ahead of NASA in execution of complex long term "operations". Hell, NASA hasn't even been effective maintaining its own facilities. A probable NASA argument is "there isn't enough money in the budget"! I say bullshit! If they made one or two changes and decisions on how to do Hazard Analysis for example there would have been thousands of meetings avoided, thousands of fights between thousands of personnel avoided, and hundreds of thousands of man hours of wasted discussion avoided. The resulting savings could have been used on something useful. Such hypothetical savings on Hazard Analysis format is most minimal of all potential NASA savings. NASA takes over 16,000 persons to launch a shuttle. Civilian corporations trying to pursue civilian commercial launches expected to use about 38 persons to execute a launch. Now that ratio is improved.

- A huge number of NASA's 16,000 persons per launch then did absolutely nothing to very little. Other observations of NASA and my own were that it is very difficult to ever catch a NASA employee doing anything. To make the scenario even clearer - NASA rolls on several thousand wheels. How many attempts are the American public expected to pay for? America's bank was broke already! Why should we think NASA's missions can never be achieved? Note that the truth plus lies has not been adequate to fund NASA.

What were NASA lies you ask - Here are a few:

- The Space Shuttle SRM's are reusable for 20 flights - no way! Many can show them unsafe for one flight.
- The space shuttles are only $150,000,000.00 per launch - $300,000,000.00++ each was closer then!

- NASA's original and revised shuttle launch schedules were blatant deceptions. Well documented. {Of all of the money allocated for the hundreds of flights that never happened, where did all of that money go? I believe the FBI followed the money.}

NASA started as a research organization and it should be returned to that status. In my opinion, NASA was and is totally incompetent to effectively pursue operational space flight activities as it is currently staffed, funded and exists. NASA is both victim of and contributor to the greatest banana republic of all. NASA had become a national disgrace. Now in 2017 NASA has no human flight certified rocket. Russia takes all of the various nations astronauts to and from the International Space Station. NASA is in fact out of control of operations and comprehensive safety of the / our International Space Station.

Who Is The Owner Of The International Space Station?

Nov 25, 2016

Just like the science and engineering that make it work, ownership of the International Space Station (ISS) is complicated.

Ownership is shared between the different partner organisations, namely the US, Russia, Canada, Japan and the European Space Agency. However, rather than owning a percentage of the whole ISS, different partners each own different parts of it.

One side of the station is entirely owned by Russia, while the other side is controlled by the United States, but includes modules belonging to the European Space Agency and Japan. Agreements allow partners to use sections of the space station that don't belong to them

Notice here that not only is America incapable of bringing people to and from the Space Station. Russia Owns 1/2 of the Space Station. Russia transports personnel to and from it for all nations for appropriate fees.

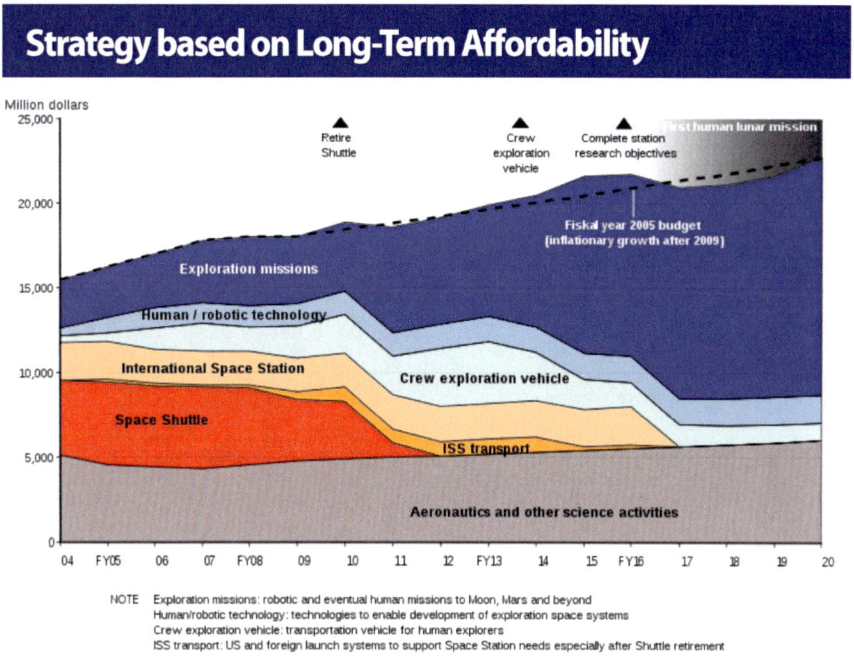

Is this chart more NASA "Spin" or is there substance behind it?

CHAPTER 40

FBI 302's

Time: May 1987 - July 1987

FBI 302's are the case write-ups which are prepared by the FBI for the U.S. Department of Justice. The U.S. Department of Justice then studies them and decides to act on the case or not to act. A simple case could have only one 302 file, a complex case could have dozens of 302's derived from a host of crimes, witnesses, sources, tape recordings, etc..

I do not speak for the FBI. I can only relate that it was my distinct impression the D.O.J. was and is a profoundly weak and impotent segment of America's legal system. My conclusion was not derived frivolously. I only had visibility to a portion of D.O.J. inadequacies and the U.S. Attorneys operations in Utah so I am therefore unable to comment on the entire scope of the D.O.J.'s global performance. If most FBI Agents refer to the DOJ attorneys as LD's (limp dicks) there must be a well informed historical reason. My general impression is that most of the FBI hates most of the D.O.J. and many of the liberal Federal Judges. I watched Special Agent Healy struggle preparing at least three more 302s on other sources and witnesses. Tim 'agonized to be fair, objective, accurate, and in proper perspective'. Tim's efforts might not have been perfect, but they were as good as any human being's could be. Tim agonized toward excellence and fairness.

Even though Tim's efforts were magnificent he was still faced with possibly insurmountable political problems. Tim's primary problem was that no one around him understood rockets. Tim had to educate the FBI, D.O.J. and plenty of others about Solid Rocket Motors, and a massively poor government contract, 14 years of events, integrate hundreds of sources and witnesses while handling 25 other separate FBI Cases. How could a smart government eight-hour-a-day employee come to understand this? A good one couldn't. A typical government drone didn't stand a chance. Particularly slow drones couldn't and wouldn't even read the 302's.

Tim had a super support team headed by Bob Bryant. They were not government drones. They were dedicated American Patriots. They strained at the task. After Tim prepared a group of 302's he offered his 302's up the FBI totem pole for their perusal. Shocked, Tim found that his 302's were too complex for his FBI supervisors. He was 'mind boggled!' He thought he had them reduced to "breath in, breath out" simplicity.

Tim had to boil his conclusions down and down and down. No small task. When Tim's learned supervisors understood, then the 302's went off to the D.O.J. Tim cannot and has never told me anything other than bits and pieces. My assembly of bits and pieces is that when the D.O.J. received the first 302 write ups on Thiokol (approximately August 1987) the D.O.J. Officer / attorney (LD) refused to read Tim's 60 page write-up. Tim was pissed off - crazed in fact. Tim and I had boiled over 40,000 pages, and thousands of events down to a paltry 60 pages and the lazy bastard was too lazy and arrogant to read it!!

I was angered that the D.O.J. were so fucking lazy. I then and there concluded the D.O.J. was a fucking joke and disgrace. Over the last 30 years my opinion of the DOJ has eroded much further. They had a lot in common with Malcolm and others - take the money and run. Perhaps they weren't lazy, maybe they just weren't going to work on this case? Perhaps U.S. Attorney General Ed Meese had his hands on the Thiokol pulse??

When the D.O.J. refused to read 60 pages, shudders of profound insecurity ran through me. The LD's just might be as fucked up as NASA. In fact it occurred to me the DOJ might also be guilty of De Facto Treason. Just where did the greatest Treason lie? OH FUCK! Why me God? Where did my responsibility end? Does our responsibility ever end? Our answer lies in our hearts and our minds.

Tim reworked, refined and finally derived a ten page document from the crucible of his keen mind. This document was understood and finally the MTI Case became a D.O.J. responsibility and problem. Tim and I read 40,000+ pages, the LD's finally accepted 10 pages initially. I asked myself - what in the hell do we pay the LD's for? A bunch of useless prima - donna bastards!

In summary, the FBI 302's are the FBI's product. The conclusions of the 302's are what determines if cases are valid or not. If cases are not valid, then the case agents or FBI managers file the 302's and drop the cases or they are dealt with in other ways . If the cases are valid then the 302's guide the D.O.J. through the case prosecution.

Tim prepared the D.O.J. well for his MTI Cases. His efforts had to be above and beyond the call of duty and he had to do huge parts of the LD's jobs for them. I know for a fact the 302's were powerful and 100 % indicative of severe and extensive criminal prosecutions. Tim's 302's were armed and launched legal 'nuclear weapons'.

CHAPTER 41

CASE ERRORS

Time: 15 Jan. 1987 to present

Several FBI Agents told me that most FBI cases have some errors prior to their conclusions. The FBI's Case # 207C-77 (Morton/Thiokol Fraud against U.S. Government Office of Origin Salt Lake City was no exception. Undoubtedly I will never know all of the case errors, but I can tell about some of them in a general way.

Probably the first case errors were made my me. Being tired, under pressure, and inexperienced I had taken my former friend Malcolm Phillips partly into my confidence. We shared the same exasperations over MTI suppression and apparent crimes. We discussed the crimes as we saw them, we accurately predicted Thiokol cover-ups in advance. Malcolm helped me carry documents out of Thiokol by the briefcases full. Malcolm was 'greased' for the FBI to recruit him. He indicated to me he would cooperate willingly, but he also thought all of the Thiokol problems were none of our business. It was not ours to reason why, it was ours to grab the money and run. I never once felt that way.

Almost a year and a half later I was told by a well informed FBI source that my former friend Malcolm Phillips had sold me, the FBI, and his new country out. Information was furnished to me that on the 6th or 7th of March 1987 Malcolm had cut a deal with Don Thompson.

The sense of his deal was that by sacrificing my FBI cover to Don Thompson, Malcolm received a very long term contact with MTI and it is reported that he also received other substantial pecuniary benefits. This is why the FBI sends some Agents to a special undercover school. I of course was not schooled in 'undercover covert operations'.

I couldn't believe Malcolm could be anything but patriotic to his new country, America. He had immigrated from Australia. I was WRONG! Malcolm, when originally visited by the FBI, sparred with the FBI, was condescending to them and lied to them. Fortunately, some of the FBI knew he lied. Even though this happened he knowingly carried documents for me when I was overloaded, but he had his ulterior motives. Malcolm was manoeuvring for a longer term contract next to his Florida home, more money, more of everything for Malcolm. Malcolm had told me he was manoeuvring to be 'Thiokol's man' at the Kennedy Space Centre. The FBI and I thought that was great at first. Good to have Malcolm's eyes on the launch vehicle at its assembly and launch site.

When I left Thiokol I was polite to Malcolm. Malcolm raised more hell inside Thiokol than I did and he was accepted - I wasn't. I was literally forbidden to raise any hell by Don Thompsons orders delivered via Kent McKinnon. Malcolm was working on SRM handling equipment rather than Flight hardware. I sensed collusion but could not prove it early on. The FBI was aware of my thoughts along the way. Also the FBI knew much more than I did bout Malcolm. The FBI had their own thoughts, wire taps, Security Clearance Investigations, etc. I suspect the FBI 'derogatory' file on Malcolm exploded with new entries as time progressed. Then nearly a year and a half later, Malcolm Phillips and Kelly Guiles were MTI's contract appointees to Kennedy Space Center. So, while nearly all of the original contractors were gone, Malcolm had maintained an extraordinarily lucrative contract (nearly a third of a million dollars per year) through it all, and would live in his Florida home next to the Kennedy Space Centre, and no doubt felt proud of himself.

Other than myself the Indian Engineers were the only persons I knew who fully appreciated or understood Malcolm's incredible farsightedness and ability to plan for his own personal position. So I was wrong, I failed, I had faith where there should have been none. Malcolm had taken up an accelerating new American attitude, "What's in it for me?" I hate to fail, I hate to be betrayed, I hate to be lied to, and am ashamed my judgement slowed down the FBI and DOJ.

Probably the next case errors were when the FBI cornered NASA. A disastrous fuck up! Had the FBI kept the Hazard Analysis in hand which I provided them, not told NASA, and later subpoenaed documents their cases against Thiokol and NASA would have been dirt simple. Once NASA was involved, the FBI had to overcome NASA and Thiokol lies and mutual cover-ups. A sad but true state of affairs. Literally willing partners in TREASON!

When the FBI went to the D.O.J. to work the case, a DOJ attorney, David Long, had to cover his ass due to some technical legal requirements. When he covered his ass he exposed mine. I also eventually concluded he was not keen to do much work or precipitate long hours of work. I suspected he was actually lazy.

One night, just after I departed Thiokol, Tim called me and told me to 'be cool'! An article would probably be out in the next days' paper which would spill some FBI beans, but not name me.

The article is as follows and is Appendix 14a:

Quite an exposure just to cover David Long's DOJ ass! David Long was smart, dedicated, and a real good guy, but he really fucked up and endangered me. Letters from Thiokol went out to all Thiokol employees instructing them on the fine details of how to bullshit the FBI.

Another of my grandiose errors was VR (Very Reliable). She was my friend, my confidant, one of my sources, and much more. Miss VR was partly screened and the FBI was ready, in fact late, to make her an official FBI Source. She was ready, willing and able to help the FBI. VR had internal MTI access of incredible value. Indirectly VR had already helped the FBI massively through me.

Everything regarding VR eventually went wrong. Tim was absorbed by cops and robbers and simply didn't get to her as soon as he wanted or needed to. I did not know she had some 'lurking' emotional instabilities. The Mormons had her by the 'short curlies'. Her Mormon girlfriend turned on her. Thiokol then turned on her, etc, etc. I'm not sure how much damage she caused to Tim's investigations. Perhaps none, perhaps profound? Perhaps she enhanced the FBI necessary 'fixes'. The revelation to Thiokol that VR might be 'FBI associated' no doubt scared the hell out of all Morton Thiokol management. I cannot evaluate. When VR's girl friend told Thiokol Security Miss VR had traded sex for Thiokol secrets there was a huge commotion. I am certain Miss VR traded no sex for anything, but beyond that I'm confused. So much data came available then that I couldn't sort the wheat from the chaff. I had to assume the FBI could and did.

Even though Miss VR was gathering facts, figures, and documents for me to pass to the FBI she had very restricted knowledge of my activities. Did she spill any beans to MTI Security? Unknown. Did her girlfriend go to Thiokol Security or did Thiokol Security make a desperate lunge at one or the other of them? I personally believed that Thiokol Security and management were scared to the point of irrationality at that point in time. I no longer saw that as a bad thing. Rumours and bullshit ran through Thiokol and Brigham City five feet deep because of articles such as Appendix 15a, 15b. I couldn't figure out what was truth or lies, who knew what and who didn't. Tim couldn't put all the fires out. Troubling and depressing. Miss. VR was mad at me for marrying someone else, but I still do not believe her capable of throwing National Security around lightly. She was fully aware of the dire ramifications. Personally, I suspect Thiokol was desperate. I think they pounced Miss VR and did their usual corporate and Mormon threats routine. I was and am concerned that Miss VR's life was destroyed. I am at least partially at fault and I

Appendix 15a, 15b: FBI Probing Morton Thiokol

am sad for that. I tried my best, however, I did not always achieve the desired result.

Bob Bryant, (Utah FBI #1) wrongfully anticipated NASA upper management would help. Even Bob Bryant FBI SAIC could fuck up - though rarely. Bob Bryant was more than displeased with Jim Thomas when NASA refused to meet me because Jim didn't want to upset Thiokol! Jim Thomas said to FBI No. 1, "What would the managers at Morton Thiokol think if they knew I had meetings with an FBI Source or the FBI? I could lose my credibility with MTI!". What more do we need to know about Jim Thomas from NASA? Jim Thomas later verified his conversation with the FBI to the Seattle Times. Astounding! "What will MTI think?" First it wasn't MTI's business to ask and second why should any NASA official care?

There were case disasters that weren't really errors. The case was against the largest employer in Northern Utah and could put Thiokol out of business and seriously damage UTAH. The NASA side of the case was against a primadonna government agency gone wrong.

Utah Senator Garns had displaced an important D.O.D. project to fly aboard the Space Shuttle. He was both Mormon and an ardent supporter of the Shuttle. It suited his political purposes to keep on supporting Thiokol and NASA. Who know what deals had been cut? Everyone wondered. Senator Garns flew on Shuttle Flight STS 51 D.

The case against MTI involved many Mormons, including Bishops, and was in the Mormon stronghold. The Director of NASA, Dr. James Fletcher, was a Mormon. Judges and juries would most likely be Mormons. The lists of disasters and problems went on and on. Tim overcame most of them. Did the D.O.J.? Did Ed Meese suppress the case? Did an entourage of Senators and Congressmen slow or stop the case? Did the President of the United States suppress the MTI Case? An entourage of Mormon and Thiokol attorneys reportedly visited #1 (Bob Bryant) early in the case. Did the Utah D.O.J. Prosecutors act in good legal faith? I know much more than I can tell, however I do not know all. Other than Special Agent Tim Healy, #1, #2, #3 and I, did anybody act in good faith? Damn good question. The FBI had at least three 302's from me, and a bunch of others at various stages of completion. The FBI knew the breadth, depth, and scope of the crimes, their direct and indirect implications. The FBI had always had a priority as follows:

1. Get the Shuttles up and flying safely again
2. Protect FBI sources and affiliates
3. Prosecute the Cases

I personally concurred with the FBI's priorities. I also frequently wondered what the D.O.J. priorities might be as opposed to the FBI priorities? Other peripheral things happened which were interesting to watch unfold. Mahalon Hollway put pressure on Wayne Tilley and scared him off from talking to the CBS West 57 (TV) crew, however Mahalon wasn't slow to get in front of their cameras. However his performance wasn't considered credible and was not used / published / or shown on TV by the CBS reporters. CBS must not have liked what he had to say - they used none of it. Not surprising, they undoubtedly knew who and what he was from an array of sources.

After MTI improperly fired Dr. Tony Laine they sick'd a bunch of U.S. Government Agencies on Tony for improper acquisition of Dept. of Defence and Dept. of Energy Classified Nuclear documents. He had in fact acquired them via legal and correct procedures. Word was out and about that MTI had also accused Tony of being a drug addict. A curious twist is that supposedly MTI had pressured System Safety Contractors to acknowledge his drug use. IT turned out that Dr. Line was using the same antihistamines as Kent McKinnon needed for the hyper allergenic Utah desert. Also Dr. Laine held a Security Clearance far higher than 'Top Secret'. He had been checked for illegal drugs as part and parcel of the Security Clearance processes. MTI personnel had not. No one can ever afford to forget - *"No Good Deed Goes Unpunished"!*

I was at MTI when the shit hit the fan over Tony's requested documents arriving at MTI. Of course the documents were not classified, they were obtained by simple ordinary phone calls. MTI was looking for a way to throttle Tony's insights into MTI"s and NASA's discrepancies. Dr. Laine was essentially under' house arrest' for a few hours for receiving documents helpful to MTI. Souls were bought and sold, corruption of the ugliest kind metastasised to hidden places. How far was the metastasis in MTI, in NASA, in the D.O.J.? Had it spread to the WHITE HOUSE? Was the disease global and 'terminal'?

CHAPTER 42

GOVERNMENT PROCRASTINATION

Time: Anytime

When a great nation's national defences are damaged - how long can the 'fix' be delayed? It depends on the nation, many variables, and the segment of history. Must the fix be coincident to prosecutions? Does the fix demand prosecutions? Obviously there are no simple, universal answers, however, there still remains wisdom, timing, and insight.

I only had piecemeal visibility of what happened inside NASA after the FBI informed them of known problems and crimes. When I went to the FBI in January 1987, NASA was progressing in its inept ways as usual. NASA would not tolerate the truth. NASA apparently only wanted to hear positive 'can do' or better yet 'we did' answers. No aerospace contractor would be rewarded for telling NASA anything negative or for delaying launch schedules. The truth would be punished as usual. This is the same flawed thinking the U.S. Armed Forces had early this century by punishing persons with venereal diseases. Punishment caused the problem to go underground and later surface on a broad front and as a more advanced infection - often fatal.

When the FBI called in the NASA Inspector General's Office to meet me, NASA apparently had entered a radical new phase. Both the NASA Inspector General's Office and Mr. Jim Thomas's office went into high gear to cover up NASA's errors and publish NASA's credibility. The FBI came back to me time and again asking about details; the resounding reply, half under their breath, was 'Those NASA bastards!' I won't name the FBI Agents who said it - there were many!

Mr. Jim Thomas's written reply to the FBI was ludicrous and devious. It was a detailed point by point cover-up and whitewash. When I read it for the FBI I ranted and raved. After the FBI understood it they ranted and raved. This occurred in the first week I was full time with the FBI. Jim Thomas's letter was clever, learned, and informed. It was also spin and bullshit. The FBI and I talked about it word by word for days. The FBI went over the cliff in disgust with NASA at that point. After the proof I showed the FBI of serious deviousness in Jim Thomas's letter the Utah FBI saw NASA as virtually public enemies and treasonous from then on.

The Utah FBI had to deal with a willing rape victim - NASA. Yes, we at NASA and in fact we the American people were raped. It was just wonderful. Our (NASA) rape kept 'front end' SRM project costs down. Usually the equipment worked well enough. Many near catastrophes occurred, a few we even catalogued but they ignored them. NASA folks that liked to hunt, fish, and ski, loved to 'reside' or 'visit' in Utah. And, besides, the Mormons are nice people and have good picnics and great hospitality, and consider themselves 'God's Chosen People.' However, 'board rooms' are different in Utah. The FBI told me that Utah Mormons had the highest 'white collar crime ' rate per capita in the USA. Also the Utah Mormons had the highest divorce rate in the USA. Certainly something to ponder. There is absolutely no question that the Thiokol rape of NASA continued on after the Challenger Accident, continued after the Rogers Commission Report, continued after I went to the FBI, continued after the FBI alerted NASA and the DOJ.

In fact the rape no doubt continues as I'm writing this book 30 years later even though MTI no longer exists. How long can the United States and the free world allow this rape to continue? In my opinion we were already too late then. The first time I met Bob Bryant (FBI #1) he instantly seized the concept that curing the criminality on this case was obviously secondary to fixing NASA and replacing Thiokol and getting the Space Shuttle back up into space. Bob Bryant came to the instant realization that we were dealing with "*Treason*". He was in a 'state of shock'. In an instant he jumped up at a Bountiful restaurant and screamed "that's Fucking Treason" and then he acted accordingly. He sat down, pulled a note pad out of his pocket, and started writing. Next AM he was going to put a 'poison document' into the case file to protect against 'Presidential case interference'. Right there at the table he set many things into motion, against MTI and NASA. He was not a procrastinator and he was very experienced. He soon became FBI no. 2 over the entire agency and held that position for some time.

From the beginning of this dilemma to now the FBI swears to me that enormous NASA overhauls have taken place. Many of the problems were being worked and some FBI Agents are satisfied the equipment was vastly improved primarily thanks to me. These are honest and good men but I assure you the problems were at best bandaged.

Consider the following realities:

1. The field joint O-ring continued in use with a modification thousands of engineers saw as stupid or insane. On the first three launches after Challenger many of the SRM joints leaked hot gases like sieves as they had for years.

2. The SRM cases were the same old cases. Not safe for one flight in many opinions. They failed to meet the required 1.4 safety factor for a total use of 20 flights or for one flight. They 'oil can' bend when they hit the ocean swinging after being parachuted from space, thus permanently weakening them.

3) The same old 1950's dangerous Safe and Arm device.

4) There were no new SRM nozzles development programs. Same old Rembrandts (Nozzles all different and unique 'works of art'.)

5) Countless and ongoing Thiokol false statements to the government.

We have to face reality. The FBI broke NASA's arms to "shape NASA up"! Neither NASA nor the FBI could get the STS program refunded, so America continued with a genuinely poor, obsolete, dangerous, and unreliable space transportation system. Approximately 22 billion dollars wasted on the Solid Rocket Boosters alone up to 1987 Since then untold amounts of money has been wasted. Finally the last shuttle (Endeavour) was retired in July 2011 Some educated opinions view American launch windows into space as no longer paramount. America's space program and its world leadership in space exploration crashed and burned just as Challenger died, and for the same reasons. Conclusion - the FBI acted within its delegated powers to push for fixes. Our U.S. Congress have failed to face up to perilous realities. My personal opinion is that Congress never understood the questions, let alone the answers on space related issues - civilian or military. The U.S. Department of Justice took over two years to act. My perception of the D.O.J. in general is a lazy, uninterested attitude - mostly political drones. D.O.J. attorneys usually want to know what is in it for their careers. The D.O.J. has foiled all attempts by myself and others for freedom of information release of documents. NASA had in fact broken more laws by not furnishing "all" of the requested documents but that is the world we live in today. "Which Law", or "Who's Law". Who really knows? Were deals and settlements made between the FBI / MTI / NASA/ DOJ/ the Presidential Investigative committees, the Mormon Church of Latter Day Saints, or other players? Were sanctions agreed to? Is the D.O.J. into treason via protecting very high level criminals?

I won the war, if not all of the battles. Morton Thiokol Inc. was stopped from bidding on America's forthcoming advanced Solid Rocket Boosters. The two following articles told me I had won the basic war by persistent 'ways and means' rather than straight out 'legal remedies'. As a result of Thiokol getting out of the Space business America would be safer in many ways. [16]

The aggregation of delays has forced American military and defence space launches to use small or geriatric missiles. Their initial failure rate was incredible, but has improved over the years. After a few years we ran out of our old, retired missiles. Our newer rockets are not soundly funded nor are they likely to be. NASA, the National Reconnaissance Office, National Security Agency, and the U.S. Air Force have divergent requirements. Directly or indirectly, U.S. government procrastination has irrevocably damaged our sentinels of freedom, America's gambit into space has nearly terminated American space exploration and space based defences. Tragedy beyond measure. Is the 'super secrecy' to hide our successes or our failures?

Much later the following article showed that Morton (of Morton Salt) was being so damaged that they moved for divesture. [17]

Appendix 16a, 16b: "Thiokol Getting Out of Space"
AP and Times Staff, Seattle Times, 1988
"Thiokol Wont Bid Again"
AP and Times Staff, Seattle Times, June 7, 1988
Appendix 17: "Morton Thiokol Plans to Split..."
Chicago Tribune/Washington, February 27, 1989

CHAPTER 43

PRESIDENT REAGAN AND PENTAGON FOLLY

or 8 years of Ronald Reagan's presidency.
Time: 8/16/89

President Reagan had spoken benevolently about space. He even had his Space Policy published. It all sounded learned, and looked good. The only problem was that his policy, words, and actions were Hollway and empty. Many wondered if he ever read it? For eight years under President Reagan, America's space program had been without any meaningful presidential support. As President Reagan's successor, George H. Bush, entered the White House the national bank accounts were empty and overdrawn. There was not even enough money around to fill the pork barrel, let alone run it over. A repeat of California. Reagan came in during abundance and exited in bankruptcy. A political genius we are still told 30 years later.

There have been articles over the years that President Reagan really did not understand "space stuff" at all. I don't know what he understood, but I know that the office of President gave inadequate and improper support to NASA. When President Reagan was elected the Governor of California it had a large 'savings' account. When he left office California was bankrupt and is in shambles 30 years later. When he was elected President of America, America was in the best financial shape it had been in for decades. When he left it was bankrupt! [18]

As you can see the above comprehensive Pentagon problems are our nation's problems. Our entire industrial military complex was and is faltering.

Wasn't the bottom line of the above article that America's entire governmental system was in failure? More crime than ever before, more foreign and industrial spies than ever before, escalating issues in every sense.

More failure than ever before, and finally more apathy than ever before. I was appalled when the FBI was moving and shaking early in the case to prevent the upper echelons of the White House and the DOJ from squelching their case against MTI. I'm glad the FBI made their moves, I'm disillusioned that they had to. I question that the FBI's poison pills in the files

---

Appendix 18: The Pentagon Probe Poops Out
An August 1988 article from U.S. News:

were entirely effective? I'll never know? 'Poison Pills' in files are placed there in order to prevent Presidential or other political meddling. Another fascinating facet of the Challenger Disaster story is why there was so much pressure on NASA to launch Challenger. President Reagan was scheduled to talk with Christa McCauliffe, America's teacher in space, while she was in orbit as part of his 'State of the Union' address to America. 'State of the Union' addresses are traditionally 'Glossy' since George Washington set the precedent. There were subsequent allegations from various sources, investigations by various agencies and the usual bureaucratic bullshit. In the end NASA claimed it couldn't be rushed, even by the President of the United States!

The White House claimed they would never rush NASA. All of the government agencies and their drones fell right into line and immediately determined that there was 'no collusion' and 'no problems'! Presidential pressure, and NASA launched in the freezing weather amongst the icicles of its own free will and against the advice of knowledgeable engineers, scientists and even some managers. A deadly disgrace!

For nearly three minutes after the Challenger exploded at least four of the astronauts lived. They knew they were dead, they were recorded screaming, saying their prayers, saying goodbye to their relatives. Their 200 G impact with the Atlantic Ocean terminated their suffering and anguish. The other three were probably already dead. These wonderful astronauts were wasted in an act of total NASA, Morton Thiokol Inc., and White House stupidity.

Notice below that the U.S. Government did not make Thiokol pay all the bill for the deceased Astronauts. American tax payers picked up 40% of the bill. At one point the U.S. Government tried to suppress the total amount of beneficiary pay out to $250,000.00. Personally I found that an obscene low amount for the valuable highly educated Astronauts. [19]

After the Challenger accident, President Reagan ordered the Rogers Commission on the Space Shuttle Challenger Accident formed. The Rogers team found plenty wrong with both NASA and Thiokol and they published the Rogers Commission Report on the Challenger Accident.

In my estimation I felt the Rogers commission did a phenomenal job of probing the problems in a short time. Also, in my opinion the Rogers Commission suggestions were appropriate and generally sound though not

---

Appendix 19: "$7.7 Million Paid to Challenger Kin"
Washington AP, July 3, 1988

entirely complete, that would be impossible considering the limited time and hidden documents and conspiracies.

The Rogers commission failed in the same manner as the FBI originally did. Both the FBI and the Rogers Commission trusted NASA to fix their own problem in a responsible manner. NASA could not, and cannot be trusted to fix its own problems. NASA hid behind parallel ghost activities and subcontracts to secondary aerospace contractors not to necessarily repair the problems, but to certify NASA blameless, not at fault. NASA is big on CYA (cover your ass), and herds of government drones!

President Reagan's final blow to me and America's space program was when he vetoed the "Whistleblowers" Bill. The new "Whistleblowers Bill" was unanimously passed by both the House of Representatives and the Senate. It was vetoed by President Reagan.

The following article is from the Seattle Times -

'WHISTLE-BLOWER BILL DEAD

Pocket Veto Exercised by Reagan

by Ruth Marcus Washington Post

WASHINGTON: In a surprise move, President Reagan decided to veto legislation that would have given major new protection to federal workers who expose fraud and mismanagement.

The move to kill the legislation which passed the house, 418 to 0 and the Senate by unanimous voice vote this month, came after four years of intensive negotiations in which congressional leaders thought they had achieved a compromise acceptable to the administration.

Congressional negotiators expressed outrage at what they viewed as the administration's betrayal.

'I think there's going to be bipartisan upheaval around here because the administration gave its word, made a commitment worked out a compromise, signed off on this bill,' said Sen. Carl Levsin, D-Mich.. "If the administration doesn't keep its commitment, then it's going to fundamentally change the way processes works here."

The fact that the veto came after the end of the congressional session, when there was no opportunity for a vote to override, "just rubs our faces in it" he said.

Reagan, in a two-page statement issued just hours before a midnight deadline for signing the legislation, said he was withholding his approval because the bill raises "serious constitutional concerns" by creating a special counsel's office removed from presidential supervision and would essentially mean that the federal government would end up suing itself.

Sponsors said 'the whistle-blower bill was needed because, the Office of Special Counsel, established to represent the interests of whistle blowers, was not doing its job'

The bill was designed to reverse decisions of courts and the Merit Systems Protection Board, a Federal Judicial Agency, that made it difficult for whistle blowers to win their cases.

The bill would have made the Office of Special Counsel an independent agency and would have defined its role clearly as one for protecting whistle blowers. The office now is part of the Merit Systems Protection Board and at times it has acted against whistle blowers who came to the office for help.

Critics accused the special counsel under the current system of attacking whistle blowers rather than protecting them by using the information they gathered in investigations against the whistle blowers.

The legislation would have made it easier for government whistle blowers to prove they were disciplined or fired for exposing the government wrongdoing, requiring them to prove only that their whistle blowing was a "factor" in the action. Currently, they must show that it was a "significant" or "dominant" reason.

In his statement, Reagan said he had directed the attorney general and the directors of the Office of Management and Budget and of Personnel Management to prepare "effective" whistle blower legislation for Congress to consider next year.

Rep. Patricia Schroeder, D-Colo., chairman of the House Civil Service subcommittee, said "I'm really shocked. There was not one Republican or Democratic vote against this. "We worked with everybody, we had (acting OMB Director Joseph Wright Jr.) in there all the time. Supposedly he was speaking for the administration. Obviously the new head of OMB couldn't deliver anything"

Rep. Frank Horton, R-N.Y., a member of the House Operations Committee, accused the administration of "reneging on what was agreed to. I've had no explanation " he said, adding that he only agreed to bring the legislation to the House floor after the administration had signed off.

"All I've heard is that they were going to veto it and that's very shocking to me," he said. "The message now apparently is that we don't want the whistle blowers and we're not going to protect them. I think it's atrocious".

White House spokesman Marlin Fitzwater defended the decision. "OMB did negotiate to try to get a bill that would be acceptable," he said. "The fact is there were major problems registered with the bill by the justice department. Their constitutional concerns were many and grave, and the president just had to reflect these constitutional concerns in making his judgment."

"The presidential veto is the same as an official endorsement of attacks on federal workers who challenge bureaucratic corruption," said Tom Devine of the Government Accountability Project, which has represented a number of whistle blowers in lawsuits and lobbied for the legislation.

Reagan's decision to pocket veto the bill followed a White House meeting yesterday at which Attorney General Dick Thornburgh argued strenuously that the bill was unconstitutional. A senior administration official said Thornburgh's opposition was the decisive force in Reagan's decision."

My blood ran cold as I'd read the article. *Bastards*, or rather whistle-blowers like me, are essentially unconstitutional, and a clear cut danger to *Pork Barrels, Feather Bedding, The High Priest Of Graft And Corruption*, and we can (and probably will) be punished by a bunch of federal drones from the Merit

System Protection Board. So Fucking be it! I finish what I start. I started these cases *"For The Right Reasons"*, I was already destroyed as a consultant to high tech industries, and would not back off. No wonder the D.O.J. was so slow on these cases! This article confirmed to me in a final sense why Roger Boisjoly's Qui Tam case was thrown out of the Utah Federal Courts. I'm still American. We'll just see if the drones would throw my case out of court and stop me!

The President of the United States didn't give a damn about those of us who would stand up and be counted. The President of the United States didn't want the government managers to be bothered! Hell, it's the managers who need the whistle blowers to blow the whistle and inform them what the problems are or the whistle blower needs to blow the whistle on the managers who have failed.

The Presidents' veto was a "swift kick in the nuts" to each and every whistle blower in the United States. A betrayal of enormous magnitude, and a sprinkling of Holy Water on crimes against the U.S. Government, the free world, and our development as a species.

There was yet another horrendous anomaly during President Reagan's tenure. America's space programs had always been billed for the American people, industry, commercial endeavour, and to support the military. Now, all of a sudden, after more than a trillion dollars has been spent, after NASA fails, after the pork barrel is broke, President Reagan wants to give space back to the private sector, i.e. let the citizens and private sector buy and operate it again! Now 30 years later, a number of civilian commercial space launch companies are moving into Cape Kennedy, and also developing elsewhere. The best funded U.S. Intelligence Agency is the NRO, or National Reconnaissance Office which serves many needs of many organizations. The National Reconnaissance Office (NRO) is one of the 17 U.S. intelligence agencies and considered, along with the Central Intelligence Agency (CIA), National Security Agency (NSA), Defence Intelligence Agency (DIA), and National Geospatial-Intelligence Agency (NGA), and other evolving names to be one of the "big five" U.S. Intelligence agencies. The NRO is headquartered in Fairfax County, Virginia, 2 miles (3 km) south of Washington Dulles International Airport. It designs, builds, and operates the Reconnaissance Satellites for the United States Government, and provides satellite intelligence to several government agencies, particularly Signals Intelligence (SIGINT) to the NSA, Imagery Intelligence (IMINT) to the NGA, and Measurement and Signature Intelligence (MASINT) to the DIA (Defence Intelligence Agency). The Director of the NRO reports to both the Director of National Intelligence and the Secretary of Defence and serves as Assistant Secretary of the Air Force (Intelligence Space Technology). The NRO's federal workforce consists primarily of Air Force,

CIA, NGA's, NSA, and Navy personnel. A 1996 bipartisan commission report described the NRO as having by far the largest budget of any intelligence agency, and "virtually no federal workforce", accomplishing most of its work through "tens of thousands" of defence contractor personnel. Personnel like myself. All American Tax Payers support NRO, nearly none have heard of it. It operates at a very high level of secrecy. It obviously has a symbiotic relationship with NASA. After President Reagan, George W. Bush occupied the White House. What would he do? In just a few weeks there was shaking and moving to cut some parts of military budgets. Congress could be led, several presidents have proven that. Without a presidential leader America had no hope of recovering our space race explorations or national security status. [20]

In other words, you have paid for it and you may now do it all over again for yourselves - pay for it all over again!

Appendix 20: "Reagan Sees Space as Private Place"
By Harry F. Rosenthal, AP Washington Feb 12, 1988, Seattle Times

CHAPTER 44

GOING PUBLIC - THE SEATTLE TIMES

Time: December 1987 onward

By late December 1987 I was going crazy that the FBI and D.O.J. had not yet brought indictments against Thiokol and NASA. I simply couldn't believe it! After all, the FBI's 302's were long since complete, the FBI ranted and raved about the Fucking criminals at Thiokol and the scumbags at NASA, and the LD's at the D.O.J..

My wife Fay was born in Rhodesia, Africa which fell and became Zimbabwe. Then she lived in Durban, South Africa and witnessed apartheid problems. Next she lived in London. For the last seven years, Fay lived in Sydney, Australia. Fay said, "Steven, the D.O.J. will never prosecute!" I couldn't believe her. It would be un-American to let these cases pass. I believed Fay was wrong, but I had a creeping agony and terror that the FBI had been silenced from above. There were occasionally strange words and looks from FBI Agents. They were clearly frustrated. Also a lot about the cases had gone secret/silent. The crimes continued on!

On December 15, 1987 I called the Seattle Times. It was on my mind to quietly start having a responsible group of reporters start preparing a publicity spectacular to hit when the Fed's "hit". Instinctively it seemed like a good idea. When I called the Seattle Times I was shuffled off to the special projects editor. He didn't seem terribly interested but said he was leaving and in a few days. His replacement, Mrs. Kathy Tisch would be on board to replace him. He thought I should throw the story to her. I was in no hurry so I thought that was fine. If I didn't like her there were a few thousand other papers to talk to.

On January 15, 1988 I called Kathy Tisch and had a talk. I then became known to her as "FBI/NASA". I didn't want to give up my name yet. Didn't trust her and her friends yet. I didn't know enough about them. I was afraid of them. At that time I had established Trans - International Corporation. My office was directly under the old Central Control Tower at Boeing Field. I was reaching all over the world to do arbitrage business: buy low, sell high.

Kathy and I talked briefly and decided I would call her again on January 16, 1988 and set up a face to face lunch. I suffered serious self induced stress - to go public, not to go public? Are there cases against Thiokol or aren't there? Adding to my frustration were the signals I was getting from my frequent discussions with the FBI. I felt that they were bullshitting me and "handling" me. They still wanted me to do odds and ends, but there was a definitive change. I felt pressured and uneasy. Special Agent Tim Healy had so many cases going on he couldn't possibly spend any time on NASA and MTI cases. Tim was buried in cops and robbers, prostitutes, drugs, murders, bank robberies, and kidnappings.

When do you go to the Grand Jury Tim? Beat around the bush, ho hum, fiddle diddle, B.S.. B.S. He could never legitimately tell me most of what was currently going on, but what I read between the lines was that Tim was on hard times with the D.O.J. and other forces. He never said, but I knew him well enough to know all was not well on our cases and yet the cases were in fact "made". Tim was clearly pressured and frustrated. *Force Majuere!* What if I went public and fucked the cases up? That would be DeFacto Treason! What if I didn't go public? Where did my responsibilities end? What would my conscience let me live with?

I was more driven and more troubled and very unhappy. Fay knew I was troubled. There was no high technology consulting in sight in America and I didn't know what to do or where to go. My mind was on fire and I could not escape the all engulfing uncertainty. The ball had landed in my court some time ago! Every person has their 'cross to bear' and some are heavier than others. Additionally I was creeping my new corporation, Trans-International around the world.

As usually happens - our bodies only take so much. I had a deep cough which soon became pneumonia again. My health, emotions, fears and anxieties were all wrapped around the axle. Lost career, lost friends, lost communications, lost faith and new corporate aspirations. My mind directed me to start a new business. I had created Trans-International Corporation. I was to be a global arbitrage organization - buy low, then sell high. In a relatively short time it was trading all over Asia, the Middle East, Europe, South America. I acquired a group of employees who were divergent, and all up read, wrote, and spoke about 30 languages between them. I landed a USD$38 million Letter of Credit (L.C.) from a Korean firm for No. 2 corn for human consumption. We had the rights to sell a used 747 which had been repossessed in Florida. We had gold from Norway to sell - approximately USD$200,000,000.00. We had ghee going to Africa.

Trans-International did extremely well intrinsically, but global conspiring and enemy forces acted against me! No one would load a ship. Those who contracted to deliver then refused or defaulted. I was under siege and I didn't understand the scope or power of the problems surrounding me.

After some time the facts emerged. The FBI brought some to my attention, police brought others. General Banks enlightened me. Trans-International and I were under attack from the Utah Mormon Industrial Complex theocracy and from Morton Thiokol! I was being systematically destroyed by a team or teams of professionals!

My personal bank account with several thousand in it bounced two rent checks for $600! My business accounts were not operating. I was blocked from all angles. I had proof of both Mormon and Thiokol collusion, but I had to wonder if Senators Jake Garns and Warren Hatch were also targeting me. The answer was "Yes". Were offices or agencies of the U.S. government involved? Was NASA involved?

My company's staff drifted away and Tran-International was isolated and destroyed - with a USD$38 million dollar L.C. in my bank! Should I file bankruptcy, I pondered. No, I didn't. I felt I had to depart the USA. I had no protection there. The FBI admitted they knew I was being destroyed but they could not and would not intervene. Would I be safer in another country? What to do? What should "destroyed" people under siege do in a situation where the President of the United States hated whistle - blowers?

On January 16th, 1988 I called Kathy Tisch and we sat a lunch meeting at the Blue Max Restaurant in a couple of days. It was a restaurant on Boeing Airfield just south of downtown Seattle. We were set for 11:30 AM.

I was punctual as usual since my office was in the same building. Kathy Tisch and Peter Lewis were also punctual. Introductions went around and we were escorted to a quiet table. They asked me about my story and with that my next or continuing ordeal began. Kathy was controlled and dignified. Peter was clearly very smart and also a touch abrasive. He was to the point and wanted to know if I was "clean". I assured them I was. Sometimes the FBI uses "turned" criminals - I wasn't one of those. I have no criminal record of any kind and I had carried various Dept. of Defence Security Clearances for most of my adult life. They seemed interested. They wanted to go back to the Seattle Times Offices and get their high technology big gun, Eric Nalder. Then he and Peter Lewis would come to see me at home.

See me at home, hell! They damn near lived there for the next three months. The Seattle Times spent a fortune on the story, around $250,000.00 they told me later. Peter and Eric went to Utah several times and took other trips to NASA facilities. My story checked top to bottom as I knew it would. The truth will always stand. They interrogated me as severely as the FBI did. On certain aspects they dug deeper than the FBI it seemed to me. Other issues were closed to them. They did not dig quietly however - they stirred up a hornets' nest!

Toward the end of my time consulting to the Seattle Times my health was really going downhill. Also I was really alarmed by what the FBI wasn't saying. There was clearly some heavy bullshit going on between the FBI and several other agencies. Also, as Eric and Peter twisted the arms of knowledgeable persons the FBI wanted to know why I was going public. At first I didn't tell them because I wasn't completely sure why. It seemed partially instinctive, a deep inner voice spoke to me, hell it was screaming!. Simultaneously, my new international arbitrage business was actively developing deals on used locomotives, used aircraft including 747's, gold, etc. all over the world. I was a busy man. Trying to save my self and family.

Now Peter Lewis and Eric Nadler had stirred up a hornets' nest of trouble in Utah, Washington D.C., and all through NASA. In fact I had heard from my sources I was in physical danger. Before, I suspected trouble, now I knew. So, I went to a police station and applied for a concealed weapons permit. A few weeks later I had been harassed sufficiently to start 'packing a cannon'. No sense relying on a 'pea shooter'. Just before and just after the Seattle Times newspaper article was published I was paranoid and felt continual fear. Why shouldn't Thiokol or the Utah Mormons shoot me? The FBI and the D.O.J. were not getting into court. Special Agent Tim Healy called me several times - crazed! Steve, we (the FBI) hear you are going public!". Yes Tim. The D.O.J. is too slow for the good of the country Tim and you know it!. Did the FBI stop me from going public - No! His pleadings were ambivalent -"Oh shit"! "No," as well as "Might not be a bad idea"! I had Tim in a tough spot, but I did it *For The Right Reasons*. I didn't feel that Tim's or my personal needs counted for much, in fact for nothing in the scheme of things important. At the end of the day I felt the FBI wanted me to go public, however there would be more than hell to pay if they said so or were conceived to be part of it.

I knew Tim had FBI internal political pressures on him including the White House. So, I told him I would send him a letter of 'why'. I sensed he had to appease the Gods and his Superiors. What the hell! I might as well take the heat, I was the one going public. My letter was a written 'tool' to take the pressure off of Tim - as follows:

4/4/88

Dear Tim,

As promised this is a letter to tell you what I have been doing with the Seattle Times. My first contact discussions with the Seattle Times began mid-January 1988. I told the Seattle Times 'Special Projects Investigative' Research Staff the Broad outline of the Story.

I anticipated it would take months of the Research Staff to study the MTI, NASA, and National Space Program problems. They agreed not to publish anything until the D.O.J. had indicted the appropriate Case subjects. They were quite interested in the U.S. Space Program since the Seattle economy is driven by high technology and aerospace industries.

I anticipated the D.O.J. to strike nearly eight months ago, and kept continuing to expect the D.O.J. to strike. Now I feel sick inside that the D.O.J. has not struck. I am acutely aware from my continuing contract search all over the U.S. that the U.S. Space Programs are in more desperate shape that ever. I am not aware of everything going on in the industry, nor is anyone else. However, I could write a book on what has not improved and what foolishness is continuing at Morton Thiokol Inc., NASA, and with associate contractors on the Solid Rocket Motors, and the Space Transportation System. Perhaps I am idealistic. Perhaps I am a fool. Perhaps I am improperly informed. Every part of me is screaming FOUL!

I am not a law enforcement person, nor an attorney. I have kept the utmost respect for Bob Bryant, #2, #3, and especially you. I have not ever tried to tell you how to do your jobs and never will. I have counted on the D.O.J. to prosecute your findings effectively. I have long understood a few of the forces on your MTI and NASA investigations.

1. *D.O.J. probably has to help defend NASA from Astronauts families.*
2 *D.O.J. has internal problems (Ed Meese)*
3 *FBI has discontinuous Directors*
4 *NASA continuing poor cooperation with D.O.J. and FBI*

5. *Mormon / Astronaut U.S. Senator from Utah*
6. *U.S. Attorney is a political animal and Mormon*
7. *D.O.J. and FBI both could not act fast enough on Roger Boisjoly's case to meet initial legal time limit.*
8. *Local (and not completely understood) incestuous relationships between the FBI, D.O.J. and NASA, and Mormons.*
9. *D.O.J. probably unwilling to tangle with mismanaged NASA*
10. *President Reagan wants to leave office untarnished*
11. *President Reagan wants to pass the exploration of Space to FREE ENTERPRISE after the U.S. Government has spent roughly the equivalent of the U.S. National Debt on Space Exploration and now can only launch small pay loads into space and not astronauts.*
12. *D.O.J. attorneys are not funded nearly as well as MTI attorneys*
13. *MTI attorneys will be the best in the world, an awesome foe*
14. *MTI and NASA management have incestuous relations*
15. *NASA Inspector General's Office provided poor cooperation to the FBI*
16. *MTI is largest employer in Utah*
17. *D.O.J. Prosecutors would probably face a Mormon Judge*
18. *D.O.J. Prosecutors would probably face a predominately Mormon jury*
19. *D.O.J. difficulty separating and prosecuting false statement issues, technical issues, civil issues, criminal issues, prejudicial issues, theocratic issues.*
20. *1988 is a Presidential Election year*
21. *Numerous other restrictive and contributory issues*

Since the D.O.J. clearly had its hands full I decided it would take a crack newspaper team and make a serious effort to unravel the issues. I felt eventually the issues and truth as well as the D.O.J.. I felt eventually the *comprehensive truth of the collapse of the U.S. Space Program should become public. My book (**For The Right Reasons**) will be my story, but not the whole story. I felt evidently a proper in-depth media vehicle was proper, in fact American.*

The Seattle Times went to work and we all awaited prosecution and still do 2 years later! {Now 30} It has become evident to me this Case has either been shut down or suppressed. The lack of aggressive action forced me to examine my conscience. Originally my conscience directed me to bring in the FBI. Then my conscience directed me to aid the FBI and the DOJ and I have to my maximum capacity. Later my conscience is screaming "FOUL" and my decision was to go public in the most responsible way I know to do it.

My decision to go public is not because I am mad, or displeased with the FBI. All I know about the D.O.J. is that they haven't acted. It is clear to me that the D.O.J. has been bogged down or completely stopped by any of a number of forces or an aggregation of forces. Something is radically wrong. I cannot say what exactly is wrong, but I can bring the problems to a position where they will be dealt with and the FBI and the D.O.J. are no longer restricted from 'acting'! Working with the FBI and the Seattle Times has been very hard on me. It has consumed vast amounts of time and emotional energy which could have been better spent elsewhere. It is a personal embarrassment to me to need to be involved - why hasn't someone else fixed the problems?

NASA was in its most critical budgeting year ever and when the process was over there were changes. NASA, and therefore the United States are budgeting the U.S. Space Program into oblivion (you can watch some of it on TV). My conscience is screaming ! Whenever the Seattle Times articles go public they will be comprehensive, and they are expected to remove political obstacles and free the FBI and the D.O.J. to do their jobs freely and in the best interests of the United States.

U.S. National Defence is severely damaged, over a million high tech and aerospace persons are displaced, our national economy reduced, and our world image diminished. We are not a leader in Space, we are not a follower, we are barely a participant in Space Activities. Tim, your goals of fixing the problem, protecting the sources, and prosecuting are proper. Now, over a year since I left MTI the problems still rage. Another launch is scheduled soon which will no question endanger lives, National Defence payloads, and another orbiter.

It makes me feel sick to embarrass you, FBI #1, #2, #3, sources and the innocent and even the guilty, but it has to be done **For The Right Reasons**! *Now my antagonists will be MTI, NASA, the FBI and the D.O.J. and probably others unthought-of. I never wanted any of this, but it has fallen to me and I have acted in the best way I know to do. You can only tell me what you are directed. And the same for the rest of the FBI. I cannot talk to the D.O.J. I wish there was another way to circumvent the complexities immediately, but I believe no other means exists.*

I and others have risked it all. The D.O.J. has not responded in any meaningful way. MTI still lies to NASA as a matter of course. NASA continually lies to government agencies and the public. Five more persons are dead unnecessarily. The FBI is concerned. The entire free world is at stake.

I could no longer be concerned if the DOJ or the Seattle Times went public first. I would continue to aid the FBI or the D.O.J. as needed in preparation for prosecution if it will help. The Seattle Times was preparing a huge group of articles. They are intelligent and responsible people. The three main personnel were :

Mrs. Cathy Tisch Special Projects Editor Mr. Eric Nadler Investigative Reporter Mr. Peter Lewis Investigative Reporter These Newspaper personnel are not antagonistic toward FBI and are willing to assist you.

I feel sorry for your position Tim. You have worked extraordinarily hard and have done well in my opinion. I also feel FBI #1, #2, and #3 have been excellent support to you. You are clearly in an uncomfortable position and surrounded by extraordinary forces. Your emotional plea to me to be 'controlled' was noted and forgiven. I am certain there have been improvements in MTI and NASA, however it is an insult to my intelligence when you tell me the problems are fixed! The problems haven't even been all listed yet!

Kent McKinnon will have to face his responsibilities as a man, a Mormon, and an American either at a Federal Trial or News Media exposure. I called Kent five times and talked to his daughter, and to date have had no response from Kent. I have confidence Kent will do his duty when it is critical.

No matter what you have to do or say I will always be your friend, retain respect, and in fact call you family. Please no more emotional plea phone calls (actual case work or personal - glad to hear from you). This case has dominated my life and nearly ruined my health. I must now start a new business and save myself financially (in a collapsing economy). There is absolutely no consulting for me in the U.S. Currently nor is there expected to be in the foreseeable future. This letter is meant to be both personal and business. I have felt in danger for some time and am licensed and carry. My idealism is the same as MTI and NASA "everything will be alright". My (our) reality is that the U.S. is at war in space currently (Russians target practice on our satellites). You may show this letter to FBI #1, #2, #3 as necessary.

GOD'S SPEED AND BEST WISHES,

Steven W. Agee
Apt. 1302
10407 S.E. 174th
Renton, We. 98055

If I was dead maybe the D.O.J. and FBI couldn't prevail in the minds of the guilty or fearful. I was not the only dissident. There were many others. [21]

Appendix 21: Space Shuttle "No Good"
By Rocky Mountain News, Jan 29, 1988

Oh shit - another paranoid thought! Thiokol, NASA, the Utah Mormons, and the D.O.J. probably all hated me, but who was my most serious threat? The KGB, the GRU, and the Russian hierarchy must be laughing their asses off at American stupidity regarding its space program. Russia and the KGB probably don't want our problems fixed. Was I only paranoid or could the Russian block intervene? Good question? Just look at MTI's track record on safety! See the Seattle Times article [22]

When the Seattle Times started digging around in Utah, Special Agent Tim Healy was jolted. He was confused by my actions. I could tell he was under pressure to keep his "Goddamned source" (me) under control. I had put him and his career in a tough spot. The Seattle Times eight and a half page article was published on March 22, 1988. It was written by Eric Nalder and Peter Lewis. Nalder went on to receive two Pulitzer Prizes. [43]

The following 8 page article was thermonuclear and changed subsequent history. Here it is:

The Seattle Times Seattle Post-Intelligencer

MOSTLY SUNNY
Sunny this morning. Increasing clouds this afternoon. High, 70s; low, 50s.
DETAILS, D16

SUNDAY

$1.00
May 22, 1988
412 pages

WASHINGTON'S LARGEST NEWSPAPER ■ COPYRIGHT 1988, SEATTLE TIMES COMPANY

TIMES SPECIAL REPORT

In the nation's race to get back into space, some see the same problems resurfacing

Beyond Challenger

IS THE SHUTTLE SAFE?

by Eric Nalder and Peter Lewis / Times staff reporters

School gunwoman left trail of poison

5 sickened by arsenic-laced snacks

Times news services

WINNETKA, Ill. — A woman who went on a shooting rampage at a Winnetka school also delivered arsenic-laced snacks to two fraternities at Northwestern University in Evanston, Ill., and poisoned containers of juice to at least six homes, police said yesterday.

Two deliveries were made to fraternities at Northwestern University in Evanston, Ill. Five people who ate the snacks at Alpha Tau Omega were made seriously ill and taken to Evanston Hospital, Police Chief Herbert Timm said.

WHAT IF . . .

■ The bizarre case of Laurie Dann left acquaintances and authorities wondering: "What if . . ." A 18.

A Northwestern spokesman, Ken Wilds, said last night a laboratory analysis conducted by the university confirmed the presence of arsenic in the snacks.

A subsequent analysis by a police laboratory showed the snacks "contained a very heavy concentration of arsenic," Winnetka police Sgt. Gene Kalvaitis said.

"The situation with Laurie has not ended," Timm said.

The chief said Laurie Dann, 30, dropped off CapriSun juice drinks at six homes in Glencoe, Highland Park and unincorporated Winnetka where she once worked as a baby sitter.

One Glencoe girl felt ill after

Please see POISON on A 18

Appendix 22: "Fatal Shortcut Puts Spotlight on Safety..." May 22, 1988
Appendix 43: Reporters' Relentless Pursuit Of Facts Won {Seattle} Times 2 Pulitzers
Email Article Eric Nalder received Two Pulitzer Prizes
Email Article: Tuesday 8 April 1997

FOR the RECORD
The first of five excerpts from Donald Regan's kiss-and-tell memoir looks at the Reagans and astrology. Scene/K1

COLUMNIST HINTERBERGER MAKES A MOVE
Arts/Entertainment/L1

WINNING COLORS WEREN'T
Sports/B1

The Seattle Times Seattle Post-Intelligencer

MOSTLY SUNNY
Sunny this morning. Increasing clouds this afternoon. High, 70s, low 50s.
DETAILS, D16

SUNDAY

$1.00
May 22, 1988
412 pages

WASHINGTON'S LARGEST NEWSPAPER ■ COPYRIGHT© 1988, SEATTLE TIMES COMPANY

TIMES SPECIAL REPORT

In the nation's race to get back into space, some see the same problems resurfacing

Beyond Challenger

IS THE SHUTTLE SAFE?

by Eric Nalder and Peter Lewis / Times staff reporters

NASA / 1986

COPYRIGHT, 1988, Seattle Times Company

When the failure of a small piece of hardware brought the shuttle Challenger and the U.S. space program to a shattering halt 28 months ago, the nation resolved to find out what went wrong and to fix it.

But while space-agency officials vowed that safety would remain their No. 1 priority — and significantly improved the testing and inspection procedures for assuring that safety — a 3½-month Seattle Times investigation focusing on one major shuttle contractor shows that the system is still flawed.

The investigation of Morton Thiokol Inc., primary maker of the solid rocket booster that failed on Challenger, suggests that in the rush to recover from the disaster, Thiokol and NASA, its boss, are repeating some of the patterns that a presidential commission said contributed to Challenger's loss.

Among those patterns: A desire to maintain a "can-do" image and a reluctance to hear the warnings of those closest to the work.

Kent McKinnon, a systems safety supervisor at Thiokol in the months after the accident, described the atmosphere there in a letter dated Jan. 10, 1987, a year after Challenger:

"If you could see the pressure that management is now under to try and look good to the world and to meet the recovery schedule, it would make the pressure around here prior to the 'incident' seem insignificant."

Many of the concerns that McKinnon and others felt then remain unresolved as the National Aeronautics and Space Administration prepares to launch the shuttle Discovery in late August.

The Times' review of stacks of government and NASA contractor documents, as well as interviews with 200 people involved in the shuttle program, reveal that:

■ Reports designed to identify and address vital safety and quality problems have been delayed and even, some critics say, shelved.

■ Thiokol employees say they have been discouraged, and in some cases prevented, from raising their concerns.

■ In the effort to build a safer booster, important decisions about its design have been compromised by budget and time constraints.

■ And in creating a complex system that combines the brute power of a bomb and the delicate precision of a fine watch, the critics say NASA has strained the process with confusing procedures, a blizzard of paperwork and an overambitious schedule.

NASA / 1986

One consultant, called in to troubleshoot at Thiokol 10 months after Challenger, was so disturbed by what he found that he went to the FBI, which opened an investigation that is continuing.

Top NASA officials insist they have a far better machine today, and some wonder if America is becoming afraid to fly.

"If you don't try to reach for the stars sometime, you really don't know what you can do," said NASA engineer James Thomas Jr.

Meanwhile, a public impatient to see America back in space has begun asking why it's taking so long. Within the space program, others are asking: Has it taken long enough?

STORIES IN SECTION C

School gunwoman left trail of poison

5 sickened by arsenic-laced snacks

Times news services

WINNETKA, Ill. — A woman who went on a shooting rampage at a Winnetka school also delivered arsenic-laced snacks to two fraternities at Northwestern University in Evanston, Ill., and poisoned containers of juice to at least six homes, police said yesterday.

Two deliveries were made to fraternities at Northwestern University in Evanston, Ill. Five people who ate the snacks at Alpha Tau Omega were made seriously ill and taken to Evanston Hospital, Police Chief Herbert Timm said.

WHAT IF . . .
■ The bizarre case of Laurie Dann left acquaintances and authorities wondering: "What if . . ." A 18.

A Northwestern spokesman, Ken Wilds, said last night a laboratory analysis conducted by the university confirmed the presence of arsenic in the snacks.

A subsequent analysis by a police laboratory showed the snacks "contained a very heavy concentration of arsenic," Winnetka police Sgt. Gene Kalvaitis said.

"The situation with Laurie has not ended," Timm said.

The chief said Laurie Dann, 30, dropped off CapriSun juice drinks at six homes in Glencoe, Highland Park and unincorporated Winnetka where she once worked as a baby-sitter.

One Glencoe girl felt ill after

Please see POISON on A 18

Peter Liddell / Seattle Times

Paul Evans, of Federal Way, tries out a five-valve double C tuba at a gathering in Tacoma.

Splendor in the brass

by Kit Boss
Times staff reporter

Julie Castiglioni did not so much choose the tuba as the tuba chose her.

She picked up the cornet in sixth grade in Butte, Mont., and played it well enough to sit in the first chair of the horn section.

But the high school band needed lower brass. So she switched to euphonium — a slightly smaller and higher-pitched relative of the tuba — then learned to play tuba for the orchestra.

"I couldn't fathom playing something like that. I thought the parts would be boring," remembered Castiglioni, now a junior music-education major at Montana State University. "You can't get me away from it now. There's no more beautiful sound in the world."

Castiglioni and three classmates loaded two euphoniums and a pair of tubas into a station wagon and drove to Tacoma this weekend for the first annual Pacific Northwest Regional Conference of the Tubists' Universal Brotherhood Association — or TUBA.

She was joined by more than 100 tuba, euphonium and baritone players who came from schools and professional orchestras around

Please see TUBA on A 12

Gorbachev plans summit offer of joint unmanned Mars flight

by Jim Hoagland
Washington Post

MOSCOW — Soviet leader Mikhail Gorbachev will ask President Reagan to approve a joint Soviet-U.S. unmanned flight to Mars as the symbol of an ambitious new era of superpower cooperation on Earth and in space when the two men meet at the Moscow summit later this month.

In an extended interview here, Gorbachev also expressed hope that he and Reagan will be able to sign an agreement before the president leaves office early next year that would require the superpowers to cut their strategic nuclear arsenals by 50 percent.

CLOSE-UP / A3
■ The summit beginning next Sunday in Moscow is not expected to be as productive as the previous ones.

"I would certainly welcome that," the general secretary of the Soviet Communist Party said, strongly implying that he is willing to meet Reagan in a fifth summit to sign a new arms-control accord even if it has to be left for Reagan's successor to send it to the Senate for ratification.

Gorbachev rebuffed informal U.S. suggestions that some key

Please see SOVIET on A 12

Tug of war ensues behind 'deal' for M's

by Terry McDermott
Times staff reporter

ANALYSIS

The best thing David Sabey has going for him at the moment is not who he is, but who he is not: He is not George Argyros.

At least not yet.

"Listening to Sabey talk, I close my eyes and sometimes I swear I hear George Argyros," says one man who has had dealings with both. "There are the same vocal inflections, the same vocabularies. They even look a little alike. Both try to intimidate their opposition; both have the same football mentality."

Argyros and Sabey, the owner and would-be owner of the Seattle Mariners, are engaged in a war of words, press releases and news leaks over which of them ought to be the owner of the Mariners and why. Argyros owns the team. Sabey would like to.

Argyros, claiming possession is ten-tenths of the law, says the team is his. He's right. He has contracts to prove it.

Sabey, pointing to Seattle pride and a lease that allows a qualified local buyer to purchase 20 percent of the team at an agreed-upon price, has yet to agree to the price but would like soon to buy 20 percent of the team and, at some point, all of it.

Sabey, too, is right. The lease allows him to buy one-fifth of the team for $11.7 million.

In their private actions and public statements, Argyros and Sabey have exhibited what seems to be a disregard for one another.

Please see DEAL on A 13

INDEX

NASA / 1986

When the failure of a small piece of hardware brought the shuttle Challenger and the U.S. space program to a shattering halt 28 months ago, the nation resolved to find out what went wrong and to fix it.

But while space-agency officials vowed that safety would remain their No. 1 priority — and significantly improved the testing and inspection procedures for assuring that safety — a 3½-month Seattle Times investigation focusing on one major shuttle contractor shows that the system is still flawed.

The investigation of Morton Thiokol Inc., primary maker of the solid rocket booster that failed on Challenger, suggests that in the rush to recover from the disaster, Thiokol and NASA, its boss, are repeating some of the patterns that a presidential commission said contributed to Challenger's loss.

Among those patterns: A desire to maintain a "can-do" image and a reluctance to hear the warnings of those closest to the work.

Kent McKinnon, a systems safety supervisor at Thiokol in the months after the accident, described the atmosphere there in a letter dated Jan. 10, 1987, a year after Challenger:

"If you could see the pressure that management is now under to try and look good to the world and to meet the recovery schedule, it would make the pressure around here prior to the 'incident' seem insignificant."

Many of the concerns that McKinnon and others felt then remain unresolved as the National Aeronautics and Space Administration prepares to launch the shuttle Discovery in late August.

The Times' review of stacks of government and NASA contractor documents, as well as interviews with 200 people involved in the shuttle program, reveal that:

■ Reports designed to identify and address vital safety and quality problems have been delayed and even, some critics say, shelved.

■ Thiokol employees say they have been discouraged, and in some cases prevented, from raising their concerns.

■ In the effort to build a safer booster, important decisions about its design have been compromised by budget and time constraints.

■ And in creating a complex system that combines the brute power of a bomb and the delicate precision of a fine watch, the critics say NASA has strained the process with confusing procedures, a blizzard of paperwork and an overambitious schedule.

NASA / 1986

One consultant, called in to troubleshoot at Thiokol 10 months after Challenger, was so disturbed by what he found that he went to the FBI, which opened an investigation that is continuing.

Top NASA officials insist they have a far better machine today, and some wonder if America is becoming afraid to fly.

"If you don't try to reach for the stars sometime, you really don't know what you can do," said NASA engineer James Thomas Jr.

Meanwhile, a public impatient to see America back in space has begun asking why it's taking so long. Within the space program, others are asking: Has it taken long enough?

by Eric Nalder and Peter Lewis
Times staff reporters

Near the northern edge of Utah's Great Salt Lake, alongside the road that carries workers to and from America's biggest rocket factory, an optimistic sign shows a space shuttle roaring toward the heavens.

"Morton Thiokol . . . Things are looking up," it proclaims.

But while Morton Thiokol Inc. and its customer, the National Aeronautics and Space Administration, look heavenward toward recovering America's space program, there are signs that at least one foot is mired in the tragic past.

In the nearly 2½ years since the shuttle Challenger exploded in the cold blue sky over Florida, NASA and its contractors have been scrutinized by a presidential commission, by expert scientists and engineers at the National Research Council, by Congress, and by NASA's own safety organizations.

Each has recommended changes that would minimize the risk in what all agree can never be a risk-free enterprise. And each has been assured by space program officials that the improvements have been made, and that everything looks good for the first post-Challenger launch, now scheduled for late August.

Despite those assurances, the post-Challenger record shows disturbing evidence of a space agency and a major contractor, Morton Thiokol, sometimes unable or unwilling to carry out the chief mandate from the investigating panels: an orderly, thorough and timely review of safety issues, as well as more authority and independence for the people doing safety work.

The presidential commission made it clear that the Challenger disaster might well have been avoided if a "properly staffed, supported and robust safety organization" had been at work.

Among the evidence suggesting that the system remains flawed:

■ Safety reports that should have been completed a year ago were not turned in to NASA until this March. By that time, the equipment they analyzed had already been made and shipped to Kennedy Space Center in Florida. The importance of such reports was made clear by the presidential commission and by the National Research Council — a private, independent adviser to the government — which concluded that the Challenger accident might have been averted if reports and other safety data had not been so out-of-date and in such poor shape.

■ Louis Polaski, a former NASA engineer who served on a special NASA panel created after the presidential commission, said Morton Thiokol and other NASA contractors have been "covering their tails" by adding inspections rather than fixing problems. A top NASA official agreed that a tenfold increase in inspections since Challenger is making them too routine. Air Force inspectors on the shuttle rocket in Utah say NASA has gone so overboard with "inspection points" that they sometimes occur in the wrong order.

■ Though audits are typically a mainstay of government oversight, outside audits of Thiokol's safety and quality have been infrequent. There was a NASA management review in March, the first since 1985, but a full-scale Air Force audit scheduled for June has been postponed until Oct. 17, after the shuttle is scheduled to fly again.

The reason for the rescheduling, said Alex McCool, NASA safety director at Marshall Space Flight Center in Huntsville, Ala.: "We were just concerned they were going to tie up not only the Thiokol people but also my resident (NASA) people out there."

Besides the faults in the safety process, there is evidence that it has been further compromised by officials who, facing extremely tight schedules and budgets, have shunned and even suppressed the safety professionals.

■ Copies of a 400-page report critical of Morton Thiokol operations, written in mid-1986 by George Schick, a Thiokol engineer, were gathered up and allegedly destroyed because Thiokol executives worried it would spur more "Thiokol-bashing."

■ Kent McKinnon, a systems safety supervisor at Thiokol until March 1987, was warned by his boss not to tell a NASA committee about safety problems. McKinnon drafted a letter earlier that year saying, "Thiokol management doesn't really want me to identify anything that might make the company look bad."

■ Steven Agee, a Renton engineer called to Thiokol in November 1986 to help troubleshoot safety concerns there, was banned from meetings after raising questions about the safety of the parts he was examining. The company's unwillingness to address vital safety problems, he said, drove him to bring evidence to the FBI in January 1987. The case remains open.

■ James Thomas Jr., a top assistant to the chief of NASA's solid rocket booster program, was invited by the FBI to talk to Agee and other concerned workers at Thiokol. He declined, he said, because "if Thiokol thought I was part of an investigation into this, I would not be welcome in their plant."

There is more. And though no one can say whether it means history will repeat itself, the evidence is enough to raise again the question: Is the shuttle safe?

The first major examination of shuttle safety after Challenger was undertaken by the presidential commission on the accident, convened within a month after the explosion. Led by former Secretary of State William Rogers, the 13-member panel of experts, including three astronauts, heard weeks of testimony and received hundreds of pages of documents.

Its five-volume report concluded that the primary cause of the Challenger accident was the failure of a rubber O-ring designed to seal a joint holding together the rocket made by Morton Thiokol. The O-ring failed, the commission said, because of a faulty design that made it unacceptably sensitive to factors such as temperature. NASA had launched Challenger when the temperature was below Thiokol engineers' recommendations.

The commission called for changing the faulty joint and seal. But it went beyond those particulars to examine — and criticize — the space program's procedures for analyzing and addressing safety issues. In the case of Challenger, the commission observed, NASA officials making the final decisions to launch knew nothing about O-ring problems that the Thiokol engineers had identified.

Panelists urged that NASA start over again, beginning with a complete review of all safety documents and the reliability of shuttle parts to "identify those items that must be improved prior to flight."

They called for changing NASA's management to expand safety programs and make them more independent. And they called for more rigorous tests and inspections.

The commission also criticized NASA management for maintaining a driving schedule and a "can-do" attitude that discouraged people from raising concerns that might cause delays or damage public confidence.

In the 28 months since the accident, the space program has made significant changes to address those criticisms.

NASA reorganized its safety program to make it more independent. In April 1986, engineers and scientists began redesigning numerous parts of the shuttle's components: the solid rocket boosters, main engines, orbiter and external fuel tank. New hardware has been made, and the whole system is undergoing more testing than ever before.

George Rodney, a respected engineer and former test pilot, was appointed in August 1986 to the new position of associate administrator for safety, reliability, and quality assurance. He reports directly to James Fletcher, NASA administrator.

In October 1986, at the urging of Congress and the presidential commission, Rodney ordered creation of a committee to investigate the status of safety and quality operations at NASA since Challenger.

Two months later, NASA brought in The Boeing Co. to review NASA safety reports and Martin Marietta Corp. to independently review solid rocket safety.

Many experts say the new shuttle equipment is far superior to the old, and there is a heady confidence that the first launch since Challenger will go well.

"If we didn't think the damned thing was going to work, we wouldn't play with it," said Wiley Bunn, NASA quality assurance director at Marshall. "We just couldn't withstand another accident."

Making sure that an accident doesn't happen is a very tall order. The shuttle involves some of the most innovative and sophisticated technology in the world.

More than 100 contractors now share the nearly $3 billion-a-year job of redesigning and making parts for the shuttle, as well as preparing everything and everyone for launch day. Three of NASA's centers oversee different parts of the work: Marshall, which monitors rocket-propulsion systems, in Huntsville, Ala.; the Johnson Space Center in Houston, which handles mission control; and the Kennedy Space Center at Cape Canaveral in Florida, which assembles and launches the shuttle.

NASA's job is complicated by the fact that the shuttle's key manufacturers are scattered all over the country.

The fat, winged orbiter that carries the astronauts is built in California by Rockwell International Corp. The main engines for the orbiter are built by Rockwell's Rocketdyne division, also in California.

The bulky external fuel tank is made in Louisiana by Martin Marietta Corp., and the solid rocket motors for each of the shuttle's two giant boosters are built in Utah by Morton Thiokol.

NASA officials agree that the most hazardous components are the complex liquid-fuel main engines on the orbiter and the powerful solid rocket boosters.

Once investigators concluded that Challenger's loss was caused primarily by a flawed design in the booster joints, the pressure was on Morton Thiokol to modify or replace them.

The redesign work at NASA and Thiokol began in April 1986. NASA signed a memorandum of understanding with Thiokol in February 1987, agreeing that Thiokol would do more than $400 million worth of redesign and testing on the rockets — at

Please turn to C 2

Continued from C1

government expense but without profit. Two weeks ago, NASA and Thiokol agreed to a contract amendment that raised the amount to $505 million.

Those astronomical costs reflect the size of the task. When assembled, the rocket boosters are the largest ever flown, 149 feet tall and loaded with more than 1 million pounds of explosive propellant. Together, two boosters provide about three-fourths of the thrust needed to get the 4.5-million-pound shuttle off the launch pad.

Morton Thiokol's Wasatch Plant, sprawling over 20,000 acres in the arid foothills of the Wasatch Mountains in Utah, is the only source of solid rockets for the shuttle, and for many of the nation's strategic military weapons.

On the day Challenger exploded, Thiokol employed 6,500 people, making it second only to the Mormon Church as Utah's largest private employer. The disaster that took seven lives and destroyed the 25th shuttle flight on Jan. 28, 1986, brought home the enormity of their job as never before.

"It was very rough. Everybody was in a panic mode," recalled Cindy Ferrari, a structural engineer who left the company eight months later.

During the months of finger-pointing and investigations after the Challenger explosion, Thiokol was thrown into upheaval, as people were laid off and then rehired.

The push to recover was frantic because virtually the entire space program — everything from telescopes to spy satellites — had been left waiting on the ground. Forced overtime was the rule, morale dipped to an all-time low, and NASA directives changed more quickly than the Wasatch winds, several Thiokol veterans say.

Les McFee, a tooling engineer who was a Thiokol consultant until last December, said NASA was driven by a new launch date, and that forced the company to comply.

NASA "said we are going to fly, and everything was based on that date," said McFee. "They were picking numbers out of the sky that were unrealistic. There was no way in the world they could meet the schedule. I felt like the people who had the information were never consulted with by the people making the decisions."

Ferrari wrote an angry note of resignation in September 1986, summarizing what she had seen.

"Now at a time which is so critical in the space program," she wrote, "the company has succeeded in reducing morale to an all-time low. The entire engineering workforce is upset, demoralized, and unconcerned with retaining their jobs. Their work output level is falling despite, and probably because of, mandatory overtime. Employees are frustrated since their recommendations are so often ignored. They are faced with unreasonable workloads and totally unrealistic schedules developed by management personnel with no technical training."

After copies of her note spread around the plant, "strangers were walking up to me in the hallways and commenting. Everybody was saying they felt the same way," said Ferrari, who now works for a consulting company in New York.

The burden of the work, the confusion and the frustration were mounting, and the effects on the safety effort were being felt.

'Heart of the safety process'

Perhaps no single element of the recovery effort had been more emphasized in the analyses of the Challenger accident than the system for reporting and resolving safety concerns.

The safety reports are so important that the National Research Council, in a January report requested by the presidential commission, called them the "heart of the safety process." A review of all reports was clearly the presidential commission's mandate to NASA and its contractors.

From these documents, NASA makes the critical decisions about shuttle safety: whether a part or a component needs redesigning or fixing, or whether it is safe enough for flight.

The system for preparing and reviewing these documents is complex, and deliberately redundant. NASA's contractors have primary responsibility for preparing them, and they are reviewed both at the contractors' plants and again by NASA, up the chain of command to headquarters in

Washington, D.C.

NASA rules require that contractors subject shuttle equipment to rigorous safety analyses. Two are key:

The first is a review of each part inside each piece of equipment, to determine every way that a part or system could fail. Always assuming the worst case, this review is known in engineering jargon as a "Failure Mode and Effect Analysis," or FMEA.

From those analyses, contractors prepare a Critical Items List, identifying how serious the failure of an item or system would be. Any item on the top level of that list must be reworked, or must be granted a waiver explaining why it can be used.

The other basic review, called hazard analysis, takes a broader view. It pessimistically considers all natural disasters, hardware failures and human errors that could destroy the shuttle and its crew. Every cause from meteorites to metal fatigue should be cataloged and examined, according to NASA guidelines.

The Thiokol contract with NASA says "a listing of all hazards identified during the analysis shall be provided," and that a report on all hazard work sheets shall be submitted.

While these basic procedures were in place before the Challenger accident, the presidential commission and other panels found that vital safety information was not getting relayed up the channels.

The commission also noted that in the years of "unrelenting pressure" to meet launch schedules, the safety checks-and-balance system had been degraded. It was imperative, the commission said, that as work began on redesigning and improving the shuttle, it be done with an expanded, more independent safety program.

The National Research Council said safety reports should be done at least parallel to the redesign work — soon enough to be a "tool to provide impetus for design change."

Despite those recommendations, the new hazard reports had little effect on the way the rocket was redesigned, according to the research council and the Rodney panel, which issued its first major report in August 1987.

Morton Thiokol began working on the safety analysis of the rocket in the fall of 1986, months after redesign began in April.

The redesign work was exhaustive. NASA offices, including one representing astronauts, submitted lists of parts needing work. Sweeping changes — such as a switch from solid fuel to liquid in the boosters — were out of the question because they would take many more years and billions of dollars to accomplish.

Primarily, the redesign people concentrated on modifying the existing systems. They consulted old safety reports, test results and blueprints to make changes. Informally, engineers working on new FMEAs did affect redesign by raising issues, said Mahlon Hollway, a former Thiokol consultant who now lives in Renton.

But the new FMEAs on Thiokol's rocket, scheduled for completion between August and December 1986, weren't finished until mid-1987. Some were still being reviewed this month.

The hazard-analysis reports were due in July 1987, but Thiokol didn't get them finished until March this year. A NASA board that normally uses hazard reports to decide whether the redesigned rocket meets all safety criteria met last January, four months late, and without benefit of hazard reports.

And while cranes were lowering the rocket segments into place at Kennedy last month, Thiokol safety engineers were meeting with NASA officials explaining their hazard reports for the first time in detail.

"Yes, we are doing it late," said James Thomas Jr., the solid rocket project manager's technical assistant. "Yes, the chance we are taking is that we will have to change or scrap some hardware. If we find we have done a good job, then we will meet the schedule. If we haven't, then it will cost us that hardware."

The rocket booster segments might have to be unstacked. "That's the biggest fear they have, of finding some concern all buried up. That's a real, real possibility," said one Thiokol engineer.

Why were the reports so late?

Before Challenger, Thiokol had two people writing hazard analyses and safety reports. By late 1986, the company had hired more than a dozen consultants to help get through all the additional work NASA was requiring.

Those consultants, as well as other workers at the plant, found plenty of reasons for the delays. The most obvious was volume: As attention to safety mounted, so did the paperwork. But there were other, more troublesome reasons.

'A lot of useless paperwork'

Steven Agee, the Renton consultant who ultimately went to the FBI, says that when he arrived at Thiokol to work on hazard reports in November 1986, there was only one phone for the whole staff, and "desks were on a steal-your-own basis."

He and other consultants say they couldn't get their hazard reports typed, and when two overworked secretaries did complete some, they were full of errors.

And each time a report was finished, NASA changed the format, Agee and other consultants say. Sometimes a column on a form would be moved from one side of the page to another. Sometimes a numbering system was changed.

"Confused and angry worker bees, good consultants and dedicated employees changed and rechanged their documents," Agee said.

The changes delayed their reports for nearly a year, Thiokol and NASA officials confirm.

Thiokol safety consultant Kelly Giles said he feels the blame belongs firmly in the lap of NASA, not Thiokol.

"It's been a frustrating environment," he said. "Just kind of the aimless situation we've been in. They really didn't know

Please turn to C 4

Continued from C 2

what they wanted, NASA didn't."

Bunn, NASA quality assurance director at Marshall, said the delay wasn't intentional. He said the "thing got wrapped around the axle" because each NASA center wanted different formats.

A format was finally settled on in the late summer or fall of 1987, safety officials say, but hundreds of the hazard reports that were written before then could not be used.

Dale Myers, NASA deputy administrator, said if the shuttle were a completely new piece of equipment, the FMEA safety reports would be done before redesign was complete. But because the system was 8 years old, designers used the older FMEAs to help determine needed changes.

The National Research Council, in its January report, anticipated Myers' arguments. If that is the case, the report said, then NASA may have "overemphasized the importance" of the re-evaluation effort ordered by the presidential commission "while simultaneously not giving sufficient attention to its results."

The research council also said the waiver process for approving equipment on the Critical Items List appears to be "biased toward proving that the design is 'safe,' sometimes ignoring significant evidence to the contrary."

But Myers insists that when the new FMEA reports finally came in, "they confirmed that they did the right thing."

But whatever their differences, all agree that the work since Challenger has created a blizzard of paper.

Mark Connolly, a veteran engineer who spent a month last year working on shuttle reliability, said he was kept busy doing "a lot of useless paperwork, like 1,001 ways a bolt could fail." Connolly was recently laid off from Thiokol's strategic division.

The National Research Council urged NASA to set priorities for addressing problems so that the most important ones get the most attention and are not lost in the shuffle. NASA is working on such a program, but it isn't finished. It is also working on a computer system to track hazard reports and FMEAs so that none is overlooked.

The tracking is being done by hand, said McCool, the safety and quality chief at Marshall.

"There is so much paper . . . piles and piles," McCool said.

Although as many as 60 NASA people have been in the plant during redesign, the daily job of overseeing Thiokol operations falls to the NASA resident staff of eight, headed by Edward (Ski) Skrobiszewski, a 35-year NASA veteran.

"With eight people, it's kind of difficult to cover all these areas," said Skrobiszewski. "I think they are doing a good job, considering time constraints."

He said a large part of the work of inspecting hardware is delegated to 115 Air Force people in the plant. And from time to time, more NASA people are brought in.

But even Air Force inspectors can't guarantee quality when they are overloaded with inspections.

Polaski of the Rodney panel said that in its search for fixes, Thiokol "had so many mandatory inspections that mandatory inspections really didn't mean anything to the workers anymore." For instance, he said, Thiokol officials found that workers at their refurbishing plant had grit-blasted rocket cases too deeply — but "instead of fixing the problem, they called for another mandatory inspection point."

But disorganization, paperwork and unnecessary inspections were not the only barriers to consideration of important safety questions, say consultants and others who worked for Thiokol.

Roger Boisjoly was one of those workers — a senior engineer who testified before the presidential commission that, on the eve of Challenger's flight, he and others had adamantly opposed the launch because of concerns about the booster rocket's O-rings.

The presidential commission not only confirmed that Boisjoly was right, but that his superiors had wrongfully overridden his advice.

Six months after the explosion, believing Thiokol had isolated him from the redesign efforts as punishment for speaking out, Boisjoly took sick leave from Thiokol, and never returned. He has sued Thiokol for, among other things, fraud and negligence. And he has become a regular on the lecture and media circuit.

He says his own experiences, as well as those of contacts he maintains at the plant, have convinced him that some things at Thiokol haven't changed.

"Thiokol had a propensity to want to ignore problems, or to ignore people who brought forward problems," he said. "If something of a very severe nature cropped up, they put resources on it to put the flames out, but the embers continued to burn."

In the months since Challenger, new voices have been raised. Some have continued to work in the system. Others have left quietly; still others have spoken out. Their experiences are all different — and disputed by company officials. But they reflect the atmosphere that the Rodney panel found when it visited Thiokol in November 1986.

Rodney panelist Louis Polaski, retired assistant chief engineer for NASA's Ames Research Center, said the panel found "a contractor who was not totally committed to believing safety was an important issue. They didn't have a lot of encouragement from the government, their customer."

Tough critic or team player?

Steven Agee was hired by Thiokol in November 1986 as a safety troubleshooter — one of the consultants brought in to help Thiokol with the hazard reports.

Thiokol's response to his criticism about crucial rocket nozzles branded him a troublemaker and got him reassigned, he says.

Within a few weeks, he wrote 221 hazard reports, calling them "221 ways the space shuttle could blow up." The reports, he says, were shelved.

Believing his work was being buried, Agee went to the FBI and became an informant for the agency while still working at Thiokol.

"I hadn't been there two weeks until I found the nozzles were just an abomination," Agee recalled.

Many experts say the rocket's greatest weaknesses are in the nozzle (at the bottom of the rocket) and the redesigned joint that connects the nozzle with the body of the rocket. Agee found problems with the durability of the nozzle's carbon liner. He also found that the nozzles, which help steer the shuttle, were not uniform. Nozzles on the twin boosters must function almost exactly alike, experts agree, in order to keep them on course.

Because of his persistent criticism, Agee said, he was reassigned to a "more innocuous" part of the rocket.

John Sucher, manager of the nozzle program, said he asked Thiokol's systems assurance manager, Don Thompson, for "team players" — safety people "willing to cooperate and give us positive criticism versus negative criticism."

Thompson said he ordered Agee removed immediately. He said there was no investigation.

"If he spent the time reading the documentation, he wouldn't have had to be bothering them with questions," Thompson said.

"He was asked not to attend (nozzle meetings). I didn't have time to look into it," Thompson said. "I just assigned another man to go. We had plenty of other work to be done. The nozzle is a very complex item. Anybody with any imagination at all can come up with hazard causes."

Agee apparently had a point. In January, 10 months after Agee left, a special Thiokol team issued a 240-page report stating that tests on part of the nozzle disclosed fractures in the carbon material "so numerous, so large, so closely spaced, and so extensive that the integrity of the entire structure came into question."

Engineers recommended a number of changes, some of which were made. Others were put on hold because they were too far-reaching and controversial, and would have required more tests, NASA officials said.

Many of the other hazards Agee identified dealt with the possibility that static electricity or lightning could set off the rocket's solid propellant.

Thiokol officials concede that questions remain regarding the shuttle rocket's vulnerability to that problem, and that further studies are planned, but they say enough is known to certify the rocket for the next flight.

Agee disagrees. He said he found numerous places in the rocket that were insufficiently protected and grounded. He said explosions and fires could occur at the rocket plant, during the rocket motor's 1,500-mile rail trip to Florida, or during assembly and launch.

"It's a world-class electrical problem. Why it hadn't blown already is a mystery to me," Agee said.

A Pershing missile exploded in Germany in 1975, set off by an infinitesimally small static electrical charge. That accident alerted scientists to a danger that hadn't been fully understood before, said Rodney Perala, an expert in the field who is doing work for Thiokol.

More recently, a NASA rocket exploded shortly after takeoff because of lightning strikes. And last Dec. 29, a fire and explosion believed to have been caused by static electricity demolished a Thiokol military rocket-fuel casting plant, killing five workers.

Last year, NASA required Thiokol to install lightning protection on its railroad cars that carry the fuel-holding segments of the booster to Kennedy Space Center. A former Thiokol consultant, Mahlon Hollway, also said the company has widened a metal strap near the nozzle to better ground the rocket.

Thiokol's Don Thompson said many of the hazard reports that Agee and other consultants wrote were dismissed because they were just too incredible.

"Sabotage, for example," he said. "Does that make sense to you? Someone shooting at the Vehicle Assembly Building" where the rocket is put together at Kennedy.

The Rodney panel's Polaski, however, said if Thiokol is withholding hazard reports because they are too incredible, "that's a flaw in their process."

"Who can decide what is a credible analysis or not?" Polaski asked. "Is it done by an individual sitting at his desk, or is it done by a (NASA) panel? An individual can be prejudiced, or swayed by a point of view."

R. Ross Bowman, Thiokol's director of safety and quality assurance, said Agee's work was reviewed and rejected by another Thiokol safety professional and by a Martin Marietta engineer. He would not release their reports, however, and neither reviewer would comment.

Agee points to Thiokol's contract with NASA, which says all hazard reports, credible or not, are supposed to be relayed up the chain of command. He insists that Thiokol held back the hazard reports because if they got to NASA, Thiokol "would have to fix them, and they would probably have to pay for them out of their own profit money. So they didn't want them out the door."

Safety worry put on hold

"During the time I was there, I felt that Thiokol didn't want anybody looking over their shoulder," said Wayne Tilse, a former Thiokol consultant who now lives in Seattle.

He said he was checking the accuracy of a 1981 plan for controlling fractures on the solid rocket boosters when he found that engineers had overlooked the fact that the rocket segments sit at Kennedy Space Center for days, in a wet marine environment.

In January 1987, he concluded that the control plan was "not adequate to qualify" the rocket for future flights.

He said Thiokol made optimistic assumptions about the ability of the rocket case to withstand cracking and corrosion during repeated reuse. The cases are retrieved from the ocean after a launch, and refurbished so they can be reused up to 19 times.

But Tilse said the steel becomes relatively unstable under the heat generated by repeated flights and refurbishments. And the steel is stressed when the rocket segments are stacked into boosters at Cape Canaveral. That could force tiny, existing cracks at the joints to grow, he said, and cause the joints to fail.

Tilse also said the steel used was the least expensive that would withstand the high stress of flight — a poor choice for reusable rocket cases. Bob Omelina, vice president of Ladish Co., of Wisconsin, the case maker, said the steel is on the low end of price ranges for its type, but is very resistant to corrosion.

Tilse said changing steel could cost NASA a fortune, and delay the shuttle program for years.

A new cover for the rocket joints, designed to keep them warm, might help, Tilse said, but other parts of the rocket, such as the place where struts attach the boosters to the external fuel tank, are vulnerable.

Thompson, systems assurance manager at Thiokol, said Tilse's study was being looked at, and he predicted work would be done "before first flight."

But former consultant Mahlon Hollway said tests that were scheduled to assess Tilse's conclusions before the next launch were canceled for budgetary reasons.

While Hollway generally praised Thiokol's recent test program, he said this concerned him: "We want to make sure we can identify a flaw that would be critical."

Hollway, Thompson and Tilse agreed that problems with the cases were not likely on the first post-Challenger flight, but that they could surface on later flights as the cases get more use.

Hollway said the tests may never be done, because once an item is out of the budget, "it is hard to get back. There is always a budgetary problem."

Tossed off the 'tiger team'

After he disagreed with NASA, Thiokol safety analyst Lance Heller was removed last February from leadership of a "tiger team" investigating ground-support equipment hazards.

Reluctant to discuss details, he said it bothered him, but there was no attempt to cover anything up and he didn't "want to give the impression there is some menace."

An associate at Thiokol said, however: "He got his hand slapped real good for not accommodating the customer. People don't want to bug NASA. They see NASA as an authority figure. The way things are done here, when authority speaks, all you see is heels and elbows. Nobody dares ask questions."

Heller was assigned to lead the "tiger team" because there were serious problems completing the hazard-analysis reports at Kennedy, where Thiokol assembles and services its rockets.

At a Jan. 22 conference with NASA officials at Marshall, Heller disagreed with NASA officials over the certification and testing of tethers used to secure the Thiokol equipment at Kennedy.

Heller telephoned Thiokol several times to get answers to questions. A statement written by NASA safety engineer Dewey Channell said "a comment was made" to Thiokol personnel afterward that the meetings would "proceed smoother" if Heller

Please turn to C 6

Continued from C 4

didn't have to place so many calls.

"At no time" did the Marshall representatives recommend dismissing or removing Heller, Channell wrote.

Thompson, Heller's boss, said Heller was removed as team leader both because of a NASA complaint and because Heller disagreed with him over a procedure.

Heller took a job with Boeing shortly afterward, which he described as just coincidence.

'This is part of the cover-up'

Frank Howell — a 65-year-old reliability engineer who worked at Thiokol from November 1986 to February 1987, said his effort to address crew safety was simply snubbed. Howell, who came to Thiokol with eight years of experience in NASA's Apollo program, said he was put off when he took an interest in crew safety after stumbling onto a 1972 report about strap-on rocket boosters that could separate from the shuttle in case of accident.

"I wrote a series of letters to my supervisor. They wouldn't even type my letters or send them forward. This is part of the cover-up as far as I'm concerned," he said. "They didn't even want to talk about it. . . ."

The presidential commission report acknowledged that separating the boosters in an emergency raised nearly insurmountable technical difficulties, but urged that "new ideas and technologies should be examined."

Howell said raising new ideas was impossible: "It seemed to me that crew safety took a back seat from the beginning of the program."

Don Thompson said it was "asinine" for Howell to be bringing up issues not required in Thiokol's contract with NASA.

A long report to nowhere

In the summer of 1986, Thiokol engineer George Schick completed a report detailing Thiokol problems in communicating with NASA at Kennedy. His five-month effort resulted in a 400-page report that found, among other things, that the communication was poor.

That bad communication "caused a lot of cherished documents to end up in dead-letter offices or to get to Kennedy too late," said an engineer familiar with the report.

The engineer said the report also found that NASA and Thiokol were depending on a "buddy system" to talk to each other informally because official channels were clogged. Such an informal system, the engineer said, made it difficult to know whether all proper procedures were being followed in assembling and checking the complex equipment.

The report also questioned whether technicians who assemble the rocket at Cape Canaveral have the proper education and engineering background to implement instructions written by engineers.

A Thiokol employee familiar with the report said management "killed it" out of fear that wide distribution would cause further embarrassment to Thiokol at a time when it could not afford it.

Schick, a Thiokol engineer for 28 years, confirmed that he wrote the report at the request of Thiokol management and that it dealt with communications problems, but he refused to comment further.

Thiokol management denied knowing about Schick's report, or killing it.

Pressure to keep silent

Kent McKinnon, systems safety supervisor for Thiokol at the time Rodney Committee members visited the Wasatch plant in late 1986, also raised concerns that never got outside Thiokol offices.

McKinnon, now a Boeing employee in Utah, has refused repeated requests for comment. The account of his experience comes from Agee, from letters McKinnon wrote but never sent to the committee, and from McKinnon's boss at Thiokol, Don Thompson.

Agee said McKinnon was frustrated by the delayed new hazard reports and worried that upper management was not willing to face the problems of getting the reports completed. Those worries, Agee said, changed to fear when McKinnon met with Thompson just before the Rodney Committee visited.

In a letter McKinnon drafted for committee chairwoman Joyce McDevitt, a letter he never sent because he feared reprisal, he wrote:

"I was not given a chance to say anything at the Rodney visit, and I was cautioned several days before your visit by Mr. Don Thompson that if I said things that I felt were the case in your scheduled meeting that I just as well start looking for a job, because they (management) would get rid of me.

"This was said in front of three engineers who were sitting around listening to the discussion.

"When I realized that I was being required to be silent or face eventual dismissal, I asked to be reassigned to another assignment so that I would not continue to be a problem to them because I did not want to be part of giving a false impression" to the Rodney Committee.

Thompson tells it differently.

He said McKinnon was bothered because he wanted Thiokol's safety team to be reorganized, along lines recommended by the presidential commission, so it was independent of engineering. Thompson said he opposed the idea at first, but later changed his mind.

Thompson acknowledges that he warned McKinnon not to raise the issue.

"I cautioned him that he ought to talk to our management before he talks to Joyce McDevitt. . . . It isn't fair for somebody to stand up before the Joyce McDevitt committee and surprise our management with a statement like that. That's political suicide in any organization, and that's all I said to him. I was just talking generalities.

"You don't go telling a stranger that walks in your door that your organization is no damned good. That was just a caution to him, man on man. No threats. Just clear observation and matter of explanation. There was no threat to it whatsoever."

The stranger to which Thompson referred was the Rodney Committee.

Agee gave the FBI a copy of McKinnon's letter. He said he also told an investigator with the NASA inspector general's office about it during a joint meeting with the FBI.

Thompson said neither the FBI nor the inspector general's office has asked him about it.

Rodney Committee member Polaski said the Thiokol safety program in late 1986 "looked like a bomb had gone off inside the organization. They had all these people reporting to different divisions."

Thiokol reorganized in January 1987, creating a safety, reliability, and quality assurance office under Thompson. Thompson said it was "one of the finest things that ever happened at Thiokol."

McKinnon — who had advocated the consolidation and was expected by many to head it — was downcast on the day it happened, according to Agee and other consultants. McKinnon was concerned it would be "business as usual" under Thompson's leadership, Agee recalled.

McKinnon left Thiokol in March 1987. On Jan. 10, 1987, he had written another note addressed, but not mailed, to McDevitt. In it, he said:

"The mood and climate around here is such that everyone knows that Al McDonald and his fellow engineers (who testified before the presidential commission) were protected by national attention and even congressional support, but heaven help the next person to speak out in opposition to management."

Rodney Committee members, interviewed recently, said they knew nothing about McKinnon's concerns.

If she had received his letters, she "would have taken action," said McDevitt, who is now a private consultant.

Polaski said: "I think that's a very bad situation. It runs counter to doing things in an honest and straightforward manner."

Rodney said McKinnon should have called the NASA inspector general's employee hot line, or used a new NASA safety-problem reporting system set up after the Challenger disaster.

McKinnon did neither, and didn't mail his letters to McDevitt, said Agee, because "he was afraid NASA would come back to his bosses" and get him fired.

"As soon as the hot line was put into operation, a number of concerned MTI employees used it and were burned," Agee said. "The joke was, 'Let's call the hot line, shouldn't take more than 10 minutes to be fired.' "

In their 1987 report, Rodney Committee members said that in their visits to safety programs, "there were many comments on the hesitance of individuals to be bearers of bad news. There were some examples of people who had been such messengers and who had incurred a gamut of criticism ranging from displeasure to outright wrath."

The report also said, "The concern for safety which peaked shortly after the Challenger accident appears to be waning, and in many areas this has been translated by workers, engineers and supervisors to a 'business-as-usual' attitude."

Looking toward launch

The Rodney Committee revisited Morton Thiokol in early April. A report on the visit is due to Congress July 1, Polaski said. Both he and Rodney think things are looking better.

Nevertheless, "we are going to be struggling with how to make negative things sound positive," Polaski said. "You don't get NASA's attention by calling them bozos."

For their part, management at both NASA and Thiokol says the safety process is and will remain a sound and open one.

"I'll guarantee you one thing: There is not a hazard out there we don't want identified and brought forward," said Don Thompson at Thiokol. "It's stupid to hide hazards. . . . Anybody I found hiding hazards wouldn't have a job at Morton Thiokol very long."

"We've nearly accomplished the impossible with all the new requirements that have been levied," said Ed Garrison, president of the aerospace group at Thiokol.

But Kerry Sanofsky, who is now Thiokol's systems safety supervisor, said he wants more time to ensure that proper solutions, or "controls," for the rocket hazards are identified.

"I just think the people in my section wish they had more time. I feel good, and I think the people feel good, about the controls we've documented. The controls for the most part are out there," Sanofsky said. "The idea of having more time would give us a better feel that the controls are adequate."

James Thomas Jr., who is technical assistant to NASA's solid rocket project manager, acknowledged that within NASA there are "rumblings that we are going too fast. There are always going to be engineers who say I don't have enough data, no matter what the program is.

"Putting on my engineer's hat for a minute, I'd say I'd like to do a little bit more testing in some areas as a program manager. I'd always like to do a little more testing, get a little more data," he said.

"But at some point you've got to go try it, if you feel comfortable."

Before the shuttle Discovery is launched, a board will meet this month and next to certify its design. The hazard reports should be done then, NASA officials say.

The National Research Council, despite its criticisms, has already said the shuttle is ready to fly again.

At Thiokol, two more tests of the rocket are scheduled — one next month and one in July. That's just weeks before the scheduled launch, so data will have to be analyzed on a fast track, NASA officials say.

Two 1988 launch dates were postponed because of booster problems. And assembly problems pushed back NASA's internal launch date about another three weeks. Launch will be delayed again, if need be, says NASA Deputy Administrator Myers. "We don't have the pressure of a given schedule. We set a schedule so everybody knows what parts come in when."

All along the way, safety managers say, they'll be keeping the system honest.

At his NASA office in Washington, D.C., Rodney keeps a piece of insulation from a part that failed early in the shuttle program — his reminder to be vigilant for quality. At Marshall in Huntsville, McCool keeps a giant eight ball on his desk.

Rodney said that since Challenger he has had more people to do the job — 1,400 at NASA, helped by 440 Defense Department inspectors.

But the numbers are dropping again at Thiokol, and NASA was under a hiring restriction that started last fall and ended just this month.

The shuttle program budget dropped to $2.9 billion this year, after a high of $5.1 billion last year — which included $2 billion for a new orbiter to replace Challenger.

And while officials exude confidence that the first post-Challenger shuttle flight is getting the support it needs to be safe, they worry about what effect declining budgets could have on quality and safety later on.

Gene Thomas, the chief of safety and quality at Kennedy Space Center, said he got a third of the staff he requested this year, and Rodney said if he can't get more money for Thomas at Kennedy, he may have to delay future flights.

McCool, at Marshall, said he objects to the fact that his safety office must ask for money from the Marshall operations office rather than directly from Rodney, the associate administrator for safety, reliability, and quality assurance.

"When we have to go to the program managers on our hands and knees to give us our money, that is a conflict of interest," McCool said.

There may be some hope in proposed legislation that would provide more streamlined, multiyear appropriations for NASA.

Mahlon Hollway, the former Thiokol consultant, predicts that unless something is done, enthusiasm for continuing the recovery effort will wane.

Right now, he said, "there are no known problems. There is always the unknown. But all of the known things have been more than adequately addressed, except in those areas where the data is yet to be completed."

But, he said, "as soon as we fly again, it will be back to status quo. I don't think we are spending enough money on applied research and development."

Once the shuttle is going again, he said, from flight four on, "I think we may have problems."

If the first daring leap back into space is successful, the ever-present pressures of time and budget may very well increase as impatience to "get on with it" grows.

Already, 239 major scientific and military projects — enough for 134 shuttle trips — are waiting their turn on the "space truck," while the Soviet Union is making major advances in space.

Another shuttle accident might irrevocably damage public confidence in NASA, and with it the nation's future in space.

"My primary concern," said Rodney, "is getting the systems in place to keep us from having another Challenger accident 25 flights downstream, when people might get complacent."

The presidential commission gave the nation a sober warning of how easily that can happen: "NASA and Thiokol accepted escalating risk apparently because they 'got away with it last time.' "

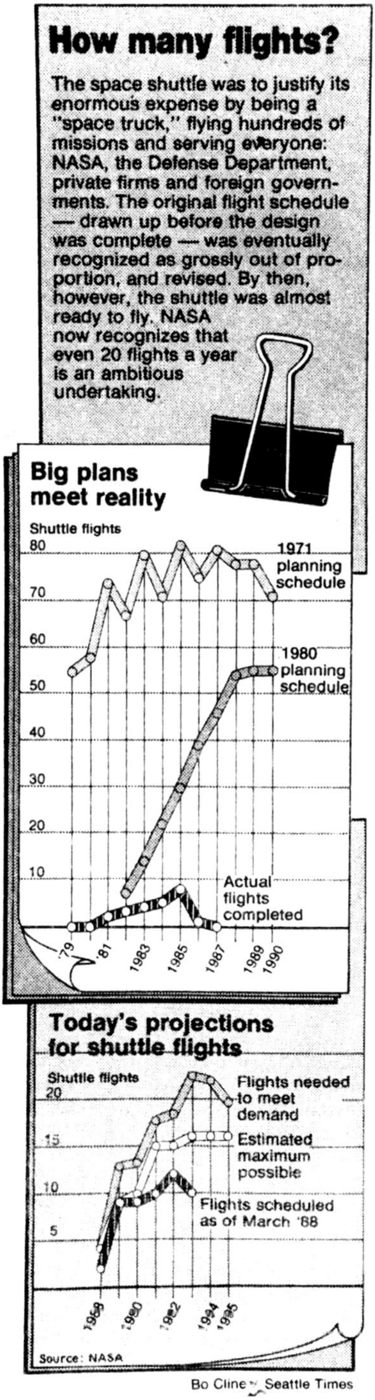

From 1981 to 2011 a total of 135 missions were flown, launched from Kennedy Space Center (KSC) in Florida.

During that time period the fleet totaled 1,322 days, 19 hours, 21 minutes and 23 seconds of flight time. So the life of the Space Transportation System was 30 years.

The average launches were 4.5 launches per year.

The average time in flight and Space was approx. 44 days per year.

All of this is a resounding difference from what was sold to the U.S. Congress.

Also where did thirty years of the money go?

At the completion of my original transcript of this book the Space Shuttle was in the order of 200 launches behind schedule. Then the schedule decayed more. Where did all of the money for the missing launches go? Neither me or my attorneys found a word on the matter. Did or does the FBI know where the missing billions went?

After the article hit the streets all over America my phone slowly began

The Seattle Times / Seattle Post-Intelligencer

BEYOND CHALLENGER: IS THE SHUTTLE SAFE?

When a consultant saw safety problems in the booster program, he called the FBI

A cry of conscience

Tom Reese / Seattle Times

Steve Agee, of Renton, went undercover for the FBI while working on the booster project for Morton Thiokol Inc. in Utah.

by **Peter Lewis and Eric Nalder**
Times staff reporters

Steven Agee's last day at the Morton Thiokol rocket plant was much different from his first.

As coolly as possible, he walked toward the exit, his leather coat and briefcase stuffed with company reports and memos — "the last documents I was going to take out for the FBI," he recalls.

As he left the factory in the high-desert country of northern Utah, where solid rocket booster components are built for the space shuttle, security guards paid Agee little attention. Armed with evidence for federal investigators, he had slipped by them many times before.

It was March 27, 1987. And it was finally over.

What had begun 4½ months earlier as just another consulting contract had become the most frustrating, unnerving and probably the most significant experience of his life. He had become the prime mover of, and an informant for, an FBI investigation probing allegations of wrongdoing at Thiokol's Wasatch Plant.

The months of stress and intrigue at Thiokol had been even tougher, he said, than his days as a combat medic in Vietnam. But what he had seen and heard at Thiokol had forced him to take drastic action — to lay his career, perhaps everything, on the line.

"Sometimes in life," he says now, "you have to stand up to the edge of the cliff and step over it."

Agee's encounter with Thiokol began innocently enough, with a telephone call from a friend who had news of a job. Though unemployed, Agee scoffed at first because the consulting contract, at $28 an hour, was less than he usually made.

A seasoned engineer with a proven record at managing and fixing "sick" projects, Agee felt he would be underpaid in — and overqualified for — the "systems safety analyst" position his friend knew of.

But reality soon took over: Prospects for senior engineers in the aerospace industry were limited. He took the job.

Agee, now 44, came to his work equipped with much of the right background — a bachelor's degree in physics from the University of Texas at El Paso, and contract engineering work on sensitive projects such as the MX missile for Boeing in Seattle and the B-1 bomber for another firm in California.

His former employers generally speak highly of him.

"He seemed to be an aggressive and go-do-it type," recalls Mel Dybdahl, who knew Agee from his work on the MX. "I had the impression he was very capable technically," says Dybdahl, now principal Boeing engineer on a program in Kent to make small intercontinental ballistic missiles.

Wes Kuhlman, an engineering manager for Boeing in Federal Way, says: "He's technically competent, maybe too competent in some ways. Mediocrity doesn't meet well with him, and if it's below that, it's terrible."

tent in some ways. Mediocrity doesn't meet well with him, and if it's below that, it's terrible."

Kuhlman, for whom Agee worked as a troubleshooter in the late 1970s, adds, "He's a supermover, very opinionated."

It did not take long for Agee to form opinions about Thiokol, his new employer.

A man accustomed to going where the work was, Agee was one of nearly 20 independent contractors, or "job shoppers," hired to bolster Thiokol's two-man systems safety program in late 1986. The Rogers Commission, appointed by President Reagan to investigate the Challenger accident, had recommended that primary shuttle contractors upgrade all safety reports and procedures.

The task of the systems safety team was specific, yet broad: Think of anything that could destroy the shuttle or force it to abandon its mission. Faulty concepts, design or human errors, even natural phenomena such as lightning were fair game. NASA requires that the team's work be written up as hazard analyses, and forwarded to NASA.

Agee started work on Nov. 9, 1986 — taking to the task with a vengeance.

At first, he was assigned to work on the nozzle at the bottom of the rocket. Almost immediately, he says, he found disturbing problems. But as the new guy on the block, he says, he was content to keep quiet and take the measure of things.

It wasn't long, though, before Agee's natural bluntness took over. He began speaking up at meetings, he says, and discovered that the systems safety engineers were not welcome.

"You didn't get Brownie points for writing hazards," he says. "You just got trouble, so a lot of people found none."

Finally, he wrote a memo detailing his concerns. That, he says, got him reassigned.

He started asking questions and poring over documents — and quickly determined that "there was almost nothing on solid rocket motors which could not cause catastrophic hazards if there was any malfunction."

Statements like that were received with mixed feelings by Agee's colleagues at Thiokol. Some agreed with him; others said he was given to exaggeration.

"I think he could be great, if he didn't fantasize too much," said Mahlon Hollway, who left Thiokol this month and now lives in Renton.

But from Agee's perspective, there was plenty to do.

Agee, who brandishes his patriotism like a weapon, said he and others in systems safety wanted to do a good job "because we knew that lives and national security would depend on solid rocket motor reliability."

But frustrating obstacles, it seemed, were everywhere. It was difficult, sometimes impossible, to get documents — including Thiokol's contract with NASA — necessary to compare designs with NASA requirements. Further, NASA kept changing the reporting format.

About the only thing going right was Agee's relationship with his supervisor,

Kent McKinnon. McKinnon would not discuss his experiences at Thiokol, but in a memo written to his superiors Dec. 11, 1986, he sought higher pay for Agee and other job shoppers, describing them as "very skilled and talented."

McKinnon, an unassuming, religious man, wasn't much like his rough-talking, world-traveling new employee, but they respected each other as professionals, says Agee. And they shared an urgent concern for getting on with the work.

So McKinnon gave him some advice, Agee says: Take a low profile and "fall out of sight of management."

Agee says McKinnon had been following his own advice, especially since November 1986, when a federal safety panel visited the plant. The panel, appointed by NASA's head safety man, George Rodney, was making the rounds of shuttle contractors to assess their progress on upgrading procedures and reviewing safety issues.

Agee says McKinnon told him he wanted to address the panel, but McKinnon's boss, Don Thompson, had told him to keep quiet or risk losing his job.

Thompson says he made no threats but told McKinnon it would be political suicide to criticize the company to outsiders.

But McKinnon, says Agee, saw it as a threat: "Kent was literally in a hollow-eyed sweat over the pressures. . . ."

The final frustration

Agee says he was reassigned to a part of the rocket believed to be relatively trouble-free — the igniter. Sitting atop the rocket, it shoots a flame down the center of the booster to light the solid propellant inside.

Agee started finding problems.

"I had asked a very simple question: Was the system electrically grounded?"

He concluded it wasn't, and started preparing hazard analyses.

As Christmas neared, the plant emptied, but Agee and some other job shoppers worked on. He wanted to pin down hazard analyses and send them to NASA.

It was not to be.

Because NASA was still changing the reporting formats, the two secretaries who were typing hundreds of pages of reports were turning back nothing usable, say Agee and others who were there.

Then, in early January, came the final frustration.

Thiokol announced, Agee says, that McKinnon was to become the manager of a new safety organization free of internal pressures, answering directly to NASA and to one Thiokol vice president.

Agee and the others were thrilled, he says. But the exhilaration was short-lived.

The day after the appointment, Agee recalls, McKinnon announced to his staff that "he had just been told by Don Thompson that even though there was an organizational-chart change, 'nothing had in effect changed.' The Thiokol operation would run as it always had."

McKinnon was dismayed. He "wanted to make sure everything was done right, so that nobody else would have been killed" in another shuttle disaster, said one systems safety employee. "He may have felt that was being put in jeopardy, because nobody would listen to him."

Agee decided it was time to have somebody else listen. He chose the FBI. While the agency will not confirm anything, an FBI letter replying to a request for documents he had given them denies the request on grounds they are part of "a pending investigation."

Agee had acted as an FBI informer once before, when he believed a California contractor was supplying defective goods and overbilling the government on the B-1 project. This time, through a third party, he contacted the agent he had worked with earlier.

On Jan. 15, 1987, another FBI agent, Tim Healy, visited Agee's apartment in Brigham City.

Agee, a man whose allegiance to friends is as passionate as his disdain for foes, took an immediate liking to Healy.

"I can't remember how long we talked, but it was a long time, and a lot of coffee went by the wayside. . . . He indicated he would like my help and would be talking to FBI No. 1," the special agent in charge, Robert Bryant, the next morning in Salt Lake City.

Soon after, Agee says, he started meeting with Healy three or four times a week, five or six hours at a time.

"Tim told me that the Salt Lake City office had gone off like a bomb over this thing," Agee says.

Early on, Agee says, Healy told him the FBI had filed court papers giving him permission to take Thiokol documents.

A short time later, the FBI set up a meeting at its Ogden office, where Agee and Healy met with two investigators from the NASA inspector general's office.

Agee says he asked one of the investigators what he thought of his allegations, and the investigator replied:

" 'Steve, I can't believe what you're telling me. I believe you believe it, but I can't believe that MTI (Morton Thiokol Inc.) would withhold safety information.' "

Agee started pilfering documents in earnest. "I could have stolen that whole plant."

Once committed to working with the FBI, he also started looking around for other Thiokol employees who might help.

Sometimes his intuitions were right, other times wrong.

Agee would take someone to dinner and ask a few questions. If things felt right, he'd set up another dinner, and bring along Healy, whom he would introduce as a friend just out of the Marine Corps.

Agee declines to specify how many Thiokol employees he helped turn into spies, or how many of them remain on the job. But recruit them he did, he says.

Before going to meals, Agee sometimes was wired to record the conversation. Healy assured him that activity had also received court authorization, Agee says.

The tape recorder fit under his armpit, a radio transmitter in his shirt pocket, an on/off switch halfway down one arm, and a microphone inside the other sleeve. An agent would monitor the radio transmission from nearby.

The equipment was cumbersome, but besides the physical discomfort there was a psychological unease: He was recording not only colleagues, but people he considered friends.

Toward the end of February, the FBI arranged another meeting with NASA in Salt Lake City.

Present, among others, was James W. Thomas Jr., a technical assistant to NASA's solid rocket project manager, as well as the FBI's Bryant and Healy. Waiting in the wings was Agee.

But Agee was never introduced because Thomas refused to talk to him, Agee says.

"If Thiokol thought I was part of an investigation into this, I would not be welcome in their plant," Thomas said later in an interview.

It was after their meeting with Thomas, Agee says, that FBI agents expressed "enormous disappointment in NASA's covering its tracks and not cooperating." They initially viewed NASA as "a victim and an ally," but came to regard it as an agency that "could never again be trusted," Agee recalls.

'Today is to be your last day'

Back at the plant, Agee tried to make himself invisible to management. "I could not continue my job or help the FBI if I was fired, so I made every effort not to get fired," he says.

At the same time, he was "ready to explode with hazards," and pressure was growing on McKinnon.

Soon after, McKinnon took a job with Boeing in Utah. On McKinnon's last day at Thiokol, March 20, 1987, Agee was called into the office of the man picked to succeed McKinnon, Kerry Sanofsky.

Sanofsky, says Agee, told him: " 'There are forces in MTI which really don't want you around, Steve. Today is to be your last day.' "

In a later interview, Sanofsky criticized Agee's work but declined to say why he fired him.

Agee says Sanofsky told him he could not divulge the reasons, but was authorized to give Agee a final week of work on condition he "promise to do MTI no harm."

Agee then "went into extremely high gear" to produce more hazard analyses and smuggle out more documents.

By the time he walked out Thiokol's door for the last time, he had turned in 221 hazard analyses — reports that Sanofsky assured him would be reviewed and forwarded to NASA.

Agee says the reports were shelved.

He had planned to set aside two weeks to help the FBI before pushing on to another job. But on April 6, he says, Healy told him the FBI had approved Agee's coming to work for the agency at $50 an hour. He was assigned the code name "Brains."

Agee's job was to read the NASA-Thiokol contract and to point out to the FBI everything he thought was not being done by Thiokol or was being done wrong.

Agee provided the FBI with lengthy lists of Thiokol and NASA personnel to be interviewed. He also compiled a list of allegations — 56 involving Thiokol, 42 related to NASA. Those lists were boiled down to 18 items of initial FBI interest, Agee says.

Thiokol management acknowledged the FBI investigation in an April 22, 1987, memorandum to employees urging them to seek a company lawyer if contacted by the FBI. In recent interviews, management indicated it had no information regarding the status of the case.

Agee says he worked full time with the FBI until July 24.

Since then, he has looked for other work as a senior engineer in the aerospace industry, but found nothing.

Last November, he returned to Renton. He has kept in touch with both the FBI and the informants he had helped recruit.

Today, Agee is writing a book about his experiences and setting up an international business to buy and sell big-ticket, high-tech items.

He is concerned that the FBI's investigation has taken so long, but he remains hopeful that something will come of it.

In a long letter to Healy recently, Agee tried to explain why he went public:

"Originally my conscience directed me to bring in the FBI. . . . Now my conscience is screaming 'foul.' . . I never wanted any of this, but it has fallen to me, and I have acted in the best way I know to do."

to ring. Eric and Peter's article was a slow cooker. It caused a huge commotion but in a quiet and methodical way. There was no Senatorial or Congressional outcry to me, but I guess the FBI heard from a flock of them.

Now 30 years later the Seattle Times no longer has this famous article on its website. I have to wonder why? It is still germane, it was a pivotal article, and now it appears to be gone! Who would have tried to remove it?

I had to have a new career to save myself. I hadn't gone on unemployment so as to not cause the D.O.J. / FBI Cases problems. I didn't' want to tell the Mormon Judge and Mormon Jury I was on unemployment benefits? Thiokol attorneys could use it as proof I was the 'worthless bastard' they wanted me to be. I in fact needed a new country to live in.

About a week after I went public I received a call from Channel 11 TV, Tacoma, Washington. A Mr. Jerry Kemp, reporter, would be to see me at 11:00 am. He wanted to shoot footage for evening News. Jerry and his cameraman came. Jerry said this was "red hot", it'll get 1 minute and 45 seconds on air! So that's what America's space program is worth - one minute and 45 seconds of TV news time! "What's the fucking use?"

Some of the other local Seattle TV Stations picked up the story. Since a local newspaper broke the story some of the local TV stations shunned it. If they hadn't found it, then it couldn't be worthy news. More political bullshit. The general Seattle Public consensus was that the Seattle TV stations wouldn't know a story if it bit them on their ass.

So I had gone public and entered American archives. Soon I was on two 10 minute segments of ABC News in Salt Lake City, Talk Radio in Salt Lake City, and numerous other programs such as CBS's West 57th Street, Good Morning America, Night Line, and literally hundreds of others. My message was disbursed globally without question. Was I technically dead or alive?

My letter to the FBI prior to going public seemingly set the D.O.J. in motion. Now the DOJ were running up and down the hallways I was told by those who were trampled. Their excuse of no performance could now be blamed on Agee going public as the reason for their non-performance for years. The LD's were hanging in the vacillating political winds. Going public was traumatic and life endangering. As soon as I went public one FBI Agent whispered to me to keep "watch on your 6 o'clock very carefully." I loaded my 'blaster' and took care. I had feared having made a mistake by going public. Then, I was glad that I did. The articles and media splash put the fear of God into Thiokol, NASA and the D.O.J.. The FBI needed all of the help it could get.

Soon after, the D.O.J. finally entered the FBI 302 Cases into the Grand Jury process. They did it with the usual secrecy and the media had a hard time prying information out of the FBI, D.O.J., and Grand Jury. That was of course the 'Mormon Grand Jury' no doubt. To this day I wonder why I was not one of the witnesses subpoenaed to the Grand Jury?

I found out years later the DOJ found the cases 'lacking evidence'! Well what that meant is the FBI Agent who went on and on about the theatre of it all was absolutely correct. "You Mormon witnesses, you Mormon jurors, you Mormon Judges", you Mormons ad nausium was absolutely correct and true. Accurately Insightful Agent Hagey! What actually happened was the religion drove the politics, which drove DOJ, etc... American Justice failed and America lost her 'high ground' in the space race. Tim Healy's lips have been sealed on this disgraceful segment of his career every since. The 'FBI Code of Silence' and sometimes Ordered Cover Up 'from above.' A Girl Scout could have made the cases! Obviously the DOJ and Congress managed the NASA and MTI problem another way, 'Ex - Law', or 'Cover-ups, or both? Whatever happened was not the standard practice of law. At a minimum the coverups had to have been touched by Congressmen or Senators. Did a US President touch it?

CHAPTER 45

FBI, NASA, MTI RESPONSES

Time: 22 March to Present

The FBI was officially non-committal formally about my going public. Some of the FBI top personnel thanked me and hoped the prosecutions would now get underway. FBI sources (Not Agents) told me I may have hurt the cases, but probably not. After all, the FBI and D.O.J. did have a lot of MTI crimes and NASA crimes to choose from - before my time at MTI during, and after. Also, at this time, the FBI was watching MTI destroy documents. At first I was shocked that the FBI would allow continuing destruction of documents to continue, but then I realized in many cases the absence of documents or forgery of documents later was proof enough. I had very restricted visibility of this aspect of the Thiokol cases. The FBI Agents seemed to be laughing themselves silly watching (monitoring) Thiokol. MTI couldn't destroy the copies held by the FBI.

There were rumours that high ranking FBI officials were pleased to get this case out in the open. The FBI reaped a harvest of immediate and direct benefits from the Seattle Times article. After MTI employees passed the article through work areas, the FBI received a landslide of new calls and correspondence enumerating other crimes and critical information not previously known to me or the FBI. A windfall of new FBI investigations.

There was limited official response from the FBI. The FBI finally publicly admitted there was an investigation. Then the following article ran. [23]

So! The Utah Grand Jury was in motion at last, and of course then 'shut down'. Thiokol had a lot to say about how "no good" I was, but they continually failed to ever mention that the FBI had them by the balls.

Bear Bryant (No. 1) laughed with Tim Healy as they had never seen a corp. in such a bind. If they killed me it was proof of guilt. If they didn't kill me it was proof of guilt. MTI and NASA were both in serious pain.

Soon, the Ogden Standard Examiner ran the following article: [24]

Appendix 23: " Subpoenas Issued For Makers of Shuttle Part"
Peter Lewis and Eric Nalder, Times Staff Reporters, Seattle Times
Appendix 24A, 24B: Thiokol Shoppers Deny Ex-Colleague's Allegations
By Jennifer Paul, Utah Standard Examiner

Then on 24, 1988 the following article showed up in Contract Engineer Weekly in rebuttal to Ogden Standard Examiner article. Not only did Ron Clarey stand up and be counted, he was non-solicited, but a little later on the television show "West 57th Street" he appeared on a pre-recorded spot with me. He had written approximately 400 Catastrophic Hazard Analyses himself at that time! Ron Clary's resounding rebuttal was as follows:

Monday 7 / 13 /88
Dallas, Texas

"Dear Ron and Jerry

I just read your 7 / 13 /88 issue with the article "Thiokol Shoppers Deny Ex-Colleagues' Allegations" and felt compelled to throw my two-cents into the commentaries.

Firstly, this 'colleague" of Steve Agee does not deny any of his allegations. That's because I, and every other shopper in his group (Systems Safety) to my knowledge, were never informed either of his allegations or his "221 ways the Space Shuttle could blow up." By my definition, this in itself constitutes evidence of suppression of Safety relevant information within the very special group of professionals specifically charged with System Safety analysis.

I did work with Steve for a period of about two weeks when he suddenly and mysteriously disappeared from his desk. I soon heard that he had been laid off' - and I mean fast. I had no opportunity or occasion to discuss any technical issues with him, but can say that he was very helpful and cooperative in my apartment search efforts while he was there. I did, and still do, hear rumours casting negative aspersion on his competence by a few, always in the absence of Steve. So I filed them in the trash and wondered what the source might say about me in my absence. As for suppression of safety concerns, I support Steve. I witnessed, and was involved in, and was the object of several such actions. At least the effect was suppression. Nearly all could be explained or defended as management prerogative, lack of resources to pursue, information unavailable, decisional (s) pending, analyses pending or not provided for, restrictions on access, and the list goes on. One of the most effective was continuous " of the System Safety group as a whole. Many millions of NASA's budget seemed to be spent on reworking, over and over, the several hundred Hazard Analyses (HA's) developed by the group. Administrative support (secretaries, stationary, computers) was so minimal as to approach invisibility. One had to spend much time and resources dealing with 'The System'. Relatively little was available whether to update old HA's or initiate new ones., and the basis for new ones was limited only by the time and resources available to study the design, testing, processing, handling, assembly, and disassembly of the Solid Rocket Motors (SRM's).

I know that one of my favourite, and apparently endless, sources of Safety concerns (the basis of all HA's) was field observation of actual testing, processing, assembly, etc.. MTI requires all observers in proximity to live propellant to wear special safety clothing. MTI does not furnish this equipment to 'shoppers'. We always had to beg and borrow it from the directs, who were loath to lend such 'personal' possessions. It got to be such a hassle that I finally was compelled to abstain from all field trips.

Most of my colleagues didn't even try'."

Months later I received a copy of Ron Clarey's complete rebuttal, and here is the remainder of his letter which was not published at that time.

"is this another form of suppression?

I could continue with many more examples. I don't see anyone yet with the gumption to discuss the influence of the Mormon Church at MTI. Or how about the obvious fact that those who deny Steve's allegations have no personal knowledge of either Steve or his safety concerns, and no competence in safety engineering. However, this is 'standard practice at MTI".

For the record a few parting titbits' that are not common knowledge:

1. *On or near March, 1987, the ex-supervisor of System Safety resigned to accept a position at Boeing. In parting, he submitted a letter to the Rogers Commission citing broad safety concerns with SRB's. Why does MTI withhold this information?*

2. *MTI does not allow safety concerns to be expressed or documented that have no clear way of being resolved.*

3. *MTI imposes resolutions of safety concerns to not be recorded, with no participation or even notification allowed to the engineer.*

In a wrap of this ramble, I thank God that none of my children are likely to be astronauts. To me they are an endangered species and I am not about a repeat flight of V51-L, the Challenger Disaster.

Sincerely,

(signed)

Mr. Ron Clarey "

Shortly after the Ogden Standard Examiner article was published the FBI called and asked 'who in the hell is Ford?' We have been investigating for two years and we never heard of him! I told the FBI I knew him hardly at all. He had been a new arrival as I was exiting. The FBI was jumping mad. They, I, and anybody with the intelligence of a pet rock could see that MTI and the Mormon mafia had control over the Ogden Standard Examiner. KZZI Radio in Salt Lake City told me that both MTI representatives and the Mormon Church representatives had been to visit them and requested they never deal with the subject of FBI or Steve Agee. Later on, the reporters from CBS West 57th Street told me Thiokol and the Mormon Church had each sent an entourage to NBC to try and suppress anything on Steve Agee or the FBI cases against them. Mr. Glen Silber of CBS told me about that. Glen also told me a similar entourage had been to ABC and CBS. Sheila Hamilton of ABC TV Salt Lake City told me a similar entourage had visited ABC's Salt Lake offices and ABC national headquarters. She also told me an MTI and Mormon entourage had been sent to NBC. She told me the entire media industry was aware of MTI's public relations efforts to suppress my and all FBI notoriety. The entourage was composed of officials, public relations firms, and attorneys. I was pleased they were so frightened by Glen Silbner's article: [25]

Appendix 25a, 25b: " NASA Overhauls the Shuttle" New York Times

Everyone but NBC showed news and TV footage on me and the FBI case. I have wondered if the Mormons or Thiokol really got to NBC? I think they may have or, possibly NBC ran a segment on me and I and my friends missed it?

Generally, the media response to me and the articles about me were very positive. The media generally was down on both Thiokol and the Mormon mafia. No one from the TV media treated me improperly other than some perhaps ignored me which was OK. The only exception to proper treatment from the newspaper media was the Ogden Standard Examiner by Jennifer Paul. Her article was so pitiful, slanted, uninformed and prejudiced that I felt sorry for her. Stupidity and ignorance must be painful. For weeks after her article, she called me at my office. I politely told her she was too stupid and prejudiced for me to even consider being interviewed by her. Political incorrectness!

The Seattle Times and I sent copies of their article to Senators and Congressmen all over America and Mr. Carlucci of the Department of Defence. I don't know who came down on the FBI and the D.O.J., but I heard rumblings there was an avalanche of government officials wanting answers. So much the better. That's why I went public - *"For the Right Reasons"* - FOR CASE EXPOSURE! When I read the following article I was slightly encouraged. Perhaps a few souls out there are getting at least part of the message, NASA Overhauls Shuttle. [26]

The word from my sources told that NASA was finally getting tough on Thiokol - about fucking time! About 14 years late. Was anybody getting tough on NASA? Yes! The FBI had and was continuing to do "attitude adjustment" on NASA. I have reason to believe the D.O.D., C.I.A., and NSA (National Security Agency) were also - possibly others undetermined. The dominoes were beginning to fall. The Russian star was rising, then the Japanese Space Program, the European Space Programs and more recently the manned Chinese programs. Even the evil empire of N. Korea now has primitive ICBM's and nuclear weapons 30 years later. About a month later the Russians quite successfully launched their unmanned Space Shuttle. It is interesting to note the Russians didn't continue with their version of the Space Shuttle. They remained with rocket technology for higher orbits.

On May 22, 1988 I sent the following letter to the HONORABLE Judge William Sessions, the new Director of the FBI at that time.

Appendix 26: " Panel Voices Concern About NASA Safety" Sunday 24, 1968, The Arizona Republic

Dear Judge, Sessions,

I am FBI code name BRAINS, Salt Lake City office, case agent Mr. Tim Healy. Case subject - Morton Thiokol Inc., and NASA.

I wanted to send this to you and tell you personally Mr. Tim Healy has done an incredible effort on these cases and I know for a fact that he has fought against all odds and for all of the right reasons. I also want to be clearly on record that so far as I could tell #1 (Bob Bryant), #2, and #3 have also been totally supportive. I admire their support of Tim Healy on this case and especially their genuine concern for U.S. national security.

I only went public when it was apparent there were political forces obstructing both justice and national security. After long anguish I realized I was probably the only one who could push the cases forward without losing their career. I have already lost mine. So I went public to clear a lot of blatant political road blocks to the FBI.

I hope you and your staff note in the articles that the FBI is treated with the greatest respect. I have never thought the hold-up and delays were in the FBI and I insisted the Seattle Times did well by you.

I sincerely hope these articles clear the way for appropriate disposition of these cases. Our American space program has suffered a near fatal blow, and our economic losses over Morton Thiokol Inc. and NASA failures are beyond measure.

I will of course continue to support any FBI activities on these cases.

Best Wishes
Steven W. Agee
Apt. 1302
10407 S.E. 174th.
Renton, Wa. 98055
Telephone: (xyz) ---

I wrote the above letter for a number of reasons. First and foremost I wanted to make sure Tim Healy didn't get beat up on by FBI hierarchy for letting his "damned source" get out of control. Losing control of a SOURCE is a cardinal sin to an FBI Agent - they lose career Brownie Points for it. Also, I was concerned that the FBI might feel insecure about my continuing support. Nothing would keep me from supporting their investigations. Also there were a number of other monumental political log jams. I knew the log jams were there by what the FBI and DOJ didn't tell me more than from what they did tell me.

I soon received a letter from the Director of the FBI as follows:

U.S. Department of Justice

Federal Bureau of Investigation

Office of the Director

Washington D.C. 20535

July 15, 1988

Mr. Steven W. Agee
Apartment 1302
10407 Southeast 174th
Renton, Washington 98055

Dear Mr. Agee:

It was kind of you to write on May 22nd to express your support of our investigative efforts and to commend Special Agent in Charge Robert Bryant and Special Agent Timothy Healy. We in the FBI take great pride in our personnel, and it is always good to have that pride reinforced by commendations of our professionalism. I will share your remarks with Messrs. Bryant and Healy, and I am sure they will join me in thanking you for writing. Your continued support of this Bureau is appreciated.

Sincerely yours,

William S. Sessions
Director

True to his word Tim received a nice letter from Judge Sessions. This letter from Judge Sessions may appear as a small morsel to you readers, but it is one of my only emotional rewards during the entire operation. My FBI source work literally almost killed me, and might yet. It ain't over until it's over, and it ain't over yet - 30 years later!

The word came back to me from all over America that the Seattle Times Article had really put the fear of God into both Thiokol and NASA. A lot of scared castrated corporate sheep thought some starved wolves were after them and it was thankfully so. NASA response to the Seattle Times article was as usual -- There were no problems! We fixed everything they said! As these cases evolved I came to learn that NASA was in fact primarily a public relations operation. Very long on public relations, at that time short on technical and managerial expertise. NASA was now beginning to feel the heat of no major launches for nearly two years. I hoped the Seattle Times article "Beyond Challenger" made NASA squirm and thrash as hard as possible.

NASA had failed and the results were tragic. Geopolitical economists would be required to assess the scope of the comprehensive planetary impact. Was / is recovery possible?

And the news continued - Test revamped, Part of Rocket Cracked? SNAFU or was it FUBAR (Fucked Up Beyond All Recognition) in military terms? [27]

Appendix 27a, 27b, 27c: " Safety Comes First..."By Washington AP
"Test of Revamped Shuttle" August 29, 1987, National, Briefs
"SNAFU Wouldn't Have Hurt Liftoff" National Briefs, July 1, 1989

CHAPTER 46

THE D.O.J. FAILS

Time: Who knows?

Appendix 28 are articles that came out soon after I went public. They characterize the GAO's (Government Accounting Office) investigation, and the investigation of Ed Meese, Attorney General of the United States, and others. Some of this was directly or indirectly related to the procurement crimes at Thiokol. [28]

So the FBI closed their cases and passed the crimes on to the U.S. Senate and Congress via an instrument known as a "Letter Head Memo" to the U.S. Senate and Congress. A powerful tool I am told, but not 'due legal process'. There were indications this route was taken to save massive legal prosecution fees, huge amounts of time, and most of all to 'save face' for America and NASA. There were also insinuations that there were issues related or entwined with the NSA regarding important launches of very highly secret intelligence satellites.

My efforts continued to retrieve more documents legally available to me via FOIA (Freedom of Information Act). Time after time the FBI told me the documents would be forthcoming. Time after time the D.O.J. evidently held them up. Restricting the FOIA's documents from me was illegal and aided and abetted both M.T.I. and NASA. I had and have a truck load of documents still, but some I was too busy to acquire for myself during these troubling and busy times. We all thought it was also protecting the guilty at the D.O.J.. [29]

So not only MTI and NASA failed. Now the D.O.J. turned out to be rotten to the core in my view. Fay had been closer to the realities of U.S. government failure than I. I was a misguided honest American - very naive and surrounded by out of control clandestine forces. I was out of my depth. I became cannon fodder! No wonder there were no indictments. Now I thought there would be some indictments on minor charges and agencies and officials would move to hide the serious stuff. Fay felt there would be no trials. Time would tell. Could the D.O.J. just ignore these cases and they would go away? Could be. Had I wasted my career, personal security, and family safety for nothing? Probably so!

Appendix 28a, 28b: "GAO Criticizes Justice Dept...."
Bu Phillip Shenon, New York Times
Appendix 29a, 29b: Agee letters to FBI SAIC Robert Bryant

The D.O.J. trudged along and failed. Eventually the new, vital D.O.J. attorneys in Utah put all of their heads together to define the Crimes and the best approach to the prosecutions. The disgusting result of the DOJ legal consortium was that they couldn't find a way to prevail in a court case. The Legal Consortium literally covered the crimes of others and in fact joined and consummated the original perpetrators' treason. The FBI cases were clean and profound. D.O.J. failure was no doubt requested and orchestrated because a 3 year old could prove or verify the crimes. *The FBI had been castrated* it appeared. Wouldn't be the first time or the last. What was the percentage of political, theological, economic, and other factors. I can't help but feel 'theological / political' was the major percentage. I had and have in my possession a huge horde of documents which the FBI has copies of. However there were some I didn't have copies of due to lack of time, exhaustion, etc.

Here is FBI response to my FOIA request. [30]
Fair enough, the rules are the rules and the LD's had their asses to protect.
Does it sound to you like the cases against Morton Thiokol were over?
Does it sound like the U.S. Government did a side-step? Did anyone other than the Mormon 'spin-doctors' move off of dead centre?

Appendix 30: "High Risk of New Shuttle Disaster..."
William. J. Broad, New York Times, April 9, 1989

CHAPTER 47

THE UTAH GRAND JURY 8/16/89

Time: Near First Launch after Challenger

Close to the launch of the first shuttle to return to space after the Challenger accident the Media was really chasing me night and day. Also about that time it became known that there was a grand jury convened in Utah and Mr. Bill Richards of THE WALL STREET JOURNAL was pushing for information. It turned out he had just been assigned to the Space Shuttle safety irregularities by WSJ. He talked to my new attorney, Edwin McLean, and Edwin pointed him in my direction. When Bill Richards called me at my office I was horrified by his lack of knowledge. He was two days late and fifty dollars short, but yet there was something about him that was compelling. We talked and the next day the following story appeared.

THE WALL STREET JOURNAL 58

Thursday, September 15, 1988

U.S. OFFICIALS SAID TO REVIEW THIOKOL PROJECT

by Bill Richards - staff reporter of the WSJ

CHICAGO - A former space program employee said Federal officials are investigating allegations by him and others that Morton Thiokol Inc. officials suppressed reports of safety problems with the redesigned space shuttle rocket motor.

The former employee, Steven W. Agee, an aerospace engineer, worked for a special safety team set up by Thiokol at its Brigham City, Utah, plant during the 1987 rebuilding of the rocket motor.

"We had nothing but problems on the project. I did everything a human being could to warn Morton Thiokol and they wouldn't listen," Mr. Agee said.

Mr. Agee, who now runs his own exporting company in Seattle, said that it was his understanding that a Federal Grand Jury in Salt Lake City has begun taking testimony in connection with the investigation. Mr. Agee said that at the request of the FBI he wore a tape recorder for three months and recruited others as undercover informants while working at the Brigham Plant in 1987.

The latest controversy over the Morton Thiokol's redesigned rockets comes only weeks before the expected launch of the space vehicle. The shuttle is now expected to go into orbit by early October. Earlier this week, the National Research Council, a scientific watchdog group, approved Morton Thiokol's redesigned booster rocket and said it should result in "considerably enhanced safety and reliability."

Anthony W. Laine, a Phoenix, Ariz., physicist who worked on the project in 1987 also said yesterday in an interview that the FBI taped Morton Thiokol employees as part of an investigation into suppression of safety information. Edwin B. McLean, a Phoenix attorney who represents both Messrs. Agee and Laine said they have provided information to the FBI alleging Morton Thiokol executives hushed up safety concerns.

Messrs. Agee and Laine filed separate lawsuits in federal court here Monday charging they were wrongfully dismissed by Morton Thiokol. Mr. Laine said senior Morton Thiokol officials ordered him to stop investigating a potential hazard involving the rocket motor's electromagnetic "plume." Mr Agee said he wrote 221 hazards reports that Thiokol officials kept from the National Aeronautics and Space Administration.

Morton Thiokol released a statement yesterday calling the charges in the law suits "without merit." The company said the two former employees were "job shoppers" hired by ARC Technical Services, a subcontractor set up to hire for the Chicago-based company's Brigham City project. A Morton Thiokol official declined to confirm or deny that federal investigators have been looking at the Utah project.

Spokesmen for the FBI in Washington and Salt Lake City didn't return telephone calls about the allegations. The FBI customarily declines to comment on pending investigations.

Earlier this month, a Federal Judge in Salt Lake City dismissed two claims against Thiokol totalling $3 billion by another engineer who said he had been defamed by the company after he told Federal investigators about problems with the booster rocket.

Thiokol made the rocket who's failure caused the explosion of the Space Shuttle Challenger Jan. 28, 1986. NASA subsequently paid Thiokol more than $500 million to redesign the space vehicle and improve safety procedures.

A NASA spokesman couldn't be reached for comment late yesterday on the lawsuits filed by the former Thiokol employees.

Mr. Agee, who has worked in the aerospace industry more than 20 years, was hired by Morton Thiokol in November 1986 to work on the projects System Safety Analysis Unit. The unit was supposed to analyse each component of the rocket motor to determine potential points of failure and report directly to NASA. "We found nothing but problems", Mr. Agee said. "We tried to get our information to NASA and Morton Thiokol suppressed it."

Dr. Laine, who also has a lengthily aerospace resume said in his law suit that he was told by Thiokol officials to stop his work on the potential hazard of electromagnetic plume and concentrate only on "cut and paste" rehashes of old reports.

I was pleased with Bill's article. I was surprised when he called me soon after the article came out. He seemed bright and honest so I agreed to another. Then his second article on me appeared as follows: (Note that it was indeed very peculiar and irregular I was not subpoenaed to the Grand Jury. Who didn't want me there?)

"THE WALL STREET JOURNAL

Friday, September 16, 1988

MORTON THIOKOL IS SUBJECT OF U.S. PROBE OF ALLEGED SUPPRESSION OF SHUTTLE DATA

by, Bill Richards - WSJ CHICAGO -

A Federal Grand Jury in Utah is investigating allegations by former employees of Morton Thiokol Inc. that company officials suppressed memos and concerns about the safety of the redesigned space shuttle space rocket motors.

A Federal Bureau of Investigation spokesman in Salt Lake City said Chicago based Morton Thiokol turned over documents to the Federal Panel within the last few weeks but that no witnesses have been called yet. Meanwhile, an official of the National Aeronautics and Space Administration said he had been contracted earlier by the FBI in connection with the investigation.

Earlier this week two former employees complained in separate lawsuits filed here that Morton Thiokol officials had wrongfully dismissed them after they raised safety concerns about the rocket motor in 1986 and 1987. Earlier this month a Federal Judge in Utah dismissed claims for $3 Billion by a former senior Morton Thiokol engineer that company officials defamed him after he voiced concerns about the safety of the booster rocket that exploded.

The Grand Jury activity comes just weeks before the National Aeronautics and Space Administration is scheduled to launch the redesigned space vehicle. Their last manned Space Shuttle launch ended in disaster when the Challenger exploded after liftoff Jan. 28, 1986, killing seven astronauts.

NASA administrator James Fletcher, said the FBI investigation wouldn't affect forthcoming launch of the Space Shuttle Discovery. "That (booster) design is in good shape" he said. He attributed the allegations to "disgruntled employees." Nevertheless, he said he believed NASA's Inspector General would investigate the charges too.

A spokesperson for Morton Thiokol said the company doesn't discuss matters "in litigation or with the potential to be litigated." But he said that if a Federal Investigation is under way Morton Thiokol would cooperate. The spokesman said Morton Thiokol is confident the redesigned Space Shuttle is safe and that the company didn't suppress any safety concerns from its employees on the project. "The redesign effort is the most scrutinized effort that ever went on in the rocket propulsion industry" the spokesman said.

But several former employees who worked at Morton Thiokol's Brigham City, Utah, plant for a special safety review unit said yesterday that they had trouble getting senior Morton Thiokol officials to pay attention to their concerns.

Ron Clary, a Structural Engineer who worked for seven months with the Systems Safety Analysis Unit at the Utah plant, said in an interview yesterday. "There was a lot of suppression of hazards and safety concerns" raised by members of the unit. "Intentionally or for whatever reason, those things never got into the record", said Mr. Clary. Rakeh Chopra, another engineer who worked with the unit said he relayed some of his concern about the safety of the rocket casing to the FBI investigators after they were ignored by Morton Thiokol officials. Mr. Chopra who now lives in Maryland, said his recommendations involved additional study for the long-term durability of the casing, which is supposed to be reusable for up to 20 launches total. "I believe that was never carried out", he said. "I think there was a public relations concern" about taking such action.

J.R. Thompson, Director of the Marshall Space Flight Center in Huntsville, Ala., said in an interview yesterday he was told a year ago that the FBI was looking into allegations Morton Thiokol had hushed up employees' safety concerns. Mr. Thompson said he met briefly with FBI officials at the time, but has had no contact since. A senior Morton Thiokol official who declined to be identified, said the company has been providing the FBI with documents for the last year

Mr. Thompson said NASA has had a team of 25 to 30 engineers and others working in Brigham City for the past two years. "We've got so many people looking at things out there that I'm extremely confident there's no piece of data that someone has not looked at.

Wayne Tilse, an engineer who now lives in Seattle, said in an interview yesterday that when he worked on the booster rocket safety unit in early 1987 he raised a number of concerns about the durability of the rocket casing. "I couldn't even succeed in getting a draft of my report typed" he said. Mr. Tilse said he ended up taking a hand- written draft of his report to Morton Thiokol Officials. "The initial reaction was, "well, this isn't a problem" he said.

Many of the concerns raised by the former Thiokol employees were first reported in a lengthily article by the Seattle Times in May. Other than the continuing FBI investigation and Grand Jury inquiry, however it isn't clear whether the controversy has had any impact on the continuing effort to prepare the shuttle for the next launch. The employee concerns were raised anew in lawsuits filed here this week by Steven Agee, a former Thiokol engineer, and Anthony W. Laine, a physicist who worked on the shuttle project.

In his lawsuit Mr. Agee said engineers with the safety unit were supposed to make their evaluation directly to NASA but MTI officials squashed most of their reports. Mr. Agee said he personally filed 245 Catastrophic Hazard Analyses related to the booster rocket, none of which made it to NASA. In all he said the units engineers wrote more than 2000 reports, mostly Catastrophic, Hazard Analyses, and MTI forwarded fewer than 50% to NASA {12 it actually turned out later}.

During part of his five-month tenure with the unit Mr. Agee worked as a paid informant for the FBI, wearing a hidden tape recorder. He also turned over hundreds of project documents to the Federal Investigators and recruited additional informants before he was fired by Thiokol in March of 1987. {{Was paid by the FBI after fired and out of MTI for later services. Was not paid while undercover at MTI by the FBI}}

In Utah the FBI spokesman said the Federal Investigation could last for some time. "We're looking at an even longer investigation than we anticipated" he said. "Running through all these documents is a laborious process.".

Judeth Valente in Chicago and
Bob Davis in Washington contributed to this article."

After Bill Richards of the WSJ came out and visited me he wrote the following article. Bill was a very bright man and he was no longer late to connect or short on comprehensive depth of understanding.

"THE WALL STREET JOURNAL
Thursday, September 22, 1988
Morton Thiokol, Issued on Shuttle Data.
Says it studied 1, 027 possible hazards

By Bill Richards and Rob Davis, Staff Reporters of the Wall St. Journal
Morton Thiokol Inc. Officials said trouble-shooters found more than 1000 problems during the redesign of the Space Shuttle booster rockets. A handful of hazards were fixed while others are being treated as unavoidable risks, the company said. Thiokol officials were responding to allegations last week by former company engineers that it had suppressed their reports about safety hazards associated with booster rockets. The officials denied the allegations. The trouble shooters were part of a team of engineers and others who went over the redesigned booster rockets piece by piece after the explosion of the Challenger shuttle in 1986. Two former members of this same review group sued Thiokol in Federal Court in Chicago last week, charging the company wrongfully fired them and suppressed their concerns.

Several other engineers who were part of the unit have raised similar suppression allegations against Thiokol a Federal Grand Jury in Salt Lake City is looking into the allegations and has subpoenaed documents from Thiokol.

Richard Davis, a Morton Thiokol Vice President, said officials of the Chicago company reviewed 1027 possible hazards raised by the troubleshooting team that some had been corrected, others were discounted and 11 were judged acceptable risks. "In some cases there was no possibility of designing around them". In the quotation Mr. Davis said Thiokol safety officials of the company the National Aeronautics and Space Administration agreed to allow 11 "accepted risks" in quotation marks stand uncorrected. Ross Bowman, Thiokol's Vice President for safety reliability and quality assurance said those uncorrected risks included such items as lightning hazards and the possibility that the booster rockets might not ignite simultaneously. NASA officials have listed some 1500 parts on the shuttle as "criticality number one" risks, meaning that if they fail the shuttle could be destroyed. The parts range from the booster rockets themselves down to tiny parts. Most haven't any backups.

Steven Agee, an aerospace engineer who worked with the safety review unit in the late 1986 and early 1987, said some of the Hazard Analyses he filed hadn't been corrected. Mr. Agee said he reported concerns with the booster rockets nozzles and with the grounding problems that could cause rockets highly volatile solid fuel to explode.

Mr. Agee was one of those who sued Thiokol. In his suit he said that he personally filed 245+ Catastrophic Hazard analyses and that altogether members of the project System Safety analysis unit filed some 2000 hazard reports. Mr. Agee claims Thiokol forwarded about 50 of these reports to NASA.

At the request of the Federal Bureau of Investigation Mr. Agee wore a hidden tape recorder during part of the time he worked at Thiokol's Brigham city, Utah, Plant. He also turned over to the FBI copies of hundreds of documents related to the booster rocket project and recruited others to act as informants. FBI Robert Bryant, Special Agent in charge, of the FBI Salt Lake City office and 5 States, called the inquiry "a priority investigation" but declined to give details. "If any information came to our attention that would endanger the safety of the launch we would have passed it on to NASA" He said.

> Mr. Davis said the company assigned a review team from its Elkton, Maryland Division to study Mr. Agee's reports. He said the company had acted on nearly a dozen of the current concerns raised by the other engineers or were deemed to be far fetched and were discounted he said.
>
> Several former members of the safety analysis unit raise questions about the durability of the rocket steel casing, which is supposed to be used for as many as 20 launches. A Thiokol spokesman said the company had subjected to casing to the equivalent of at least 20 Launches. "They function magnificently" he said. Morton makes propulsion systems, specialty chemicals, and processes and sells salt. {No limit to the lies and coverups.}

I talked to Bill Richards about many aspects of the pending DOJ cases. Unfortunately I could only tell him a small fraction of what I knew. I wanted to tell him the Seattle Times, CBS-TV and a bunch of other good folks about everything, but I just couldn't at that time.

I asked Bill to tell me what I really was? I was tired and confused and was frustrated by the slow DOJ activities. After all, since this case had come into existence, S. A. Tim Healy had rounded up many illegal drug abusers, a couple of bank robbers, a group of American Indian murders, and other criminals, etc. They had been tried and sent off to prison. Is America's priority murder and not space exploration or military high ground? I could argue the other side also the MTI and NASA cases were no doubt infinitely more complex, infinitely more political, infinitely harder to explain to the DOJ prosecuting attorneys (LD's).

My frustration climaxed on 9 December 1988 when my worst fears, predictions, and insights became true. The first Space Shuttle flight after Challenger had numerous hot gas leaks in five of the SRM joints! The next flight (Atlantis number 27) which returned had similar leaks. $500, 000, 000 and 32 months later and NASA has purchased more failing O-rings and it has a continuance of the NASA 14+ year history of ignoring O-ring problems and failures. Look at this: Wall Street Journal 9 December 1988

THE WALL STREET JOURNAL

Friday, December 9, 1988

B for technology

Space shuttle shows problem with booster

Gases burned through insulator layer during Discovery flight

By Bob Davis staff reporter for Wall Street Journal

Washington - as the U. S. Celebrates the Space Shuttle second successful mission this year, there is evidence that its redesign of the second rocket boosters didn't work the way they were intended to during the earlier launch. Royce Mitchell, a National Aeronautics and Space Administration Booster Manager at Marshall space flight Center, said gases improperly burned through a layer of insulating material in five of the 10 nozzle joints and Discovery's Boosters in September. In two joints, he said, the gases approach at the primary O-ring, a seal designed to keep hot gases from destroying the nozzle. The nozzle also has a secondary O-ring to capture any gases that burn through the primary seal.

Hot gases escaping from an O-ring led to the January 1986 Challenger Disaster, which burned a hole from the gases which killed seven Astronauts. Rocket boosters using the Atlantis launch. NASA will try to figure out better ways of applying the insulating material, he vowed. He said that NASA expected to find similar leaks in the Solid Rocket Motors that killed seven astronauts.

In the recent Discovery mission, however Mr. Mitchell said the gases didn't burn the O-ring or threaten the shuttle. NASA didn't consider the leak significant enough to hold up the launch of Atlantis last week, he had in order to delay future launches.

"This is a damn good motor", he said. NASA discovered the evidence of the leaks when it opened and dismantled the Discovery boosters. Disassembling the Discovery boosters after the end of the mission they discovered soot between the segments that make up the booster nozzle. The booster - nozzle segments are sealed by two O-rings, which line the nozzle. As added insulation, NASA injects a rubberized caulking material into small gaps between the nozzle segments. The soot (ash) was found in the caulking material. "Gases got past the thermal barrier," Mr. Mitchell said.

Mr Price, an aerospace engineering professor from the Georgia Institute of Technology said the soot indicated the leak occurred too late in the launch to harm the O-rings. If the leaks had occurred early in the launch of the gases flowed continuously past the O- rings, he said, the gases could've destroyed the seal.

Mr. Price is a member of the National Research Council solid rocket booster review panel set up after the challenge that Challenger accident to monitor the booster redesign.

Mr. Mitchell said NASA believes the hot gases may have burned through the insulating layer because there was a gap between the insulation and the metal nozzle segments, or because an air bubble in the insulating material opened a hole for the gases.

He said that NASA expected to find similar leaks in the solid-rocket boosters used in the Atlantis launch. NASA will try to figure out better ways of applying the insulating material, he vowed.

I thought to myself "What is the fucking use?"

The D. O. J. dropped the case against MTI. Well informed rumour had it that the FBI were driving MTI out of Utah . That appeared to be true. Would the SRM facility become the new national SRM facility? Information from Financial Week magazine was that the US government had allocated $1.2 billion for the new national SRM facility and MTI was lobbying to get in excess of 1/2 of it. An enormous rip off demonstrating the incredible criminality and incestuous practices of our American space programs and much more crime and cover-up pending!

It was looking to me as though I had precipitated a small national step forward. [37]

Appendix 37: "Space Shuttle shows problem with boosters ..."

Chapter 48

THIOKOL STRUGGLES AND LEGAL CONFUSIONS

Time: 1989 Onward

The D. O. J. - LD's had failed, sold out, quit, deferred? One of the saddest failures in American history up to that time. A turn of events the free world will never fully recover from. What options were left to the FBI? Not much! The then Utah FBI SAIC decided all that was left was for the FBI was to write an "FBI Letter Head Memo" to the U.S. Senate and the U.S. Congress. The FBI LHM (letterhead memo) was written in earnest sincerity by the Case Agents, subsequently neutered by lawyers of the FBI hierarchy, diluted by the DOJ, and was finally quietly entered into the last paragraph of page 2240 3321 of US Congressional record. After massaged by all concerned it was probably pushed from isle to isle, ineffective, and covered the DOJ LD's asses! [31], [32],

One of my fauvorite articles is where Mr John D. Thirkill, one of the most dangerous men imaginable decided to retire, or possibly the decision was made for him? [42]

Did he retire or was he assisted by the FBI, NASA, and / or others?[38 39]

See Appendix 31: "Bush Comes Down Hard on Wrongful Military Spending
" By Judy Wiessler, Herald News Services, Norfolk, VA
Also see Appendix 32: "Space Shuttle Delayed"
By William. J. Broad, New York Times, Feb 2, 1989
Appendix 42: "Thiokol Official Who Led Booster Design Team Announces He Will Retire"
By Seattle Times, May 25, 1989
Appendix 39: "Risky Business ..."
Miami Herald Dennis E. Powell 9 April 1989
Appendix 38: Two Minutes Financial World Magazine
June 27, 1989. Author was flown to New York by the Financial World Magazine and this book was the basis of this article

Chapter 49

EMOTIONAL DISSASSOCIATION FROM AMERICA

Time: July 1988 onward

As I was completing writing the first draft of this book 30 years ago in its original form it was March 1989. I filed a huge lawsuit (September 10, 1988) against MTI wrongfully dismissing me, racketeering (RICO), antitrust, and blackballing me. I would have filed a Qui Tam action against MTI on behalf of the United States government for several billion dollars. Illegal actions by a Federal Judge in Chicago against my attorney made it clear to my attorney and myself that the US government didn't want to recover their lost billions. Also only one Qui Tam can be filed on any single issue. The U.S. Government didn't care - why not? Seriously, why not? George H. Bush, and Ed Meese were in the hog trough with both trotters most likely. Possibly two presidents, President Reagan and George W. Bush? I was told by high level Gov. officials there well could be Ed Meese issues. I never knew for certain. [40]

I was physically and emotionally wasted. In seven months of business Trans-International Corporation brought over $90 million into America from overseas. Every single American supplier defaulted on their contracts with my new business Trans-International Corporation. You figure, TWENTY SEVEN defaults in all! Our contracts at Trans-International Corporation were crafted to International Chamber of Commerce standards or better yet twenty seven consecutive American companies defaulted, most of them when I had Letters of Credit received and cleared by my banks. In other words the funds amounting to a total of $90 million passed through my bank and I never had a spendable cent from any of it!

My two very learned vice presidents and I lamented-how many companies can we afford to sue? Two of our attorneys thought I should make a business of brokering out my potential lawsuits against American corporations in default. That was a joke of course. We all knew the Mormon

---

Appendix 40: "In Resigning as Attorney General, Edwin Meese /Did Nation a Favor ..." 17 July 1988 and Thiokol and other forces of the universe had attacked us. Did the U.S. Gov. attack us? I would never know but I suspected. Ed Meese, ex - US Attorney General, and Senators Jake Garns and Orin Hatch were dangerous.

One day one of my vice presidents literally sat down and cried. He was a distinguished, international banker type, spoke nine languages, had done business all over the world. He sincerely asked "Steven is someone tampering with our business?" We had foreigners millions' in our bank accounts in Letters of Credit (LC's) and couldn't touch a penny due to the defaults of all of our domestic suppliers. It was contemporary history that foreign money contracts were more viable than American contracts? As time went on my personal bank accounts were tampered with, my corporate and personal and business telephones were tampered with, our mail was tampered with. All of these problems were turned over to the FBI as they occurred. Who - MTI, the Mormons, the DOJ, possibly the U.S. Government, or other players? I went with the Peter Miller Agency, publishing agents, in New York. Peter Miller liked my story and felt it needed a strong edit and I agreed. He had a buyer and a strong deal on the table. The sale was coming to consummation when the United States Internal Revenue Service walked into his office and killed the sale. The IRS threatened Peter Miller if my book (this book) was ever published (by him) there would be no end to his tax troubles. And, the second largest employer in Utah was the US Internal Revenue Service at that time. Who worked at the IRS offices, pervasively Mormons of course.

Now 30 years later:

IRS Guilty of Fraud.

If you ever had any doubts concerning the morally depraved and illegal activities of some IRS employees, the U.S. Court of Appeals case Dixon v. U.S. [91 AFTR 2d 2003-569 (9th Cir. 2003)] filed January 17, 2003, *should strike fear in your heart.* The U.S. Court of Appeals remanded this case twice and finally, on the taxpayer's fourth appeal, the U.S. Court of Appeals decided to reverse the U.S. Tax Court's decision based on *intentional acts of fraud, perjury, witness tampering, secret IRS payoffs, and secret IRS deals that favoured some taxpayers and damaged others.* All of these acts were perpetrated by corrupt IRS attorneys.

Even more reprehensible is the fact the top brass at IRS awarded the corrupt attorneys large monetary bonuses for their efforts to win this case at all costs. If this had happened in the private sector, the above acts would have been referred to as *payoffs or bribes.*

The Dixon case is quite long, but I have listed a few excerpts for you to get a feel for the scathing decision handed down by the U.S. Court of Appeals.

1 Counsel for the IRS committed intentional fraud on the court.

2 IRS counsel entered into secret agreements with certain taxpayers in exchange for false testimony and cooperation in the government's case.

3 IRS attorneys corrupted the adversarial nature of the proceeding, the integrity of the witnesses, and the ability of the court to judge with impartiality.

4 IRS attorneys violated the rights of taxpayers who agreed to be bound by the U.S. Tax Court decision.

5 Factual findings of the U.S. Tax Court support the conclusion that fraud plainly designed to corrupt the legitimacy of the truth-seeking process was perpetrated on the trial court by IRS attorneys McWade and Sims.

6 There can be no questions here - the actions of IRS attorneys McWade and Sims amounted to fraud on both the taxpayers and the U.S. Tax Court. The proceeding (in U.S. Tax Court) was a charade fraught with concealed motives, hidden payments, and false testimony.

The Appeals Court ordered sanctions against the U.S. Tax Court and the IRS attorneys. Get this - IRS attorneys McWade and Sims were suspended for two weeks without pay and transferred to another district. McWade took early retirement and kept his cash bonus. Sims returned his cash bonus, but continues to work (administrative duties only!) for the IRS Regional Counsel's Office in San Francisco. "

My book should have gone on to its destiny at that time. There were substantial communications with the FBI. There was still incredible uncertainty and stress involved in the preparation of this book and living through it.

I am still an American patriot above much else, but I'm very disappointed and dissatisfied with current aspects of the evolution of our current American legal system for the last 30 years. Hopefully, I contributed to some of the cures necessary for America to survive and prosper - yet to be determined, and uncertain. Donald Trump was just elected President of the United States. Perhaps there will be a beneficial transformation - TBD?

My aerospace career was apparently over. My gut perceived my 'edge' was lost forever and that I was simply one man in a long string of historical patriots who have spent their energies mostly in vain. Another drip wearing on the drip stone. Deformation takes centuries and frequently millennia.

Let's face it, Thiokol prosecutions at that very late date meant very little. NASA prosecutions meant very little. All that mattered was to establish American pre-eminence in space, high technology and be masters of 'the high ground'. America was out of space for some time on and off, and had lost huge amounts of advantage. The bottom line of it all to me as a person was an emotional disassociation from America. The America I left when I was went to Vietnam did not exist when I returned from Vietnam. America today is not a land of security and plenty. It is in chaos on many levels. It is now massively corrupt, it is massively broke, it is disastrously mismanaged. I am now more of a citizen of the world if I like it or not.

America will never look as pure to me again as it did before. Sure, I was idealistic. Idealism and striving for idealistic standards is what made America a great nation, it fuels entrepreneurs, and drives great risks, great explorations.

To die and have failed or made only minor contributions
means you have had a worthwhile life.
To die and not have even failed at an enterprise
means you were probably a sad waste.

Now my life has been threatened numerous times. Now I am witness of many who sold their souls to Thiokol against me, the FBI, and America, and indeed the free world. Now I see clearly justice in America is sensitive to nearly everything but the simple facts and realities of truth and crimes. Now 30 years later, nearly everything is corrupt. I clearly see our great America slipping into the pages of Earth and space history as a late great nation.

The two greatest conceivable setbacks America can incur are nuclear war, and the absence from winning Earth's space race. Just look at any major newspapers front page on any typical day then and now.

Example, the Los Angeles times,
December 22, 1988 headlines:
"When one's rights made a wrong"
"Drexel admits guilt; to pay 650 million"
"Army officer held for selling secrets to Soviets"
"At least 258 dead in Scotland crash of New York bound 747"
"US targets defence firms for prosecution"
"Gallo brothers battle over the family wine business name"

31 July 2016 (30 years later) New York Times headlines are:
"Trump belittles Muslim Family of slain soldier"
"Hot air balloon in Texas kills 16"
"Real-estate dispute on front gardens"
"Illegal to grow vegetables in your yard"
"China's Nuclear Power ambitions sailing into troubled waters"
"Tunisia ousts Prime Minister"
"Malaysia's leader, dogged by a billion dollar scandal proves untouchable"

No new space news, no great achievements, no great expectations - just signs of crime, failure, and anarchy.

The following editorial cartoon perhaps says it all.

Knight Ridder

Chapter 50

America Out Of The Space Race

Time: Challenger Accident Onward

A statistical chronicle of the failures of Thiokol, NASA, and America's space program is simply too long to list here. ABC television's (NOVA) ran a special on space. In 1989 *Ganett's Media* 'Roundtable on Shuttle Safety' and many other programs all point to the sad conclusion - America's space programs had wound down to nearly nothing. The remaining life of the Space Transportation System ran to Nov. 2008. It only launched light pay loads after the Challenger Disaster, and in particular, it was trapped in 'low orbit' and in and of its self it could not go far from earth. It had a few successes, but only a fraction of what were possible or paid for.

Congress and the Senate didn't appear to understand or care. The White House didn't appear to understand or care. Some of the American public seemed to care, but didn't understand much about why, where, or how "space stuff" was important to them and all of our futures.

America went on to share projects with many nations per NASA:

Many nations now recognize the strategic value and practical benefits of space assets and are pursuing space capabilities. By the end of 2010, government, commercial or academic organizations in at least 52 nations - including 19 member states of the ESA - were operating one or more satellites, or planning to launch a satellite before the end of 2012. Some of those nations include: Argentina, Australia, Austria, Bolivia, Bulgaria, Chile, Colombia, Czech Republic, Denmark, Egypt, Greece, Indonesia, Kazakhstan, Laos, Luxembourg, Malaysia, Mexico, the Netherlands, Norway, Pakistan, Portugal, Romania, Saudi Arabia, Spain, South Africa, Sweden, Thailand, Turkey, Venezuela and Vietnam.

Obviously public relations is more important than science or technical leadership. The Russians began launching their space stations in 1971.

First-Generation Stations (1964-1977)
First-Generation Russian Stations

Salyut 1	Civilian	1971	First space station
Unnamed	Civilian	1872	Failure
Salyut 2	Military	1973	First Almaz station
Cosmos 557	Military	1973	Failure
Salyut 3	Military	1974/75	Almaz Station
Salyut 4	Civilian	1974/77	
Salyut 5	Military	1976/77	Last Almaz Station

First-generation space stations had one docking port and could not be resupplied or refuelled. Almaz military stations and Salyut were civilian capable stations. Notice the first were all Russian. Now the International Space Station which is over 70% financed by the USA has Russia as a 1/2 half owner. America is less than a half owner.

Our American Space Station took 10 years and 30 missions to assemble. It began in 1984 and continues in space to this day. It was originally scheduled to be shut down in 2016, but now it is scheduled to be deactivated in 2020. It's first components were launched via a Russian rocket. The USA's International Space Station has been deemed a 'Financial Disaster'. NASA does not currently have a rocket capable of conveying astronauts to or from the International Space Station since the Space Shuttles were retired. That job now falls to Russia. There is currently contention on the ISS between Russia and American personnel and operations. Also there are tentative agreements for Russia and America to proceed on with a new space station after the ISS.

So, our American Space Shuttle basically put most of our International Space Station up, but America cannot completely service it without Russian assistance. How sad is that? Look at this UPI article on America's early thinking of joining with Russia to go to Mars. [33]

Appendix 33: "U.S.-Soviet Trip to Mars Long Way From Agreement", William Howard

Over a year after I left MTI Ron Clary wrote the following shocking transcript - one of his MTI trip reports of the worst kind.

Note: Ammonium Perchlorate is notoriously explosive in any amount and under nearly any conditions.

MORTON THIOKOL, INC.

Wasatch Operations

Space Division

Interoffice Memo

3 June 1987
8512-FY87-M213

TO: K. G. Sanofsky, Supervisor
SRM System Safety

CC: D. E. Thompson, M. J. Phillips

FROM: R. Clary
SRM System Safety

SUBJECT: Trip Report - Kennedy Space Center (KSC), 26-29 May 1987

This visit was the third in a series of visits planned in conjunction with Operations and Support Hazard Analysis (O&SHA) activities. The objectives are the continuation of familiarization with personnel, organizations, schedules, operations, equipment, facilities, and documentation available only at KSC.

Facilities visited on this trip were Complex C (MTI/K), VAB, Hanger AF, NASA Headquarters, Complex E (Bamsi, Inc.), and the RPSF Area. The VAB required observations over two shifts because of the hazardous character of operations then being conducted.

The priority objective was first-hand observation of demating of segments containing live propellant. In this case, the LH SRB of Flight 61-G was being demated, and progress on arrival had reached the Aft Center Segment. Potentially serious incidents were observed on both Wednesday and Thursday nights in the process of scheduled demating.

On Wednesday night a sudden rainstorm occurred, with the top two Highbay No. 3 doors open and the Aft Center Segment end exposed. A low rain cloud entered the Highbay and rained directly on the exposed end grain, and down through the bore of the remaining assembled segments of the LH SRB. All exposed faces of live grain propellant were quickly saturated, and little could be done. Consternation and confusion reigned for several hours. This observer was quite negatively impressed.

K. G. Sanofsky, Supervisor
SRM System Safety

3 June 1987
8512-FY87-M213

- 2 -

Then on Thursday night all but one of the pins of the field joint were successfully removed after a series of sequentially higher lifts of the Aft Center Segment. It was then discovered that the remaining single pin was jammed in place. And all this with exposed grain propellant still wet and actively leaching Ammonium Perchlorate (AP) from the night before. Remedial action required the participation/approval of LSOC engineering, who finally approved a series of 1/8-inch sequential lowerings, each followed by attempts to extract the problem pin. After five such increments the pin was successfully extracted, apparently with no damage either to it or the mating holes. After protracted delays the Segment was finally raised to the required 18-inch separation where this observer was able to discern stalacites of AP actively forming around the bottom edge of the bore hole. There were also several puddles of water on the top face of the Aft Segment Inhibitor and exposed grain propellant. Shortly thereafter all nonessential personnel were cleared from the work platform as demating continued routinely and without further incident.

All of the above were second and third shift operations in the VAB. First shift activities were not nearly as dramatic. The largest part of the time was devoted to visiting Complex E, Hanger AF, and RPSF facilities, and meeting resident MTI/K personnel. Without exception, these personnel were extremely cooperative and informative.

The most difficult task was tracking Bamsi, Inc., to Complex E. After being mis-directed to NASA Headquarters and the LSOC Complex, they were finally discovered almost by accident with a simple telephone call. Once located, they proved cooperative in delivering several USBI drawings ordered previously, and accepting another extensive order for supplementary documentation.

On Wednesday morning Al McIntosh and myself were given an extensive tour of Hanger AF by Mr. Pringle, who proved to be a virtual historian of the facility and its myriad operations. We observed several different routine operations then underway involving both ships and the facility itself. For example, the sea channel through which the SRB's are towed was then being surveyed for depth. And an active test was underway of the EJAF mounted on an empty segment. Most interesting was the side tour through the Hydrazine Processing Facility adjacent to the main site.

K. G. Sanofsky, Supervisor
SRM System Safety

3 June 1987
8512-FY87-M213

- 3 -

On Friday the RPSF Facility was toured and all available GSE noted and inspected. No Segment handling operations were then underway and the building was unoccupied. I was particularly interested in the manner in which the facility was grounded. Apparently a large degree of dependence is placed on grounding to the surrounding network of railroad track, but little documentation could be discovered to support whatever kind of grounding system is provided. The track was "walked" both inside and outside the fence, with no resolution of concerns about the grounding system in place. No time was available to investigate further and request for all documentation pertinent to the existing grounding system was left with the Manager of Operations.

A large variety of safety concerns were identified, of which the above are only examples. Nearly all of these would have been virtually impossible to identify from review of documentation presently available at MTI/W. My parting impression was that the number of hazards identified was limited only by the length of the visit.

Ron Clary
R. Clary

This memo by Ron Clary was final proof to me that NASA has learned nothing from me, the FBI, or anyone else. In fact it is strong evidence that NASA was un-trainable at that time. NASA was a dangerous organization and not up to the responsibility of overseeing complex projects and vendors such as Thiokol. MTI definitely was not responsible enough to entrust with billions of dollars, national security and national space exploration.

Mr. Roger Boisjoly's Qui Tam lawsuit was essentially ignored forever by our American legal system. The DOJ knew his allegations were true, but they declined to act on behalf of the American public. They publically claimed his allegations were not specific enough. Rogers eleven large boxes of documents were deadly specific. The first judge to deal with Rogers cases in Washington DC chose to throw Roger's Qui Tam case to Utah Federal Court. The first Utah Federal Judge disqualified himself for Mormon prejudice. The second Utah Federal Judge chopped the case up and then in the end the Utah Mormon judge threw out Roger's cases on his 'opinions only' which were not substantiated by documented law. Roger Beaujolais was a victim of injustice by several American Federal Judges. It would appear he was also a victim of his own preferred religion. It takes a lot of gumption for a man to stand against his religion when it is tearing him apart.

The good news is that America still has men and women such as Roger Boisjoly, and Dr. Tony Laine, and others who did (and will) take the risks and who pushed *For The Right Reasons.* America is a great nation because the majority of its aggregate efforts are *For The Right Reasons.*

The following cartoon best describes how I felt as I was writing my original manuscript of *For The Right Reasons.* It was not published earlier as U.S. Presidential obstacles and threats were placed to prevent publication.

GOOD NEWS ABOUT YOUR BOOK... IT'S A TWO MILLION DOLLAR CONTRACT WITH YOUR NAME ON IT.

Letters to the editor

Mr. Ralph Jarvis wrote the following article about the renowned bad treatment of Dr. Tony Laine. Similar harassment was then harassing Faye and me. My family probably has been under siege every since. There is documentation and evidence of it. Mr. Ralph Jarson, a journalist wrote an article "Safety Wars".

Some of those who read Ralph Jarson's article demanded action. The following is an examples of further definitive concern. [34]

United States Senate
COMMITTEE ON APPROPRIATIONS
WASHINGTON, DC 20510-6025

September 23, 1988

The Honorable James C. Fletcher
Administrator
National Aeronautics and Space Administration
400 Maryland Avenue, SW
Washington, DC 20546

Dear Administrator Fletcher:

In an article published in the September 1988 issue of "Arizona Living", a number of allegations are made which, if correct, seriously threaten the safety of our astronauts, as well as the future of our space program. I have enclosed a copy of the article for your review.

I would like to have your response to these allegations. If you have previously investigated this issue, I would like to be advised of the results of that investigation.

I look forward to an early response to this request.

Sincerely,

Dennis DeConcini
DENNIS DeCONCINI
United States Senator

DDC/lsi

Appendix 34: "Safety Wars ..."Arizona Living, Sept. 1988 Ralph Jarson

As a result of that article the magazine was put out of business. Soon after that, the Utah Governor's office was trying to dissuade the moviemakers from "damaging Utah "! Edwin McLean, Tony's and my attorney, had a 'rule 11' hung on him by a Chicago judge - in other words he was fined $3,700 for a frivolous lawsuit in Dr. Laine's case! Tony was financially destroyed, blackballed, under continuous threats, and under psychiatric care and his adversaries had committed treason. Is this frivolous? Everyone should read the following article which shows how Dr. Laine was in fact destroyed and killed by his enemies which included Mormons who had strong relations with Maximus, Inc., MTI, NASA, and I have no doubt others. [35]

See Appendix 35 for the article: "How To Destroy A Nuclear Physicist" by Charles E. Corry, Ph.D.. It is astounding.

CHAPTER: 51

WHAT HAPPENED TO SOME OF THEM?

Malcolm Phillips:

My former friend Malcolm sold his soul. As he helped me carry many pounds of documents out of the MTI for the FBI he looked after himself. Informed witnesses and observers such as the Indian engineers, Dr. Tony Laine and FBI sources reported that Malcolm cut a deal with MTI to tell them all he knew about me and the FBI and in turn he was rewarded with a very long term contract and became MTI's man at NASA's Kennedy Space Center - where his home was. So, he continued his very substantial $250,000 to $350,000 a year income which was very good at that time though it required many hours of work.

Malcolm was not completely a bad guy. He did direct the FBI to the fact that out of the 1200 hazard analyses MTI admitted they delivered to NASA (under FBI pressure) there was a discrepancy of 12 which were not put forward that should have been. The last I heard about Malcolm was that the FBI was going to coax more information out of him to find out more details they knew he knew about and was involved in. Malcolm was a profound disappointment to me. His new country meant nothing to him. He had evolved to be an American yuppie - he wanted to know "what was in it for him". Let there be no question, he was a very intelligent man.

Mahalon Hollway:

Mahalon sold his soul. Mahalon reportedly returned to 'the bottle' just before I was thrown out of MTI. Even though I was instrumental in bringing Mahalon to MTI, Mahalon hated my guts for whatever his personal reasons were. I was always courteous to him, and helped him a great deal hoping he could recover from his long-standing alcoholism. Reports came to me from reliable sources that Mahalon was paid $10,000 to do a deposition against me by MTI when they found out I had probably been with the FBI. He stayed at MTI for a while and was then terminated for reasons I do not know. He had reportedly not worked since (one year plus) and was in desperate shape living in the Seattle area according to Seattle Times journalists. Could it have been he now had Security Clearance Problems? While I was at MTI Mr. Don Thompson and Mahalon were doing investigative work against Dr. Tony Laine to discredit Tony. How strange is that? Tony held highest Security Clearance, and the highest CIA income, but the Brigham City Tax man was having him investigated! Sounded a bit desperate to me? Mahalon loved this kind of work since when he put reasonable productive people down he envisioned his position rising. Reports came in that he also did an investigation on me. I wonder if NASA expected Mahalon to be doing invasive investigative work on personnel rather than redesign of the SRM's?

I wonder if MTI was aware that Mahalon's university degrees were bogus? How had the USA tedious Security Clearance processes allowed him to hold any level of Security Clearance? The FBI wanted to know? Mahalon and MTI perhaps deserved each other.

Dr. Tony Laine:

Dr. Tony Laine came to MTI sane, rational, exuberant, capable, and above all, a willing and qualified man. He was improperly fired from MTI for doing his job and sticking to his guns and not capitulating to direct MTI suppression. I talked with Dr. Tony Laine just before he was fired and just after. The FBI and I had sensed he would be sacked. Just after he was fired he was acutely bright red in the face and emotionally distraught. He then left MTI and started his own campaign to straighten out MTI and NASA. His efforts fell on mostly deaf ears! Then Tony became so distraught he went emotionally 'out of whack' for lack of a better description. Tony wrote a complaint to the President of the United States, the VP, hundreds of Congressmen, Senators, high government officials, and others. Dr. Laine was investigated as a result of MTI's slandering him to the FBI, and during that investigation he got mad and told an FBI Agent if he didn't get someone in the USA to work the MTI problems he would take the case to Russia"! This was a desperate outpouring of frustration and precipitated another FBI investigation. In the end, I believe the FBI cleared Tony of any and all wrongdoing. However his frustration and radical behaviour were problems to the investigations of MTI and NASA. Dr. Laine then filed a huge Civil Federal lawsuit against MTI for wrongful termination, RICO (racketeering), Antitrust, and other germane issues. Dr. Tony Laine was driven to poverty. There were continuing threats to him and his wife and harassments until his death. In the end it appeared to me and others that Tony was literally systematically killed. He died in November 2010 of a stroke. He was over charged hundreds of thousands for child support and repeatedly wrongfully charged and jailed. There is a documented connection between the Mormons and the company Maximus that illegally destroyed Dr. Tony Laine. He was literally hunted down by a corporation specializing in 'failed justice'. My deepest respect for Tony. He may have collapsed under his various loads, but he was a dedicated and great American. If you are interested more in what was done to Dr. Tony Laine go on line and look up "How to destroy a nuclear Physicist", by Dr. Charles E. Corry at http://www.ejfi.org/family/family-38. It is shocking and could easily precipitate a book on its own. Tony also turned on Special Agent Tim Healy. Tony erroneously thought Tim had 'done him wrong'. I am certain that is not the reality. Tim felt sorry for Tony and his situation.

Special Agent Tim Healy:

Tim was and is an extraordinary man. Tim cut his FBI teeth on this case and grew and matured profoundly before my eyes. Tim and I witnessed one of the greatest cover-ups in United States history up to that time. Tim had to balance forces of the FBI with the awesome failures of the U.S. Senate, Congress, Department of Justice, NASA, and others - no small adjustment but good training for a new and enthusiastic FBI Agent.

Tim has university degrees in Spanish and History. He married his university sweetheart. He joined the U.S. Marine corps and became a Naval Aviator and flew Harrier Jets for the U.S. Marine Corps. For whatever reasons he wanted to be an FBI Agent. So he took the exams and made it against massive competition at the time. Then he began an illustrious career. He was beyond honest and fair. He was beyond diligent. And so he amassed a career taking him to the very top echelon of the FBI. Tim was not a hard ball political player.

When he reached the upper echelon of the FBI he focused on his job, not politicking which made him a rare commodity.

Later, long after MTI, Tim and I created the FBI Internet Fraud Complaint Center (FBI IFCC). It was the FBI's first venture into Cybercrimes. Tim had the vision to see the FBI need to pursue cyber crimes. He had to campaign to get the funds. Later yet, he and I created the FBI Terrorist Screening Center (FBI TSC). Tim bent over backwards to get me the incomes I commanded on civilian projects. For my last three years I made approximately 300% of what the FBI Director made while I was contracted to the FBI creating the FBI Terrorist Screening Center. In truth, I earned it. We changed world history. However, I worked very hard and long hours for it and I truly delivered the goods. I literally killed myself. Our product has saved hundreds of thousands of lives. Our work in fact changed world history. It is certain I will fall over dead from the Vietnam War, or from working myself to death at for the FBI. The TSC is astoundingly successful. From my scratching it out on a piece of butcher paper in pencil the FBI TSC now has its own large building and facility at Freedom Park in Northern Virginia across the Potomac River from the Lincoln Monument. It is roughly a 124 million dollar facility. It integrates all terrorist data from all USA Agencies and many nations integrated efforts. It operates well above 'Top Secret'. I can't say any more.

Tim is 18 years younger than me and he probably won't last long. He has worked longer hours than me and has raised five children and fought so many 'internal' Wash., D.C. wars we all lost count. He believes in and supports America. We started as Special Agent and Source, and we have ended up brothers. Our minds and efforts were synergetic for the good. Tim is now a President of an important company serving the U.S. Intelligence Community.

To this day Tim looks at me in his embarrassment and he sticks to his 'FBI Code of Silence" and says "the DOJ could not make the cases against MTI". He still thinks the sun shines out of the FBI's ass. He knows the DOJ and the FBI were throttled! He of course insults my attorneys and my personal intelligence. We forgive Tim, we do not forgive the DOJ, the Attorney General, or the President. Tim's lips are sealed and he is clearly not allowed to tell 'ever'!

Tim and the FBI had to be satisfied with an FBI Letterhead Memo to the U.S. Congress to save U.S.A. national face that NASA had failed. So much for the normal 'law'. It was disgusting. I was disgusted. My various attorneys were disgusted. The Seattle Times was disgusted. Washington, D.C. 'business as usual'!

Special Agent In Charge: Bob Bryant:

He fought many battles and went on soon after the MTI cases to become second in command of the FBI. What more it there to say, he was and is a consummate winner.

Roger Boisjoly

Roger knew the US government suppressed his cases. Everyone who knew anything about these cases was shocked the U.S. Government did not attempt to recover its losses. Was that 'political spin' thinking the rest of the world should not know NASA was a mess? I really don't know. It does however turn my stomach. After President Ronald Reagan ordered a Presidential Commission to review the disaster, Boisjoly was one of the witnesses called. He gave accounts of how and why he felt the O - rings had failed. After the commission gave its findings, Boisjoly found himself shunned by colleagues and managers and he resigned from MTI.

Boisjoly became a national and global speaker on workplace ethics. He argued that the caucus called by Morton Thiokol managers, which resulted in a recommendation to launch, "constituted the unethical decision-making forum resulting from intense customer intimidation." For his honesty and integrity leading up to and directly following the shuttle disaster, Boisjoly was awarded the Award for Scientific Freedom and Responsibility by the American Association for the Advancement of Science in 1988.

When Boisjoly left Morton Thiokol, he took 14 boxes containing every note and paper he received or sent in seven years. On May 13, 2010, he donated his personal memoranda—six boxes of personal papers, including memos and notes from congressional testimony—to Chapman University in Orange, California. Rand Boyd, the special-collections and archival librarian at Chapman's Leatherby Libraries, said the materials will be catalogued and archived. It was to be about six months to a year before library visitors would be able to view the materials from the time of his donations.

Boisjoly died on January 6, 2012, of cancer of the colon, kidneys, and liver. He was a great man who stood his ground with dignity. He has had substantial impact for the good of man.

Judge Sessions (FBI Director beginning at 2 Nov. 1987):

A highly qualified Texas Judge, not a former Special Agent. So by definition none of the top end FBI managers liked him as he had not ever been an FBI Agent. He was not an FBI 'good old boy'. Also his wife came to work with him and watched 'soap operas' on TV while he commenced FBI Business in the FBI Directors Office. Word was that she did not of course hold a Security Clearance, and so when various 'classified' matters came to his office, which was frequently, she would be pissed off when she had to go outside! FBI managers were not amused!

To his credit he wrote me a very nice letter thanking me for my efforts. I never met him. He ended in disgrace when he had additions to his home for his family protection and safety. Some of the things added were acceptable, and some were misuse of government funds for general renovations. I never knew the details, but he was 'shifted out of office'. Here is a synopsis of his FBI History:

On November 2, 1987, former federal Judge William Steele Sessions was sworn in as FBI Director. Prior to his appointment as FBI Director, Sessions served as the Chief Judge of the U.S. District Court for the Western District of Texas. He had previously served as a District Judge and as U.S. Attorney for that district. Under Director Sessions, crime prevention efforts, in place since Director Kelley's tenure, were expanded to include a drug demand reduction program. FBI offices nationwide began working closely with local school and civic groups to educate young people to the dangers of drugs. Subsequent nationwide community outreach efforts under that program evolved and expanded through such initiatives as the Adopt-A-School/Junior G-Man Program.

NASA:

The NASA IG's office evolved to be angry about contract supervision of MTI. Many NASA officials lied profusely to the FBI. NASA came under attack from the USAF regarding other major rocket and missile systems and projects. NASA budgets were on shaky ground with Congress. Over 3000 HA's were ultimately written concerning the Space Shuttle and NASA and MTI only confessed to 1200, and only acted on a small handful! The Solid Rocket Motors of America's Space Transportation System were veritable 'flying trashcan' bombs. The FBI had copies of nearly all HA's.

For the last 30 years NASA has had astounding budgets but no rockets capable of high manned orbits or returning men to the moon or completely maintaining our International Space Station. NASA has been under contracts related to the ISS, International Space Station, and also a lot of secret defence related work. Various observers will have different views, however my view is simple. My view is that NASA has squandered vast amounts of money, been poorly run, and has severely diminished America's space related explorations and defences from what they could have been. NASA is in fact a disgrace in my view. Now NASA is developing the following, 30 years later:

> CAPE CANAVERAL Fla. (Reuters) - NASA's new heavy-lift rocket, designed to fly astronauts to the moon, asteroids and eventually Mars, likely will not have its debut test flight until November 2018, nearly a year later than previous estimates, agency officials said on Wednesday. NASA is 70 percent confident of making a November 2018 launch date, given the technical, financial and management hurdles the Space Launch System faces on the road to development, NASA associate administrators Robert Lightfoot and Bill Gerstenmaier told reporters on a conference call.
>
> NASA estimates it could spend almost $12 billion developing the first of three variations of the rocket and associated ground systems through the debut flight, and potentially billions more to build and fly heavier-lift next-generation boosters, a July 2014 General Accountability Office report on the program said.
>
> While the rocket might be ready for a test flight in December 2017, as previously planned, the new assessment showed the odds of that were "significantly less" than the 70 percent confidence level NASA requires of new programs, Gerstenmaier said.
>
> "We want to commit to this (November 2018) date and show that we can meet it," added Lightfoot.

The schedule assumes flat annual budgets of about $1.3 billion for the SLS rocket and another $1.5 billion for Orion crew capsule and associated ground launch systems at the Kennedy Space Center in Florida.

The GAO report found that NASA's SLS rocket program was about $400 million short of meeting its December 2017 target.

The rocket is a modified version of the shuttle-derived, heavy-lift booster developed under NASA's previous exploration initiative known as Constellation.

The U.S. space agency spent about $9 billion on Constellation, which included the Orion capsule, from 2005 to 2010, before President Obama axed the program. Its goal was to return astronauts to the surface of the moon by 2020.

Instead, the White House and Congress approved a flexible path toward Mars, including a visit to an asteroid that will be robotically relocated into a high lunar orbit.

NASA did not say if the 11-month slip in the new rocket's debut flight, which will be an unmanned test run around the moon, would impact the second mission, slated for 2021, with a two-member crew.

Initially, the SLS rocket, which uses leftover space shuttle main engines and shuttle-derived solid rocket boosters, will be able to put about 77 tons (70 metric tons) into an orbit about 100 miles (160 km) above Earth. Later versions are expected to carry nearly twice that load. Ultimately, the rocket is expected to be used to launch astronauts and equipment to Mars.

"Our nation has embarked on a very ambitious space exploration program and we owe it to the American taxpayers to get this right, " Lightfoot said." {This is disgusting to me. SpaceX, and other civilian corporations will lead the way to Mars, back to the moon and other great explorations. NASA is currently a little research and a lot of 'talk and Spin'. Just today another SpaceX rocket was successfully launched.} {Agee}

Just yesterday (7 May 2017) the USA and NASA were able to return an X-37-B Orbital Test Vehicle (OTV) from space after a 718 day mission in space. Operations were classified and unmanned, however a later version, the X-37-G, is anticipated to carry a few persons. Speculation is they will be military. How long to develop is unpublished.

Morton Thiokol Inc.:

This book fairly well tells the story of MTI gone wrong. Following online sources concerning later history and their corporate dissolution are fascinating . A long and twisted departure from when I first encountered MTI. A departure to save and protect the guilty and perpetuate what President Eisenhower called " the Industrial Military Complex". He was a great General, his insights were deadly accurate, and he was a great President and man. [47]

John D. Thirkill: [42, 22]

Did he resign of his own volition?

Dick Giaque, Attorney:

He later became counsel for the Utah Cold Fusion project. Dr. Fletcher left NASA to head it up and it all turned out to be a fraud. He also represented the families of the men killed in explosion of the MTI Ammonium Perchlorate plant. He did not represent me as my suit was simply the same hassle as larger suits, and not enough millions in it to keep him interested. His firm sued MTI, Kerry Sanofsky and others at Thiokol regarding those more recently killed. I never tracked the outcome of those suits.

Dr. Edwin McLean (Attorney):

Teddy had literally been a U.S. Intelligence Officer doing 'Overseas Operations' in his younger years. Enough said. He was a profoundly well educated attorney. He is both an attorney and a Ph.D. He specialized in massive cases and won the majority. When he was representing me and Dr. Laine he was newly accepted to the California Bar. Teddy was fined under rule 11 by Chicago Federal Judge Tanner on Dr. Tony Laine's case for $3,700. In Chicago Federal Court my suit against MTI and Dr. Laine's were consolidated for discovery and thrown to Utah Federal Court. In the Chicago Federal Court my attorney, Edwin McLean was not allowed to plead properly. He was only given short notice and rushed through, was always called to very early am hearings in the Chicago Federal Court. Essentially Teddy was given the 'bums rush'. "Who's Law" "Which Law"?

Teddy was astounded at the Mormon Mafia and the U.S. Government collusion as they affected our cases. In addition to Teddy five other attorneys told Fay and I there was no honest way my cases against MTI could be lost. I had

---

Appendix 47: SOME HISTORY OF THIOKOL
Appendix 42: THIOKOL OFFICIAL WHO LED BOOSTER REDESIGN TEAM ANNOUNCES HE'LL RESIGN 26 May 1989 Desert News

the documents, signatures of the criminals, and the witnesses, top attorneys, and the FBI. What could possibly go wrong? Morton Thiokol attorneys, about 20 of them, sat facing me during my deposition (interrogation in effect). The first day of deposition went for a full 12 hours and the second day the same. The deposing lady attorney tried every trick known to throw me off, defeat me, to do anything to me. At the end of 24 hours of gruelling depositions I had won every last word, point and concept! She literally said laughingly, "I'm afraid to ask any more questions". I had delivered another powerful tool to Teddy to help our cases. Again, what could possibly go wrong? I murdered them in depositions!

Teddy asked me what I thought would happen as we drove home from my depositions. I told him after the defeat of facts and truths I had just beat them over the head with, it was inconceivable they would go forward to court, they would have to settle out of court or buy a Federal Judge. He thought about it and agreed. What else could they possibly do? They didn't have one deposition fact on their side. I simply destroyed them. It is a matter of record.

Then we lost in the Utah Federal Court, literally trumped by Mormon Theocracy! Then we proceeded to the Federal Appeals court where we were very strong and again virtually unbeatable. We lost again! So much for un-losable 100% documented and witnessed cases.

At this point in my life I finally realized there is no limiting depth to comprehensive corruption. Our case Judges, the DOJ, and ultimately the FBI had simply capitulated to "Mormon Political Power' and authority. MTI could not have pulled it off by themselves. The new Federal Judge that sat on our Appeal Case was literally red in the face, in a sweat, and stunk in 'fear and manipulation'! He had just been appointed. We had no alternative except to wonder if he had just been bought? Teddy still practices Law, though now in California rather than in Phoenix, Arizona.

Eric Nalder (Reporter):

He received two Pulitzer Prizes, what more is there to say? [43] He was told he would have received a Pulitzer for "Beyond Challenger ...." if he had not shared the research the credit with Peter Lewis, a much younger and less experienced reporter.

Appendix 43: Reporters' Relentless Pursuit Of Facts Won {Seattle} Times 2 Pulitzers Chuck Taylor, Seattle Times Staff Reporter

Steven W Agee:

I had been raised and educated in a lie. Honesty is frequently in fact damaging and limiting. I have been through hell several times. First in the Vietnam War. I was with U.S. Reconnaissance Marines in the I - Core of Vietnam with the 3rd Marines also known as 'Swift, Silent, Deadly' and indeed we were. Once at MTI before I went to the FBI. Once for the FBI. Once for the Seattle Times. Once for "Trans-International Corporation". Twice for trying to settle into Australia and New Zealand. It was Hell to re-establish myself after "massive life injuries'. And hell in every sense because there seemed to be no justice in America or in fact much of the world at that time or since. Dorsey Clement, a top U.S. National Security Advisor whom I had started a company for, Recursive Solutions Corp, told me much later that the FBI and DOJ should have placed me into "Witness Protection" for my own good. Interestingly it was never once discussed by the FBI. I suppose none of us clearly saw what was coming.

My lawsuit against MTI was moved from Illinois to Utah and was pending at a Mormon influenced Federal Court in Utah, and half my case issues had to be dropped because the 'related law' in Utah doesn't even exist as it does in most other states and Illinois (Chicago, where filed).

My book *'For The Right Reasons'* was under serious consideration by several major publishers. I signed a contract with a movie group for an 'option' to make a major motion picture on my original manuscript of *'For The Right Reasons'*. I was and am tired, lonely, frustrated, and very beat up. I am owed millions in addition to my former lawsuit and realize that even if I collected I never really could comprehensively recover. Honesty has cost me nearly everything except my satisfaction that my integrity is still intact. I was not allowed to hold any substantial employment in Australia. I did not fit into Australia. I was miserable there and the only reason I capitulated to stay there as long as I did was to get my two younger and beautiful children, Mariel and Evan, to a better age to transpose to some where better. My marriage was a lonely disaster and my life was simply my children. Then it was time to move on and we shifted to New Zealand.

NZ was a much more honest nation than Australia. NZ people are much nicer in my view, but I really didn't fit into NZ either. NZ is provincial and I am not. Mariel and Evan were straight into school at Omokoroa, N.Z. Fay was PA (Personal Assistant) for a hospital owner. I was absolutely unemployable for anything. New Zealand is real short on high technology jobs. Ultimately I had to return to America to make any substantial income.

Ultimately my marriage disintegrated. One of the most contentious and expensive divorces in N.Z. history. When it was about all settled Fay grabbed our children and abducted them to Australia. She broke the Hague Convention on Child Abduction lock stock and barrel. Years and a fortune were spent trying to extradite the children from Australia. It was a waste of time as Australia has never even read the Hague Convention rules even though they are signatories. I have not seen those abducted children for 17 years and never expect to see them again. They have been brain washed and are gone forever. I do not, and would not know them, they are 'GONE'. In my heart I wish them well and all the best. They are emancipated to a new and changing world.

Now 30 years after my collision with MTI and NASA criminals, DOJ LD's, a disgraced Attorney General (Edwin Meese), the Mormon Mafia, and 17 years after divorcing my wife, Fay N. Agee, there are still remnants of hostilities and issues. Fay and Australia are still trying to usurp money from me which is clearly not owed to any of them. My top NZ attorney keeps them held at bay. A local physician friend and several others speculate that Fay, my ex-wife sold out to the Mormons, and she and the Mormons have possibly capitalized on the twisted Australian Family Court system. She certainly changed. I really don't know and I really no longer care. Fay N. Agee was initially a decent person and wife and it ended in what I view as a psychotic meltdown. No one can ever fault her in her ability to manipulate the well documented shabby Australian legal and governmental systems. She is a 'Black Belt' at 'street smart' and 'rat cunning'. I can't believe I was ever married to a person such as she became.

Contributions from my aerospace career can be found on the Moon, Mars and USA Ballistic Missiles in silos.

While consulting in Washington DC to the USA Intelligence community it cost me a fortune to have a hotel room there, so one day I calculated it to be far less expensive to buy a substantial sailing yacht to live on, so I did. I lived on it for well over a year and then sailed her back to New Zealand with crew. Better an asset for one of the childrens trusts than money spent on hotel rooms. On sailing back to New Zealand we stopped at 16 countries

In my old age I think on a university paper I wrote 46 years ago. When I went to capitalize on it 46 years ago it was the wrong time in history. Now is the right time and the 'Green revolution' and my project is worth US$ Trillions. I am working to launch that project and have made progress. I also do pro-bono global connections and fund raising for scientists seeking financing. I am orchestrating Physics Scholarships at a University in America. I also administer my two children's Trusts. In my spare time I carve wood and am a shameful Spanish classical guitar player.

Carving of my oldest son, Eric W. Bullock. He is now a Lt Colonel.
Approx. 3 ft. by 3 ft in English Walnut

Senior RN with two Masters Degrees.
Was stationed in Mosul, Iraq for some time with U.S. Army
He remains in the US Army

Carving of my esteemed and learned Physician who has kept me going.
Approx. 2 1/2 ft. by 3 ft., Silver Beach Wood

Dr Anthony H. Townsend MNZM MGP MBChB BSc FRNZCGP(Dist) DipObst,
Order of New Zealand

I am at peace. I have done all I can for everyone. I never hear from any of my children and it used to bother me. It no longer bothers me. They have made their decisions and I have made mine. They are emancipated. We are all committed to our individual destinies. I am occasionally asked if I would do it all over again. The answer is a resounding "Yes!" Human excursion beyond Earth is that important in my vision of human achievable expectations.

I do not feel humans will remain a 'one planet species'. I have done all I can to project humans into space. [36] The X-37-B landed after 718 days on 07 may, 2017. It was a USAF vehicle and project. I think it was heard loud and clearly.

President Donald Trump:

Newly elected only a few weeks ago and there is shaking and moving. Who knows at this time what will happen or what he or the United States will do? [41]

Appendix 36: "Crucial Flight" Wall Street Journal By Bob Davis & Laurie McGinley
Appendix 41: "Donald Trump asks NASA to explore putting crew on rockets debut flight" Thomson Reuters, Irene Klotz, Feb. 24th 2017

CHAPTER: 52

TRANS-INTERNATIONAL CORPORATION

TIME: 1989

After returning to Seattle after MTI I couldn't score a contract there or anywhere. It was very harsh times for high technology work and I was apparently 'Black Balled'. So after frustrating and depressing failures to score I decided it was my time to do something else. I had a list of business possibilities but in the end I decided to start an arbitrage company. Buy low, and sell high, pay taxes, live off of the difference. An ancient form of business.

I rented the office directly under the 'old tower' at Boeing Field. It was a beautiful office looking over all of the planes and runways. Outside my windows were usually a P-51, a Phantom jet or two, several corporate jets, and a few puddle jumpers like I flew. I held a U.S. Commercial Pilots Certificate. It was a nice and dynamic environment. I was energized, recovering from too many hours, and looking forward to new successes coming forth. When I focused I nearly always won, so why would this be any different? Morton Thiokol was in the past, my contact with the FBI was diminished, and I was now 'just another businessman'. I was 46 years old. What could possibly go wrong?

I acquired a Vice President, Dr. Jerry Curtis. He had been a Professor of French at Rice University, and he had a very gifted set of communications skills. He was a 'class act' to say the least. I had bought an early personal computer, and the civilian internet had not been instituted yet. Then business was mostly by FAX and snail mail, telephone and couriers There was a lot of international business done via telex at that time. Slow and limited, but it worked. So we set off trading. A number of people wanted to join my firm as they thought we were on our way. It appeared they were right. We had deals going in Saudi Arabia, South Africa, South Korea, Hong Kong, and Thailand, etc.

As time went on we had a small group accumulating around us. One was Joe Kolo. He was the grandson of the last High Chief of Cameroon. He spoke a dozen languages, had a Masters Degree from a university in Oregon, and he was smart. Another showed up, Mr. Wallace Simpson. He was a 6'4"

TRANS-INTERNATIONAL CORP.
7277 Perimeter Rd., Suite 213, Boeing Field
King Co. International Airport
Seattle, Washington 98108, U.S.A.
Telephone (206) 762-9343
Telex No. 798440
Fax No. (206) 762-9345

Steven Agee
President

Pakistani, Christian. He was an expert in bank 'Letters of Credit', which roughly translates to 'bank to bank money'. He had been with Hong Kong Shanghai Bank. He was a powerful asset to my company "Trans-International Corp." and he became my second VP. He was a class act, many languages, and great dignity.

Once the team was up to speed you could come into my office spaces any time of day or night and one team or another would be negotiating in any of 20 languages spoken in house. Finland gold to Canadian buyer, repossessed '747 combi' (half cargo, half passengers) in Georgia to a French buyer, Ghee (lard) to West Africa. And then finally with 25 balls in the air relating to many millions we were invited to cut a deal with South Koreans. So Jerry Curtis and I went to lunch with about 6 Koreans and we shook hands and were awarded a deal and we were issued a USD$38,000,000.00 LC (Letter of Credit) for No. 6 Human Corn. As each of about 24 ships would be loaded the cash funds would be released into my accounts in the normal manner for Letters of Credit. It appeared Trans-International Corp. had made it! What could possibly go wrong? Our "i's" were dotted and our "t's" were crossed. We were diligent. Our work load was increasing asymptotically. We were working more intensely, we had signed contracts, we had LC's in from all over the world, we had sellers, we had buyers, yet there was something wrong? None of our commodities were shipping, none of our LC's were being signed off - not a cent changed hands? Simultaneously my home phone rang five times a night with no one answering?

Next, Wallace and I were visited by, let's say 'A Gentleman'. He showed up on time for his appointment. He was missing a couple of teeth. His suit was old and a tad dirty, poorly cut. We asked him what was on his mind? He was very polite and courteous and told us his employers were aware we had 'millions' in LC's in our bank. They were in particular interested in the $38,000,000.00 for 'Number 6 Corn.' He went on to explain he was sent to offer us 'help' acquiring a supply and getting the ships loaded. Wallace and I were stunned and sat there in amazement! How did he know we were in a thermonuclear melt down with our suppliers? How did he know the amounts and details? This was clearly a seasoned man, but not an intellectual giant. He politely refused to tell us who his employers were.

We politely sent him on his way and told him we would consider his employers kind offer of assistance. After he left we sat in 'stunned silence' and looked at each other. Finally, I asked Wallace, "Was that the Mafia?". Wallace said he was sure it was! We were in shock. Were our phones bugged? Was our Fax bugged? Did we have a spy amongst us? AAaRrggGGHhh! Were the Mormon's messing with us? Was Thiokol messing with us? Was the U.S. Government via Attorney General Ed Meese messing us? Were we simply targeted by the Mafia? We knew there was trouble and danger in the air. Our incomes and lives were clearly center stage. Time to strap on my shoulder holster and wear my cannon and so I did.

Then within a day or two I asked Wallace, what is happening with the USD$220,000,000.00 gold sale from Finland to our Canadian buyer? Has been quiet for a bit long I thought. At those amounts our commission was a paltry 1/4 of 1% or USD$550,000.00! He agreed it was time to check on the buyer and seller and see how it was progressing and nurture it along. Next day Wallace came into my office and he was clearly 'devastated' and upset. Oh, oh I thought? What's up? I knew Wallace was honest and profoundly competent and dedicated. Wallace said "Well Steven, I just talked to Angel in Canada. He laughed at me and thanked me for bringing his attention to that particular gold deal. He said he had just returned from France where he and the Fin's had consummated the deal without our services or assistance!" HOLY SHIT!!! I thought we had been arranging delivery and settlement at the Federal Reserve Bank in New York? Wallace said that's what had been decided. He had arranged for it. Wallace said somehow Angel, a fine Canadian Italian business man, had found out who the Fin's were. Probably by simply contacting one of the banks in Finland? Then he simply went directly to the Fin's and expunged us from the deal so they could split the extra 1/4 of 1%! Oh *!!#4%##&&##++! Then I was as devastated as Wallace.

Simultaneously I received a $17,000.00 movie 'option' on this book. Over almost two years the Hollywood guy failed to sell the movie. I wondered if he was a truly trying to sell it or was he an agent to prevent it being sold? I will never know. He finally disappeared.

In a matter of a week and half every deal and contract at Tran-International Corp. went 'tits up' in a similar manner. Our Corn contract with South Korea epitomized all of the deals. Money in my bank via an 'accepted', 'bonafide' LC, a ship ordered, contract for Corn in place and not so much as one kernel of Corn moved one inch with in U.S borders. Wallace, others with us, and myself ran all day and all night to get the Corn on a ship. The Corn would not move! The suppliers were laughing and didn't seem to be concerned about the massive law suit looming against them. Under U.S. law a thirty eight million suit plus damages represents a serious concern to anyone? Whatever was up was from institutions or individuals who were not afraid of a law suit. These people were playing 'for keeps'. Then in a moment it hit Wallace, me, and my other VP that we were under siege and in danger. Both of my VP's were virtually in tears.

In front of them I picked up the phone to Special Agent Tim Healy and called him in Utah. I told him what was up and his answers were astounding. He said the FBI knew I was being fucked with big time. So I asked who, he said he could not say! I asked was he going to help my company or just let me sink. He did his famous 'schmooze' routine and made it clear to me he couldn't interfere. He did however suggest I get a hold of a U.S. Senator or Congressman to assist! I asked, 'Tim what in the hell would they know about my business?'. He said they often took consulting fees for smoothing things like loading ships, etc..... What about the night calls Tim, same. In our own personal 'speech codes' what I heard was that I was being messed with 'big time', and there were FBI cases wrapped around it. Also Tim was emphatic for me "to watch my six o'clock!"

Over the next few weeks all contracts failed which was a statistical impossibility if they were unmolested. Gerry's wife filed for a divorce, Wallace went into depression, I felt fear and threat, my staff slowly bled away. At the end two of my fine young men, one of them Joe Kolo, helped me pack everything into a shipping container. Trans-International Corp. was a corpse. I could easily have been a corpse soon I though at the time.

Simultaneously, Edwin McLean, my new attorney was suing Morton Thiokol for millions and my case and Dr. Tony Laine's had been 'conjoined' by a Chicago Federal Judge and referred back to Utah. Edwin McLean was also 'sanctioned' for filing these cases which is an extraordinary dirty and inappropriate move by a U.S. Federal Judge. We found out much later that Chicago Judge was married to a Mormon wife. The long arm of Mormon business had reached out.

Mariel was born, so now we had a daughter. A wonderful thing but another pressure and responsibility, and she needed protecting. At this time

Fay worked for a Renton, Washington. Travel Agent. The owner donated us tickets to Australia as a gifts. The crimes and improper actions of Thiokol had just came out in the Seattle Times about me being an undercover source for the FBI at the Utah plant. The crimes were so well documented no one doubted their accuracy or veracity. Many lives had just exploded. NASA had exploded. Thiokol exploded. Where were the DOJ's LD's???

As Dr. Laine and my multimillions Federal Case came to a conclusion there were issues. Near the end of the case Dr. Laine had a mental melt down. The Utah Federal Mormon Judge wrote an opinion, not a learned conclusion of law. He stated that Thiokol could do anything they wanted. He of course ruled in Thiokol's favour.

The case had been completed for four days before my attorney called me and told me we had lost. My impression was that he had consumed a lot of grog over the decision and he also told me he had already filed in the Federal Appeals Court in Denver - a state which was not dominated by Mormon Theocracy.

Our appeal case went on for some time and finally culminated and was in the hands of a newly appointed Federal Judge. He ruled in a sweat and with shaking hands. He told my attorney, Teddy, and the opposing attorney he found for Thiokol. Thiokol's attorney was shocked. He knew he had been beaten, and he won, Thiokol won, the Mormons won, and the NASA and Mormon fix was in! America and the Western World had lost!

We packed, left my nearly new Toyota station wagon with friends. Our other possessions went into a 20 foot sea freight connex box. Just before we left America I went to check on something in our connex sea freight box. When I opened it I was astounded to find it had been opened, and clearly unloaded and gone through. I reported it to the police, Tim at the Ogden FBI, and Teddy my attorney. Which of the villains had rummaged in our possessions?

Friends took us to the airport. Our plane's departure was held up because two CIA male passengers were late. Fancy that! Finally we departed and flew to Cairns, Australia and a new life there. We had USD$17,000.00 in our pocket that was an 'option fee' for a movie on this book (yet unpublished) at that time. My consulting was ruined, my American life was ruined, and I had to seek a new life for my family. Not an untypical status for an American Whistleblower.

CHAPTER: 53

EXILE IN AUSTRALIA

Fay and I hit the ground in Queensland and it was interesting. I had travelled enough to know it would be a huge change in culture. Indeed it was. I am not a heavy drinker and many Australians are serious 'blotto' drinkers. I immediately did not fit. It would be wrong to go into all of the ways I didn't fit. I am what I am, they were what they were. It was like mixing oil and water, materials not miscible.

Various legal contracts remaining from Tran-International Corp. came to 'performance dates' and they failed by some ones design? One was a massive Saudi Oil 'Petromen' Contract. The Saudi Prince simply cancelled it. Nothing I could do to retrieve it. Then the option for a movie on this book expired with no performance by the Hollywood promoter. Also I was notified by friends where my USA Toyota Station Wagon was stored that a 'meter maid' had queried them about it and then it was towed and disappeared! We never found out what happened to my car? Friends and a local attorney couldn't find what happened? Gone forever!

I tried and tried to get a respectable job, but the fact is there were no white collar jobs there except for petty criminal activities. Work in Queensland is blue collar at best. Over the next few years I had interviews in Sydney, and Perth.

I finally took a job in Perth. I became No. 2 (General Manager) of Uni-lab Telecommunications. To make a long story short the founder and CEO and President was Mr. Terry Robinson, the then current 'Western Australian Executive of the Year'. When I interviewed I could see he had 'criminal' etched on his forehead. I didn't want the job because of how I read him (con man), but my family needed to eat. I got the job and enjoyed it when Terry was away which was often. As a publically listed company he was at the 'hog trough' with both trotters.

In a few weeks Terry Robinson was in Poland schmoosing the Polish Police to buy equipment yet un-invented. While he was gone I called in our company Financial Officer and asked him to enlighten me as to why the company had just posted AU$24,000,000.00 in profits with the Australian Stock Exchange and Australian Tax office. Was I confused or perhaps had a stroke? Any way I calculated it the company was AU$8,000,000.00 in debt with no profits. Being a Physicist and Engineer there was of course a possibility I couldn't count! He seemed nice, helpful, competent and I quite liked him. He was refreshingly direct.

So, why the difference? Please enlighten me? Enquiring minds need to understand. He said, "Well Steve, the company is absolutely bankrupt. There are two sets of books, the company books here and the 'real books' in my basement at home!" I thought and said NO, NO, NOT AGAIN, OH AAaRrggGGHhh - or something along those lines!!

And so we made an appointment with the Australian Tax Office and Investment Commission. We went and we told them what was known. I told them he had been 100% up front to me and I didn't think they should be prosecuting and jailing him. I did however think Terry Robinson, as the Western Australian Executive of the Year was a problem. The Securities and Tax office asked me if I had a plan to save the 160 poor souls who worked there. I told them I hadn't had time to work the problem but would look into selling it off to a competitor. That idea seemed good to them. The Securities and Investment Commission went to our Financial Officers home and retrieved the 'real books'. I then learned that two to four sets of books is common practice in Australia. Australians are no longer 'transported prisoners' from England, but the 'shackle marks' seem to lurk in their genetic profiles. Malcolm Phillips might have 'shackle marks' I was thinking!

During eight years in Australia I never found a single business activity that was not corrupt to my American mind. I acquired Australian Citizenship as my second nation. Our friends were primarily Europeans. I was miserable and unhappy there. They didn't like me and I didn't like them. We had one leftover contract from Tran-International. It was a Petrochem Allotment Wallace and I had brokered. It's due date came due and that particular Saudi Prince defaulted everyone. That was then standard procedure and risk for doing any business with any Saudi at that time. I heard from those who would know it is not any better now. By Australian standards I made serious money at Uni-Lab Telecommunications.

While the Stock Brokers, and the Board of Directors jumped hoops and bull shitted each other and pointed fingers I quietly made a few phone calls and arranged to sell the company. There were some personnel from British Aerospace to be saved and they were excellent. Also a lot of Rhodesians, and South Africans. I had some emotional problems being kind to the North Vietnamese on staff. After all I had 'dispatched' a bunch of them during the war. It was now in my blood.

After selling the company to one of the three buyers I found, I flew home to our Trinity Beach home just north of Cairns. I paid sixty two percent to the tax office, built a fence around the house to contain children, and had small change in my pocket. I was in awe of the Australian's too drunk and stupefied to revolt over the tax codes there. A friend pointed out that none of the Commonwealth had revolted like America had. He was of course right. 'English Disease' had overwhelmed all of the Commonwealth. They had all been taxed to death. I could see no light in the tunnel for me except for the oncoming train! We needed to either do a drastic move forward or get the hell out of Australia. I was beyond miserable.

Over a few months period there had been a documentary movie project in the winds from the Australian Film Corp. and Australian ABC. One movie maker was awarded the rights and funds on my story, but in a relative Australian manner when he received the first AU$35,000.00 he and his film crew went on a serious binge. So when I and his manager complained to the Australian Film Authorities he was sacked immediately. Much to their credit they enlisted Titus Films, of Sydney to replace him and make the Documentary.

They came to Trinity Beach, Queensland and met me and started to make the documentary. They were very nice and understanding and insightful. After some time around where we lived they were off to America to talk to others I had interfaced with to include in the documentary. Soon they contacted me and asked if I minded if they added a number of other persons to what was supposed to be entirely my documentary. They had sat in on some astounding U.S. Congressional testimony concerning several whistle blowers. Also they wanted to add a Queensland policeman for Australian content. I was not jealous or frustrated, after all the problems of whistleblowers and their impact is a valuable part of all societies.

Before they returned to the Cairns, Queensland area I had quite an incident. One day I was in my beat - up Land Rover and drove into the local shopping center parking lot. Very slow as usual. A car was behind me and followed me a short way to where I parked. There were a few people loading groceries or walking to the store. The car following me parked right next to me.

As I got out of my car two men got out of the car parked next to me. Before I had a clue as to what was about to happen I was in an almighty fist fight. They both attacked me. I had no time to think or consider anything. I was scared and about to be severely hurt. They both landed some painful punches and not only was I frightened, but then I was really pissed off. My old U.S. Marine Corp commando training snapped into place fortunately. I graduated No. 2 from the USMC Amphibious Recon. School. I really was a trained commando in my Vietnam days.

When my fear, anger, and training kicked in, which didn't take long, I up and kicked and slugged those bastards senseless. 'Bystanders' were in shock and horror because there was quite a bit of explicative's, screaming, shouting and flat out 'ass kicking'. They were clearly intent on attacking me but somehow I don't think they or I realized what they were dealing with. Both had bashed in faces, and severely damaged lower legs, and more. One was dazed and one was unconscious.

I don't quite remember all of the brutal details, but not knowing what to do I was in a panic and a quandary. There is nothing on earth more corrupt than the Queensland, Australian police in the view of many. So I was loath to ask for police help. Who in the hell were these assholes? Were they the Mormon Mafia? Were they the Morton Thiokol Mafia? Were they NASA Mafia? Were they US agents as part of the 'cover up' of the crimes and the obvious theft of billions?? Oh shit, what should I do???? Whatever I did had to be done instantly or I was going to have a police problem or possibly worse?

As I stood there looking back at onlookers in shock I kicked the unconscious one under the car and dragged the 'dazed and out of it' one next to him. I wondered about cameras on the parking lot? I got into my Land Rover and drove off about five blocks, my heart was pounding out of my chest, and parked under a tree and just sat there for a half an hour and tried to regain my composure. Yes, go to the police? No? Oh shit what to do. My marriage was in serious trouble and Fay wasn't going to take this well. It was clear I couldn't tell her. But could I hide it, I had been hit hard on my sides and chest. My knuckles were bleeding, strained and bruised.

I decided if the police showed up from camera information the cameras would be my best witness. I had not started a fight. However it would also show I had 'ended a fight' and not arranged medical assistance. So I decided to wash up and go home and say nothing. So that is what I did. I pondered the 'on-lookers!' In Australia punch-ups are so common that no one much cared.

Fay really didn't pay much attention to me and she never noticed the purple bruises on my ribs and chest. She never noticed my damaged hands and fingers. We had really reached the point where I really was not on her radar. She was changing one way and I was changing another way. We both realized it, but we never faced or confronted it.

When Titus returned from the USA with great footage I told them about my attack. I told them and I tell you I have no idea who they were. I was and am a very courteous driver so I am certain it was not road rage. Also the tone and vernacular coming from them while I beat the shit out of them was clearly American until they were unable to communicate any more. To this day, thirty years later I wonder? Though I took care of that attack I had to wonder if there would be more, from who, why, or could it escalate? No one in Australia can legally carry a concealed weapon for protection, but of course all of the 'bad guys' do.

Titus Films completed the documentary. It was "Nice Guys Finish Last". Australian TV showed it many times. It was on Public broadcasting all over Europe, N. America, and surprisingly all over Africa. I believe it can be obtained from Australian ABC if anyone is interested. It is a shocker. I felt Titus Films did whistle blowers and society in general a lot of good.

Since that time I lived in quiet resolve and fear which will never completely subside or evaporate. I refused to let fear rule my life.

My next move was to start what would be the most advanced robotics company in the world. We would use the Australian Tax Office 100% Tax Rebate to Research and Development company investors. So I ran around and found AU$22,000,000.00 in investors, which would fund the facility and 56 high end jobs. Android Protocol Ltd. was, I believe, the first company on earth to be registered to develop and manufacture human like robots, Androids.

Computers were just beginning to be small enough and powerful enough.

I applied for the simple Tax Rebate and was eventually and curiously called to Canberra. I wondered why the Tax Office had armed guards? Well, I found out soon. At our meeting I had a top Australian Tax expert with me. Then two Australian males and a Sri Lanken male (a biologist) came in. The Sri Lanken laid it out to me in just a few words. The Honourable Senator Chris Schacht and the Tax Office were happy to issue the document as soon as I delivered U$120,000.00 in cash virtually in a brown paper bag! I came very close to hitting him, and just then I saw through the glass wall the armed officers were there and were anticipating a 'beating'. They all clearly knew what was going on. He was possibly offended when I told him to "*Fuck Himself*" and the two of us got up and marched out under the eye of the 'armed' guards eying us and extensively 'herding us out the door'. This was clearly standard procedure at DIST (Dept. of Industry, Science and Technology). Once anyone paid them off then the Australian IRD (Tax Office) owned them forever. Those Australian tax bastards / politicos were not going to own me. I would not pay them off!

Over the next few days Fay and I went to Senator Bill O'Chee in Cairns, Queensland to get it sorted. It did not sort. Australian business was and I suppose is still corrupt. I was the wrong man in the wrong place at the wrong time. I was destroyed. I was miserable. I was cornered. I was tired to the bone. I loved my children very much, but my marriage was in strife. Fay had been hurt by all of this but so was everyone. She thought I should now join the criminals and participate as that was how it was done where she grew up, in Rhodesia, South Africa, and England... She was losing respect for me because I would not capitulate and turn to local criminal income enhancements. I lost respect because she would even emulate to it! There was honesty in my world, and unfortunately there had not been in hers prior to marrying me. I did not want my children to grow up with an Australian mentality, i.e. shackle marks. I had evolved to prejudice against Australian 'ways and means'.

Soon we planned a trip to New Zealand to see if life would be better in New Zealand for us. There had to be somewhere compatible.

CHAPTER: 54

ESCAPE TO NEW ZEALAND

TIME: Approx. 1996

After being told by the Honourable Senator Chris Schacht's representative in Australia at the CSIRO (Commonwealth, Science, Industry, and Research Organization) that they required AU$120,000.00 illegal payoff for what was supposed to be a free Australian Tax Benefit for R&D investors I knew I had to leave Australia. Since our Australian Home was nearly paid off we took advantage of a special bank deal to take some cash back from the Commonwealth Bank of Australia. We took the money and charted a 'look and see' trip to New Zealand.

When we came to NZ it was a vastly different nation to Australia in nearly every way. We liked it. We toured Northland of the North Island - very beautiful. We didn't want to be in Auckland. Then we were toured around Tauranga by Ian Madden, the Development Officer for that area. Tauranga had a great deal going for it and we liked it. Then on down to Wellington via other possibilities. It was windy and cold in Wellington. When I drove our rental car to the Ferry we walked around and I found I had locked my keys in the car. So a local from the Ferry company came and burgled our car and got our keys. He was very nice and an expert car opener. I joked with him - which was his first or second job?

Then on to the Interisland Ferry to South Island. The seas were very rough between the islands. Neither Fay nor I got sea sick. During the last few minutes of the Ferry ride I went to the Casino on board. Per usual, as a non - gambler, I risked five dollars worth of small change and Lo and Behold I won NZ$327.00! First and last time I ever won anything. At Picton {Northeast of the South Island} we disembarked and stayed overnight.

Next day, we looked over Blenheim. Beautiful, but it didn't feel right. Then on to Nelson, full of artists and a huge beautiful Cork Oak Tree. Much too remote and isolated we thought. Then on to Christchurch. Beautiful gardens, Alzheimer's drivers, cold. Then on to friends further south at Timaru. Beautiful home, swimming pool frozen over, sheep, and the coldest I have ever been or want to be! OMG it was cold.

Our minds were focused on Tauranga it the Bay of Plenty. So it was we flew back to Cairns, Australia and drove home to Trinity Beach to the north. In a nut shell we packed up our home, put it on the market, and departed for

New Zealand soonest. I was done with Australia.

On arrival we rented a van and drove to Tauranga. Our hunt for home and work was on. Fay had two jobs offered soon. I was beating my head against a stone wall again. Many people said I was far too old to work! After all I was 53 or so - ancient. Executive 'Head Hunters' told me no one in NZ was paid anything like I had been. Finally I had a 'face to face' with the top Executive Search person in N.Z.. He was kind, polite, and to the point. "Steven, live in NZ and enjoy it, but if you want to make a living get on a plane, and go make your money elsewhere. No one in NZ will ever trust or feel comfortable around you after your former USA career!" He had checked me out and knew I was for real, but a terrible fit to New Zealand. I had dealt with projects and funding far beyond anything conceivable in New Zealand. I was devastated and felt adrift.

I knew he was pragmatic and correct. So my marathon to recovery was not over yet. Would it ever be? I was tired to the bone and beyond. What to do? He told me that the week before he had a similar Dutch Executive who was simply 'overqualified' for New Zealand as well. He told that fellow the same he told me.

My wife Fay didn't take it very well. She was then the PA to the owner of a small hospital. Then out of nowhere a call came from an old friend, Alan Salzmann, in Colorado. He told me there was a job cooking at Ft. Collins, Colorado with my name all over it. It was to go fix a high technology company doing digital geographic information system applications. Long story short is I got the job and fixed many of their problems. I could not however fix the owner, who originated in Puna, India, but that is another story. When done there I consulted briefly for a small aircraft company in America.

When I came home Fay had a terrible cold. I was glad to be back on the beach again. Glorious to see and be with Mariel and Evan again. Fay was resoundingly rude and seemed to have lost the plot while I was gone. In a few days I flew off to Trinity Beach, Australia and put up a car port for our Australian Home which we still had up for sale and it was then rented and had been for some time. After a couple of weeks I returned to New Zealand. Fay had gone 'out of her mind' in my view and was screaming and yelling continually.

One evening she came home and had another 'scream out' and I told her, if that was the way she felt then we should get a divorce, the kids (7 and 9) did not need to hear her screaming and her rudeness. I was sorry she was so unhappy, but I had been unhappy also and I was travelling half way around planet earth to make a living and we were in fact in much better financial shape than most. The next day sadly I filed for a divorce.

At the time I hadn't given it a thought but many educated persons such as attorneys and others felt that just possibly the Mormon Church or Thiokol made a deal with her? It was one of the most acrimonious divorces in NZ history when there was no need for the dramas. Instead of making any type of settlement she ultimately abducted our two children from NZ to Australia and she wanted the sun, the moon, and the stars and the asteroids! Nothing was negotiable with her, she had become provocateur on all matters. She was absolutely guilty of breaking the Hague Convention on Child Abduction. I could easily write a book on the Criminality of the Family Court in Australia playing income generating games with children but I won't bore you with that reality. Suffice it to say she cut her own throat with me, she destroyed two beautiful children and she was aided and abetted by the Australian Government and their famously slanted family court system.

A very sad story of a malignant and toxic woman whom I no longer knew. I bent over backwards to do right by her and it was a waste of my time and money. Whatever her demons were they consumed her and our children. I had my own demons. Against me was MTI, the Mormon Mafia, the USA DOJ LD's, possibly Ed Meece {Attorney General of America}, the destruction of my marriage, the pending destruction of my children, my team of attorneys in New Zealand, my team of attorneys in Australia, and not least the absolute criminality of the Australian Family Court System regarding the Hague Convention On Child Abduction, Fay the provocateur, re-establishing my USA consulting, and re-establishing my USA Security Clearances just to name some of my demons at the time. Several editors of this book thought I should leave these details out, however I have followed my usual internal voices. Whistleblowers have to understand and recognize the risks and forces that may come against them. This is especially true when your President vetoes a Whistleblower Bill passed by every single member of the U.S. Congress and the U.S. Senate. Whistleblowers have to have it clear in their minds that what is right, good for the country, important, and imperative may just be a beneficial effort and good deed that will not go unpunished! Absolutely "No Good Deed Goes Unpunished" and you have to decide if losing everything is an affordable price for what is to be saved. Every person has their cross to bear.

Fay had created a quagmire for all concerned including herself. My filing of a Hague Convention Suit and later an Appeal to the Hague Suit all up cost me NZ$200,000.00. The court actions were 'theatre', were pitiful jokes at my children's expense. The only human beings that didn't think my children were kidnapped and transported over international lines illegally were the morons of the Australian Family Court. They saw my children and my loss as a grand opportunity for their profit.

However, thanks to some great attorneys in NZ it has not panned out so well for a few criminals in the Australian Legal system. Enough said concerning pestilence, vermin, and the Australian Family Court system. Incredibly the NZ Family Court System is not especially 'flash' either.

Sixteen years later Fay and Australia were still conspiring and campaigning for money. My children are now about 26 and 24, they don't speak to me and they have apparently been 100% brain washed by Fay. Doubtful I will never see them again and I do not know them.

To her credit Mariel obtained a scholarship in Mechanical Engineering from the University of Sydney and she graduated and works in Australia. Evan is a plumber in Australia. They both are highly intelligent and could be anything they desired to become. I have absolutely no communications by their choice. I have loved them and wish them well in their lives. It is obviously very sad their lives have been orchestrated and controlled by their mother. I hope in due time they completely reclaim control of their lives.

CHAPTER: 55

EXTRACTION FROM QUAGMIRE

TIME: 1999 to 2016

I was consulting at the U.S. Federal Reserve Bank when Fay fled to Australia from New Zealand and abducted our children to Australia. My attorney and Fay's were both astounded. While my Hague Suit on Child Abduction smouldered for two years and the attorneys 'played with them selves' I returned to consulting in the USA as I wasn't employable in either Australia or New Zealand. A number of things came along, and bit by bit I reacquired my USA Security Clearances. That is not a simple process for an international person. I put the FBI's Internet Fraud Complaint Center together for Tim Healy who had gone to the then Director of the FBI, Louis Freeh for authorization and funding. It was and is a smashing success. Its operations processes began by me making 'process diagrams' for Tim on a napkin over lunch the day he acquired funding. Currently on line at www.ic3.gov

Then my health took a dive, most likely from the Vietnam War, and my colitis escalated to desperately serious. Then I lost 100 lbs., then I was bleeding to death. Then I had surgery and became infected with MRSA (which can kill you all by its self) and it took 18 months to heal up and get used to an ileostomy and put some weight back on. I was in the hospital for four months, in ICU 68 days, and nearly died.

My friend, Ern Reynolds, a Harvard and Kentucky Attorney visited me every day except 4 days when the snow was too bad. None of my 5 children ever visited. After a few months out of hospital I returned to New Zealand. When I returned I still had a massive open surgical wound from my MRSA infection which took another 18 months to close leaving a massive scar.

Sometime later I received a call from Tim Healy. Tim said, "pack, go to airport and fly back to D.C. immediately. Classified and can't talk on phone." In less than 48 hours I was at Dulles Airport in N. Virginia and Tim picked me up. It was about 8:00 pm, raining and wet and cold. We went to an Italian dinner in a quiet corner and I asked him, "So what's up?"

Tim told me a new organization was to be formed and there was "much trouble at Flat Rock". The new FBI Director's ass and job were on the line if it wasn't functional in 30 days. The new Director was Robert S. Mueller III, a very respected man and the best the FBI Director they ever had up to that time. So what is the project Tim? Well it is the integration of all of the U.S.

Intelligence Agencies data on terrorists and more in one place. I asked why not at the CIA or several others? Simple answer, no one 'on the hill' trusted any of the other agencies, so it was to be the FBI. OK, that makes sense from my experience. So how many Agencies? Over 40 you say. Well Tim this is terrific. I thought they should have been integrated 50 years ago. The existence of the FBI TSC is not classified now, however its internal processes, information, and products are highly classified. It holds a substantial place on the FB's website, www.FBI, TSC So how long do we have to do this Tim? "30 days Steve!" *"Oh Shit! That Is Crazy Tim!* George W. Bush has completely lost his mind!"

"What time are you picking me up tomorrow morning Tim? "I'm not, we are going to work now........." This is going to cost you Tim. He said "I know, and it is already signed off by the FBI Director"! How do I end up in such demanding positions? Because - I am a results person. There have been many of these. I created the FBI Terrorist Screening Center (FBI TSC) from scratch, and I didn't sleep for the next 3 years really. On its first day open we had eight possible terrorist contacts globally. Four of them turned out to actually be terrorists. Now the daily terrorist contact count is huge and shocking. I am not at liberty to share the numbers.

Both Tim and I had to have our Security Clearances upgraded to far past ordinary Top Secret. We created the official *'no fly list'* processes and many other things I cannot list or go into. I hit the ground there about 1 Nov. 2003 and my consulting retirement luncheon from the FBI TSC (Terrorist Screening Center) was on 30 April 2006. An extraordinary success. By now you realize my time in the Utah Desert at MTI did not completely define my life, though it tarnished it. I was staggering from exhaustion for years.

Tim Healy, Donna Bucella, and an entourage of CIA, FBI, DOD, State Dept. of Homeland Security, and other managers and officials were there at my 'going away luncheon'. All but one of my yacht crew preparing to sail to New Zealand were there. The missing crew member, Colin Keppel, had a terrible cold. Dohn Taylor from Perth, Australia, and Bill MacDonald from Western, Australia. My oldest son, Eric and his wife Elizabeth were there, and their very young daughter were. Rachel, my oldest daughter, and her then husband were there. It was emotional to me as my systems and all of our work at the TSC had already saved thousands of lives all over the world. By then the lives saved could well have been in the hundreds of thousands. The FBI TSC is a three letter organization hardly ever mentioned in the news. It is a quiet and powerful contributor to the saving of lives from terrorism and other arenas on a daily basis. At this time there is no doubt hundreds of thousands of lives have been saved. Numerous 9 / 11 type of incidents have been interdicted. I can't discuss this.

Two weeks later my crew and I sailed out from Friendship, Maryland on my 44 ft. La Fitte Yacht *"Mariel"* named after my abducted daughter. We were headed to Bermuda, the Caribbean, Panama, the Galapagos, the Marquis, Tonga, and to my home in New Zealand.

It is now 2017 and those children are adults and Fay and Australia are still screaming for money. Those two children's minds have been destroyed towards me. I have seen them for 2 hours since their abduction from NZ on 28 Jan. 2000. They are 26 and 24 now. I am of no interest or value to them. They were wasted like cannon fodder to a woman's and several courts insanities.

My American children from my first marriage are all well educated. None of them can be bothered to communicate with me in any meaningful manner. For many decades I spent my energies foolishly trying to re-engage with them. All was and is lost there. I never expect to see any of them again. They are all simply gone, I assume forever. Now I am old and infirm and cannot possibly last long. I have always fought the good fight and I sleep at night knowing I have done all I can for everyone. I am at peace, but I am sometimes lonely. I wish all of them the very best. I am comforted by a loving partner and a loving Papillion, 'Indi'.

To fill my time I am writing books, carving wood, and mentoring students and business persons. I am nurturing a project based on a university paper I wrote 46 years ago That project is now underway and you will see much more about it in due time. I have handed that project over to a much younger man who is running with it. The project will be a pivotal point in human evolutionary history as humans change from a petrochemical based economy to a better more flexible, global Green economy. I walk several beaches daily and contemplate my 'disgraceful' Classical Guitar Technique.

The following articles clearly spell out the continuing confusion. I find it very sad. [45] [46]

Appendix 45: "Will Trump Go to Mars" WWW article
Appendix 46: "Donald Trump's first 100 days" WWW article

CHAPTER: 56

EPITAPH: DEAD DREAMS DYING NATION STRESSED PLANET

Time: 1987 Onward

Under adult mature supervision and proper fiscal management America could have initially colonized both the moon and Mars by now (2017). The funds were available over the years but were all squandered, misallocated, or simply stolen. Human beings are a species fraught with confusion and irrational belief systems. Long term focus seems to elude our species. Even educated humans are myopic. Humans' are not yet far from dancing around our nightly cave fires after the days hunt.

Prevailing consensus is that Earth took approximately 4.3 billion years to evolve to the condition we experience today. Evolution of the earth into a habitable planet was a great natural and rare event. Some estimates of man's evolution on earth at approximately 3 million years. Humanoids as we know them about 250,000 years. Democratic civilization has existed since the sixth century B.C., and then only flourished in limited geographic areas and societies briefly. Democracy is relatively new and difficult to find on earth. During the evolution of human history entire nations and millions of lives have been extinguished experiencing the evolutionary cost of civilized law. The machine revolution is now about 250 years old. The electronic age is maturing and about 80 years old. The computer age is 40 years old and still an embryo. Yet there has not been an iota of change of the human political heart or emotions for millennia. Is the human heart eternal or rational? Will human beings exist long? I think not. The question is how will we evolve if at all? It is 2017 now and wars and terrorism are raging all over our planet, probably worse than ever in history. Modern technology has expedited wars and discontent.

Each of these ages has independently levied both deteriorating and synergistic effects on the lives of all subsequent human beings. All of these revolutionary and progressive ages and contributions enjoy a symbiotic relationship with mankind. No one can say which human achievements are greatest, but few can deny that America's astronauts exploration of the moon has been one of mankind's most spectacular achievements. They opened the door on human 'space migration'.

From 1969 to 1973 America sent seven flights and 12 men to the moon and returned them to earth. The Space Age vaulted from simple rockets to the moon and beyond in just 20 years - a phenomenal feat. Those great feats were sponsored by desire, dedication, hard work, governments, and businesses working together - even with the 'cold war' raging. The 'cold war' beneficially contributed to the speed, excellence, and success of these great achievements.

After large rocket vehicles came America's Space Transportation System (STS also known as the Space Shuttle). The STS was designed and built in a few years and flown 25 times over 14 years at that time. Since then many more have flown and now all of our Space Shuttles are retired. The STS was far behind schedule per the Seattle Times graph below.

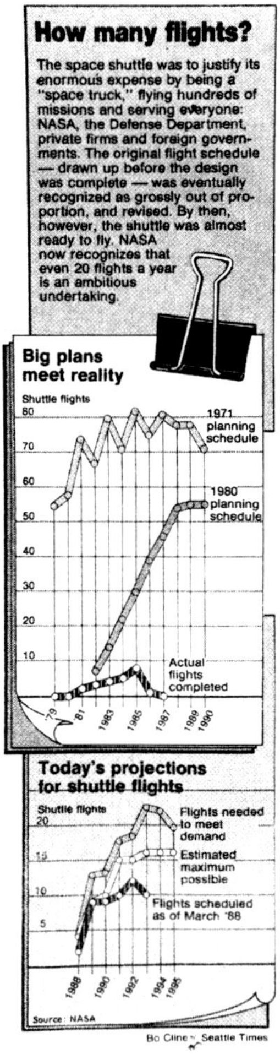

The above charts show the Shuttle Launches were about 200 launches behind original schedule sold to the U.S. Congress. After the Challenger Accident NASA fell even more behind.

The 25th flight ended in what came to be known as the Challenger Disaster. After the Challenger Disaster America was essentially grounded from space for 32 months. After returning to space the next three space shuttle flights were very light loads. The light loads were below contract specifications. The solid rocket motors continued to exhibit massive potentially fatal and unacceptable faults. The Space Shuttle was comprehensively a disaster.

> "From NASA,
>
> On February 1, 2003, the space shuttle Columbia broke apart while re-entering the atmosphere over Texas, killing all seven crew members on board. The disaster occurred minutes before Columbia was scheduled to land at Florida's Kennedy Space Center. An investigation later determined the catastrophe was caused by a problem that took place shortly after launch on January 16, when a piece of foam insulation broke off from the shuttle's propellant tank and damaged the edge of the shuttle's left wing. The Columbia disaster was the second major tragedy in the history of the space shuttle program after the space shuttle Challenger broke apart shortly after launch on January 28, 1986, and all seven astronauts on board perished."

America began its space exploration with difficult, noble and attainable goals. America was bold and one of two leaders of human invasion into the vast and unimaginable space that earth and our lives are surrounded by. Logic and human efforts have found no evidence that there is any life beyond earth. No proof has been found to indicate that we humans are not the most incredible species in history. We achieved great feats.

We have now failed. I doubt that America will be able to repeat what it did. If our greatest freedom, our boldest exploration, passes into history then our species is finished. Contemporary treason then culminates into past greatness. Boisjoly and his attorneys fought valiantly to repair the problem. Many great people fought to repair the failures.

As a person I am wary. "Gone in fact". I have been professionally harassed. I have risked everything and passed the responsibilities to the FBI, and to America. I have done the best I know to do and I am at peace. Only you and history can judge the full impact of my efforts. My efforts are irretrievable in the grand scheme of life. We have to determine if we are part of the problem or part of the solution. I am satisfied with my efforts. I have been steadfast in perspective.

We were going to have our first child in a few days when I began my initial manuscript. In preparation we toured our hospital's newborn nursery. Miracles! New borns', the fine young men and women of tomorrow! Then I realized that no matter if I have succeeded or failed, no matter that Morton Thiokol failed, no matter that NASA failed, the DOJ failed, no matter if the United States of America fails - most of these young miracles would know right from wrong, would seek freedom, and would follow their genetic instincts to explore the great unknowns and seek their destinies. I had been but a 'drip on the drip stone of human effort and evolution'. They will be a new and continuing wear on the drip stone of human evolution and progress to where ever it leads. The drip stone wears and alters to time, force, intelligence, effort, care, circumstance and even failure and despair.

In a flash I realized why I was compelled to do what I did:

I did it FOR THEM and their best long term outcomes!

I did it 'For Inevitable Human Planetary Expansion'

I did it For Unlimited Resources and Infinite Possibilities

CHAPTER: 57

CONCLUSIONS

TIME: 2016

When I was young I felt there was much honesty in the world. I knew there were manners, justice, and also there was frequent fairness. Now at this time in history I am disillusioned.

There is nearly no honesty or integrity, and as a result I feel the quality of all of our lives are grievously diminished. I am saddened that nothing can be done in an honest way anymore.

Dishonesty now rules Planet Earth. Financial collapse also rules Planet Earth. The fact is that people have not changed as individuals over millennia, however our Planetary Societies have changed.

America should currently have a substantial research base on the moon. We do not. America should have sent humans explorers to Mars by now. We have not. Instead planet earth is quagmire'd in legal chaos, economic chaos, political chaos, and global wars of several types with no possible winners. [44]

There are strong arguments that World War III has begun. If so it will be truly a lethal and terrible World War in both the short and long run. Over the last 30 years NASA has accomplished much, but not nearly anything appropriate to what they have been paid to do. I would never believe anything NASA 'spun' or said and with good reason.*

If not Space X and Elon Musk then it will be someone else. NASA will be standing there to take a lot of the credit but look at this article's comparisons of costs and look at the comparison of achievements. NASA could never compete with well run private enterprise.

The wearing of the drip stone has been slow and costly. Human beings have sufficient intelligence and energy to do much better. I suspect human beings will not fully understand the beauty, bounty, resources, and comfort of earth until such time as they have to establish human life elsewhere and look back at Earth over their shoulders. Then the value of Earth's, Peace,

---

Appendix 44: "Joint Russian - USA Planning For Manned Mars Mission"
Internet July 29, 2015, Joseph P. Farrell
* See Post Log: "With or Without NASA SpaceX is going to Mars"
By Rich Smith on Internet 2018

Abundance, and Tranquillity will finally become universally valued and cherished. Expansion from Earth is actually much more of a beginning than an end destination.

I asked myself the following questions:

Did I Do Enough?
What Should I Have Done?
Is There More I Can Contribute?
What Would You Have Done?
What Nations Will Emerge To Dominate Space?
Does It Matter Any More Or Is It Too Late For America?
Is It Too Late For Planet Earth?
How many lives did I save?
Did my efforts save America's future in space?

Think About It

END

POST LOG

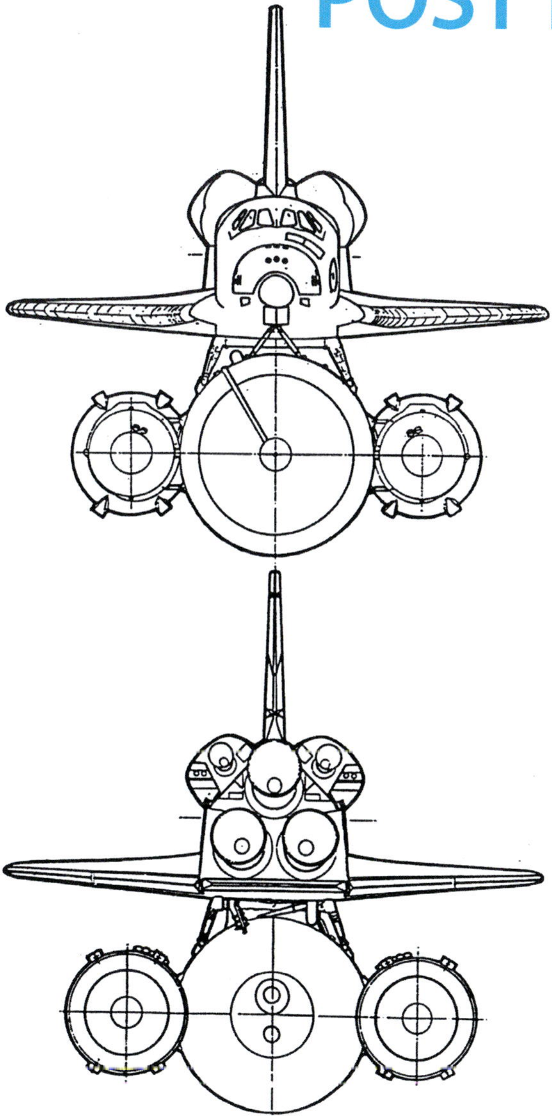

POST LOG:

Dynamic Possibilities cooking to humans travelling to Mars.
From the Internet:

"With or Without NASA, SpaceX Is Going to Mars
SpaceX just changed the world with one tweet.
Here's what you need to know about "Mars 2018."
Rich Smith

Start the countdown. In just two years, SpaceX is blasting off for Mars.

It began with a Tweet. On Wednesday, April 27, SpaceX posted two photos on Twitter, announcing plans to send a Red Dragon space capsule to Mars "as soon as 2018."

That, of course, got everyone wondering what the heck a "Red Dragon" was. And so, in due course, SpaceX CEO Elon Musk clarified that the capsule in question will be SpaceX's new Dragon 2 space capsule. This fact was then further clarified when SpaceX added that the Dragon 2 will be equipped with a "SuperDraco" propulsive landing system.

Questions, questions, so many questions

Of course, even then we had questions: Why is SpaceX going to Mars? What will it do when it gets there? And is SpaceX going alone, or is this some as-yet unannounced NASA mission it will be fulfilling (seeing as Neil deGrasse Tyson has assured us that SpaceX would have to be "delusional" to try to go to Mars without help from NASA)?

Mars

IMAGE SOURCE: NASA

You've got questions. We've got answers

To try to clarify these nagging questions, I reached out to SpaceX earlier this week and got a bit more color on the company's upcoming mission to Mars. Basically, the story goes like this:

SpaceX is laser-focused on putting humans on Mars, and the sooner the better. In fact, Elon Musk himself has stated a 2025 deadline for his first manned mission to Mars, opining that "nine years" from now "seems like a long time." That said, there's a lot of prep work that must be done in the meantime.

To name just a few likely issues, the company will want to survey the terrain, and pick a landing site -- preferably one with access to resources useful for sustaining life, and producing rocket fuel. The company will also probably want to prepare the beachhead, so to speak, pre-landing such useful items as a prefab habitat, fuel, food, and water for astronauts to use upon arrival. All this will require a series of unmanned missions to Mars before the first manned mission can take place.

The first such unmanned mission will be in 2018, launching probably in April or May. It will begin with a launch of SpaceX's new Falcon Heavy lift vehicle -- due to make its maiden voyage in November of this year -- and end with a Dragon 2 landing vertically atop its own rocket exhaust on the Red Planet.

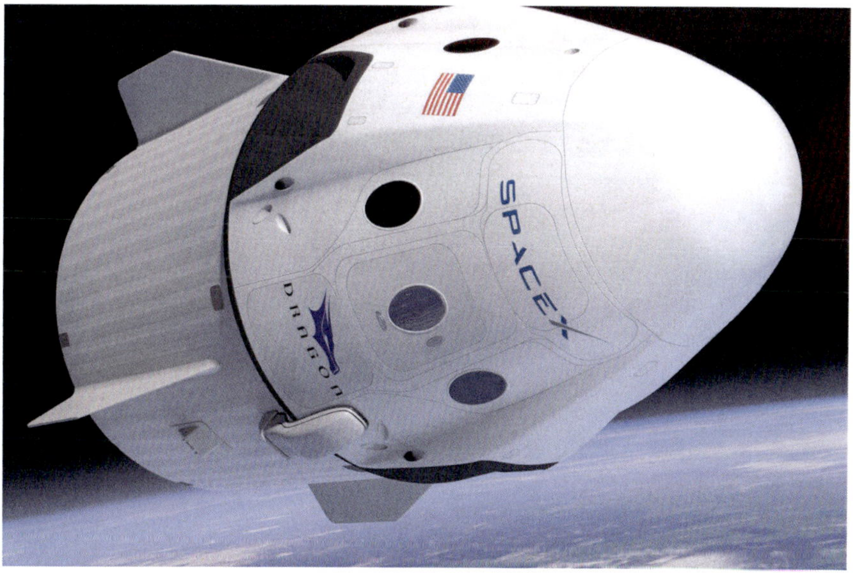

Artists conception

None of this, by the way, will be sponsored by NASA. As SpaceX advised, it will foot the bill for this 2018 Mars mission entirely on its own. More than that, the company will share any data generated by the mission with NASA -- free of charge.

How much will it cost?

Speaking of charges, as investors looking forward to the day when SpaceX ultimately IPOs, we're naturally interested in knowing how much SpaceX thinks a mission to Mars will cost. The company was entirely mum on that point. But here's what we know.

On the high end, NASA spent $2.5 billion to send the Curiosity Rover to Mars in 2011. That's about $2.65 billion in today's dollars. The cheaper Phoenix Mars Mission put a lander on Mars for just $386 million in 2008 ($427 million today). Both those missions used extremely high-cost launch vehicles from Boeing (NYSE:BA) and Lockheed Martin's (NYSE:LMT) United Launch Alliance, however, to get them to their destination -- Lockheed Martin's Atlas V in the case of Curiosity, and Boeing's Delta II in the case of Phoenix.

India's recent Mangalyaan orbiter, on the other hand, went to Mars for a grand total of just $74 million (including $25 million for the orbiter). According to new reports, the launcher on that mission, a domestic Polar Satellite Launch Vehicle, cost only $15 million -- roughly an order of magnitude cheaper than anything Boeing or Lockheed have to offer.

That's a price tag I suspect even SpaceX would be hard pressed to match. According to its published prices, the cost of launching a Falcon Heavy into Earth orbit will average $90 million. Add the cost of extra fuel to escape Earth orbit and reach Mars, add a bit more cost for equipment aboard the Dragon lander -- we're still probably only talking a total mission cost in the low $100 million range.

That's a bill Elon Musk should be able to foot easily.

What it means for investors

Unless and until SpaceX goes public, most of the above probably seems academic. We can't invest in SpaceX today; perhaps we never will. Be that as it may, one thing is clear: Mankind is going to Mars, and sooner than you think. That this will open up the possibilities of new investments -- literally out-of-this-world investments -- seems almost certain.

Hold on to your hats, folks. The world is changing as we watch."

Postscript

New Zealand launched its first substantial rocket on 25th May 2017. It seems the smallest of nations realises the necessity of space exploration.

APPENDICESE 1-47

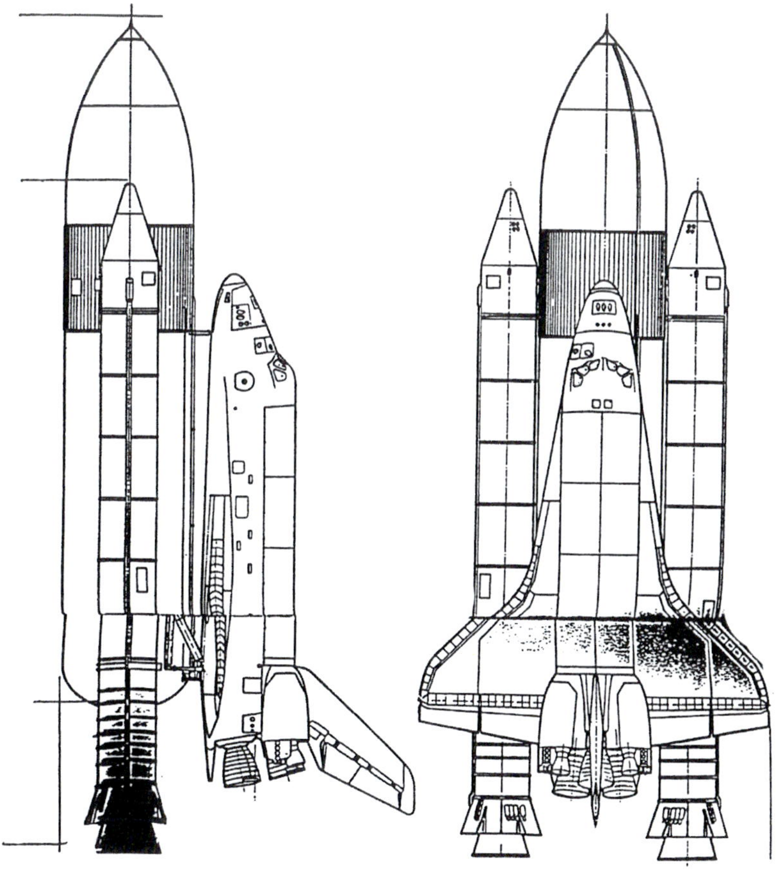

APPENDICESE

Appendix 1

Soviet lasers 'zapping' U.S. satellites?

$1.0
January 24, 19
288 pag

First 'Star Wars' skirmishes are reported

by Richard Sale
United Press International

WASHINGTON. — U.S. intelligence agencies are convinced that Soviet laser attacks have damaged super-sophisticated American spy satellites deployed to monitor missile and spacecraft launches at a major Soviet space center.

Administration sources believe the Soviets fired ground-based lasers to cripple optical equipment which was trying to scan launches at the center in Tyuratam, east of the Aral Sea in the south-central Soviet Union, to obtain sensitive military information.

The sources, most of whom requested anonymity, said they fear other vital U.S. reconnaissance satellites will soon be endangered because six new Soviet laser battle stations are under construction in the Caucasus Mountains.

However, one intelligence analyst told United Press International that experts disagree as to whether the U.S. satellite malfunctions are due to Soviet attacks and noted that some skeptics in the intelligence community put the blame on "natural causes

"Proof of what has happened to something in space is awfully hard to come by," the analyst said.

Nonetheless, other sources believe the suspected Soviet laser "hosings" — or illuminating the optical sensors — of costly satellites, details of which remain classified, occurred over the past several years and have left U.S. scientists scrambling to shield the space surveillance system.

"There is no way you can protect the optical sensors on satellites" from laser attacks, an Air Force official said.

Intelligence sources said the United States, responding to the Soviet attacks in what appear to be the first skirmishes of the "Star Wars" era, has electronically jammed Soviet satellite transmissions of surveillance data from U.S. weapons tests.

The sources acknowledged that the Pentagon also has trained

Note: Subsequently there have been numerous zapping issues involving many parties.

Appendix 2A

OVERALL DIMENSIONS

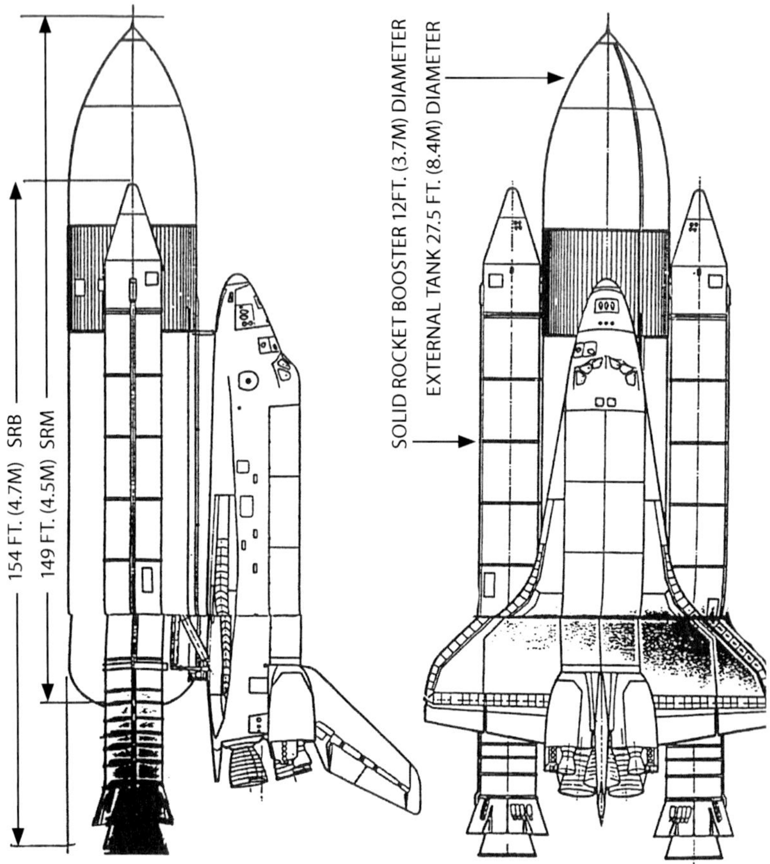

WEIGHTS
ORBITER (Empty) 165,000 pounds (75,000kg)
EXTERNAL TANK (Empty) 78,100 pounds (35,500kg)
SOLID ROCKET BOOSTER (Empty) 185,Ø00 pounds (84,100kg)
ENTIRE VEHICLE FUELED FOR LAUNCH 4.4 million pounds (2 million kgs)

General View Drawing of STS, Space Transportation System. The Solid Rocket Boosters are on the left and right of the Shuttle and the Main Tank.

Appendix 2B

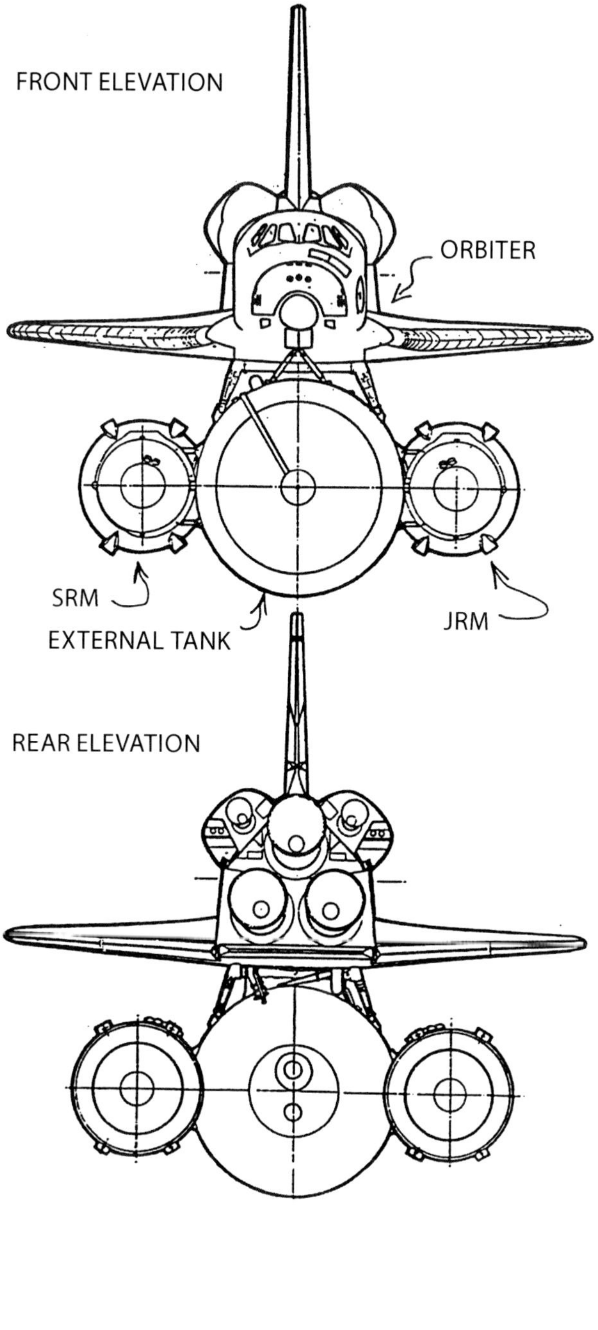
FRONT ELEVATION
ORBITER
SRM
EXTERNAL TANK
JRM
REAR ELEVATION

Appendix 2C

Solid Rocket Booster Separation

Exhaust gas plume still exiting both SRM's, however their force contribution is now near zero. The vapor above the SRM rocket exhaust and the vapor at the top of the rockets is where 'ordinance' has been exploded by a system to jettison both of the SRM's simultaneously.
Both SRM's then drop back to earth, parachutes open, and they are subsequently recovered from the Atlantic Ocean for recycling. This recycling was to be a vast cost savings operation, however it turned out the SRM cases were not safe for a single launch.

Appendix 3a

Introduction to the space transportation system

The reusable Space Shuttle is the most versatile launch system ever devised. Every other launch vehicle in the world is expendable, that is, used for only one flight. The Space Shuttle is the only launch vehicle that can be reused. In addition, it can retrieve and return payloads from space.

Because of its unique design, the Space Shuttle can serve as a launch vehicle, platform for scientific laboratories, orbiting service center for other satellites, or return carrier for previously orbited spacecraft. In short, the Space Shuttle is a multipurpose vehicle, a "space truck" which carries cargo to and from space.

Three main segments make up the Space Shuttle: a reusable Orbiter, a pair of solid-propellant boosters, and a large liquid-propellant tank.

The Orbiter is the principal part of the Shuttle. Designed to last for one hundred flights, this winged vehicle is part spacecraft, part aircraft. It carries payloads in a 15-by-60 foot (4.5-by-18-meter) cargo bay that dominates the middle of its fuselage. Forward of the cargo bay, you find the crew compartment where you live while in space. Behind the cargo bay, in the aft fuselage, are three engines that propel the Orbiter during launch. Reusable insulation on the craft's exterior enables it to survive the searing heat of atmospheric entry. After entry, the vehicle becomes a glider as it makes an unpowered approach and landing on a runaway. Onboard auxiliary power units provide power for the hydraulic system that operates the control surfaces during atmospheric flight. Hydrogen and oxygen are combined in fuel cells to generate electricity and water.

The three engines on the base of the Orbiter, called the Space Shuttle Main Engines (SSMEs), are the most advanced liquid-fuel rocket engines ever built. They burn liquid hydrogen and liquid oxygen under high pressure to create as much thrust as possible. Each SSME has a rated thrust of 375,000 pounds (1.6 million newtons) at sea level. The thrust can be varied from 65% to 109% of this rated value.

Propellants for the SSME's are contained in a large tank, called, appropriately, the External Tank (ET). The tank, made of aluminum, is 154 feet (47 meters) long and 27.5 feet (8 meters) in diameter. It is the largest single Space Shuttle component and the only part not reused. When empty, it is allowed to enter the atmosphere to break apart and burn up over the Indian Ocean. The ET attaches to the belly of the Orbiter. Two ıarge conduits feed propellants into the Orbiter's aft fuselage.

Appendix 3a: Space Shuttle Operational Description

Appendix 3b

THE SPACE TRANSPORTATION SYSTEM 1.

At launch, two large solid-propellant rockets are attached to the tank, one on each side. These are the Solid Rocket Boosters (SRBs), which provide most of the power to lift the Shuttle off the pad and propel it during the first two minutes of flight. After their propellants are consumed, the empty boosters separate from the Shuttle. They descend on parachutes and land in the ocean. The empty motor casings are recovered and reused. These are the largest solid-propellant rockets ever flown—149 feet (45 meters) long and 12.4 feet (3.5 meters) in diameter—and they are the first ever designed for reuse.

The Space Shuttle, an amazing flying machine, provides a means for reaching space, but it cannot open the "space frontier" by itself. The Space Transportation System (STS) has been devised to support all types of space missions. The Space Shuttle is the backbone of this system, for it is the conveyance "truck" that gets things there. Accessories for the Shuttle complete the flying portion of the STS.

Most communications satellites are positioned 22,300 miles (35,900 kilometers) above the earth, where it takes 24 hours to complete one orbit (see 3.7). However, the Orbiter, the central part of the Space Shuttle, is limited to orbits below 690 miles (1,100 kilometers). Once the Shuttle delivers a satellite to a low orbit, a rocket motor attached to the satellite can boost it to the desired higher orbit. Such rocket motors are built as part of the STS so that the Shuttle can launch these satellites.

In addition, the STS also includes ground facilities where the Shuttle is assembled, launched, and recovered. Together, all portions of the Space Transportation System provide an economical, efficient way to launch all types of satellites—scientific, application, planetary, or military—into space.

Finally, there is one more task the Shuttle can perform: it can retrieve satellites already in orbit. This means that a satellite that breaks down can be returned to earth for repair. This is the first time we've been able to do such work and the capability is unique to the Shuttle.

Appendix 3c

FLIGHT PROFILE 1.7

altitude of 28 miles (45 kilometers). The empty motor casings separate from the vehicle and land in the ocean via parachute. They are retrieved and later reused.

The Shuttle now consists of the Orbiter and External Tank. It continues to gain speed and altitude; 6½ minutes into the flight, you are traveling 15 times the speed of sound at an altitude of 80 miles (130 kilometers). If one or two of the main engines fails now, you have several abort modes, depending upon when the failure occurs. First, you can make a Trans Atlantic Abort (TAL) to one of several landing sites which include Benguirir, Morocco; Casablanca, Morocco; Zaragoza, Spain; Moron, Spain; Rota Naval Station, Spain; Bonn, West Germany; or Dakar, Senegal, depending upon your launch azimuth and ascent profile. The next option is an Abort once Around (AOA) which would bring you in to White Sands Space Harbor (also known as Northrup Strip) in New Mexico. Finally, there is the Abort to Orbit (ATO) mode where you have enough energy to reach a lower than planned orbit with an extended Orbital Maneuvering System engine burn. The ATO was used during STS 51-F on July 29, 1985, when an engine shut down due to a temperature sensor failure.

About 8½ minutes after launch, the main engine cut-off (MECO) command is given. The External Tank is discarded 20 seconds later. The Orbiter maneuvers down and to the left of the tank which will splash down in a remote ocean area. Remember—throughout the ascent, you travel "upside down" with your head toward the ground. The Orbiter will not assume an "upright" position until you are in orbit—and then it doesn't matter, since you are weightless.

Following ET separation, the **orbital maneuvering system** (OMS), consisting of two 6,000-pound-thrust (26,700-newton) engines, fires to place the vehicle in a low elliptical orbit. Half an orbit later, about 45 minutes

Appendix 3d

A Solid Rocket Booster in the Atlantic Ocean being recovered for return to MTI. When these large and heavy burned out rocket casings are swinging from parachute arrays they often hit the ocean with a side blow which causes 'oil canning' or substantial bends in their skins. This was a serious problem to recycling them safely.

Appendix 3e

Solid rocket booster segment move

I found these rail cars to have been unguarded, un-electrically grounded and unlocked at Corine, Ut.. The FBI found them to have moved from Brigham City to Vandenberg AFB in California for storage and then to KSC in Florida for launch. Many were not electrically grounded. Lack of electrical grounding didn't bother NASA until NASA realized they, not MTI, were responsible for damages and liability if they blew up enroute.

Appendix 4a

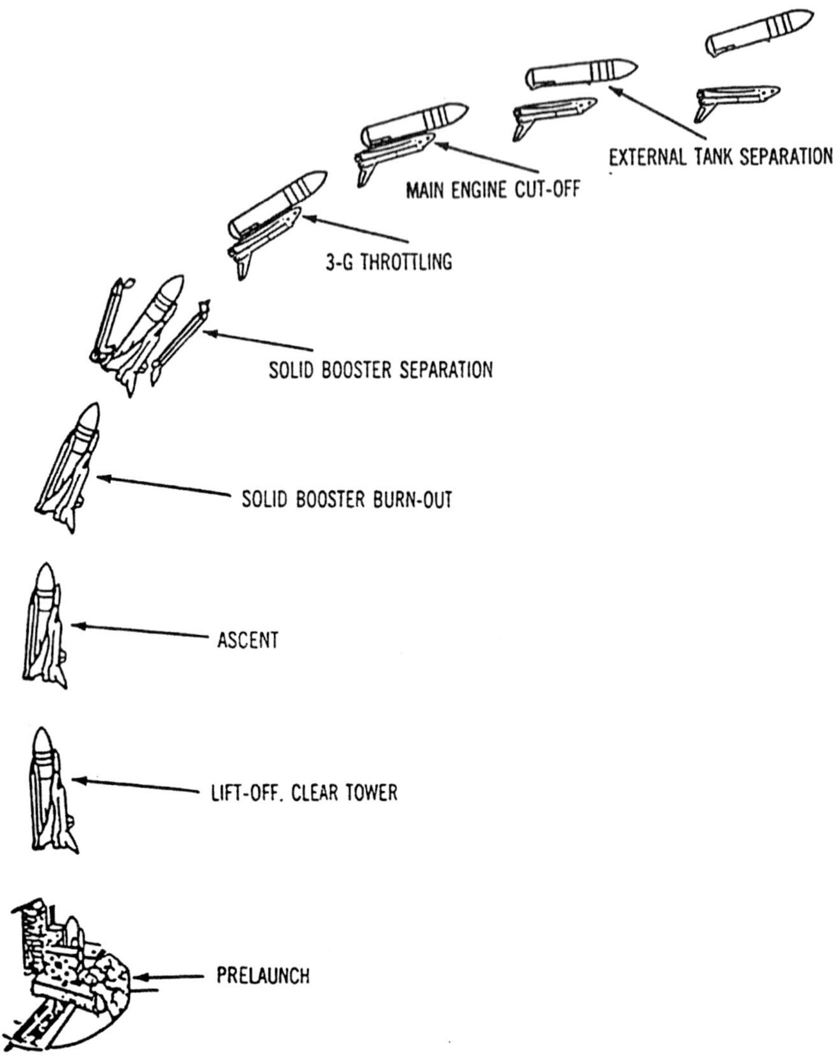

Space shuttle (STS Space Transportation system Launch sequence

Appendix 4b

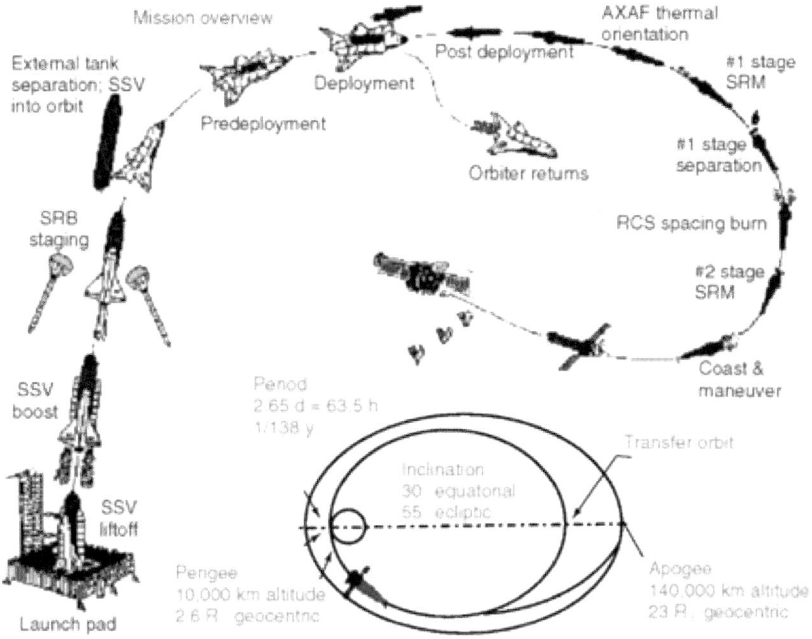

Appendix 4c

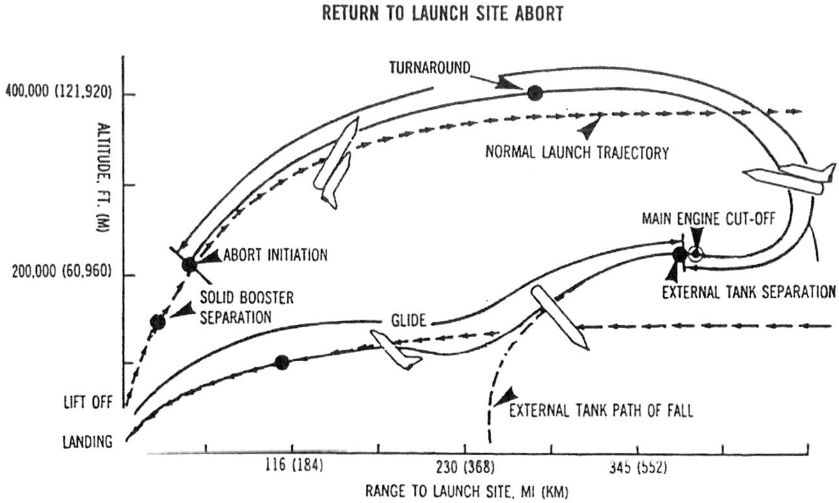

Appendix 4d

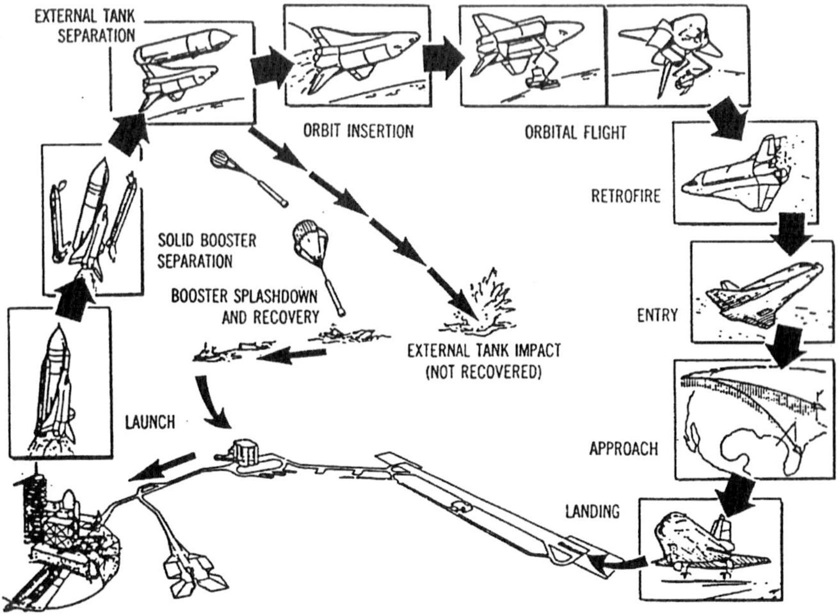

Mission Cycle

Appendix 4e

Solid Rocket Booster Recycling

The Shuttle's Solid Rocket Boosters (SRBs) are the first solid-propellant rocket motors designed for reuse.

During the first 2 minutes of flight, two SRBs provide most of the power needed for the Shuttle's ascent. After they burn out, four small motors on each end of the SRB casings fire, pushing the cases away from the External Tank and Orbiter. The nozzle extensions separate 70 seconds later, as the casings fall to earth. After 2 more minutes, the nosecaps are jettisoned. Small drogue parachutes pop out, followed by three 115-foot (35-meter)-diameter main parachutes on each booster. The boosters land base-first in the ocean.

Two ocean-going tugs locate the boosters with tracking and sonar beacons. Each tug recovers one booster. Air in the casing keeps it floating upright. After the casing is "safed," a towline is attached. A remote-controlled nozzle plug seals the open end of the casing. When the water is pumped out of the casing, it floats horizontally like a log, and is then towed to port.

In port, the casing is broken down into four sections. The sections are washed clean of any remaining insulation and propellant, using water jets with a pressure of 6,000 psi (41,000 kilopascals). These four pieces are then broken down into the original eleven segments, which are degreased and grit-blasted. The segments are then tested for cracks. Then they are pressure-tested with oil and once more checked for cracks. The segments are then put back together and repainted, and are then ready to be filled and reused. It is hoped that these will last for about twenty missions, but that depends upon test results after each flight.

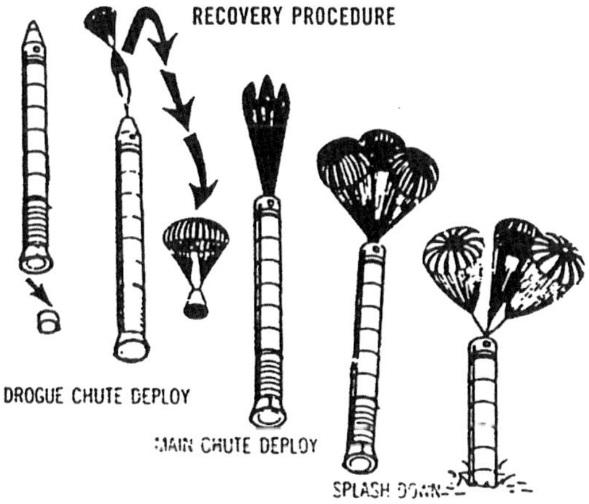

Appendix 4f

Propellants

Fuel:	Aluminum powder, 16%
Oxidizer:	Ammonium perchlorate, 69.83%
Catalyst:	Iron oxide powder, 0.17%
Other ingredients:	Binder and curing agent, 14%

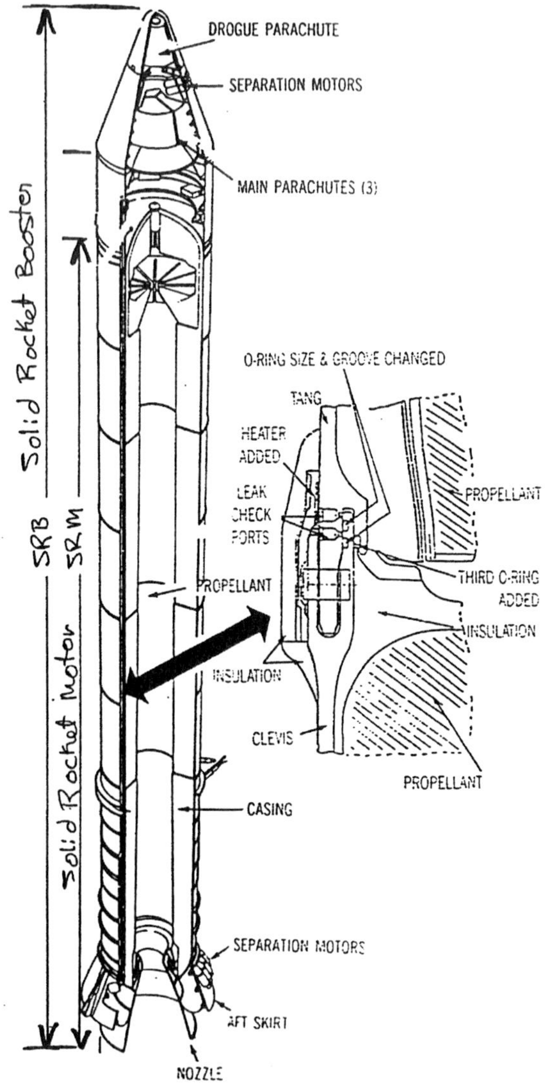

Solid Rocket Motor

Appendix 4g

Shuttle Carrier Aircraft

For cross-country movement, the Orbiter rides piggy-back atop a Boeing 747-100 jumbo jet. This mode of transport is used to ferry the craft to and from various support facilities.

After NASA purchased the 747 from American Airlines in 1974, they outfitted it with more powerful engines and modified it to accommodate the 75-ton (68,000 kilogram) Orbiter. All passenger equipment—seats, galleys, etc.—was removed. Additional bulkheads and structural supports were added inside the aircraft. Three supports for the Shuttle were placed on the jet's top, and tip fins were added to the horizontal stabilizer.

A tapered tail fairing fits over the Orbiter's aft end during transport. The fairing smooths the airflow around the Orbiter, reducing aerodynamic drag while it is in flight.

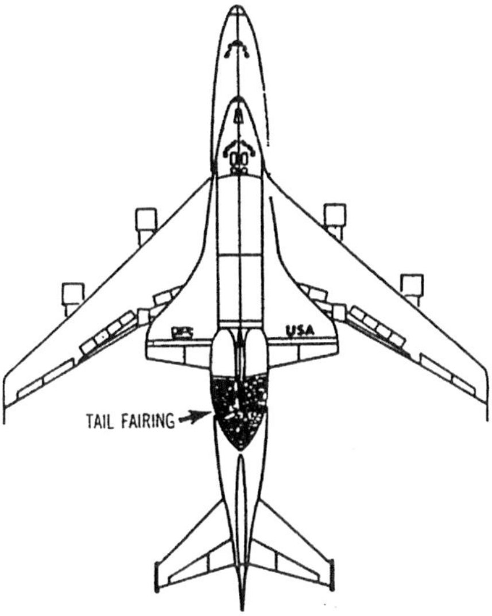

Cycling of Space Shuttle if it lands away from KSC.

Appendix 4h

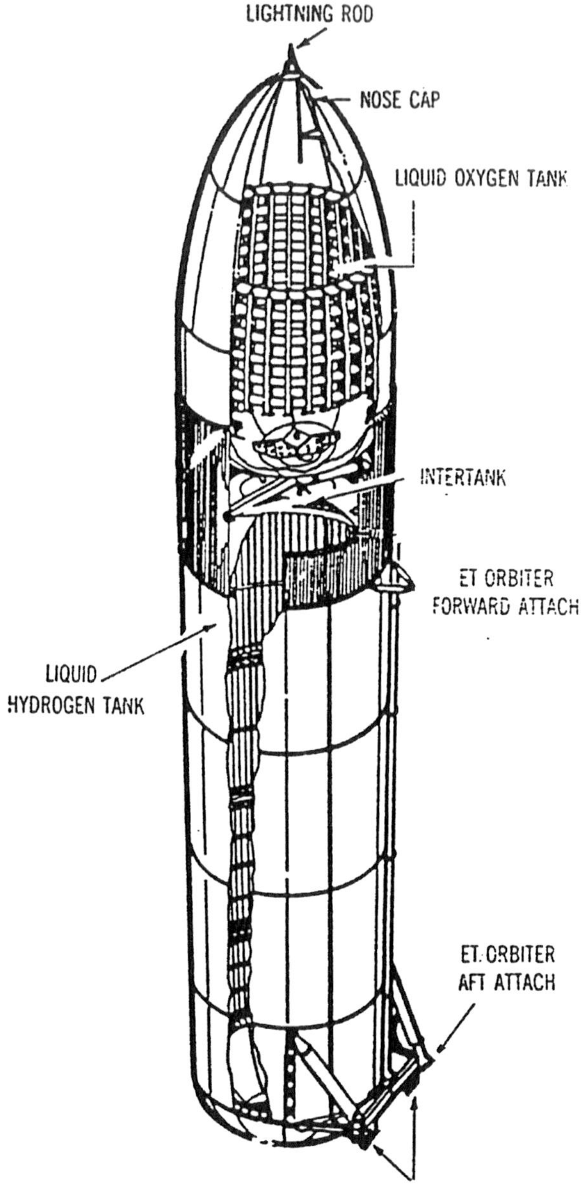

External Tank for Hydrogen fuel supplying the Shuttle Maine Engines.
Used one time only. Jettisoned from space and fall back to earth and sinks into the Atlantic Ocean.

Appendix 4i

Two large solid-propellant rockets operate in parallel with the main engines during the first 2 minutes of flight. The Solid Rocket Boosters (SRBs) provide most of the power needed to lift the Shuttle during this part of the flight. The SRB is the first solid rocket motor designed for reuse and the largest solid motor ever flown.

The SRB **nozzles** swivel (or **gimbal**) up to 6° to direct the thrust and steer the Shuttle.

The cylindrical portion of each booster is made up of four segments. The joints between these segments must contain the hot internal explosion that occurs when the booster is firing. The joint design used on the first twenty-five launches was made up of a tongue-shaped extension which fit into a U-shaped rim on the adjoining section. Two rubber sealing O-rings on the inside surface were to contain any gases that burned through the insulating putty that separated the solid fuel from the joint. This design had serious problems that led to damaged O-rings on several flights, and a complete failure on **Challenger** flight 51-L. The new design adds an extra flange with an additional seal, new materials for the O-rings, and an unvented insulation gap that should provide further protection for each joint. In addition, the requirements for safe launch environmental conditions will be strictly adhered to.

Overall Dimensions (Each Booster)

Length:	149 feet (45.5 meters)
Diameter:	12 feet (3.6 meters)
Weight at launch:	1,300,000 pounds (590,000 kilograms)
Thrust at launch:	2,650,000 pounds (11,800,000 newtons)

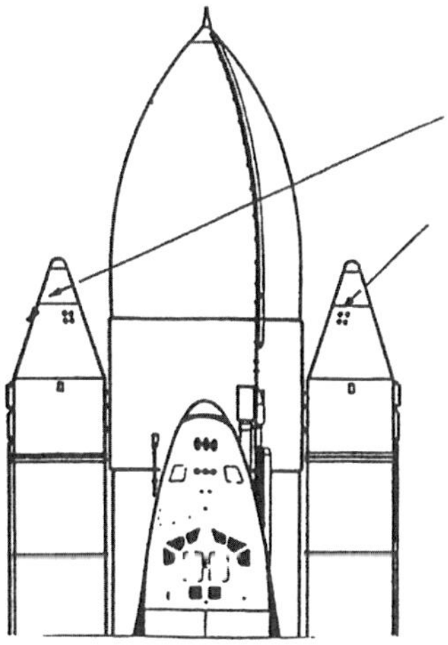

Appendix 5a

Space Shuttle Contractors 1987

Aerojet Liquid Rocket Company Sacramento, Calif. 95813	Orbiter maneuvering system engines
AiResearch Manufacturing Company of California Torrance, Calif. 90509	Air data transducer assembly and computer Safety valve (cabin air pressure) Solenoid valve (shutoff, air)
Albany International Company FRL Dedham, Mass. 02026	Nomex felt (for the thermal protection system)
Ball Aerospace Systems Division Boulder, Colo. 80302	Star tracker
Bendix Corporation Navigation and Control Group Teterboro, N.J. 07608	Airspeed altimeter Vertical velocity indicator Surface position indicator
The B.F. Goodrich Company Akron, Ohio 44318	Main and nose landing-gear wheel and main landing-gear brake assembly
Boeing Aerospace Company Seattle, Wash. 98124	Carrier aircraft modification
CCI Corporation The Marquardt Company Van Nuys, Calif. 91409	Reaction control system thrusters
Conrac Corporation West Caldwell, N.J. 07006	Engine interface unit (main propulsion system)
Corning Glass Works Technical Products Division Corning, N.Y. 14830	Windows, windshield, and side hatch window Glass-ceramic retainers (for TPS tiles
Cutler-Hammer, Inc. AIL Division Farmingdale, N.Y. 11735	Microwave scanning-beam landing- system navigation set
Fairchild Republic Company Farmingdale, N.Y. 11735	Vertical tail
General Dynamics Corporation Convair Division San Diego, Calif. 92138	Mid fuselage
Grumman Corporation Bethpage, NY 11714	Wings
Honeywell, Inc. Avionics Division St. Petersburg, Fla. 33733	Flight-control system displays and controls
Hughes Aircraft Company Space and Communications Group Los Angeles, Calif. 90009	Ku-band radar/communication system

Appendix 5b

Hydraulic Research Textron Valencia, Calif. 91355	Servo actuator elevon-electro command hydraulics Four-way hydraulic system flow control pressure valve
IBM Corporation Federal Systems Division Oswego, N.Y. 13827	Mass-memory and multifunction cathode-ray-tube display subsystem General-purpose computer and input-output processor
Instrument Systems Corporation Telephonics Division Huntington, N.Y. 11743	Audio distribution system (voice and tonal signals)
Lear Siegler Instrument Division Grand Rapids, Mich. 49508	Attitude direction indicator
Lockheed-California Company Burbank, Calif. 91520	Static and fatigue testing of orbiter structure
Lockheed Missiles and Space Company, Inc. Sunnyvale, Calif. 94088	High- and low-temperature reusable surface insulation
Martin Marietta Corporation New Orleans, La. 70189	External Tank
McDonnell Douglas Astronautics Company Huntington Beach, Calif. 92647	Solid Rocket Booster structure
McDonnell Douglas Corporation St. Louis, Mo. 63166	Aft propulsion system
Menasco Burbank, Calif. 91510	Main and nose landing-gear shock struts and brace assembly
Northrop Corporation Precision Products Division Norwood, Mass. 02062	Rate gyro assembly
Rockwell International Rocketdyne Division Canoga Park, Calif. 91304	Space Shuttle Main Engine
Rockwell International Space Systems Group Downey, Calif. 90241	Space Shuttle Orbiter System integration
Rockwell International Tulsa Division Tulsa, Okla. 74151	Cargo-bay doors

Appendix 5c

Spar Aerospace Limited Toronto, Ont. M6B 3X8 Canada	Remote manipulator system
Sperry-Rand Corporation Flight Systems Division Phoenix, Ariz. 85036	Automatic landing Multiplexer-demultiplexer
Sundstrand Corporation Rockford, Ill. 61101	Auxiliary power unit Rudder-brake actuation unit Actuation unit (body flap) Hydrogen recirculation pump assembly (main propulsion system)
Thiokol Corporation Wasatch Division Brigham City, Utah 84302	Solid Rocket Booster motors
TRW Electronic Systems Division Redondo Beach, Calif. 90278	S-band payload interregator S-bank network equipment Network signal processor Payload signal processor
United Technologies Corporation Chemical Systems Division Sunnyvale, Calif. 94088	Solid Rocket Booster separation motors Propulsion for inertial upper stage
United Technologies Corporation Hamilton Standard Division Windsor Locks, Conn. 06096	Atmospheric revitalization subsystem Freon coolant loop and flash evaporator system Water boiler, hydraulic thermal control unit Shuttle spacesuit
United Technologies Corporation Power Systems Division South Windsor, Conn. 06074	Fuel-cell powerplant
United Technologies Corporation United Space Boosters, Inc. Sunnyvale, Calif. 94088	Solid Rocket Booster assembly (checkout, launch)
Vought Corporation Dallas, Tex. 75265	Leading-edge structural subsystem and nosecap, reinforced carbon-carbon Cargo-bay door radiator and flow-control assembly system
Westinghouse Electric Corporation Aerospace Electrical Division Lima, Ohio 45802	Remote power controller Electrical system inverters
Westinghouse Electric Corporation Systems Development Division Baltimore, Md. 21203	Master timing unit

Appendix 6a

Sample of Hazard Analysis

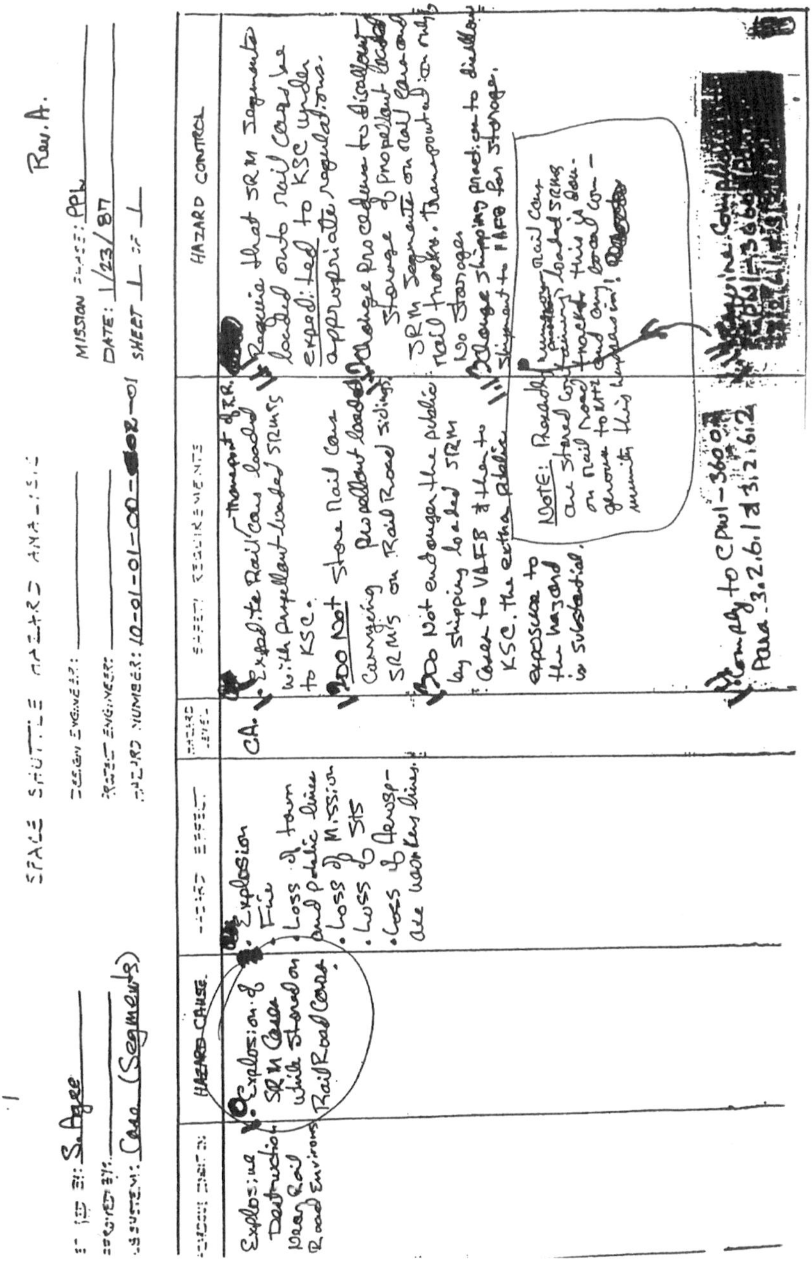

SPACE SHUTTLE HAZARD ANALYSIS Rev. A.

SUBMITTED BY: S. Agee
APPROVED BY:
SUBSYSTEM: Case (Segments)

DESIGN ENGINEER:
PROJECT ENGINEER:
HAZARD NUMBER: 10-01-01-00-[redacted]02-01

MISSION PHASE: PPL
DATE: 1/23/87
SHEET 1 OF 1

HAZARDOUS CONDITION	HAZARD CAUSE	HAZARD EFFECT	HAZARD LEVEL	SAFETY REQUIREMENTS	HAZARD CONTROL
Explosive Destruction Near Rail Road Environs	1. Explosion of SRM Cases while stored on Rail Road Cars.	· Explosion · Fire · Loss of town and public lives · Loss of Mission · Loss of STS · Loss of Aerospace workers lives.	CA.	1. Expedite Rail Cars loaded with propellant-loaded SRM's to KSC. (transport of R.R.) 1.2 DO NOT Store Rail Cars carrying propellant loaded SRM's on Rail Road sidings. 1.3 Do Not endanger the public by shipping loaded SRM Cases to VAFB & then to KSC, the extra public exposure to the hazard is substantial. 1. Comply to CPW1-3600A Para 3.2.6.1 & 3.2.6.2	1.1 Require that SRM Segments loaded onto rail cars be expedited to KSC under appropriate regulations. 1.1 Change procedures to disallow storage of propellant loaded SRM Segments on rail cars and rail tracks. Transported in only. No storage. 1.1 Change shipping practices to disallow shipment to VAFB for storage. NOTE: Presently rail cars containing loaded SRMs are stored on rail road tracks; this is dangerous to MTI and any local community this happens in! 1. Require Compl[redacted]

Appendix 6b

SPACE SHUTTLE HAZARD ANALYSIS

Rev. A. SHT 1 OF 1

EPARED BY S.W. Agee
PPROVED BY
BSYSTEM: Operations & Support

DESIGN ENGINEER
PROJECT ENGINEER:
HAZARD NUMBER 10-08-00-02-01-01

MISSION PHASE PPL
DATE 2/1/87

ZARDOUS CONDITION	HAZARD ~~CAUSE~~	HAZARD EFFECT	HAZARD LEVEL	SAFETY REQUIREMENTS	HAZARD CONTROL
No Knowledge of Egress Route by Personnel	• Personnel not instructed or trained to know emergency egress Route	1.0 • Unnecessary loss of life. • Other major Losses	CA	1.1 • Revise OMI, No. B5141, Rev. A, Para. 1.6.2.9 to require training & instructions for personnel egress method/Routes in the event of accident (fire, explosion, Critical Condition). • Comply to CPW1-3600 Para. 3.2.6.1 & 3.2.6.2	1.1.1 • Require Revision of OMI, No. B5141, Rev. A, Para. 1.6.2.9 to require egress training of personnel. • Require Compliance to CPW1-3600 Para 3.2.6.1 & 3.2.6.2

NOTE:
this is a
Fundamental
Safety Activity

Appendix 6c

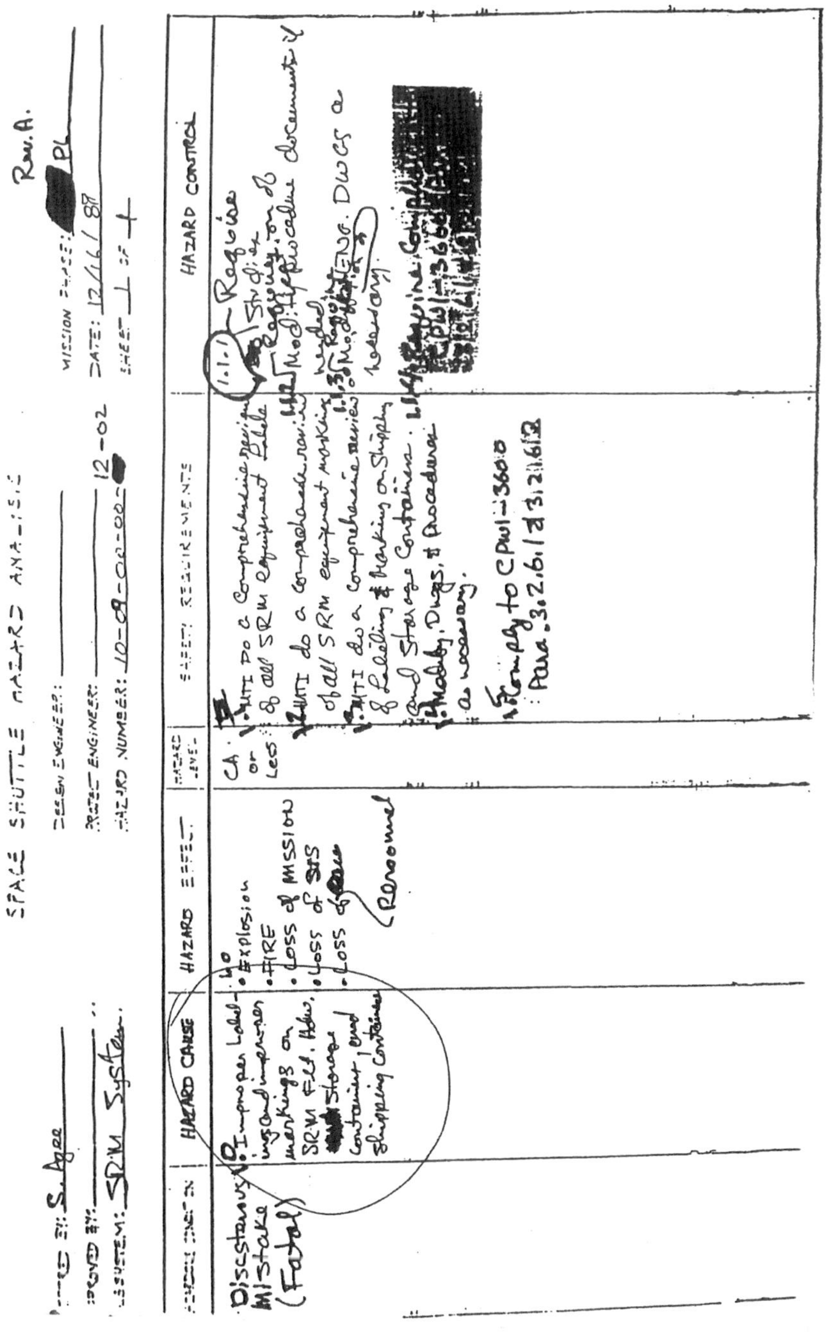

SPACE SHUTTLE HAZARD ANALYSIS

Rev. A.

PREPARED BY: S. Agee
APPROVED BY:
SUBSYSTEM: SRM System.

DESIGN ENGINEER:
PROJECT ENGINEER: 12-02
HAZARD NUMBER: 10-09-00-00-[redacted]

MISSION PHASE: [redacted] PL
DATE: 12/16/87
SHEET 1 OF 1

HAZARDOUS CONDITION	HAZARD CAUSE	HAZARD EFFECT	HAZARD LEVEL	SAFETY REQUIREMENTS	HAZARD CONTROL
Disastrous Mistake (Fatal)	1.0 Improper Labelings and improper markings on SRM Flt. Hdw. [illegible] Storage container, and shipping containers	1.0 • Explosion • Fire • Loss of Mission • Loss of S[redacted]S • Loss of [illegible] Personnel	CA or Less	1. MTI do a comprehensive review of all SRM equipment labels. 2. MTI do a comprehensive review of all SRM equipment marking. 3. MTI do a comprehensive review of labeling & marking on shipping and storage containers. 4. Modify Dwgs, & procedures as necessary. 5. Comply to CPW1-3600 Para. 3.2.6.1 & 3.2.6.2	1.1.1 Require Studies [illegible] 1.1.2 Modify procedure documents if needed. 1.1.3 Require Modify [illegible] ENG. DWGS as necessary. 1.1.4 Require [redacted]

Appendix 6d

SPACE SHUTTLE HAZARD ANALYSIS Rev. A. SHT 1 OF 1

PREPARED BY S. Agee
APPROVED BY
SUBSYSTEM: Igniter Adapter.
DESIGN ENGINEER
PROJECT ENGINEER:
HAZARD NUMBER 10-03-04-[illegible]
MISSION PHASE BT
DATE 2/4/87

HAZARDOUS CONDITION	HAZARD CAUSE	HAZARD EFFECT	HAZARD LEVEL	SAFETY REQUIREMENTS	HAZARD CONTROL
• Gas leak paths through around Igniter Adapter • Tension on Igniter Adapter bolts at launch rather than have the igniter adapter pressure/compression seal its self at ignition • Tension on S&A device bolts at launch rather than have the S&A device pressure/compression seal its self at ignition. Shatter of igniter adapter from bolt pattern rather than complete surface seating.	• High impulse propellant ignition force spike could shatter, or break the igniter adapter bolts or warp the igniter adapter. • The igniter adapter & S&A bolts are in tension rather than the sealing component being under compression.	1.0 • Explosion • Fire • Loss of Mission • Loss of STS • Loss of Crew • Loss of VAB	CA	1.1 • Do a trade study for Block II igniter adapter. Characterize advantages of a pressure/compression seal rather than a bolted-in-tension seal for the igniter adapter. 3.2.6 • Comply to CPW1-3600 Para. 3.2.6.1 & 3.2.6.2	1.1.1 • Require a trade study to improve igniter adapter sealing & seating properties. • Require compliance CPW1-3600 Para. 3.2.6.1 & 3.2.6.2

Appendix 6e

SPACE SHUTTLE HAZARD ANALYSIS — DC — Rev. A. — SHT 1 OF 1

PREPARED BY: S. Agee
APPROVED BY:
SUBSYSTEM: Systems Tunnel

DESIGN ENGINEER:
PROJECT ENGINEER:
HAZARD NUMBER: 10-01-03-~~04~~-01-05

MISSION PHASE: [redacted]
DATE: [redacted]

HAZARDOUS CONDITION	HAZARD CAUSE	HAZARD EFFECT	HAZARD LEVEL	SAFETY REQUIREMENTS	HAZARD CONTROL
Systems tunnel covers blow off	1. "Temporary" Nylon Shipping Bolts (1U82472-02) are not removed & replaced by metal flight certified bolts prior to launch. • Temp. nuts, bolts, & other equipment not Audited & accounted for prior to launch.	1.0 • Systems tunnel cover attachment fails. • Explosion • Fire • Loss of Mission • Loss of STS • Loss of Crew.	CA.	1.1 • Careful Audit of all temp. Components taken off of STS & specifically the Systems tunnel • Do not fly STS with temp. hardware installed. • Modify all appropriate Dwgs. & Procedures as appropriate. • Careful & Complete Safety audit of both ~~~~ Cft & tem. items. • Amend OMI No. 5141, Rev. A, Page 98, Para. 03-172 to account for temp. Hdw. prior to flight. • Comply to CPW1-3600 Para. 3.2.6.1 & 3.2.6.2	1.1.1 • Require Audit of temp. Parts being used on SRM/Systems tunnel. • Require Audit of flight Parts being used on SRM/Systems tunnel. • Require Modifications to all appropriate Dwgs. & procedures required. • Require amendment of OMI, No. 5141, Rev. A, Page 98, Para 03-172 to account for temporary Hardware prior to flight. • Require Compliance to CPW1-3600 Para. 3.2.6.1 & 3.2.6.2

Appendix 6f

SPACE SHUTTLE HAZARD ANALYSIS — Rev. A.

...ED BY: S. Agee — DESIGN ENGINEER: ______ — MISSION PHASE: [redacted]PL

...YED BY: ______ — PROJECT ENGINEER: ______ — DATE: 1/26/87

...YSTEM: Safe & Arm Device — HAZARD NUMBER: 10-03-[redacted] 02 00-17-02 — SHEET 1 OF 1

DESCRIPTION	HAZARD CAUSE	HAZARD EFFECT	HAZARD LEVEL	SAFETY REQUIREMENTS	HAZARD CONTROL

...fe & Arm Device Mechanical/Electrical Malfunction Failure	• Contaminated (Impure) Nitrogen & hydrogen gas mixture
	• Water Contamination of Gas Mixture
	• Petroleum distillate Contamination of Gas mixture.
	• Other Contamination of Gas Mixture.
	• Improper ratio of Nitrogen & hydrogen (Note: too much hydrogen could precipitate hydrogen embrittlement of S&A metal part or weaken
		1.0
		• Explosion
		• Fire
		• Loss of Mission
		• Loss of STS
		• Loss of Crew.
		
		• Comply to CPW1-3600 Para. 3.2.6.1 & 3.2.6...
			CA.	1.1
				• Specify the purity of the gasses by spec. requirement.
				• Specify the ratio of Gasses (Mix.)
				• Specify Moisture Content of gasses
				• Do a Study on the gasses to Characterize the Gas Mixtures Ageing phenomena, i.e. in situ Contamination.
				• Analyze Gas Mixture for more & possibly destructive or conductive impurities.
				• Take gas samples at the same time the S&A Cavity is filled. Immediately test one sample, save the others for future reference.
				- Modify all dwgs, specs, procedures, and OMI's as needed,
				• ENFORCE Strict QA.
					1.1.1
					• Require gas purity specifications on the Nitrogen & hydrogen
					• Require gas Ratio (Mix) specifications (Nitrogen/Hydrogen).
					• Require Moisture Content of Gas Mixture specification with Max. Min. parameters specified.
					• Require Gas in Situ (S&A) ageing study for —
					  • Degradation of Mixture
					  • Contamination of Gases by electro static products from Armature.
					  • Contamination of Gases by Corrosion products in the S&A Cavity.
					  • Contamination of Gases by Contact with rubber and/or petroleum products.
					  • Others as needed.
					• Modify all Dwgs. and Procedures to include gas samples where applicable.
					• [illegible] Strict QA.
					• Require Compliance to. [illegible]

The mark ups and mark outs were because the HA form changed from day to day and was in deliberate total chaos from MTI and NASA.

Appendix 7

SPACE SHUTTLE
SOLID ROCKET MOTOR
CHRONOLOGY

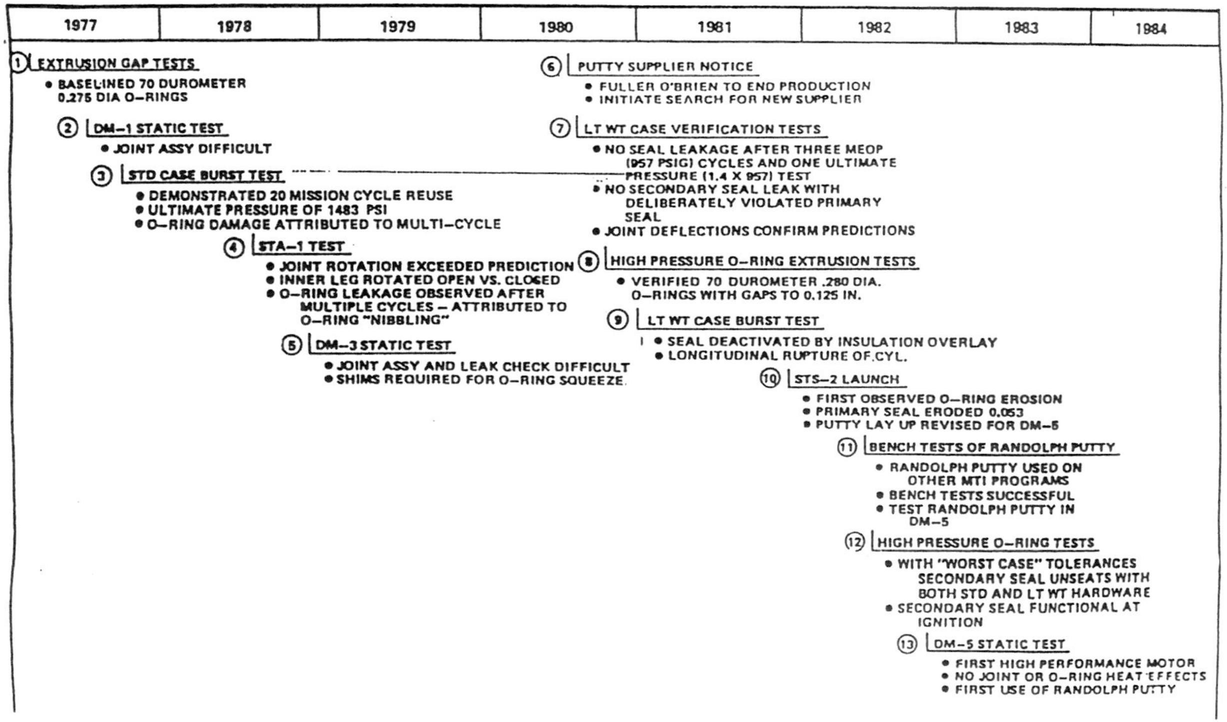

Appendix 8a

Thiokol Docked $1 Million a Month For 'Unsatisfactory' Work on MX

By Norman Black
AP Military Writer

WASHINGTON — The Air Force, disclosing a new problem with production of the giant MX nuclear missile, said it began withholding contract payments from Morton Thiokol Inc. on Wednesday because of "unsatisfactory workmanship practices" on the weapon's first-stage booster.

The service said the disciplinary measure entailed withholding 10 percent of the monthly contract payment normally made to the Chicago-based company for work on the stage-one rocket motors.

That amounts "to approximately $1 million per month," the Air Force added.

The MX, which is called the Peacekeeper by the Reagan administration, is a long-range, four-stage intercontinental ballistic missile capable of carrying 10 nuclear warheads. The first stage of the missile is made by Morton Thiokol's Wasatch Operations in Brigham City.

The remaining rocket stages for the MX are made by other defense contractors. The Wasatch division of Morton Thiokol is the same one that made the solid-fuel boosters for the ill-fated space shuttle Challenger.

An Air Force statement said the contract-payment penalty was imposed "because of unsatisfactory workmanship practices relating to the production of the Stage I rocket motor for the Peacekeeper missile system."

"Deficiencies in contractor perf mance in this case are characteri: by inadequate work instructio poor workmanship and a lack proper quality-control managem during the production process for t Stage I rocket motor."

The Air Force asserted that as now, the problems at Morton Thio were not expected to affect t schedule for deployment of compl ed missiles.

"However, internal Morton Thio production schedules have be missed, which, if left uncheck could impact on future delive schedules," it said.

At another point in its announ ment, however, the Air Force

See A-2, Colum

Appendix 8b

Air Force Penalizes Thiokol for Work on MX

July 16, 1987 | Associated Press

WASHINGTON – The Air Force, disclosing a new problem with production of the MX nuclear missile, said Wednesday that it has begun penalizing Morton Thiokol because of "unsatisfactory workmanship practices" on the weapon's first-stage booster.

The service said the disciplinary measure entailed withholding 10% of the usual monthly contract payment made to the Chicago-based company for work on the first-stage rocket motors. That amounts "to approximately $1 million per month," the Air Force added.

The MX--called the Peacekeeper by the Reagan Administration--is a long-range, four-stage intercontinental ballistic missile capable of carrying 10 nuclear warheads. Its first stage is made by Morton Thiokol's Wasatch Operations in Brigham City, Utah, the same unit that made the solid-fuel boosters for the ill-fated space shuttle Challenger.

The remaining rocket stages for the MX are made by other defense contractors.

An Air Force statement said the penalty was imposed "because of unsatisfactory workmanship practices relating to the production of the Stage I rocket motor for the Peacekeeper missile system."

It added: "Deficiencies in contractor performance in this case are characterized by inadequate work instructions, poor workmanship and a lack of proper quality-control management during the production process for the Stage I rocket motor."

The Air Force asserted that, as of now, the problems at Morton Thiokol are not expected to affect the schedule for deployment of completed missiles. "However, internal Morton Thiokol production schedules have been missed, which, if left unchecked, could impact on future delivery schedules," it said.

At another point, however, the Air Force appeared to contradict the statement that only internal production schedules had been missed, saying that there have been "some short delivery delays."

The Air Force's public affairs office said it could not immediately answer such questions as when and how the service became aware of the workmanship problems or explain why it felt confident in saying that the problems "have had no impact on the quality of units already accepted by the Air Force."

Telephone calls to Morton Thiokol officials were not immediately returned.

The disclosure of the disciplinary move against Morton Thiokol followed similar action earlier this spring against Northrop, which makes the guidance system for the MX. The Air Force has so far withheld about $90 million in contract payments to Northrop because of delivery delays.

The Air Force acknowledged last month that because of the problems on the Northrop production line, seven of the 21 MX missiles installed at F. E. Warren Air Force Base, Wyo., have not been brought to alert status.

Responding to congressional concerns, the Air Force last week ordered a technical review by a scientific advisory board to ensure that the guidance units that have been delivered will work properly.

Northrop has acknowledged production delays but dismissed suggestions that the guidance units might malfunction.

Appendix 9

Will Tragedy Delay MX Program?

"It is the only facility equipped to do Peacekeeper casting at this moment," said Capt. Louis Figueroa, Air Force Systems Command. "However, Morton Thiokol does have other casting facilities that I'm told with some modification could be set up for the Peacekeeper stage one casting, and keep the Peacekeeper program right on track."

Castings are molds for the missile motor's solid fuel.

Capt. Figueroa said he did not know the amount of time that will be needed to modify the other Morton Thiokol facilities. "We just don't know what the impact will be yet," he said.

Systems Command is responsible for research, development and acquisition of weapons systems.

"Right now the Air Force is assessing the impact of the loss of this facility on the Peacekeeper program," Capt. Figueroa said. "To this point in time, Morton Thiokol has been right on time for their delivery of stage one."

The MX program is to be completed by December 1988.

Hercules Inc. manufactures the third stage of the rocket motor at its Magna facility. Hercules spokesman Dave Nicponski said it is "too early to tell" whether the fire will cause a delay in the production of the MX.

Capt. Ron Joy, a spokesman for the Air Force Ballistic Missile Office, said, "This shouldn't affect the other contractors at all."

Morton Thiokol has already met its January schedule for shipment of the first stage of the MX missile, Capt. Joy said. The four stages of the MX are manufactured separately by the contractors and are assembled in their silos in Wyoming.

The Ballistic Missile Office is responsible for the Air Force's ICBMs.

The MX is a four-stage intercontinental ballistic missile, with a range of 6,000 miles, designed to deliver 10 warheads to independent targets, said Air Force spokesman Capt. Jay DeFrank. The destructive power of

See A-2, Column 1

Test Failure Sets Back The Shuttle

By Dawn Tracy
Tribune Staff Writer

Just hours after fire destroyed an MX missile facility Tuesday, Morton Thiokol received another setback when NASA announced that a nozzle component failed during a test firing last week, forcing the nation's space shuttle launch -- originally scheduled for June — to be postponed indefinitely

The nozzle's failure and the flash fire Tuesday that killed five workers are unrelated, said Thiokol spokesman Rocky Raab. He added, however, that Tuesday "was a particularly bad day for us."

Late Tuesday, NASA officials at the Marshall Space Flight Center, Huntsville, Ala., said "the STS-26" launch date will be delayed, referring to Discovery's planned June 2 blastoff, the nation's first attempt to launch a space shuttle since the Challenger disaster two years ago.

Thiokol's redesigned rocket was test fired Dec. 23 at the company's Brigham City facility, the second of four full-scale test firings required before the nation's next space shuttle can be launched

Initially, NASA and Thiokol officials said the test was a success. But Tuesday NASA officials said further examination of the motor showed that a nozzle anchored to the rocket case came apart during the test.

NASA officials said they and Thiokol engineers are developing contingency plans, but in the meantime, shipments of flight motors to the Kennedy Space Center, Fla., will be halted.

"The extent of the delay will not be known for the next several days," said the statement. "Additional information will be released as analysis continues."

Appendix 10

G. CONOVER	D. ROBSERTSON	K. SONOFOSKY
S. RODGERS	B. HORWICK	C. SWOPES
*L. DUNTHORN	C. SHARP	P. GREENHALGH
*S. SIEMION	M. FORBES	L. HELLER
*L. TOUGHAW	* J. SEILER	R. LARSON
*B. GARMA	* D. ALEN	* M. PHILLIPS
*V. SMITH	* K. SMITH	* M. HOLLWAY
*R. JAQUEZ	* R. DANIEL	* K. GILES
*R. MAZZLE	* C. GRABIN	* T. BRADLEY
*E. MOON	4 RECS - PERM.	* S. AGEE
*J. HOEPFER		* A. MURTHI
*R. ENLOW		* R. CHOPRA
*W. BARLOW		* H. HINKLE
6 RECS - PERM.		* J. TAYLOR
		* R. AVE
		* D. DANALDS
		* F. BROADHACKER
		* D. JONACEK
		* S. MASQUEREZ
		* T. LAINE PH.D.
		4 RECS. PERM.
TOTAL = 20	TOTAL = 16	TOTAL = 25

GRAND TOTAL = 63 AT THAT TIME

* = JOB SHOPPERS (CONTRACT PERSONNEL)

After this list of contractors many more contractors came to MTI

Appendix 11

Tragic Day for Utah's Thiokol

Blast Claims Five, Levels Building

By Eric McMullin
Tribune Staff Writer

BRIGHAM CITY — Nearly 100,000 pounds of rocket propellant in an MX missile section ignited at Morton Thiokol Tuesday, killing five men and destroying a casting facility.

Killed were Morton Thiokol employees Michael H. Smith, 24, and Lawrence W. Rothenberger, 47, both Brigham City; Del M. Loosli, 42, Nibley; Richard J. Goodsel, 36, Clarkston; and Jeffory W. Ellis, 28, Brigham City.

Mr. Ellis survived the initial fire but died Tuesday afternoon at the Intermountain Burn Center.

It was the worst disaster ever at the northern Utah aerospace plant and the first since a March 1986 explosion.

The explosion — technically an extremely fast-burning fire — occurred at 6:20 a.m.

"It lit up the whole valley," said Morton Thiokol fire crew chief Marcus Petersen, who was with the first crew to respond.

Firemen could see the flames "50 feet in the sky" as they raced to the scene, he said.

Even so, by the time they arrived two minutes later, the propellant had burned itself out and only twisted, burning pieces of metal and wood were left to extinguish, he said.

"Myself and a safety man went around with air packs. We found one survivor [Mr. Ellis]. He was lying about 25 yards to the west of the building, conscious.

"He said he was cold."

Two of the victims were found inside the building and two "just outside," he said.

The five men had been working to remove a metal cylinder that creates the hollow core in the finished rocket's propellant, said spokesman Rocky Raab. The situation is volatile, but is not the most volatile process at the company, he said.

"All I can say is we've done thousands and thousands of these and each one goes through the same process and we've never had one explode before."

The company has suspended propellant casting until it determines the cause of Tuesday's fire. Mr. Raab said the Peacekeeper motors in production will remain in production. If they get to the casting stage and the problem has not been solved, work will halt, he said.

Part of the casting procedure is automated, just as many of the company's other procedures are automated so that workers will not be in the area during volatile operations. Even so, workers must enter the area to connect parts or set instruments in anticipation of the next automated step, said Mr. Raab.

In a general sense, that's what Tuesday's workers were doing, he said, although he did not know what point they were at in the procedure.

The motors are filled with 100,000 pounds of propellant, which consists of aluminum as the fuel, ammonium perchlorate as an oxidizer and a synthetic rubber binder for stability. The liquid propellant is poured into the motors and dries around a metal cylinder. The cylinder is then removed, creating a hollow core through which the propellant is ignited during a firing.

The 23-foot motor is filled underground in a 30-foot pit. The building that was destroyed Tuesday rises about four stories above the pit.

Reporters were allowed to tour the site Tuesday afternoon. The building's steel girder frame stood blackened and twisted. Debris was scattered over a half-mile area and the remains of the motor were still smoking. Morton Thiokol firemen patrolled the area.

The company staffs its own fire crews and had a station about a mile away from the fire.

The company is on a holiday standdown as all non-essential employees were let off Christmas Day and are not due back until Jan. 4. Mr. Raab said he did not know how many employees were at Morton Thiokol Tuesday.

An investigation team will be formed of Morton Thiokol and Air Force personnel "at a minimum," said Mr. Raab. Officials met through the day, but not as a formal team, he said.

Utah Occupational Safety and Health Compliance Supervisor Don Anderson said his office sent one person to the site Tuesday and expects to send more Wednesday for an investigation.

"We'll try to put together what happened and why," Mr. Anderson said. He said the task will take "a couple weeks" as investigators conduct interviews and reconstruct the procedure used for the missile's production.

Morton Thiokol has the contract for the first stage — or lowest segment — of the Peacekeeper missile.

Appendix 12a

Washington Post Thurs. 1/29/87, pg. A-15

THE WASH

Engineer Sues Thiokol For 'Criminal Homicides'

Health, Career Said Ruined After Testifying

ROGER BOISJOLY
... retired after shuttle explosion

By Nancy Lewis
Washington Post Staff Writer

A former Morton Thiokol engineer who tried to stop last year's fatal launch of the Challenger filed a $1 billion damages suit in federal court here yesterday, saying that the company ruined his health and career after he testified before the presidential commission investigating the disaster.

Roger Boisjoly, 48, of Willard, Utah, claimed in his unusual civil damages suit filed on the first anniversary of the space shuttle explosion that the deaths of the seven Challenger astronauts were "criminal homicides" committed by the company, which manufactured the solid-rocket boosters whose failure led to the disaster.

Legal observers speculated yesterday that while such a lawsuit might have symbolic significance, the chances of "criminal homicide" charges surviving in the federal court are extremely limited.

Boisjoly retired from the company in October on a disability pension because of "post-traumatic stress disorder and depression caused directly by the disaster" and his treatment after testifying, according to the suit.

"Roger feels very strongly about what happened," said Robert N. Levin, Boisjoly's attorney. ". . . But even today, a year later, there has not been an inquiry that focused on who was responsible for the accident."

Inquiries by the Rogers commission and Congress dealt primarily with the "technical side," Levin said, "what kind of management and structural changes were needed to prevent a recurrence."

He said Boisjoly's suit is designed to address the question of responsibility.

Officials of Morton Thiokol, which is based in Chicago but manufactured the shuttle boosters in its Utah facility, could not be reached for comment.

The suit has been assigned to U.S. District Court Judge Harold H. Greene.

Boisjoly is one of several Morton Thiokol engineers who testified before the Rogers commission and a congressional panel investigating the shuttle crash about recurrent and unsolved problems with the booster's O-rings.

The engineers told the panels that, prior to the Challenger disaster, National Aeronautics and Space Administration and the company kept information about the problems from Congress to keep the shuttle launches on schedule and to ensure that Morton Thiokol would win a new contract it was seeking for the boosters.

Boisjoly and another engineer who were transferred to lesser posts after they testified were later reinstated.

In addition to $1 billion in punitive damages, Boisjoly seeks $1 million in compensatory damages for lost income and damage to his health.

Appendix 12b

PAGE A-4/THE POST-STANDARD, Saturday, Jan. 31, 1987

NEWS EXTRA

Sometime on the night of Jan. 27, 1986, Roger Boisjoly's frustration turned to anger.

He and a team of engineers unanimously recommended postponing the shuttle launch because of frigid conditions in Florida. But four Thiokol vice presidents, under pressure from NASA in a late-night telephone conference, overruled their top technical experts.

Boisjoly saw it coming. After listening to NASA executives challenge the reliability of their scientific data and complain that they were "appalled" at a recommendation to delay. Thiokol General Manager Jerry Mason told his fellow vice presidents it was time to "take off your engineering hats and put on your management hats."

Boisjoly grabbed a set of photos showing seal erosion from the January 1985 shuttle mission and "slapped them down, not gingerly" in front of the Thiokol executives.

"I said, 'Look at the photos. How the hell can you ignore this? Look at the physical evidence.'" Instead, says Boisjoly, the vice presidents looked at one another.

That is when Arnie Thompson scooped up a pad of paper and squeezed between two of the executives. He tried to draw one more picture that might convince them where mere data had failed, a picture of a rocket joint with clevis and tang and O-rings and.... He never finished.

"They weren't listening," he would confide to a colleague later.

The vote of the vice presidents was: "Go." Boisjoly did not really hear the poll. He was too angry — "angry in body and mind."

In Florida, icicles were forming on the Challenger pad. In Utah, Boisjoly and Thompson went home to their first sleepless night.

Through the magic of videotape, schoolteacher Christa McAuliffe kept strolling across the nation's television screens over and over again all morning as launch time approached. Each time, she waved and smiled on her way to the shuttle crew module. But Thompson was not watching. When liftoff finally came, he was in his office with another engineer, still reliving the frustration of the night before. And trying to ignore the dread.

"I couldn't make them listen," he kept saying. Although he would not watch the launch, he could not avoid the awful news.

Maybe more than any of them Bob Ebeling knew tragedy. He actually held it in his arms once. That was the day his son committed suicide. Ebeling could only cradle the young man in his arms as he died — and wonder why he had not done more to prevent it.

"How many times can you stand to feel responsible for something that goes wrong?" Ebeling agonized. "I should have been more boisterous. I should have said something more."

Fortunately, perhaps, there was little time for anyone to ponder long all the implications of the disaster. An investigation was under way almost instantly. Boisjoly and Thompson were dispatched to the Marshall Space Flight Center in Huntsville, Ala., where NASA planned to conduct its own probe.

Maybe there was defective workmanship or defective equipment. Maybe something in the cargo hold exploded. Maybe the big rust-colored external tank was leaking. Maybe one of the main engines threw a rotor blade. Maybe a strand of hair compromised a seal in the rocket booster. The maybes seemed infinite.

Most engineers and managers at Thiokol felt no sense of guilt, corporate or personal, in those first days after the Challenger crash. Most doubted that the solid rocket boosters should even be considered suspects. Most doubted — but not Ebeling or Boisjoly or Thompson or McDonald or Russell.

Only hours after the explosion "ol optimistic Arnie" Thompson stood in a Thiokol office with stunned colleagues, all gathered around to speculate about what went wrong. One of them recalls Thompson swiping a big, boney finger across a picture of a booster rocket joint and saying sadly and certainly: "There. That's where it failed."

It is not clear what most startled his colleagues: the fact that Thompson could presume to solve the mystery so quickly, or the fact that he was pointing a finger right back at his own company.

The pre-launch dispute between management and the engineers might never have come out had the engineers kept silent. During the first closed session of the presidential commission investigating the crash, neither NASA executives nor Thiokol managers had mentioned any hint of the emotional debate over cold conditions or the engineers' recommendation to wait.

The panel was about to adjourn when McDonald, sitting silently to that moment, raised his hand. Al McDonald was never noted for his diplomacy. Like any good engineer he drew straight lines from Point A to Point B and he was accustomed to speaking just as directly.

McDonald's message was plain: The engineers had warned NASA and their managers that it was unsafe to launch in the cold but they had been overruled.

In short, the Challenger and its crew had been victims of cold O-rings and bad judgment.

In subsequent days Boisjoly, Thompson, Ebeling and Russell also were called to give similar testimony. Thiokol was embarrassed. Its engineers were making its managers — and prime customer, NASA — look bad.

The space agency's own investigation of the Challenger crash seemed at times to be focusing on discrediting the cold theory. An early NASA draft report spent only a few paragraphs on the effects of cold on rocket seals while devoting several pages to other potential problems, including errors in the rocket assembly process that could have caused the failure.

But McDonald's bombshell was followed by Boisjoly's bombshell, the release of his private company memo of July 31, 1985, and other papers documenting a history of warnings and "red flags" over rocket seal problems.

NASA's irritation with McDonald and Boisjoly, in particular, finally resulted in their virtual isolation. Thiokol managers cut them off from dealing with the space agency — until Congress learned about it and threatened to retaliate against the company.

Ebeling was systematically excluded from technical conferences once part of his duties. "They acted like I wasn't needed anymore," he shrugged.

Roger Boisjoly did not notice the anger that raged within him. Not at first. Everyone else did — Roberta and the girls, Norma and Darlene, and the dog. He yelled at the dog. He yelled at his daughters. He raged at anything and anyone at the slightest provocation.

For months Boisjoly was obsessed with the rocket seal problem. He could not sleep. He could not talk about anything without coming around to O-ring seals. He stopped going to church to avoid the crowds. He was starting to get headaches.

Other members of "the leper colony" made a habit of dropping by his office to check on Boisjoly's condition. Sometimes they found him staring at the wall. Then came the numbness in his left arm and tightness in his chest. One day at his desk his vision failed. The sudden onset of double vision finally frightened him into a visit to the doctor.

It was diagnosed as stress. He finally sought counseling.

Last July, he put in his last full day at the plant. He had intended to return after a two-month rest, and after Darlene's September wedding. But the anger was like a disabling poison in his system. He filed for long-term disability, which the company granted

For Bob Ebeling, 62, the crippling poison is fear. He wants no role in another space shuttle launch. He wants no responsibility that affects the lives of others.

"I couldn't stand another malfunction that I had anything to do with," Ebeling explained. "Physically, I just couldn't stand another disaster. I'd go off the deep end. All this guilt's just been piling up. I had to say, 'Take me out coach, I'm through.'"

Appendix 12c

ENGINEER WHO OPPOSED LAUNCHING CHALLENGER SUES THIOKOL FOR $1 BILLION

By PHILIP M. BOFFEY, Special to the New York Times
Published: January 29, 1987

WASHINGTON, Jan. 28— A Morton Thiokol engineer, whose anguished pleas to delay the launching of the space shuttle Challenger were ignored a year ago, filed a $1 billion personal injury and damage suit against the company today and accused it of "criminal homicide" in the death of the seven crew members.

The engineer, Roger Boisjoly, accused Morton Thiokol of fraud, negligence, manslaughter, racketeering, defamatory statements and untruthful testimony to a Presidential investigating commission and Congress, among other wrongdoings. Plan for $10 Million Claim

Robert N. Levin, a Washington lawyer who represents Mr. Boisjoly, said he would also file by mail on Thursday a $10 million damage claim against the National Aeronautics and Space Administration under the Federal Tort Claims Act. A NASA spokesman said he was unaware of that imminent claim or the suit against Morton Thiokol, which made the defective rocket that caused the shuttle explosion.

Mr. Boisjoly's legal actions today were his first detailed public comments on the behavior of Morton Thiokol and NASA before and after the Challenger disaster of a year ago today. The taciturn engineer has previously testified on shuttle rocket problems, but he has refrained from making strong public criticism.

The suit filed today in Federal District Court here states that, because of "intentional wrongdoings" by both Thiokol and NASA, Mr. Boisjoly was "severely injured in his emotional and mental health" and was disabled by depression and by "post-traumatic stress disorder," a mental disability that impairs the functioning of people who have experienced traumatic events.

The damages sought in the legal actions would be awarded to Mr. Boisjoly himself, as compensation and punitive damages. The charges of criminal behavior were offered in support of his claim for personal damages and would not result in criminal penalties against the company.

However, Mr. Levin, his attorney, also suggested that Federal or state prosecutors should consider bringing criminal charges against company officials. "There are seven dead people from that flight," he said. "I think somebody ought to investigate individual responsibility."

An executive assistant to Thomas Russell, the chief spokesman for Morton Thiokol, said that company officials had no knowledge of the filing of the suit and could not respond to the charges today.

The legal action comes after a year of increasing despair by Mr. Boisjoly over the events leading up to the accident and over his subsequent treatment by the company. The 48-year-old engineer was one of several Thiokol engineers who warned against launching the Challenger in cold weather, only to be challenged by NASA officials and overruled by Thiokol management. When he told his story to a Presidential commission investigating the disaster, he incurred an apparent demotion and reprisals from Thiokol, bringing a sharp reprimand to the company from the commission.

For the last year Mr. Boisjoly has told friends he could never work on the shuttle again because the experience was so painful, and he has given at least one lecture to engineering students on the importance of ethical behavior by engineers. He recently left the company and is receiving long-term disability payments from its insurance program.

Lawyers familiar with litigation for damages said that Mr. Boisjoly might face difficulties in proving his claim. Alan B. Morrison, director of the Public Citizen Litigation Group, a public interest law firm founded by Ralph Nader, expressed a "gut-level feeling" that corporations could be held liable for injuring a worker in such circumstances, but he said that Mr. Boisjoly might have problems establishing that his mental distress was caused by Morton Thiokol's actions as opposed to other factors in his life.

A prominent Washington lawyer who is an expert in tort litigation, asking that his name be withheld, said that Mr. Boisjoly might have a sound case at bottom but that $1 billion was such an "elephantine claim" that the courts may react skeptically, a view in which Mr. Morrison concurred.

Another Thiokol engineer who opposed the launching, Allan J. McDonald, has remained with the company and emerged as a spokesman for its redesigned rockets.

Mr. Boisjoly's suit charges that, after the accident, both NASA and Thiokol "undertook a program to discredit" him and other engineers "who had warned against the flight." As part of this campaign, the suit alleges, "untrue and defamatory statements" were circulated about Mr. Boisjoly, his job was threatened and his professional competence was attacked.

Thiokol spokesmen "consistently and falsely" portrayed Mr. Boisjoly as "a disgruntled or malcontented employee whose views should be discounted and whose professional expertise should be doubted," the suit said. It cited press interviews in which Thiokol spokesmen labeled Mr. Boisjoly a "tattletale" and an "impatient" employee who tried to hire subcontractors in violation of a contract.

The suit contends that such allegations were false and "made with the specific intent to injure" Mr. Boisjoly's professional reputation "and to punish him for testifying truthfully." The suit describes the events as a "campaign to defame" which amounted to "intentional infliction of emotional distress."

Mr. Boisjoly is seeking $1 million from Thiokol in compensatory damages to cover the earnings that will be lost by his inability to function as an engineer and to compensate for the mental health problems he has suffered.

The suit also seeks $1 billion in punitive damages against Thiokol to serve as a warning and deterrent against similar actions by the company in the future.

Under some of the relevant statutes, Mr. Boisjoly could also obtain reasonable attorneys' fees, if he wins his case. Use of Racketeering Law

Mr. Boisjoly's attorneys are seeking damages under a range of statutes. The laws include the Federal "Racketeer Influenced and Corrupt Organizations Act," which was originally designed to combat infiltration of legitimate businesses by mobsters but has since been interpreted to prohibit various business frauds as well. "

The suit charges that, on the night before the accident, Thiokol used interstate telephones, wires and mails to transmit "false certifications" that it was safe to launch the Challenger. It also charges that Thiokol filed "false and misleading statements" with the Securities and Exchange Commission, stockbrokers, and its own stockholders, thereby committing securities fraud.

Under Utah's version of the racketeering act, the suit alleges, Thiokol "committed acts of criminal homicide" and securities fraud that ultimately led to the injuries suffered by Mr. Boisjoly. His attorney, Mr. Levin, stressed that he was not accusing anyone of intentionally killing the astronauts but rather of unintentional manslaughter or criminal homicide.

The suit accuses NASA and Thiokol of a long-term "conspiracy of silence" to "keep secret" their mismanagement of safety problems arising in the shuttle's booster rockets. It charges that NASA and Thiokol sought to keep a monopoly on shuttle rocket production for Thiokol in violation of antitrust and competitive contracting laws, "testified untruthfully" before investigating panels, pressured staff people to overlook safety problems, and committed insurance fraud by withholding information on the risks from insurance companies.

It also charges that Thiokol sought to impede investigations of the disaster and tried, unsuccessfully, to get Mr. Boisjoly to alter his testimony.

The widow of Dr. Ronald E. McNair, one of the astronauts killed on the Challenger, has also sued Thiokol for an undisclosed amount.

Appendix 13

1974
1975
1976
1977
1978
1979
1980
NASA/MTI Contract
Roger Boisjoly Works at
ABBREVIATED TIME LINE CHART OF MORTON THIOKOL INC./NASA CONTRACT.
O-RING PROBLEMS LINE:

Appendix 13 cont

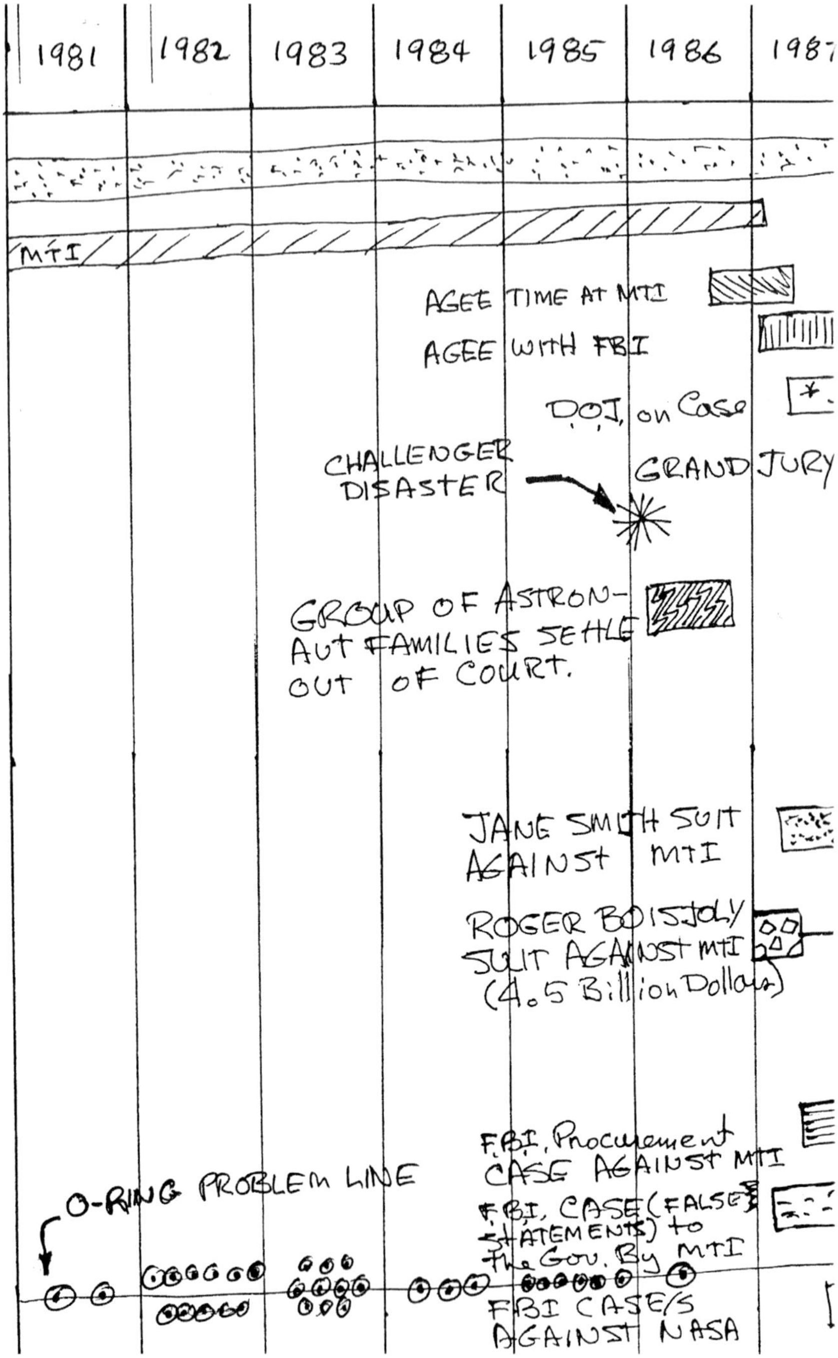

Appendix 13 cont

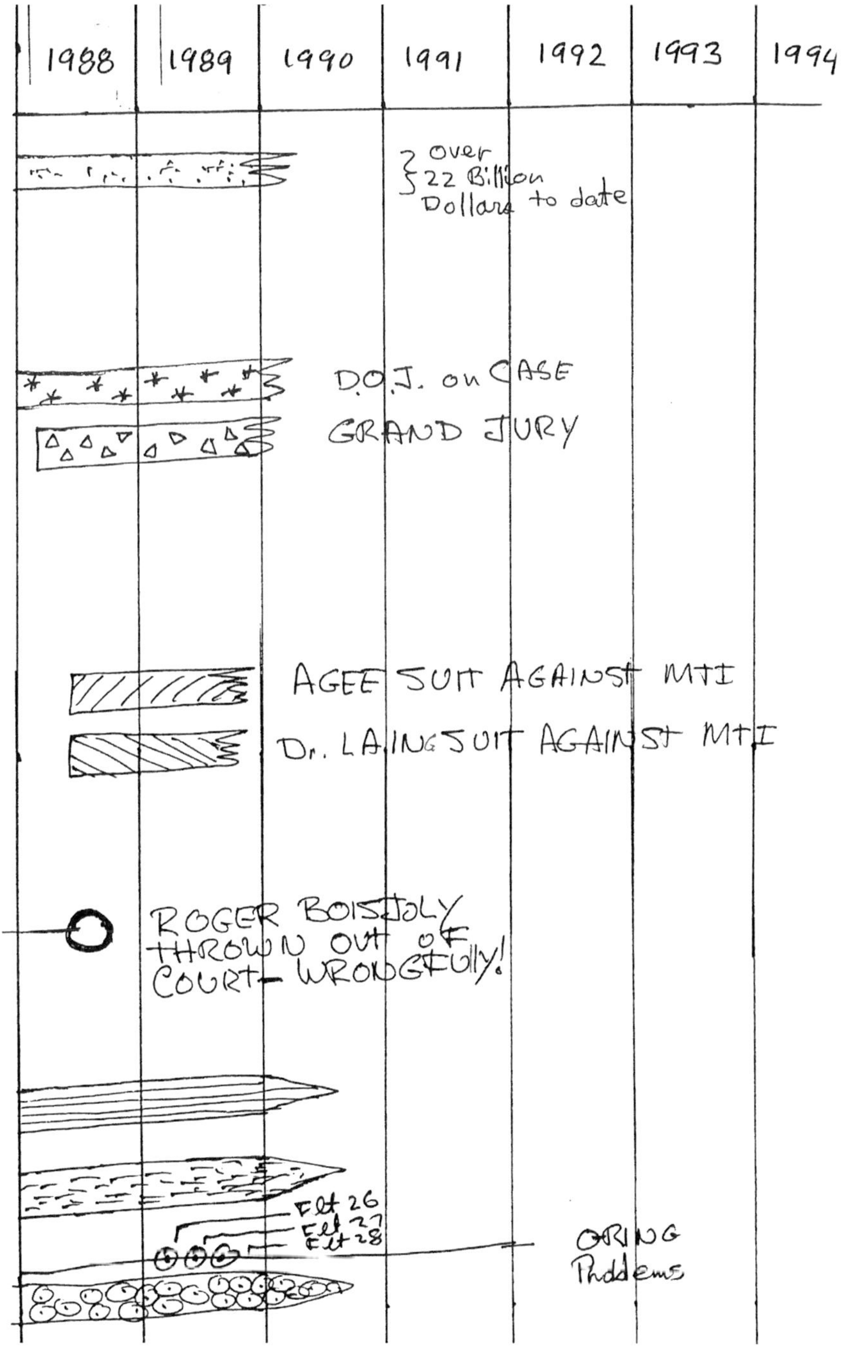

Appendix 14

Utah standoff ends with officer slain

Fri., Jan. 29, 1988, Denver, Colo. Rocky Mountain News—3

MARION, Utah (AP) — Bursts of gunfire yesterday ended a tense, 13-day standoff between a polygamist clan and police, killing an officer and seriously wounding the group's leader, who was suspected of bombing a Mormon chapel.

The shooting erupted just after dawn as police who had infiltrated the clan's compound used a police dog to try to isolate Addam Swapp and his brother from the log house they had left moments before, officials said.

In the exchange of gunfire, the dog's handler was shot in the abdomen, and Swapp fell in the snow with bullet wounds in the arm and chest, said John T. Nielsen, Utah's public safety director.

The operation didn't go as planned in part because the dog hesitated after being ordered to attack the Swapps, said FBI agent Cal Clegg.

Utah State Prison Lt. Fred House and the dog were hidden in a building about 30 yards from the Swapps when the command was given.

When the dog hesitated, House, 35, stood up in the doorway to urge it on and was struck by a bullet between the armored plates in his flak jacket, Clegg said.

Agents fired two rounds, which struck Swapp, Clegg said. Swapp stood and walked into the house and began waving a white towel through the doorway. He then walked to an outbuilding where officers had hidden and surrendered, Clegg said.

In the succeeding, following moments after the gunfire, the agents moved in an armored personnel carrier to evacuate those who were wounded, and they came under extremely heavy gunfire from the ... residence," Nielsen said.

More than 100 rounds poured from the log house for 15 to 30 minutes before Swapp's brother, Jonathan, 21, called the police command post on the compound's telephone and asked how to surrender, Clegg said.

Then Jonathan Swapp, the clan's four other adults and nine children filed from the house in two groups, their arms above their heads.

"We are very happy to report in this — that's if there's anything to be happy about in this tragic situation — that all of the children are safe," Nielsen said.

The siege had begun within hours of the predawn bombing Jan. 16 of a Mormon chapel a half-mile from the compound. That night, police talked by telephone with Swapp, who said the bombing was revenge against the church and state for the 1979 police slaying of polygamist patriarch John Singer. He told others he sought an armed confrontation to bring about Singer's resurrection.

The polygamists broke away from the mainline Mormon Church over its ban on polygamy in 1890.

Nielsen said the decision to seize Swapp, 27, on federal warrants was made after a family friend delivered to police Wednesday afternoon Swapp's response to a letter from Gov. Norm Bangerter pleading with him to surrender.

Nielsen said Swapp "declared his ranch and people an independent and free nation" and warned he would use any means to defend the 2½-acre compound.

Vickie Singer, Swapp's mother-in-law and widow of John Singer, was taken from the compound to a federal courthouse in Salt Lake City for arraignment on charges related to the Jan. 16 church bombing and gunfire that had punctuated the standoff.

As Vickie Singer, 44, was brought into the federal building in Salt Lake City in handcuffs, she was asked if her children were all right. "They're all being taken away from us," she said.

Teary-eyed and wiping her nose, she pleaded innocent before a federal magistrate.

Her wheelchair-bound son, Timothy, 21; Addam Swapp's wives, Heidi, 23, and Charlotte, 19, and brother Jonathan Swapp all were charged with being accessories after the bombing. The five were ordered held without bail at the Salt Lake County Jail.

In Marion, Nolan Douglas, an agent for the federal Bureau of Alcohol, Tobacco and Firearms, displayed a box containing what he said were 22 sticks of dynamite found in a vehicle outside the log house.

Singer had been gunned down outside the compound nine years ago by police trying to arrest him after an 18-day standoff. Singer, who had defied a court order to send his children to public schools, allegedly drew a gun before he was killed.

Swapp, married to two Singer daughters and father of six of the clan's children, was listed in serious but stable condition at the University of Utah Health Sciences Center in Salt Lake City. He underwent surgery on his left arm.

Legislators briefed by the governor were told that police had been planning Swapp's capture for several days.

Officers had tried to set a trap for Swapp by installing a booby-trapped loudspeaker designed to set off a bright flash when moved. The idea was to blind Swapp while police dogs moved in to subdue him, the legislators said.

However, they said, Swapp shot at the loudspeaker, apparently tipping it over and triggering a flash not as bright as authorities had hoped.

In the letter to Bangerter, Swapp said, "Take warning, any man of yours who attempts to cross the boundaries of this property without permission will be treated as an aggressive act on your part against us and we will defend ourselves in any manner as we see fit."

Note: Tim Healy was driving the armored vehicle. This was the first time he was shot at but not the last.

Appendix 15a

Ventura County
April 4, 1987
11th Year, No. 149
48 Pages
25 Cents
STAR ☆ Free Pres

FBI probing Morton Thiokol

Criminal charges possible for shuttle rocket firm

WASHINGTON (AP) — The FBI is conducting a criminal investigation of Morton Thiokol Inc., the company that made faulty rocket seals blamed for the 1986 space shuttle Challenger disaster, according to a document on file in U.S. District Court.

The FBI has developed informants among the employees of the aerospace contractor as part of its criminal investigation, said the document, a private communication to a federal judge by the Justice Department.

Details of the criminal allegations under investigation could not be obtained.

It is not known whether the FBI's investigation concerns the company's conduct before the ill-fated Jan. 28, 1986 launch that killed seven shuttle crew members or during subsequent investigations into the disaster by Congress and a presidential commission.

Philip Weyna, the company's director of corporate development, declined to comment when asked about the FBI investigation.

The Chicago-based company makes rocket-booster components at a plant in Wasatch, Utah.

Susan Schnitzer, an FBI spokeswoman, refused to confirm or deny the investigation, saying only that the court document that revealed the probe had been inadvertently released and resealed.

The presidential commission blamed the explosion on the failure of a joint on the right rocket booster.

The joint failed because cold weather prevented rubber pressure seals from containing gases, the commission concluded.

The FBI investigation was revealed in court papers on file in connection with a lawsuit against Thiokol by Roger Boisjoly, a former engineer for the company who worked on the space shuttle project

Boisjoly, who pleaded with his superiors and the National Aeronautics and Space Administration not to allow the launch in 38-degree weather, has accused his former employer of lying to congressional committees and the presidential commission headed by former Secretary of State William Rogers.

In his lawsuit, which was unsealed Friday in federal court, Boisjoly accused Thiokol of falsely telling NASA that the rubber seals, known as O-rings, were safe in cold weather conditions.

The lawsuit, which had been sealed since Jan. 29, charged that senior NASA officials pressured Thiokol to certify the safety of the seals holding together the sections of the solid-rocket boosters that power the space shuttle into orbit.

The suit charged that the false safety assurances were made to "preserve Thiokol's unlawful monopoly as the sole-source provider" of solid-rocket sections and to enable the company to collect a $75 million incentive award.

Boisjoly accused the company in a separate defamation suit in January of impugning his

(Please turn to A-4, Col. 1)

Appendix 15b

FBI investigating shuttle booster firm

Continued from A-1

reputation to punish him for giving truthful testimony to the Rogers Commission.

The defamation suit accused Thiokol and NASA of conspiring to give false testimony about the shuttle explosion to congressional committees and the commission.

The engineer's second suit, filed a day later was placed under seal under terms of the False Claims Act.

U.S. District Judge Harold Greene signed an order this week releasing papers filed in the suit, which seeks to recover $2.075 billion Boisjoly claims Morton Thiokol owes the government for its alleged false claims.

The figure includes the $75 million bonus incentive that NASA decided to pay Morton Thiokol after the disaster. NASA has paid Morton Thiokol roughly $40 million of the bonus, said Robert N. Levin, an attorney for Boisjoly.

Under the False Claims Act, private citizens can bring claims against government contractors on behalf of taxpayers. The government is required to investigate the validity of the allegations.

It was not known if the FBI investigation was prompted by Boisjoly's lawsuit.

In the complaint unsealed Friday, Boisjoly said nearly 20 Morton Thiokol engineers "unanimously recommended against the flight" because freezing temperatures at the Kennedy Space Center endangered the effectiveness of the O-ring seals.

Appendix 16a

25¢
50¢ outside Seattle metro area
48 pages

The Seattle Times

WASHINGTON'S LARGEST NEWSPAPER ■

Company not bidding on advanced boosters

Associated Press and Times staff

Morton Thiokol, sole manufacturer of solid-fuel space shuttle boosters blamed for the Challenger disaster, has dropped out of the competition for a $1.2 billion contract to make boosters for the 1990s.

The decision means Thiokol is getting out of the shuttle-booster business when the new generation of powerful rockets comes into being.

The Chicago-based company announced yesterday that it wants to concentrate on redesigning the booster used on Challenger, whose fiery disintegration shortly after takeoff killed all seven aboard.

Robert Bartels, an analyst with William Blair investment firm in Chicago, said: "There's been a lot of speculation that the company will sell the aerospace business and is not willing to commit to it in the long term."

At Thiokol's rocket plant west of Brigham City, Utah, rumors have been circulating for several months that the company's aerospace division is up for sale.

But Tom Russell, vice president for corporate development and strategic planning, recently denied that Thiokol was actively looking to sell the aerospace group.

After the Challenger accident, congressmen spoke out about the need to open up competition. Recently introduced legislation regarding the Advanced Solid Rocket Motor (ASRM) reinforces that

Please see **THIOKOL** on A 10

Shuttle in mock launch

United Press International

CAPE CANAVERAL, Fla. – A make-believe space shuttle blasted off today after a successful mock countdown that served as a major test for post-Challenger management changes designed to improve the launch decision-making process.

A major goal of the simulation was to exercise NASA's "mission management team," a group of top agency managers under astronaut Robert Crippen charged with making the final launch decision.

The presidential commission that investigated the Challenger disaster charged NASA's decision to launch the doomed shuttle was "flawed."

Appendix 16b

The Seattle Times   Tuesday, June 7, 1988

Thiokol won't bid again

Troubled builder of space-shuttle boosters out of race for advanced-rocket contract

Associated Press and Times staff

Morton Thiokol, sole manufacturer of solid-fuel space shuttle boosters blamed for the Challenger disaster, has dropped out of the competition for a $1.2 billion contract to design and build new rocket boosters for the 1990s.

The decision means Thiokol is getting out of the shuttle-booster business after it finishes its work on a redesigned booster that will be used for about seven years, until the advanced generation of powerful rockets is ready.

The Chicago-based company announced yesterday that it wants to concentrate on redesigning the booster used on Challenger, whose fiery disintegration shortly after takeoff in January 1986 killed all seven aboard.

Robert Bartels, an analyst with William Blair investment firm in Chicago, said, "There's been a lot of speculation that the company will sell the aerospace business and is not willing to commit to it in the long term."

At Thiokol's rocket plant west of Brigham City, Utah, rumors

Please see **THIOKOL** on A 10

Soviet capsule launched

Associated Press

MOSCOW – A Bulgarian cosmonaut rocketed into a clear sky today along with two Soviets in the first manned space mission of 1988, a launch televised live and featuring broadcasts from inside the capsule.

Alexander Alexandrov, 36, of the Bulgarian town of Omurtag, and Viktor Savinykh and Anatoly Solovyov blasted off aboard a Soyuz TM-5 capsule from the Baikonur space complex in Soviet Central Asia. They are scheduled to dock Thursday with the space station Mir, which has two cosmonauts on board.

Soviet TV coverage was unusually extensive. The mission comes while the U.S. manned space program is grounded.

THIOKOL

continued from Page 1

have been circulating for several months that the company's aerospace division is up for sale.

But Thiokol spokesman Rocky Raab said today, "We can state categorically that we have not assembled a package to put ourselves for sale, nor have we been approached by anyone to sell the aerospace group."

The firm employs 20,000 people, of whom about 12,000 are in the aerospace group, including about 8,000 in Utah. Of the 8,000 in Utah, about 2,500 work in the space division.

Outside of the Mormon Church, Thiokol is the largest private employer in Utah.

After the Challenger accident, members of Congress spoke out about the need to open up competition for building an improved

Appendix 17

Morton Thiokol Plans to Split Into Two Separate Businesses

Chicago Tribune

CHICAGO—Morton Thiokol Inc., which once defended its diverse operations as "a portfolio of businesses," plans to separate itself into two businesses in a tax-free spinoff to shareholders.

The spinoff will create Morton International Inc., consisting of the company's commercial businesses—chemicals, salt and air bags; and Thiokol Corp., which will handle the aerospace operations.

The company and its aerospace operations still have scars from the Challenger space shuttle explosion, which killed seven astronauts in 1986. The aerospace division made the booster rocket used on the shuttle.

The company made no mention of using the spinoff as an effort to distance itself from its aerospace business.

Charles S. Locke, chairman and chief executive of Morton Thiokol, said the spinoff will separate two businesses with "distinct financial, investment and operating characteristics so that each can adopt strategies and pursue objectives appropriate to its specific businesses."

Locke said the transaction will allow investors to evaluate better the individual businesses and their outlook, enhancing the likelihood that the true value of each business will be properly reflected.

The restructuring is expected to be completed July 1, and will include a share-for-share distribution of Morton International's stock to Morton Thiokol's shareholders.

The plan is subject to board approval of agreements, stockholder approval and an Internal Revenue Service ruling that the distribution will be tax-free to Morton Thiokol and its stockholders.

"Shareholders will have a share in Thiokol Corp. and a share in Morton International," said Morton spokeswoman Nancy Hobor. "It's a one-for-one distribution."

The new Morton International will have an estimated 1989 sales of $1.5 billion and assets of $1.4 billion. It will continue to be based in Chicago and employ more than 8,000 people.

The new Thiokol Corp. will have sales of about $1.1 billion and assets of $800 million.

Appendix 18

U.S.NEWS

The Pentagon probe poops out

Law-enforcement authorities familiar with the federal inquiry into procurement fraud say it is in trouble

At the annual convention of the American Bar Association last week in Toronto, there was the usual gossip and glad-handing in the oak-paneled pubs of the Royal York Hotel. But when the standing-room-only seminar on defense-contractor cases finally convened, it was strictly business as 250 lawyers—many already on retainer from Pentagon contractors and consultants—contemplated the best course of defense in the most publicized federal investigation since Abscam. "It was a little," says one panelist, "like previewing the fall hemlines."

Strategy aside, the hottest topic among the lawyers was the uncertain status of the Justice Department investigation that burst into the headlines earlier this summer amid dire predictions of more than 100 indictments and that jeopardized Pentagon contracts valued at tens of billions of dollars. Among many of the defense lawyers involved in the case, there is hopeful speculation now: Has the Pentagon probe gone underground, or just under? Though it's impossible to tell yet, sources with first-hand knowledge of the inquiry agree that there are problems. These sources, who include Justice Department officials, private attorneys and government investigators, say the case is stalled, at least temporarily. Among the key concerns the sources identified:

▪ Intense turf and strategy battles between the lead prosecutor, U.S. Attorney Henry Hudson of Alexandria, Va., and another key prosecutor, U.S. Attorney Andrew Maloney of Brooklyn. Additionally, the sources say, since the resignation in March of William Weld, the chief of the Justice Department's criminal division, who had been supervising the investigation, there have been disagreements on strategy between Hudson, senior Justice officials in Washington and the FBI.

▪ The possibility, according to some of the sources, that the issuance of 250 search warrants in mid-June may have been premature. The FBI, which picked up obviously self-serving statements from some wiretapped suspects, fearing the investigation might have been compromised, pushed to serve the warrants early.

▪ The apparent failure, so far, of any of the principal targets of the investigation to agree to cooperate with prosecutors in exchange for leniency. The sources say at least two lower-level targets have "flipped," and without central players to testify to a jury directly about improper payments to Pentagon officials by contractors or consultants, the prosecution's bribery case could be seriously flawed.

▪ The politically sensitive nature of an investigation that could well involve some House members. Hudson told the congressional leadership earlier in the summer that no lawmakers currently were targets of the investigation, but the sources say there is sharp disagreement now within the Justice Department about how to proceed with evidence unearthed pertaining to some House members. Federal officials confirm they are investigating the relationship of Bill Chappell, the chair of the House Appropriations Defense Subcommittee, with consultants linked to the Unisys Corporation, and with defense consultant Leonard Killgore, a former aide to Chappell (D-Fla.), who has figured in the investigation. These sources also say that investigators are inquiring about the contractor relationships of Representatives Roy Dyson (D-Md.) and John Murtha (D-Pa.), Chappell's No. 2 on the Appropriations Defense Subcommittee, another Killgore contact. Spokesmen for all three lawmakers say they've done nothing improper.

Explosion at Justice. The sources say the prosecutors' turf war "blew up" at a Justice Department strategy session in December. After a "vigorous debate" in which Hudson pushed hard to retain full control of the investigation, senior Justice officials decided to let him guide it, knowing Weld would continue to oversee things. It was an unusual decision, federal sources say, because Maloney, who had more experience with big, complex investigations, "had his act together [while] Hudson's case was running dry." When Weld, who had played referee, coach and father confessor to the disputing parties, left the Justice Department in protest against the continuing legal difficulties of Edwin Meese, the battle among the various agencies and over specific legal tactics exploded.

In the midst of all this, the FBI agents working the case began getting nervous. Investigators overheard statements on the wiretaps such as "Of course, this isn't illegal," suggesting that some subjects already knew they were under surveillance. It was then that the FBI pushed to lift the taps and secure the search warrants. But while they anticipated that some suspects would rush to cooperate once they heard their words played back to them, most defense attorneys so far have been advising clients to sit back and see what else the prosecutors have. Meanwhile, Hudson has declined to comment but says he doesn't expect indictments for some time.

Defense attorneys and federal officials alike think the case may get back on track when Richard Thornburgh, the new Attorney General, takes command. "He's smart, honest," says one former prosecutor. "And knows a large messy case when he sees it."

by Alicia Mundy

24 U.S.NEWS & WORLD REPORT, Aug. 22, 1988

Appendix 19

$7.7 million paid to Challenger kin

3/1/88

WASHINGTON (AP) — The government and rocket maker Morton Thiokol paid $7,735,000 in cash and annuities and divided the cost 40-60 to settle all claims by the families of four of the astronauts who died in the Challenger explosion.

Documents released yesterday showed that Morton Thiokol, which produced the faulty booster rocket blamed for the Jan. 28, 1986, explosion, paid $4,641,000. The government's share of the settlements was $3,094,000.

A lawyer who represented other family members called the settlements "woefully inadequate."

The surviving four spouses and six children actually will receive more than $7.7 million. This is because each was given an annuity, which pays benefits larger than its cost but over a period of many years.

The total dollar amount the families will receive over time and the breakdown by family were not released.

With the release of the documents, the Justice Department settled a civil suit brought under the Freedom of Information Act by the Associated Press and six other news organizations. The government originally had kept details of the settlements and negotiations secret, saying it needed to keep its internal deliberations confidential and that the company and families demanded secrecy.

The documents, with some deletions to preserve family privacy, included the final settlement agreements with the families and the company, some correspondence between the government and the company, and several Justice Department statements concerning the negotiations.

U.S. urged to expand space-based research

BOSTON (AP) — The United States has been embarrassed in space research since the Challenger disaster and risks forfeiting its edge in high technology and other fields unless it expands space-based scientific missions, a NASA official and researchers told a congressional hearing yesterday.

Further delays in projects already backed up three years plus a lack of funding for new space experiments will allow the Soviet Union, Japan and European nations to cement the lead they have taken over the United States since the Challenger explosion, said Lennard A. Fisk, associate administrator of the National Aeronautics and Space Administration.

Fisk outlined the science-related portion of NASA's proposed fiscal 1989 budget to the Senate subcommittee on science, technology and space.

Fisk said 20 major science-related payloads are scheduled to be carried into space in the next four years, half of them aboard space shuttles and half on expendable rockets.

The projects include probes to Venus and Jupiter.

Appendix 20

SECTION B Business B 4-6

Reagan sees space as private place

President unveils plans for research

by Harry F. Rosenthal
Associated Press

WASHINGTON: — President Reagan unveiled a national space policy yesterday to turn commercial space activity over to private industry and start scientists developing the technologies needed to send humans beyond the moon.

Reagan said he will ask for $100 million in next year's federal budget to get the research started.

The policy also pledged to remove any government roadblock to allowing private industry to launch rockets and place satellites in orbit.

"It shifts the major responsibility for space development from public to the private sector, said Secretary of Commerce C. William Verity at a White House briefing.

James Fletcher, NASA administrator said the policy "clearly establishes that, for the first time, the United States has a long-range goal of expanding human presence and activity beyond Earth orbit into the solar system.

Other than the $100 million for the new technology program — called "Pathfinder" — the new policy did not spell out how much the joint federal-commercial partnership will cost. Whatever the cost, it will be borne long after Reagan leaves office.

In the budget proposal he will make to Congress next week, the President asks $1 billion for the space station, which will be built in the next decade and be in operation by 1996. NASA asked for nearly twice that much.

Jim Burnley, transportation secretary, used the occasion to announce his department had completed approval of the nation's first commercial launch license, which will permit Conatec Inc., a Lanham, Md. company, to launch a series of sounding rockets for an unnamed European industrial firm.

Fletcher said the government will lease space aboard a commercially developed space module that will be carried into space and released there by the shuttle.

A leading candidate to build the module is Space Industries, Inc., of Houston which has a long standing agreement with NASA for the "Industrial Space Facility."

But Fletcher said there would be open competition and NASA would agree to lease 70 percent of the module's space at a cost of $140 million a year for five years.

The new space directive should give NASA's various programs a sharper focus after two years of disruption following the loss of the shuttle Challenger and its seven crew members.

As a start, the administration has requested $11.3 billion for NASA in fiscal 1989, up from about $9 billion this year. That includes the money for Pathfinder which is to study technologies for establishing a scientific base on the moon and for dispatching a manned mission to Mars early in the 21st century.

NASA's space station has been slow in developing as Congress has reined in its financing. But Reagan gave the station a firm endorsement. A series of space shuttle flights is to start assembling the station in orbit in the mid-1990s.

The Boeing Aerospace Co. is one of four contractors working on the space station. It has a 10-year contract worth about $750 million to design and build the living quarters, laboratory and logistics module.

Friday February 12 1988 The Seattle Times - B

REAGAN'S SPACE POLICY

President Reagan's space proposals included:

■ Announcing the government's intention to lease space as an "anchor tenant" in a privately financed orbiting facility suitable for research and commercial manufacturing.

■ Committing to include in the space shuttle's cargo bay in the early 1990s, a commercially developed, owned and managed middeck module "Spacehab.

■ Saying the government will make available for five years the expended external tanks of the shuttle fleet at no cost for uses as research storage or manufacturing in space.

■ Saying NASA will "clarify and strengthen" the commitment to private investment in the space-station program.

■ Encouraging development of commercial remote sensing systems, including government purchases from private firms.

■ Ordering federal agencies to buy single-use rockets directly from private firms as much as possible.

■ Proposing a $200,000 ceiling on damage awards to third parties injured in commercial launch accidents.

■ Saying the administration will will consult with industry on possibly constructing private launch facilities.

Appendix 21

40—Rocky Mountain News Fri., Jan. 29, 1988, Denver, Colo.

Space shuttle 'no good,' investigator charges

WASHINGTON (AP) — The space shuttle still is "a very dangerous system" that should be flown with caution by crews limited to essential personnel, say some members of the presidential commission that investigated the Challenger accident.

Some former commissioners said Wednesday the National Aeronautics and Space Administration has yet to complete all of the testing they believe should have been done even before the first shuttle flight.

Yesterday marked the second anniversary of the Challenger explosion.

"The shuttle is a very dangerous system," said Joseph F. Sutter, an engineer and a vice president at Boeing Co. "There is no way that civilians should be allowed to fly on it. The crew should be only essential personnel and those should be single folks."

David C. Acheson, a Washington lawyer and commission member, said the shuttle "is no good" and should be replaced as soon as possible.

"It's a sloppy, loose, budget-driven design," said Acheson.

Acheson and some other former commissioners claim the shuttle development was underfunded, forcing NASA to forgo tes that should have been done during develo ment of the craft.

Since the accident, NASA has re-evaluat more than 1,000 critical parts, and has cc tinued to find flaws, such as a faulty we discovered last week in a key rocket engi part. Some commissioners wonder if the remain other, undetected flaws.

Appendix 22

Sunday, May 22, 1988. C

Fatal shortcut puts spotlight on safety issue

Last December, at Morton Thiokol's Wasatch Plant near Brigham City, Utah, 108,000 pounds of solid rocket fuel in a freshly cast MX missile segment exploded. The blast, and the violent fire that followed, killed five employees and melted a $6.2 million rocket plant.

The accident has raised serious questions at NASA and the Air Force. For the past three years, Thiokol has received awards from the National Safety Council for preventing worker injuries. However, several oversight reports by NASA and the Air Force "indicate significant and potentially serious quality control and safety problems at Thiokol," according to the General Accounting Office, a watchdog arm of Congress.

Six times in the past four years, including last December, buildings at the Wasatch plant have exploded, burned or both. The other accidents:

■ March 1984

The plant where shuttle booster segments are filled with propellant exploded and caught fire.

■ April 1985

A building for mixing propellant used to inflate automobile air bags caught fire. It was rebuilt, and exploded again in August.

■ June 1985

The building where shuttle booster fuel is mixed was struck by lightning, then exploded when the mixing machine went haywire.

■ March 1986

making, Thiokol workers operate the plants by remote control. In most catastrophic accidents, the plant operators were safely tucked away in control bunkers watching gauges, dials and video monitors.

The MX accident last December was different. Because a video camera was poorly placed, crews couldn't see what they were doing. Five workers went inside, shortcutting a delicate operation that had been done successfully 321 times before. The rocket exploded when the crew was standing right over it.

Thiokol was fined $31,700 by the Utah Department of Occupational Safety and Health for "willful" and "serious" violations of worker-protection rules.

The cause of the accident isn't known exactly, but an investigation led by the Air Force came to the conclusion that, while the steel core was being lifted out of the rocket segment, either static electricity or friction ignited the fuel.

The core and its fins are coated with Teflon to reduce friction inherent in the core-extraction process. The Air Force found that not only were there large patches of rust on the core and fins, where the Teflon had chipped, but that earlier inspections had demanded "use one time only." Thiokol could not present records to show that the rust had even been removed before the core parts were used again.

Rust, sand or grit mixed with rocket propellant makes it highly sensitive to static discharge. The Air Force found numerous ways in which MX missile fuel, a "Class A" explosive, was virtually assured of contamination as the segments were cast.

■ Times research intern Tim Hyatt contributed to this report

Appendix 23

Subpoenas issued for makers of shuttle part

FBI won't give details in ongoing probe of Morton Thiokol

by Peter Lewis
and Eric Nalder
Times staff reporters

Subpoenas have been issued in a federal investigation of the company that makes solid-fuel rockets for the space shuttle in Utah, the FBI confirmed today.

"There were subpoenas that were issued (to Morton Thiokol) for certain documents, but I don't know what the results of those are or if they will bear fruit or what," said FBI agent Cal Clegg in Salt Lake City.

Steve Agee

Clegg declined to detail the nature of the FBI inquiry.

But documents show that an investigation started in January 1987, when a Thiokol safety consultant, Steve Agee, told federal authorities he felt safety information about the shuttle rocket was being suppressed by company officials.

Agee started working for Thiokol in November 1986 and remained on the job until late March 1987. During that time he worked under cover for the FBI, bringing out copies of company documents at the direction of an official FBI agent. Agee also dined with other Thiokol employees at an Ogden restaurant, discussing company safety practices while wearing a hidden microphone for the federal bureau.

The FBI told him it had obtained court authorization for him to smuggle out documents and

Please see THIOKOL on A 6

Thiokol is subpoenaed

THIOKOL

continued from Page 1

clandestinely record conversations, Agee said.

Agee, who now lives in Renton, said that while serving as a safety analyst at Thiokol, he wrote numerous reports on hazards associated with the shuttle's booster rockets but that the reports were shelved. Company officials said his reports were not submitted to NASA because they were not credible.

After Agee left Thiokol, he became a full-time consultant for the FBI for about four months. He was paid $50 an hour and helped the FBI examine Thiokol's contract with the National Aeronautics and Space Administration to build boosters for the space shuttle. He helped the FBI track numerous areas of potential criminal conduct, including contract fraud and false statements.

This week, Thiokol announced it was withdrawing from competition for a $1.2 billion contract to design and build new rocket boosters for the 1990s. The Chicago-based company announced it wants to concentrate on redesigning the booster used on Challenger, which exploded over Cape Canaveral shortly after takeoff in January 1986, killing all seven aboard.

Each shuttle has twin, 149-foot boosters, powered by solid fuel, which provide 80 percent of the 7.7 million pounds of thrust needed for liftoff. When they burn out – in 2 minutes, 12 seconds – the shuttle is 25 miles high and going 4½ times the speed of sound.

Morton Thiokol general counsel Darryl Lee would neither confirm nor deny whether the company had recieved a subpoena. "If there is such a subpoena," he said, the company intends to cooperate fully with any investigation.

A former Thiokol consultant who now lives in Arizona said he has been in touch with the FBI several times recently, and was told he might be called upon to testify before a grand jury.

Agee, in a recent interview, said, "I have every confidence that soon there will be subpoenas and indictments, but my visibility on that issue is limited by necessary FBI secrecy."

The FBI investigation of Thiokol has been cited in court papers. It was also confirmed by others besides Agee, including the inspector general's office of NASA, other NASA and Thiokol officials. The investigation also was confirmed by the FBI itself in a recent response to a Freedom of Information Act request filed by Agee. The FBI, in that response, declined to release documents because of an ongoing investigation.

Appendix 24a

Thiokol Shoppers Deny Ex-Colleague's Allegations

Reprinted by permission of Utah Standard-Examiner

Brigham City — Four temporary "job shop" workers involved in the space shuttle booster redesign effort are disputing allegations by a former colleague and FBI informant that Morton Thiokol squelched internal criticism of the program.

Thiokol solicits questions, criticism, debate and new ideas. Checks and balances within the redesign program ensure an open, healthy environment, and managers never sweep their comments or concerns under the rug, the workers say.

"No pressure is put on me by anybody," said John Ford, one of the temporary employees.

Ford and three other job shoppers told their own stories to the Standard-Examiner about the ongoing redesign program at Thiokol. They said their experiences are nothing like those of Steven Agee, a former job shopper who was fired from Morton Thiokol about March 1987.

Agee was quoted extensively in a May article published by the Seattle Times and syndicated nationwide through wire services. The story was picked up by many newspapers, including the Salt Lake Tribune.

In the story, Agee said Thiokol stymied his efforts to look for and write up safety concerns. He said he was banned from meetings after raising questions about the safety of parts he was examining.

He said Thiokol shelved his safety report on "221 ways the space shuttle could blow up."

In the Times story, Agee said he was so alarmed about the situation that he became an informant for the FBI, smuggling out documents and recorded co-workers between January and March 1987.

Ford and the other job shoppers who spoke to the Standard-Examiner didn't work with Agee and don't know first-hand about his experiences. But they talked about their roles in the huge program to redesign the shuttle booster rocket and get the space program back in flight. They said they have never been pressured to suppress criticism or alter documents.

They said Thiokol gives them freedom to do their jobs as they should be done. They are as obligated to bring up safety concerns as those who are hired for that task. The company is accomplishing a monumental engineering feat in a short amount of time, they said. The job shoppers said that abbreviated time frame has never hindered their ability to professionally and thoroughly do their tasks.

Four temporary workers agreed to speak to the Standard-Examiner after 11 job shoppers sent a June 8 letter to the Seattle Times, local media and C.E. *Weekly*, a contract trade publication. The letter rebutted Agee's allegations.

Job shopper Wilmer Griest signed his first name and last initial to the letter and later went on the record to talk about his Thiokol experiences. He's been a project engineer at the company for 15 months. He oversees people who test components that are eventually incorporated into full-scale solid rocket motors produced at Thiokol for space shuttles.

"(My comments) have never, ever been squashed," said Griest, who's worked since 1958 as an engineer on projects for Aerojet, the U.S. Navy, Martin Marietta, Westinghouse Electrical Corp., and Bechtel Inc.

At Thiokol, Griest works for the case insulation project engineering department. That group is responsible for insulation that lines steel motor cases to protect them from burning propellant. He said he has total latitude in his job except for authority to change drawings of hardware.

He meets once a day with others in his department to talk about the work. He relays those discussions to program managers, who are responsible for all testing involved in the redesign.

"We as an engineering community work together on joint tests," Griest said. "Nobody tells you you can't bring up a problem. You don't have any problem expressing that."

Safety people aren't the only ones who bring concerns to light. Anyone working on any aspect of the redesign is encouraged to speak up, Griest said.

"There is an ongoing engineering effort at Morton Thiokol," he said. "We sit down as a community. We address what we think is a problem. We discuss solutions."

Once the group has come to conclusions and worked out a specific way to fix the problem, the package is presented to the Engineering Management Team, a group of Thiokol engineering managers who review concerns as a board.

"Sometimes I've been there until 9:30 at night to get on the agenda," Griest said. "I believe that if you document well a concern, if it's done well and it's well-substantiated with engineering logic, that there will be a response out of this (NASA and Thiokol redesign) system. We had concerns, but we have to be capable enough as engineers to document our concerns ... You don't go to any one of these redesign boards without a solution to the problem ... Nothing in this redesign program has occurred by osmosis."

Griest said he's gone to the Engineering Management Team with four concerns. One of those times, his team saw a way to make a component better. He said the board listened to the suggestions and later incorporated them into the redeisgn.

Thiokol hasn't interfered with redesign audits, said Jean Heller, another job shopper.

Heller said she has worked 15 years in technical writing and design engineering fields at various companies. At Thiokol, she and her team members in the requirements department, some of whom are full-time employees, audit all paper work created by all individuals involved with redesign efforts at the company. The documents explain materials, designs, tests, procedures and hardware needed to produce solid rocket motors that comply with specifications agreed upon by the corporation and NASA.

"At least the part that I have been auditing, we go through with a fine-toothed comb," Heller said. "I have never been asked to put something down that is untrue. Never. And I work basically on my own in a little room. I

Continued on next page

Appendix 24b

come up with my own audit results, my own compliance reports."

She said that as an auditor she would be one of the people to feel pressure to change reports on her work that she sends on to NASA officials or project engineers at Thiokol.

"Agee is insinuating that the information that is official in that company has been reworded to the benefit of Morton Thiokol. That has never been the case in my position," she said. "And as an auditor, if there is going to be pressure someplace, that is one of the places."

Ford, also in the requirements department, said he's never been pressured to speed or alter his work at Thiokol. Ford has worked 30 years in the missile and aerospace industry for companies and agencies including Boeing, Westinghouse, NASA and a subcontractor in Italy for the European Space Association.

"There are deadlines, but no pressure as far as how much I have to do in a certain amount of time," Ford said. "The only thing I have to do is meet the schedule laid out by NASA and Thiokol to verify that all requirements were met."

Since September 1987, Ford said, he has reviewed more than 5,000 drawings. He said he spotted only a few minor problems, ones that would not affect flight safety. He said he followed up as he is supposed to do, and learned the problems already had been addressed.

Heller and systems engineer Bob Enlow said they've never lacked the resources or equipment to do their jobs. They said they disagree with Agee's comments that work was done in a helter-skelter fashion without adecuate desks, telephones or typing services, and that paper work requirements bogged down safety review efforts.

There was a space crunch for a while, but it n interfered with his work, Enlow said.

Griest said, "The bottom line is, there are a l checks and balances in this whole system ... I t stood and signed, 'We are ready to launch,' (on c projects at jobs before he came to Morton Thiokol). lot of times there's a queasy feeling in the stomach nevertheless they fly it now or they fly it later. You k you've done everything you could do to make it a system."

By Jennifer Paul
Standard-Examiner staff

Appendix 25

Seattle Post-Intelligencer Sunday December 28 1986 A 9

NASA overhauls the shuttle

Changes go far beyond the redesign of O-rings

by David E. Sanger
New York Times

NEW YORK — Piece by painstaking piece NASA is revamping the space shuttle system that disintegrated, along with a good measure of the nation's technological self-confidence, with last January's explosion of the Challenger.

The next shuttle orbiter that astronauts clamber into on the Kennedy Space Center launching pad will probably have a crude but effective escape system for them to use in case disaster strikes again.

Its landing brakes will be improved. Its main liquid fuel engines will be upgraded, and for the first time astronauts may fly without harboring the seldom-spoken fear that a malfunction in a simple 17-inch valve could choke off the craft's liquid fuel supply and send it crashing into the sea.

The changes being made by the National Aeronautics and Space Administration thus go far beyond the redesigning of the solid-fuel booster rocket whose failure killed seven astronauts 73 seconds into the flight of the Challenger last Jan. 28. While the debate over how to redesign those boosters has drawn much attention, both agency and outside experts are concerned that other critical elements of the shuttle may have been neglected in the pressures to keep flying that contributed to the Challenger's destruction.

Most experts agree that a significantly safer shuttle will result. But long-term overhauls are scheduled for about half a dozen major subsystems - and most will not be complete until the mid-1990s. Thus, the craft that is scheduled to fly again in about a year will be a shuttle of transition, a 1970s model that engineers are patching up for 1988 but know they need to virtually redesign for 2000.

The hard truth, said Arnold Aldrich the NASA engineer overseeing the multibillion-dollar effort is that the really major changes take years.

How those major changes proceed may also be linked inextricably to both the politics and the future role of the shuttle. Even after the trauma of the accident the shuttle is widely regarded as scientifically valuable and essential to the more ambitious military projects of the next two decades But there is rising debate on how much the nation should rely on a shuttle system made safe for man to fly at extraordinary cost and complexity.

Most experts agree the shuttle should no longer serve as a regular freight and passenger train to space, the role NASA once considered central to selling the shuttle to Congress and the public.

The redesign effort has itself raised some new worries, chiefly

Arnold Aldrich
Overseeing shuttle redesign

from engineers concerned that a whole new class of hazards may be introduced as others are eliminated. The new escape system, for example, puts explosives in the crew cabin for the first time.

Interviews with space agency officials and outside experts over the last month about the effort to get the space shuttle flying again have also revealed:

- The escape mechanism, which can be used only after the shuttle's booster rockets have burned out two minutes into flight, will enable astronauts, each harnessed to a miniature rocket, to blast out of the cabin and then parachute to earth.
- Construction of an orbiter to replace the lost Challenger will not begin until next August, and some NASA officials say Rockwell International, which is to build the new shuttle, is thus watching helplessly as many key engineers leave the project for more immediate tasks, often with other companies.
- For reasons of safety, NASA has quietly scrapped plans to use the shuttle's main engines at 109 percent of their rated thrust to boost particularly heavy payloads

No matter how exhaustive t
shuttle's revamping, outside e
perts and even some NASA of
cials agree that another accide
may well occur in the next decad
The only real question, one seni
space agency official conced
recently is whether the next ac
dent will truly be an accide
unavoidable and unforeseeable

If instead it looks like anoth
Challenger he warned referri
to the cascade of miscommuni
tions and mismanagement ignor
warnings and startlingly bad ju
ment that the presidential comm
sion said led to the disaster
shuttle could well go the way of t
dirigible Another Challenger cou
mean the end of American mann
space flight."

From the earliest days of th
shuttle program, NASA officia
have insisted that if emergenc
strikes, the astronauts would b
far safer staying inside the spac
craft than bailing out. So the cre
has been extensively trained i
how to make an emergency lan
ing. And if all else failed, the
were instructed to ditch the cra
in the ocean and await rescue in
life raft. The shuttle does not floa

But the image of seven trappe
astronauts aboard the Challeng
radically changed that thinkin
Some individual emergency a
supplies for the crew had bee
turned on, strongly suggesting tha
a few crew members were aliv
for at least part of a terrifyin
plunge into the Atlantic And th
flattened and fractured conditio
of the recovered crew cabi
showed that "the whole concept o
putting the shuttle down in th
ocean is a fantasy," in the words o
one investigator.

Indeed, NASA has now con

Appendix 26

The Arizona Republic Saturday, December 24, 1988

Panel voices concern about NASA safety

The Associated Press

WASHINGTON — The chairman of a scientific advisory group monitoring the redesign of the space shuttle's boosters said Friday the panel has concerns about NASA's efforts to ensure continued flight safety.

H. Guyford Stever, chairman of the panel set up by the National Research Council, said he has a "broad concern, namely that the continuing program be carried out" now that the National Aeronautics and Space Administration has returned to space with two successful flights since the 1986 Challenger disaster.

"Success breeds confidence, and that's what you want, but it can also threaten you with complacency," Stever said.

"I don't see complacency developing, but I do know they have budget problems, and how they'll solve that is difficult to say."

The panel's report, delivered to NASA on Wednesday, lists nine problem areas that need to be addressed before the redesign effort can be considered complete.

The panel's recommendations for improvements include the strengthening of the conical base of the boosters, more testing of booster segments to determine their potential to be reused safely, improving materials for bonding booster parts, developing alternative materials for O-ring seals and more test-firing at cold temperatures.

In the Challenger accident, one of the O rings between the booster segments failed, allowing hot exhaust gases from the rocket to leak and cause the explosion that destroyed the shuttle and killed its crew of seven. Cold temperatures also contributed to the O ring's failure.

NASA said it will "carefully consider all the group's recommendations."

Stever said NASA officials have told him the agency has plans to carry out some of the tasks cited in the report, including a cold-weather test scheduled next month, and is considering the others on the list.

The boosters worked for 24 flights before failing on the Challenger mission, Stever pointed out. Referring to the redesigned rockets, he said, "We don't consider it a bad design, it's just that things could be improved further."

The panel's report recommends risk-reduction steps that include maintaining technically competent personnel familiar with booster design to evaluate and improve quality control and launch operations, and maintaining ground-testing and in-flight evaluation programs.

It also recommends insulating the budget for these safety activities from "competitive pressures" from other programs.

The panel of independent scientists was set up at the request of the White House-appointed Rogers Commission that investigated the Challenger accident. The commission asked the panel to monitor NASA's compliance with its recommendations.

Appendix 27a

1-25-88

Safety comes first, NASA officials say

WASHINGTON (AP) — A top NASA safety officer said yesterday that "there has been a singular change in attitude" in the space agency as a result of reforms made in the wake of the Challenger accident two years ago. Another official called the system "supersensitive" to safety.

"There is a great deal of openness in NASA operations now," said George A. Rodney, the head of NASA's new safety, reliability and quality assurance office.

THE OFFICE was created after a presidential commission concluded that an effective safety organization could have prevented the Challenger accident, which was caused by a booster rocket leak problem that had been known but disregarded in the agency.

"Somebody asked if the new system had been in place, would it have stopped the Challenger launch," Dr. Noel Hinners, NASA's No. 3 official, said at a news conference. "With the new system you would never have gotten to the point where you were even ready for launch."

He said he worries not about safety for the first post-Challenger flight expected this summer, but about safety in the long run. "I'm not so concerned about STS-26," he said. "I'm concerned about STS-47." The first post-Challenger flight is the 26th in the shuttle series.

"The system now is supersensitive," said Hinners who holds the dual titles of associate deputy administrator for institutions and NASA chief scientist. "You can say it's oversensitive and for the first mission we all buy that. But how do you come down in the future?

"We know you can't forever keep up this intensity on it," he said. "The problem is how do you gracefully go into the right steady-state"

Rodney said NASA is "in the high-risk business' and will remain so Among the reforms instituted since the Challenger explosion on Jan. 28, 1986, he said, is a system that tracks close calls.

"A close call is one that doesn't lead to an accident," Rodney said. Procedures are put in place so that any such happenings are investigated and steps taken to assure they don't recur.

The presidential commission cited gaps in communciation both within the agency and with its contractors as requiring corrections. As a result, NASA has established a "significant problem" reporting system.

RODNEY SAID communications are improved to the point that J.R. Thompson, director of the Marshall Space Flight Center - has a policy that anyone can walk into his office and receive 15 minutes of his time.

The most significant problem being worked on prior to return to flight, Rodney said, is recertification of the redesigned shuttle booster rockets. He said because of recent problems with a part in the shuttle nozzle, NASA will probably order three more full-scale tests instead of the two that were scheduled before the first flight. Astronauts have asked the agency for the third test.

Another major project for recertification is a 17-inch valve that has a quick disconnected" function when the orbiter separates from its external fuel tank. The valve had problems in earlier flights.

Appendix 27b

NATIONAL BRIEFS

Test of revamped shuttle rocket postponed over aborted firings

BRIGHAM CITY, Utah — A critical first full-scale test of the redesigned space shuttle rocket will be postponed still further while engineers sort out problems that caused the firing to be aborted three times, officials said yesterday.

In a joint announcement, NASA and rocket manufacturer Morton Thiokol Inc. said engineers will conduct several countdown dry runs to test systems that malfunctioned Thursday. This work is expected to take most of today.

A decison will be made tonight on a day for the test firing, the announcement said.

The statement said officials are working "at a measured pace and won't reschedule the test until they are satisfied that all problems identified yesterday are fully understood and have been resolved."

Thiokol engineers conducted three countdowns toward ignition of the test rocket on Thursday and were forced to stop short of the firing each time.

Appendix 27c

NATIONAL BRIEFS

Snafu wouldn't have hurt liftoff, NASA says

BRIGHAM CITY, Utah — The failure of a section of the new space shuttle booster rocket last month would not have affected an actual launch because the breakage occurred seconds after the test firing ended, a NASA official said yesterday.

Engineers have determined that half of the nozzle's 8-foot-diameter boot ring popped off after the two-minute ground test was completed Dec. 23, said J.R. Thompson, director of Marshall Space Flight Center in Alabama.

Still, engineers remain concerned about the component's design as they get closer to discovering what caused it to fail, Thompson said.

A similar failure would not have affected a real launch because the rocket is used on a shuttle for only a couple of minutes during and after launch, and is then ejected into the ocean, NASA said.

Appendix 28a

GAO criticizes Justice Dept.'s fraud program

Report doesn't cover the current Pentagon procurement investigation

by Philip Shenon
New York Times

WASHINGTON – The Justice Department's program to prosecute fraud at the Pentagon and among military contractors was criticized yesterday in a report by Congress' General Accounting Office.

The report by the congressional investigating agency does not address the current investigation into allegations of fraud and bribery in Pentagon procurement. But it comes at a time that Justice Department officials themselves are expressing concern about the management of the newly disclosed procurement bribery case that is being overseen by the U.S. Attorney's office in Alexandria, Va.

The General Accounting Office said in a report to the Senate that the Justice Department's program against fraud in procurement appeared to be understaffed and that departmental supervisors were failing to keep track of military fraud cases.

While taking note that the department might be in the midst of uncovering "one of the biggest procurement scandals" ever, the report raised questions about whether other important fraud investigations into military supplying had been aggressively pursued.

Deputy Assistant Attorney General Victoria Toensing, the Justice Department official who oversees procurement fraud cases handled in Washington, called the report "irrelevant, especially when we're in the midst of a very large defense fraud investigation."

Toensing acknowledged that there were serious problems in the department's program involving military procurement, but she said they were largely the result of inadequate financing by Congress.

She is not in charge of day-to-day management of the current Pentagon bribery investigation, in which several of the nation's largest military contractors have come under scrutiny.

It is being handled by the U.S. attorney in Alexandria, Henry Hudson, with assistance from prosecutors in Washington, as well as investigators from the FBI and the Naval Investigative Service.

A number of department officials said that concern over management of the current inquiry by the Virginia prosecutors had grown acute on Thursday when a highly sensitive court document in the case, an affidavit that detailed the history of the investigation and identified suspects, was made public in Dallas, apparently without the approval or even the knowledge, of the Virginia prosecutors.

Justice Department officials said the incident, in which valuable information may have been provided to potential defendants, demonstrated the necessity of transferring supervision of the

Please see **PENTAGON** on A 7

Justice Dept. under fire

PENTAGON

continued from Page 1

case from Virginia to the department's Washington headquarters.

The officials said recent turmoil at the department's headquarters prompted by the departure of several senior officials angered by the continuing legal troubles of Attorney General Edwin Meese, had ended.

"We have strong replacements for the people who left," one Justice Department official said. "It's time for this case to be handled out of Washington."

The senators who requested the GAO study, Charles Grassley, R-Iowa, and William Proxmire, D-Wis., said in a prepared statement that it showed that "the Justice Department's lackadaisical, careless, hands-off management goes far to explain why the government is losing the war against defense fraud."

They added, "If any major private law firm handled its affairs the way Justice handles the defense fraud program, it would be out of business within a year."

The senators, who have often had harsh words for the Justice Department's efforts against fraud in procurement, said they believed that "the recent revelations of defense fraud are only the tip of an iceberg of widespread improper and illegal activities."

The GAO report made little mention of the newly disclosed bribery investigation except to say that if news accounts about wrongdoing among Defense Department employees and military consultants proved true, "the country may have witnessed one of the biggest procurement scandals in its history."

"GAO's work," the report went on, "viewed Justice's efforts in a broader context than the current investigation."

In its study, begun about a year before the current military contracts fraud case became widely known, the GAO said investigators had heard Justice Department officials complain repeatedly that the program on military contracting fraud had been hamstrung by understaffing and by turnover among senior lawyers.

Appendix 28b and c

MEESE

continued from Page 1

fice of Professional Responsibility, in the report released today, found that Meese, who left office in August, had committed numerous violations of both departmental regulations and a 1965 executive order setting out ethical standards for government employees. Its report cites Meese's relationship with San Francisco lawyer E. Bob Wallach; his handling of telecommunications matters at a time he owned stock in regional telephone companies; and his failure to pay full taxes on time.

"We trust that this analysis will lay to rest the claims by Mr. Meese that the appropriate standard for official behavior is whether an independent counsel seeks an official's indictment," the report said. "We found that the independent counsel's report far from vindicates Mr. Meese; rather, it details conduct which should not be tolerated of any government employee, especially not the attorney general of the United States."

The report concludes that if Meese were still attorney general, "we would recommend that the president take disciplinary action."

In a 16-page statement issued in response to the report, Meese's lawyers contend that "there is absolutely no basis for criticizing his conduct" and call the report "a travesty of justice."

The Office of Professional Responsibility is the arm of the Justice Department charged with investigating allegations of wrongdoing by department employees.

The report focuses on Meese's dealings with Wallach, his close friend since they attended law school. One of Wallach's clients was the now-defunct Wedtech Corp., a defense contractor.

There were three instances "in which friendship" between Meese and Wallach "dictated government action," the report said:

■ "As a direct result of the preferential, improper efforts of Meese and his staff," the Army awarded a $32 million engine-building contract to Wedtech in 1982 while Meese was counselor to President Reagan.

■ The next year, the report said, "Meese and his staff injected themselves" into a financial problem Wedtech was having with the Commerce Department's Economic Development Administration.

At the urging of Wallach, Meese tried to help clear up the problem Wedtech was having by making a phone call to Malcolm Baldrige, then secretary of commerce, the report said.

■ Meese, again at Wallach's urging, arranged a meeting in 1985 for Wallach and his client, Swiss oilman Bruce Rappaport, with Robert McFarlane, then White House national security adviser. The meeting was an attempt to help secure an insurance package that would support a Mideast pipeline project, the report said.

The Office of Professional Responsibility also found that Meese violated the executive order on ethics and the Justice Department's standards of conduct by failing to file his 1985 federal income tax return on time.

It also said Meese violated ethics standards by participating in official matters involving the regional Bell Telephone companies while he held some $14,000 in telecommunications stock.

In a statement, the department said Attorney General Richard Thornburgh has reviewed and accepted the report and considers the matter closed.

Appendix 29

2/5/89
RENTON, WA.

MR. BOB BRYANT
SPECIAL AGENT IN CHARGE
FEDERAL BUREAU OF INVESTIGATION
FEDERAL BUILDING
SALT LAKE CITY, UTAH

DEAR BOB,

This is a letter of URGENT request for you to release all of the documents I have requested to date via FOIR. You, and all of the other members of the FBI, and myself, have to the best of my knowledge striven to pursue justice. S.A. Tim Healy and I often talked about the Morton Thiokol Inc. case priorities -

FIRST - Get America's space shuttle flying again
SECOND - Protect the sources and innocent
THIRD _ Prosecute the guilty

I agreed with those priorities then and now. However I have some very legitimate and well informed concerns. My concerns are briefly as follow -

1. The 32 months required for the redesign of the solid rocket boosters cost $500,000,000.00++ and has in no way shape or form improved the flight mission safety status of the SRM'S.

   The SRM cases are still inferior to requirements for even one flight.

   The SRM Safe and Arm (ignition) Device is still antiquated and flawed.

   The O-Ring's at the SRM field joints still leak and are incredibly dangerous (five joints leaked and each of the joints had multiple leaks on the recent Discovery flight., and more were anticipated on the Atlantis flight).

   Contract safety factors still have not been complied to on the -

   SRM cases
   SRM nozzles
   SRM field joints
   SRM to external tank attach pts.
   SRM nozzle to case joint
   SRM Contract refurbishability requirements can still never be met.
   SRM electrical grounding

Appendix 29

There are still numerous other gas path leaks in the SRM systems

MTI still maintains low professional standards.

The total testing required to ensure safety has not been even closely approximated or accomplished.

2. Even though the FBI has acted in good faith as far as I can see I am very concerned about the enormous time delay in prosecutions by the D.O.J. for the following reasons -

   A. If the D.O.J. waits much longer not even persons with good memories will remember the facts and truth. In my opinion the D.O.J. attorneys are inappropriately slow.

   B. Morton Thiokol Inc. has had a lot of time to cover it's tracks.

   C. NASA has had a lot of time to cover their tracks.

   D. America is being deceived by not being made aware of MTI crimes via our American legal system.

   E. America is being deceived by not being made aware of NASA failures and problems.

   F. Other aerospace vendors may have their problems, but none have done so much damage to so many diverse programs.

      MTI has irrevocably damaged the U.S. Space Programs

      MTI has damaged the U.S.A.F. programs such as the -
      MX MISSILE FIRST STAGE
      MX MISSILE LEGG (launch egress gas generator)
      CREST PROGRAM (ejection seats for advanced tactical fighter)
      Many others

      MTI has damaged nearly every program they have touched.

Appendix 29

Other vendors need to have the way cleared to do business with the U.S. government - higher quality business and vendors.

G. America has been robbed and deserves better.

3. The United States can not adequately solve it's space programs disastrous decline unless the current status of programs and equipment are put on the public table so the proper democratic and legal funding processes can immediately proceed and the new programs can move immediately forward, what ever they are.

4. The space shuttle is now approximately 145 launches behind NASA'S earlier published schedules.

5. The space shuttle is flying with drastically reduced pay loads and thus the cost per pound of hardware in orbit is extraordinarily higher than NASA ever told congress it could be.

6. Since the Challenger disaster the Russians have now launched approximately 700 heavy launches and the U.S. has only launched about forty launches - small wimpy launches. The Russians and Americans both had about the same number of explosions and failures, therefore the Russians are having a far better percentage of successful launches in addition to a far superior launch volume.

As you might guess I am frustrated by the slow progress of our American legal system and the fast progress of History writing the United States of America out of power, position, and national security. In no way do I criticize you or the FBI. In my book "FOR THE RIGHT REASONS" I carefully detail all of the incredible forces against the FBI on the Morton Thiokol Inc. cases. My book will probably be published soon and also a movie made. In every way I have striven to protect the FBI and laud it's efforts. I leave it to your imagination what questions I raise about the D.O.J., NASA, and others. What ever I say in the end it will be the truth.

As I saw the American Justice system wrestle with this geopolitical problem I have taken independent parallel actions which have been enormously stressful and dangerous to my self. I have been irreparably damaged in many ways and would do it all over again in a heart beat if it was necessary and if I felt my efforts would be fruitful in the end.

Appendix 29a

No matter what anyone says, or what anyone thinks my going public via the Seattle Times was a bold stroke and has done an enormous amount of good. My effort was not perfect but no doubt broke some of the political logjams. You and I both know there are a number of logjam's remaining. I won't insult your intelligence and list the remaining logjams in this letter. This case makes Watergate look like tinker toys. The Watergate burglary was about a dumb crook who had a few political documents stolen. This case is about a mostly successful conspiracy which has castrated United States National Defenses and our nations future ability to exist as a dominant global power.

PLEASE! PLEASE! Forward the documents I have requested. They are from me, they are literally "nothings" compared to the mass of data and documents your agency has compiled.

I have brought this case to you, I have brought hundreds of pounds of documents to you, I wore the "wire" for you, I helped scope out and place sources for you. Now I have parallel civil actions pending as you have criminal actions pending. The requested documents would aid my current suit against Morton Thiokol Inc.. Also it is my understanding that the FOIR request is not even necessary for documents I supplied the FBI. My efforts are not in any way against FBI efforts. My efforts are correct citizen behavior. After all, Morton Thiokol Inc. did fire me improperly, Morton Thiokol did blackball me, Morton Thiokol did steal several billion dollars, Morton Thiokol did castrate the U.S. space program, and Morton Thiokol did many other despicable things. I am trying my best to help America return to its proper global demeanor. No problem is ever solved by not being completely surfaced. This problem will be America's undoing if ignored and I know and respect that you understand that.

Please give me the tools I need to do the best job I can for America, myself, and for the FBI.

Please know my attorney Mr. Ed. McLean has a huge data base on my case against MTI and you and the D.O.J. are welcome to tap it with no strings attached if you desire or if it would benefit your various cases. There is a body of evidence emerging which shows NASA officials had more to do with Kent McKinnon's and my being pushed out of MTI than we originally thought.

Please note that strange things are happening on my case. Several federal judges have already done some very strange, irregular, and questionably ethical things in my case. It appears that in the next week or so my case will be thrown from the Illinois federal courts to the Utah federal courts. We will see what form justice takes. I intend to win. I want, deserve, and need all of the tools available. I am not asking for documents I have not seen or that are FBI secrets. What I am asking for are RUDIMENTARY DOCUMENTS.

Appendix 29b

U.S. Department of Justice

Federal Bureau of Investigation

In Reply, Please Refer to
File No.

Room 3203
125 South State Street
Salt Lake City, Utah
April 14, 1989

Steven W. Agee
10407 S.E. 174th Apt. 1302
Renton, Washington 98055

Dear Mr. Agee:

Reference is made to your letter dated February 5, 1989, and received at the Salt Lake City Office of the Federal Bureau of Investigation on February 10, 1989.

As previously advised in our letter to you dated April 25, 1988 you have made an inquiry into a pending investigation.

Investigations of a pending nature and documents contained therein are exempt from disclosure by subsection (b)(7)(A) of Title 5, United States Code, Section 552.

The decision to withhold exempt portions of our records is the responsibility of William S. Sessions, Director of the Federal Bureau of Investigation.

Very truly yours,

Robert M. Bryant
Special Agent in Charge

By:
David M. Tubbs
Assistant Special Agent in Charge

Appendix 30

High Risk of New Shuttle Disaster Leads NASA to Consider Options

By WILLIAM J. BROAD
Published: April 9, 1989

The space agency is worried that another catastrophic space shuttle accident seems almost inevitable in the next decade, so it is improving the winged spaceships still more and considering a switch to unmanned rockets for some missions.

For any single shuttle flight, the agency now estimates the risk of "catastrophic failure" in the fiery ascent to orbit at 1 in 78. Such an accident would destroy the shuttle and almost certainly the astronauts as well. The ascent is by far the riskiest period of flight, with the shuttle's powerful engines running at dangerously high speeds, temperatures and pressures.

The grim assessment of risk to the shuttle holds true despite the successes of the three flights since the 1986 Challenger disaster, which killed all seven crew members, officials of the National Aeronautics and Space Administration said last week.

The risk is becoming a major policy issue as NASA begins work on its proposed $25 billion space station, which would rely on the shuttle for transporting material and astronauts and for maintenance. Altogether, more than 100 shuttle missions are scheduled through the late 1990's, when the 508-foot-long station would be assembled high above the Earth.
'Send Eveybody Home'

So the latest estimates of risk suggest that a shuttle catastrophe is likely to occur before the space station can be completed. Another such accident, experts say, could cripple the nation's manned space program during the pursuit of its most ambitious project.

John E. Pike, head of space policy for the Federation of American Scientists, a private group based in Washington, said the political repercussions of such an explosion in the next five years or so would "basically pull the plug on the manned space program and send everybody home."

Seymour C. Himmel, a member of NASA's Aerospace Safety Advisory Panel said there was "clearly a finite probability that something will go awry over the next few years, that you'll clobber one."

"The question is whether you can do anything to make those odds longer," Mr. Himmel said. "One of the concerns that everybody has" he said, is the effect a shuttle catastrophe would have on the space station program.

James R. Thompson Jr., the head of NASA's Marshall Space Flight Center in Huntsville, Ala., said the agency needed to avoid complacency after its recent shuttle successes and to press ahead on finding ways to reduce risk.

"We shouldn't rest on any laurels after three launches," Mr. Thompson said. "There are going to be failures in the future, and we should try to minimize that" by improving the shuttle, he said. "But NASA has to avoid the trap of being a cheerleader agency, of saying all the flights are going to be successful. That's just not in the cards." New Candor by Agency

NASA's candor about the shuttle's risk of catastrophe is relatively new. Before the Challenger disaster, agency officials regularly asserted that the chance of disaster was 1 flight in 100,000. Internal Government estimates were more realistic. In 1983, a study for the Air Force found that of 14 major accidents that might destroy the shuttle, booster failure was the most likely, with chances of 1 flight in 35.

Appendix 30

After the Challenger disaster, which was caused by booster failure on the nation's 25th shuttle mission, the late Richard P. Feynman, a Nobel laureate on the Government panel that investigated the tragedy, heatedly criticized the space agency for its lack of realism about risks.

"NASA exaggerates the reliability of its product to the point of fantasy," Dr. Feynman wrote in an appendix to a Federal report on the Challenger explosion. His own estimate put the odds at 1 flight in 100. Explosions in Tests

Today, NASA bases its reliability estimates on statistical formulas, flight histories of rockets and shuttles and data derived from the ground testing of shuttle components. For example, the shuttle's main liquid-fuel engines, for their size the most powerful on earth, undergo repeated trial runs on the ground, and several have exploded in tests that mimic the strains of typical missions.

Benjamin Buchbinder, manager of safety risk assessment at NASA headquarters in Washington, said in a statement that the 1-in-78 estimate of risk of disaster was statistically the most probable, with the estimate's range of reliability extending as high as 1 flight in 168 and as low as 1 in 36.

In contrast, the Air Transport Association estimates that a single airline flight has 1 chance in 2 million of ending in fatalities. Getting Astronauts Out

James H. Ehl, a leading safety official at NASA headquarters, said a catastrophic failure of the shuttle would not always imply the death of astronauts.

"You can postulate that you'd lose the vehicle and still get the astronauts out," he said. As an example, he cited a main-engine shutdown during ascent in which the shuttle would crash but the astronauts would bail out in the escape system added after the Challenger accident. Such escapes, however, would work only during a small part of the shuttle's flight, and many accident scenarios involve engine explosions that would shatter the spaceship.

Mr. Thompson of the Marshall center, which oversees development of the shuttle's engines, said his estimate for catastrophic failure was about 1 flight in 100. If this rate was achieved, he said, it would better the flight record for the world's rockets in general, which fail on 3 or 4 of every 100 flights.

"For the shuttle," he remarked, "I'd like to say it's 1 in 1,000 but I try to keep my feet on the ground."

Taking issue with NASA's shuttle risk estimate, Jerry Grey, director of science and technology policy for the American Institute of Aeronautics and Astronautics, a professional society based in Washington, said the real odds of catastrophe were closer to 1 in 60.

"That means there's a reasonable chance of losing two orbiters by the end of space station deployment," he said. Solid Fuel or Liquid? The world's only other space shuttle, the Soviet craft, has flown once, without a crew. Nikolai A. Simyonov, head of the international department at Glavkosmos, the Soviet space agency, said it was safer than the American shuttle because of the design of its booster rockets.

"We use liquid fuel rather than solid," Mr. Simyonov said in Moscow. "With solid fuel, it is difficult to control combustion, which is far easier with liquid fuel."

Many Western experts agree with the assessment, although some experts at NASA say the solid rockets are just as safe. NASA does not compare risks in the shuttle program with those in the Mercury, Gemini and Apollo programs, which used liquid-fuel rockets.

Of the more than 100 shuttle flights scheduled until the station is assembled, 20 are devoted to lofting parts of the station, which would weigh 225 tons, orbit 250 miles above Earth and hold as many as eight crew members. The transport flights are scheduled to start in 1995 and end in 1998. More Safety Expenditures

NASA is struggling to improve the shuttle's reliability to minimize the chance of a blow to the manned space program. The agency has spent $2.4 billion on improvements since the Challenger explosion and plans expenditures of an additional $1.1 billion to further strengthen the spaceship.

Appendix 31

Bush comes down hard on wasteful military spending

By Judy Wiessler
Hearst News Service

NORFOLK, Va. – President Bush used the first out-of-town trip of his term yesterday for a dramatic landing on the deck of an aircraft carrier where he promised to "wring the last drop of waste and mismanagement" from military purchasing.

Bush, his voice reduced to a croak by a cold, continued his pattern of lauding the work ethic, encouraging individual initiative and forecasting government improvements – all without outlining specific plans.

"We want tighter controls and higher standards in weapons procurement, and we will get tighter controls and higher standards in weapons procurement," he said, in a reference to the multibillion-dollar Pentagon purchasing fraud investigations under way.

"I firmly believe that the vital first step to broadening our national consensus on defense is to wring the last drop of waste and mismanagement out of the way we buy our weapons."

Under a gleaming sun and soft breeze, about 5,000 military personnel stood on the deck of the USS America to hear the president, who did not wear a topcoat despite his cold.

Bush flew from Washington on the presidential helicopter, which landed on the deck of the massive warship.

He said the visit was the final stop on what you might call an "inaugural trip" because, since his swearing-in Jan. 20, he has been addressing groups of workers – mostly federal employees.

"I've saved for last those whose service means the most – and I mean you," he said of the personnel from all branches of the service.

Stressing that military service is "not just a job," he recalled his own years as a Navy pilot to point up the importance of honest procurement practices.

"I've been inside a submerged submarine while depth charges were going off all around it. I know what it's like to hear the vessel strain and shake and to pray to God that the people in charge of buying and building cared as much about the vessel as you do," he said.

While Bush said he thinks the overwhelming majority of procurement officers and defense contractors "do care that much," he added: "I am determined to make sure that every single one of them does."

He said the contractors and military purchasers "should think it's your son or daughter" who might be using the equipment and "remember that their lives depend on the things you make. If you're not ready to care that much and work that hard, you're not ready to do business with the United States government."

Some defense contractors, as well as military technicians, were in Bush's audience.

Bush said he would battle military cost overruns which have plagued the Pentagon. "Even if overruns are not new, they are still wrong – and hurt the national security, particularly when budgets are tight."

Bush's day out of the White House featured Chinese food on the USS America. In the ship's chow line, he loaded up on egg drop soup, pork-fried rice and a spring roll.

"I feel OK, but I wasn't sure if I could get through that brief address," Bush said of his faltering voice and repeated coughs. He said he was just "croaking along" during the speech.

Bush was examined before the trip by his physician, Lawrence Mohr, who accompanied him on the flight, said White House press secretary Marlin Fitzwater.

Fitzwater said a head cold Bush came down with last week "turned into laryngitis."

Appendix 32

THE NEW YORK TIMES **NATIONAL** THURSDAY, FEBRUARY 2, 1989 Y 11

Space Shuttle Mission Is Delayed About 3 Weeks for Repairs

By WILLIAM J. BROAD
Special to The New York Times

The space agency announced yesterday that it would delay the launching of the next space shuttle mission for about three weeks to replace suspect engine parts whose failure could cause a catastrophic explosion.

The delay, a minor blow to the National Aeronautics and Space Administration after two successful shuttle flights, is not expected to have a significant impact on the ambitious schedule of seven liftoffs for this year, including two probes to distant planets that must be launched at particular times. But the delay means there is less leeway in the schedule.

The shuttle Discovery had been scheduled to be launched Feb. 23, but the flight has now been pushed back to "mid-March," space agency officials said yesterday. They said a firm launching date would be set later.

The winged spaceship is to carry a giant $100 million communications satellite, which would be used by the space agency and other Government agencies.

Discovery of Cracks on Atlantis

The decision to delay the mission was made late yesterday at the space agency's headquarters in Washington in a two-hour meeting of top engine experts and shuttle officials.

Space agency officials had earlier decided to replace Discovery's parts when cracks in an engine pump of the shuttle Atlantis were found after its flight in December. The tiny cracks were discovered in two bearing assemblies in an oxygen turbopump.

In the shuttle, liquid oxygen and hydrogen from the giant external fuel tank are driven to high pressures, temperatures and speeds by an array of spinning turbines that feed fuel to the shuttle's three main engines.

One crack goes through the assembly on which the bearings sit. The cracks were considered too small to cause trouble in flight, but if such cracks widened and split the bearing assembly, they could cause an explosion.

At first the space agency thought the problem might be limited to the single Atlantis pump. But engineers determined that the cracks were caused by stress corrosion, apparently resulting from moisture that penetrated the metal in the manufacturing process. Since the same process was used in making Discovery's pumps, prudence dictated a delay to change parts, space agency officials said.

All three of Discovery's pumps are to be replaced with new, carefully inspected units. Their maker, the Rocketdyne Division of Rockwell International Corporation, based in Canoga Park, Calif., started assembling the new pumps last week, and the company expects assembly and testing of the pumps to take about three weeks. The three pumps are to be delivered to the space agency in late February or early March. Rocketdyne is also assembling new pumps for Atlantis, which is scheduled to carry a high-priority Venus probe into space on April 28.

Rocketdyne has traced the problem to a new plant for assembling the pumps in Canoga Park, where an oven used to cure the metal pumps apparently worked improperly.

Pump replacement poses some problems. Engine work on the shuttle is usually done when the spacecraft is horizontal. But last week Discovery was mated with its booster rockets and external fuel tank in the 52-story vehicle assembly building at the Kennedy Space Center in Florida, and the pump replacement will now have to be done with the craft in vertical position to save time.

Discovery is now scheduled to roll out to the launching pad on Friday. The repair work is to take place on the launching pad, with work to remove the first pump beginning as early as Sunday.

One option the space agency weighed was delaying Discovery's mission indefinitely so it would not interfere with the liftoff of the Venus probe, known as Magellan. If Discovery's liftoff slips much later than mid-March, the mission might be delayed until later in the year, space agency officials said.

'The Smell of Victory'?

DALLAS, Feb. 1 (AP) — A weapon manufacturer is about to give buyers a whiff of war. BEI Defense Systems is promoting its "extraordinarily lethal" Flechette rockets with scratch-and-sniff advertisements in trade publications.

Appendix 33

Shultz: U.S.-Soviet trip to Mars long way from agreement

by William Harwood
United Press International

A U.S.-Soviet unmanned mission to Mars that Soviet General Secretary Mikhail Gorbachev plans to propose at next week's Moscow summit would produce priceless data about the red planet and possibly serve as a stepping stone to more ambitious space projects, experts say.

But Secretary of State George Shultz cautioned yesterday that such a joint mission "has to be studied carefully" and may be taken up under a U.S.-Soviet space agreement, The Associated Press reported.

"That's something that's been talked about, but it's a long way from being agreed upon," Shultz said in a televised interview with ABC's David Brinkley.

James Oberg of Houston, an expert on the Soviet space program, said yesterday the United States would have to guard against technology transfer, a frequent stumbling block in the past.

On the other hand, Oberg said, a joint mission would mark "a trust-building kind of activity that could lead to greater things" and the technology required for such a mission could be applied to the exploration of other planets or moons.

Gorbachev said in an interview with The Washington Post and Newsweek magazine that he plans to propose the joint flight when President Reagan visits Moscow for their summit next week.

"Why shouldn't we try to work together?" Gorbachev said. "We have great experience, you have great experience — let us cooperate to master the cosmos, to fulfill big programs."

Gorbachev did not discuss what sort of mission he would propose, but the Soviets have said they hope to mount an unmanned soil-sample return mission by the end of the 1990s, and it is this type of mission that is frequently mentioned in the West as a good possibility for a joint flight.

The Soviet Union has made no secret of its plans ultimately to launch a manned mission to Mars, and Soviet researchers have frequently talked about the possibility of a joint manned flight.

Whether joint missions evolve or not, the Soviets are intent on exploring Mars. In July, the first in a series of ambitious unmanned Soviet probes to the red planet will blast off.

Appendix 34

SAFETY WARS

A local scientist has raised tougl questions on the safety of our astronauts. *Arizona Living* was astonished to hear of allegation that reeked of conspiracy and corruption within the space indus Dr. Anthony Laine, a consultant t an aerospace contractor, relate his thwarted attempts to remedy safety problem he perceived in tl Space Shuttle. Here is his story.

RALPH JARSON

"I call NASA the American KGB and Mo Thiokol the Western Mafia."

Those are the views of Dr. Anthon Laine, Ph.D., who is, according to resume, a former consultant on the S Shuttle, MX Peacekeeper and Trident m programs, the space station and fifty other programs.

Just who is this guy anyway? It depends on who you

According to the public relations official at Morton Thi who declined to be identified, Dr. Laine never worke them. Which is unusual, considering that there are se

ARIZONA LIVING/SEPTEMBER

Appendix 34

Morton Thiokol documents that we have seen that refer to Dr. Laine and his work for the company. We called Cathy Tinker, an attorney for Morton Thiokol, for comment on Dr. Laine. She declined all comment. Repeated calls have gone unanswered.

This is in opposition to the statements of some people with outstanding credentials in the aerospace field. We spoke with Manny Espig, a senior scientist of Command Sciences Corporation of Santa Barbara, California. Mr. Espig described Dr. Laine as "brilliant, eccentric and extremely competent." Brilliancy and eccentricity were a common theme among the people we asked about Dr. Laine.

Why would this man attack NASA, an organization highly respected by most Americans? The answer, it appears, is safety.

We met Dr. Laine at his home, a ramshackle little place in a rather unfashionable area of Phoenix. Not where one would expect a man of his world-class credentials to be living.

Dr. Laine is in Arizona — living by the generosity of a relative. He was to have started medical school this fall, but his current difficulties have made that a distant dream.

"I was hired by Kent McKinnon in late 1986," he begins, leaning back on a threadbare couch. "He was looking for the best in the world, as he was the safety director of Morton-Thiokol. I sent him my resume and after checking me out, I was hired."

"Kent said he wanted a complete redesign of the SRM (solid rocket motor, the booster rocket attached to the shuttle orbiter), and he wanted to do it by the book. I started by reading the safety and hazard analysis reports."

"I found out that since 1981, they had never solved any safety defects."

"Our duty was to go through the SRM's and address any hazards to the astronauts. I brought up Electro-Magnetic Interference (EMI) and plasma effects to McKinnon's boss, Don Thompson."

Dr. Laine was concerned for the astronauts' safety. The SRM has a range safety device on board. This is an electro-explosive device for self destruction set off by radio frequencies (RF). In Laine's opinion, the device was not adequately shielded from plasma or EMI, and could have gone off accidentally.

Dr. Laine states that there are two separate ways that the shuttle boosters could have exploded without ground-commanded instruction to do so. When the shuttle is operational, it constantly transmits a steady stream of telemetry data to ground control over radio frequencies — these transmissions are broadcast at extreme power. The resulting bleed could trigger the RF activated explosive charge that would result in the immediate self-destruct of the boosters — and the astronauts in the shuttle.

Additionally, the nozzle plume of the

They said they were going to blow me away — and blow my family away.

booster rocket is of such immense power that it creates plasma. This plasma, which is electrically charged, can flow upwards over the body of the rocket booster, thereby creating an electrical charge on the booster itself, which could then, under the right conditions, trigger the electrically charged destruct mechanism. These speculations of Dr. Laine's were confirmed in an article by the late Dr. Richard Feinman, a Nobel Laureate.

Additionally, he felt that the nozzle of the SRM was made out of inferior material which was a grave danger in itself. "In Flight 51B (of the Space Shuttle) "the nozzle burned through completely." He maintains that NASA and Morton Thiokol knew since 1981 about all safety defects and conspired to keep them secret.

"Flight 51C was Dr. Sally Ride's flight. That flight had only four seconds before it blew up."

"There is a tape called 'Spaceflight' whereby Lawrence Malloy of NASA admits on TV tape that Morton Thiokol and NASA officials conspired against the United States government and the American people by covering up all safety defects." He continues "Morton Thiokol is on record to the Seattle Times stating that safety defects are not their job."

As you may have guessed, Dr. Laine was fired by Morton Thiokol.

"Dr. Laine was hired by Morton Thiokol and subsequently fired, as he was violating security by getting access to classified information which he had no reason to have, nor did he need to know, in order to perform his duties. When he learned that he was to be fired, he claimed that someone had stolen information from him." A letter from Rep. Robert S. Walker of the United States House of Representatives dated September 18, 1987, paragraph 2.

According to Dr. Laine, security violations were never a factor in his termination.

"The last day I was there, February 6, 1987, Don Thompson said we were going to have a meeting in John Thirkill's office. He is the vice president and general manager of Morton-Thiokol's Wasatch Division in Utah. At 11:00 I went up there."

There were four or five people present at the meeting, by Dr. Laine's account. If true that's a lot of horsepower to fire one consultant. Laine said that he was informed by John Thirkill that Morton Thiokol had no problems except him, and they were tired of Laine bashing Thiokol.

After his dismissal, Dr. Laine was visited in his motel room by Roger Boisjoly, the person that had protested the launch of the Challenger just before it exploded. Laine maintains that Boisjoly told him, "Tony, they're going to go after you and make an example of you."

We asked Manny Espig of Command Sciences Corporation about the classified information and breech of security issues that had been brought up. Espig acknowledged that he had exchanged information with Dr. Laine, but said that it was totally unclassified.

Espig told us that data is commonly exchanged within the defense industry, especially between research teams. He believes Dr. Laine to be "helpful and excitable. As far as my dealings with Tony, everything was straightforward and aboveboard."

A senior engineer at North American Rockwell said that it would be impossible for anyone to get classified material without proper approval through the correct channels. "They just don't send out material to anyone who wants to see it," he reported.

Referencing Dr. Laine's difficulties with Morton Thiokol, he said, "Dr. Laine is a truthful man, perhaps a little too truthful. He works in a consultant capacity, and in this capacity you run the risk of telling the sub(contractor) something they don't want to hear."

Appendix 34

He also felt that it would be very difficult for him to believe that Dr. Laine committed any security violations. He describes Dr. Laine, as many others do, as brilliant, eccentric and ethical.

"As an ethical engineer, you report the results. You run the risk of not pleasing the rest of the company team."

Laine has been reduced to penury by his situation. Still, there are some of the trappings of a rich man. His ostrich leather briefcase glowed with the toasty warmth that good leathers do. The locks were gold and shined as if they were newly minted, "DR T" embossed on the front.

His closet, which happened to be on the way to the restroom, was filled with suits, straight out of *GQ*. Cashmere. Raw silk. Obviously custom made. Obviously expensive.

"I used to own a beautiful home, beautiful cars," he explained.

"My wife drove a Mercedes roadster and I drove a Porsche Carrera. I had excellent credit."

"Now, today, I no longer have my condo. I no longer have a credit rating. I'm being sued by the bank. they have blacklisted me."

In the corner, a custom license plate stared up at me. It said "DRS TOY 1". Dr. Laine is not a happy man. He is, however, an angry one.

"I want people to know how a good physicist can be blacklisted.

The American people have to know how a small group of men have cheated them. Is the public prepared to sell its soul for a space program that covers up?

"They destroyed me and they have destroyed others."

But why would a company with the reputation of Morton Thiokol go to such lengths to silence one lone Ph.D.? If Laine is to be believed, the answer is money.

Dr. Laine feels that there is an unholy alliance between the heads of NASA and of Morton Thiokol.

"Morton Thiokol was given 100 million dollars by NASA to set up a monopoly." It is Dr. Laine's contention that Dr. Fletcher, the director of NASA, gave Morton Thiokol the contract. And Fletcher was once an employee of Morton Thiokol.

But that's not all. Laine also alleges that Thiokol was paid 505 million dollars for a redesign of the SRM. Laine also believes that Morton Thiokol received 90 million dollars for SRM redesign research and development and spent none of it on new research.

Concerned about this information, Dr. Laine went to the Federal Bureau of Investigation and wrote the President of the United States.

"After I wrote the President, I got three threatening phone calls. They said they were going to blow me away — and blow my family away."

The calls came in February of 1987 and ceased when he informed the caller that there was an FBI tap on his line.

"The FBI does not trust NASA. The FBI took NASA (information) in early 1987; invited NASA to talk to some people at Thiokol. The FBI was going to record it. NASA declined. They didn't want to hear it."

Laine says he personally called the Inspector General of NASA to try to get an investigation and that he hung up on him. He suspects that there may be a link between that office and Morton Thiokol.

"If you called the IG's office of NASA (with a complaint), Morton Thiokol called you back within five minutes."

Arizona Living asked if he took the death threats seriously. He pointed to his two dogs (one named Christa after late Challenger astronaut Christa McAuliffe, and the other named Baron Von Messer).

"They are my ears."

We wondered what his next move was. If in fact there is such a conspiracy how can one White Knight expect to fight such august enemies? Apparently, he intends to fight them in the courts, but even that is proving a difficult task.

"Last Friday (July 15), my attorney went to Morton Thiokol. I've been under a (self-imposed) veil of silence (with regards to the press) for some time."

An appointment was reportedly made with Morton Thiokol, and their attorneys, to meet with Dr. Laine's attorneys. Laine says his attorney confirmed the details of the meeting arrangements, via Western Union, with Morton Thiokol.

"They set up the meeting. He (Laine's attorney) went in the lobby and presented his business card saying he was to meet a Morton Thiokol attorney." According to Laine, no one showed up.

As Laine's attorney prepared to leave, "security pulled up . . . in armed jeeps with blue lights flashing."

Now, Laine feels that Morton Thiokol has shown their true colors and that the gloves are off.

Another thing we heard from our inquiries on Dr. Laine was that he is a fighter, and very, very tenacious.

Laine's personal integrity seems very high. He is very intense. When he starts cataloguing the injustices that he feels have been wrongfully heaped on him, his eyes flash. He becomes agitated. The hands do not remain still, but rather constantly gesture and shift positions, as if he were a skilled fighter pilot recounting air combat maneuvers. He speaks of all of this as if it were happening, not a year and a half ago, but as if it were happening *now*. He chain smokes non-filter cigarettes.

"The American people have to know how a small group of men have cheated them. Is the American public prepared to sell its soul for a space program that covers up?"

"NASA had expertise, the best in the world. But they were scientists, not politicians, and so (were) fired because they did not say what Morton Thiokol wanted."

"There never would have been a Challenger accident if the good engineers of NASA and Morton Thiokoll were listened to."

Dr. Laine settled back into his couch, and lit another cigarette in the chain. "I am going to file suit within one or two weeks," he said. But, if Morton Thiokol thinks that I have libeled them in any way, I hope they sue me. I welcome it." AZ

Appendix 35: Extracts from article

"How To Destroy A Nuclear Physicist by Charles E. Corry, Ph.D.
Written in cooperation with Dr. Anthony and Donna Laine

Prelude

What follows is a true story involving quite disparate opponents. Though one side has suffered a series of crushing defeats, the fight continued for more than 20 years until Dr. Laine was literally driven to his death by the hounds of hell from Maximus.

On the losing side was Anthony Laine Ph.D., a nuclear physicist whose research findings on nuclear, biological and chemical (NBC) weaponry continue to be used by the Department of Homeland Security and various law enforcement agencies to help thwart terrorist attacks aimed at the United States. His security clearances were so high that he required an F.B.I. "minder" and when imprisoned has to be kept in a separate cell for security reasons.

His opponent is Maximus Incorporated whose stated mission is "Helping government serve the people." But in many ways Maximus is not like most other publicly-held corporations; the devil is in the details and the magnitude of this monster is apparent in Appendix A.

It was Dr. Laine's unfortunate circumstance to learn firsthand of the extreme methods Maximus employs to add to its coffers through one arm of its corporate structure, essentially a collection agency, but one of massive scale. Maximus administers state social programs such as child welfare, child support enforcement, Medicare, Medicaid, student loans, tax collection, and social security for many states and the federal government.

States where Maximus operates include at least Arkansas, California, Connecticut, New Jersey, Tennessee, Texas, and Missouri.

Many state legislatures view the privatization of their unwieldy social programs as a welcome relief. But Maximus almost always submits the lowest bid, a "low ball" in more appropriate street terms, and secures the lion's shares of contracts it bids. In the second year, however, Maximus follows a familiar pattern, or more correctly, a practice. Maximus simply claims it cannot operate profitably, and renegotiates the original contract for millions more than its competitors' original bids

Most of us are familiar with collection agencies in fact, or at least in

Appendix 35 cont.

theory. There are several common practices:

- They are relentless.
- They do not care about their methods except when bankruptcy is an issue.
- They use threats.
- The burden of proof is on the debtor.
- The debtor is a debtor and anything that debtor says is subject to scrutiny. Their bias is that debtors (viewed as such regardless of veracity) will do or say anything to avoid meeting their obligation.

With Maximus, debtors get this in spades.

Contract agencies seek to collect child support benefit from laws that are invasive, draconian, and far reaching. Maximus throws about the term "child support arrears" to gain the instant attention of any judge hearing the case for the first time; it's an emotionally charged term...even for the bench.

It is also common practice for Maximus and others to create paper arrearages for "state debt" after the children are grown and call it child support arrears. State debt does not have to be proven; all that is required for maximum leverage with the courts is to lump it into a "child support enforcement" claim.

It must be pointed out that this story is not simply about child support and its collection. However, Maximus would have you believe that its single-minded goal and exceptional power to extort money from Dr. Laine was only an attempt to collect from a *"deadbeat dad".*

(skipping a large amount of annotated history)

The reason Maximus's claims are so pervasive and difficult to dismantle is because all interested agencies get a piece of the action. States receive a minimum payment of $10,000 in federal funds to use as they wish for each child support case they create and enforce. Those payments are received even when DNA testing proves the man is not the father of the child.

This federal money is generally disbursed as follows:

- A percentage goes into the judge's retirement fund,
- A percentage goes to the case manager that creates the case,
- A bonus is paid to the hearing officer for ruling in Maximus' favor,
- A percentage goes to the caseworker in charge of the case as a bonus.

Appendix 35 cont.

Maximus makes an even bigger profit by imputing income and making the *"obligation"* so high that in some cases a man has to work 2-3 jobs or live in a tent because he simply cannot afford a roof over his head. Maximus has created a culture of slavery.

Any man who persists in fighting for justice in a child support case, particularly if he is *pro se or pro per*, as is Dr. Laine, can expect that eventually a bench warrant will be issued for their arrest. It will also demand that he call CSE immediately and be prepared to *"pay the purge."* This is the money that a man is forced to pay in order to stay out of, or get out of jail. If he does not have the money he stays in jail for an indeterminate amount of time until he does come up with it just as was done in more primitive times with such debtors prisons.

Most men who are hit with child support or *"arrearages"* are required to pay thousands of dollars in cash in these cases. Imagine a factory worker with a low paying, or no job trying to come up with $2,000, $3,000, or more in cash for the purge? Such cases are especially sad if he has already been thrown in jail a few times and there is no money left to pay the extortionists. And jails are now commonly charging a daily fee for their *"accommodations."*

A bail bondsman cannot help. They are strictly prohibited from providing bail in child support cases. Therefore, whatever money is used for the *"bail"* or *"purge"* has to be raised entirely by the non-custodial parent, or his parents, relatives, or friends. There is no other help available for these men.

They do not even have the right to speak with an attorney once incarcerated. Mrs. Laine learned this the last time Dr. Laine was "kidnapped." When she tried to call Legal Aid seeking an attorney's phone number she was told that once a man is incarcerated for child support he has no legal right to speak with a Legal Aid attorney,or any attorney unless he or his family can find a private attorney and pay a large amount of money to him to spend a few minutes with the "convicted" man.

If a man is unable to pay the purge he does not get out of jail, period. And whenever he is finally released months or years later his career and job prospects have been destroyed. Of course the child support arrearages continue to accrue while he is in debtors prison and one sickens to think of the effect on the children the loss of a father makes. The problems children of single parent mothers are numerous: low educational achievement, school dropouts, teenage pregnancy, gangs, drug use, juvenile detention,

Appendix 35 cont.

and many other pathologies closely correlate with the absence of a father in a child's life. Of course Maximus only profits from these cascading pathologies.

Clearly, Maximus will never give up. But Anthony W. Laine dare not give up, the principle and the consequences are far too great. Nor will the growing number of supporters who truly believe that the Constitution of The United States of America should not and will not be traded for profit on the New York Stock Exchange.

Maximus is not the underwriter of our state or federal legislative process, and We the People must take back what is ours. Until that time, under current law and practice, a man has to be functionally insane to marry and a drooling idiot to sire a child. As a result, marriage and families are becoming a thing of the past, and our civilization is facing a time of darkness.

Appendix 36

Crucial Flight

Space Travel's Future Is Riding on Launch Of 'Discovery' Shuttle

But a Success Won't Resolve All Doubts on the Value Of Manned U.S. Missions

The Powerful Lure of Mars

By BOB DAVIS and LAURIE MCGINLEY
Staff Reporters of THE WALL STREET JOURNAL

CAPE CANAVERAL, Fla.—It should be spectacular.

Next Thursday, if plans work out, the space shuttle Discovery will roar from its launch pad as a million spectators cheer its rise into orbit. The future of the U.S. space program rides on the flight.

It is the first attempt to put Americans into space in the nearly three years since a shuttle explosion here killed seven astronauts, including a young teacher. Discovery "will restore NASA as an agency that sails on the ocean of space," says Thomas Paine, once head of the National Aeronautics and Space Administration.

But another tragedy would doom U.S. manned space exploration. Even a triumphal mission won't resolve fundamental problems dogging the shuttle program.

NASA dreams of exploring the solar system by sending shuttle astronauts to launch giant telescopes, construct a manned space station and fetch satellites. But military and scientific critics complain that the shuttle is too expensive, too fragile and too unreliable for such missions. They want to turn to unmanned spacecraft instead. Even some of NASA's biggest supporters want to reduce shuttle launches to lessen the risk of another crash.

Costs Keep Rising

"The nub of the debate is: What do you do with humans in space?" says Richard DalBello, a space analyst for the Office of Technology Assessment, a research arm of Congress. "You don't need the shuttle if manned space isn't important."

The shuttle's great sophistication—Russian cosmonauts thud to Earth in charred capsules, while shuttle astronauts glide down in airliner style—also works against it. The shuttle performs more like a costly Indy 500 race car than the cheap, reliable cargo truck NASA once promised.

At a time of huge budget deficits, shuttle costs keep rising. In the early 1970s, NASA talked about launching a shuttle a week for about $10 million a launch. Now it wants to send up one a month for maybe $300 million a launch. Shuttle travel is so pricey that even an alchemist who could magically turn a cargo of lead into gold in space might lose money, says William Michael, a space expert at the National Academy of Sciences.

"The shuttle was clearly a mistake," grumbles Sen. William Proxmire, the Wisconsin Democrat who heads the subcommittee that handles NASA's budget.

Safety Questions

The spacecraft's complexity raises safety questions. A tiny flaw can send complex shuttle machinery hurtling toward disaster. Joseph Lombardo, director of the orbiter's main engines, worries that a small heating-coil leak could shut down the engine or blow it up. In the past year, he says, NASA found two heating-coil leaks during tests, although neither proved disastrous. NASA Administrator James Fletcher frets about the redesigned booster. "It's a more complicated beast than we thought it was," he says.

There is reason for concern. Already Discovery's boosters are outfitted with only half the test instruments originally planned because NASA rushed to meet an early launch date, says Myron Uman, director of the National Research Council panel overseeing the booster redesign. NASA will drop most of those sensors after flight three, instead of flight six, he adds, because of budget problems. If NASA successfully launches a few shuttles, pressure will increase to quicken the pace—and to slide a bit on safety.

NASA must become enormously more efficient, or cut safety margins, or give up on meeting its schedule. In 1985, the year before the Challenger disaster, NASA managed nine shuttle flights by working shuttle crews at the Kennedy Space Center to exhaustion and by recklessly cannibalizing spare parts from sitting shuttles. Now it takes three times as long as it once did to assemble shuttle boosters and refurbish the orbiter. But NASA still hopes to launch seven flights next year, 10 the year after that and, eventually, 14.

Hooking the Public

NASA managers swear they have learned the lessons of the Challenger disaster and won't scrimp on safety. Hot gas escaping from a faulty solid-rocket booster seal destroyed the shuttle in January 1986. Since then, the agency has redesigned about 500 parts in the orbiter, booster and external fuel tank that constitute the shuttle, and tested the changes thoroughly. NASA even riddled one booster with holes during a test firing to make sure it didn't blow up when damaged.

A successful launch will revive flagging interest in manned space exploration, the space agency hopes. "The public looks at the shuttle as the main event," asserts NASA's Mr. Fletcher. "They're interested in seeing people in space, not robots."

Initially, NASA will be content if it can launch its shuttle backlog of Pentagon spy satellites, unmanned planetary spacecraft and occasional commercial satellites without a glitch. But the agency also wants to spend $1.2 billion for a new generation of rocket boosters and about the same to build an unmanned shuttle that can

Continued . . . top of following page.

Continued . . . from following page.

trons to measure the upper atmosphere and predict how satellite communications may be affected by turbulence.

Shuttles could also be modified so astronauts could spend 30 days in orbit, rather than 10, making them guinea pigs for scientists studying the effects of weightlessness. The shuttles then could spend weeks visiting smaller, cheaper space stations that could be built for, say, producing materials. "It would be like a lighthouse tender visiting an automated lighthouse" for servicing, says Mr. Paine, the former NASA administrator.

While these missions wouldn't be as dramatic as manned journeys through the solar system, they could make sensible use of the shuttle while the U.S. decides whether it wants to commit to more grandiose goals. Even some of NASA's sternest critics admit they would like to get to Mars. "It would be fabulous," says Mr. Bahcall, the astrophysicist. "It would be like building the pyramids—something to remember."

NOTE: Discovery had multiple gas leaks in 5 joints when disassembled

Appendix 36 cont.

Crucial Flight: The Future of Space Travel Rides On 'Discovery,' but Success Won't End All Doubts

Continued From First Page

three times the load of the current model.

With these improvements, NASA figures it can more easily construct a permanently manned space station and, from there, head out to the moon and Mars.

"NASA will lead the world into a high-tech era of space exploration," predicts Mr. Paine, the former NASA chief who has mapped out a plan for colonizing Mars.

But some of NASA's most important customers say they don't need a manned space shuttle to work in space, raising the question of exactly what NASA should use the shuttle for. After the Challenger disaster, President Reagan banned the shuttle from carrying communications satellites, hoping to spur the growth of privately owned rocket companies.

Now the Air Force, spooked by NASA's earlier attempts to grab all military payloads for the shuttle, wants to eliminate as many shuttle flights as it can. It is spending lavishly to build a new fleet of unmanned rockets and has indefinitely "mothballed" a shuttle launch site it spent $3.5 billion to build at Vandenberg Air Force Base in California.

Space scientists want to be next to dump the shuttle in favor of unmanned launchers. Mr. Fletcher, NASA's chief, concedes that the shuttle has been a disaster for planetary probes because it isn't dependable enough to launch during the brief time periods—sometimes only a week—when the planets are properly aligned. An unmanned mission to Jupiter, named Galileo, is 10 years late and $1 billion over cost because of missed launch dates and other shuttle problems.

Roundabout Route

NASA once planned to equip planetary probes with powerful rockets for interplanetary travel, but after the Challenger disaster NASA discarded the rockets as too dangerous to fly on the shuttle. Galileo's replacement rockets are so underpowered that the Jupiter spacecraft must travel first to Venus and then back to Earth before it can pick up speed to head to Jupiter—adding four years to the journey.

John Bahcall, an astrophysicist at the Institute for Advanced Study in Princeton, N.J., argues that the shuttle also will cripple NASA's plans to launch four big telescopes, called "The Great Observatories," to plot the birth and evolution of the universe. The first observatory, the Hubble Space Telescope, is already six years late because of shuttle delays. The other three face similar delays unless they fly on unmanned rockets, he says. "The shuttle has set back space science in the U.S. by at least a decade," he complains.

Mr. Fletcher disagrees. He says the shuttle will be able to rescue the space telescopes should they run into trouble. NASA also plans to send shuttle astronauts to replace aging telescope parts with advanced instruments.

"You can use the shuttle to maintain [the telescopes] for decades, and even improve the electronics," adds Lennard Fisk, head of NASA's space science division.

1988-1989 Shuttle Schedule

DATE	ORBITER	MISSION
9/29/88	Discovery	Launch NASA tracking satellite
11/17/88	Atlantis	Secret Pentagon mission
2/18/89	Discovery	Launch NASA tracking satellite
4/28/89	Atlantis	Launch spacecraft to Venus
7/1/89	Columbia	Secret Pentagon mission
8/10/89	Discovery	Secret Pentagon mission
10/12/89	Atlantis	Launch spacecraft to Jupiter
11/13/89	Columbia	Launch communications satellite and retrieve long-duration exposure experiments
12/11/89	Discovery	Secret Pentagon mission
2/1/90	Atlantis	Launch Hubble Space Telescope
3/1/90	Columbia	Study the Supernova through ultraviolet and x-ray telescopes
4/5/90	Discovery	Launch Gamma Ray Observatory
5/10/90	Atlantis	Secret Pentagon mission
6/7/90	Columbia	Study effects of weightlessness
7/19/90	Discovery	Strategic Defense Initiative and other experiments
9/10/90	Columbia	Strategic Defense Initiative experiments
10/5/90	Atlantis	Launch spacecraft to the sun
11/8/90	Discovery	Launch NASA tracking satellite
12/20/90	Columbia	Study effects of sun on earth's atmosphere

Source: National Aeronautics and Space Administration

The future of the shuttle is tied tightly to whether NASA finds the money to start projects to colonize the moon or Mars, or whether it is forced to settle largely for unmanned ventures. That is a choice NASA has tried to avoid since the moon landings because it fears its budget and its public support will shrivel without astronaut heroics. "NASA has no incentive for soul-searching," says Walter McDougall, a Pulitzer Prize-winning space historian.

The Space Station

Just how the presidential election will affect the shuttle program isn't clear. Neither of the candidates says what he will do with the shuttle. The program's $3.5 billion-a-year budget will make any efforts to reduce the deficit more difficult, especially when combined with NASA's plans to build a space station.

NASA plans to ask for $7.6 billion for the space station over the next three years as part of a project costing up to $30 billion. NASA plans as many as 20 shuttle launches in the mid-1990s to haul up space station parts and bolt them together. So if the president or Congress should decide to scale back the station to reduce the huge deficit, the need for the shuttle would drop.

Even some NASA supporters urge the space agency to reduce shuttle flights because a fatal accident anytime soon could doom manned exploration. Byron Leonard, an aerospace risk-assessment consultant in El Segundo, Calif., says that NASA should cut its planned flights from 14 a year to seven to stretch out the probable time between crashes. Mr. Leonard estimates that the shuttle has about a 2% to 3% chance of a major failure in flight, but says it is impossible to predict the disasters.

Other NASA boosters say the agency should enhance the shuttle's scientific value rather than promote it as a cargo vehicle. During the Discovery mission, for instance, the shuttle's five astronauts will concentrate on releasing a NASA communications satellite—something unmanned rockets have mastered for 25 years.

Helping AIDS Research?

But the Discovery astronauts will also try to grow protein samples used in AIDS and cancer research that are difficult to make on earth, but which may blossom under near-weightless conditions. "There's a whole host of biological problems that we're approaching with this flight," says Larry DeLucas, a University of Alabama molecular biologist.

Future shuttle missions could boost scientific research even more. Spacehab Inc., of Washington, D.C., plans to outfit shuttle missions with a lab deck that would increase the number of experiments astronauts could run on the shuttle to 23 from about three. External Tanks Corp. in Boulder, Colo., proposes packing the shuttle's external fuel tank with instruments so it can take atmospheric measurements when it is jettisoned to earth.

Louis Frank, a University of Iowa physicist, sketches even bolder experiments. The shuttle could launch a small scientific satellite and circle around it while zapping it with radio waves, ion beams and elec-

Continued . . . see inset on previous page

Appendix 37

B4 THE WALL STREET JOURNAL [illegible] [illegible]EMBER 9, [illegible]

TECHNOLOGY

Space Shuttle Shows Problem With Boosters

Gases Improperly Burned Through Insulator Layer During Discovery Flight

By BOB DAVIS
Staff Reporter of THE WALL STREET JOURNAL

WASHINGTON—As the U.S. celebrates the space shuttle's second successful mission this year, there's evidence that its redesigned solid-rocket boosters didn't work the way they were intended to during the earlier launch.

Royce Mitchell, a National Aeronautics and Space Administration booster manager at Marshall Space Flight Center, said gases improperly burned through a layer of insulating material in five of the 10 nozzle joints in the Discovery's boosters in September. In two joints, he said, the gases approached the primary O-ring, a seal designed to keep hot gases from destroying the nozzle. The nozzle also has a secondary O-ring to capture any gases that burn through the primary seal.

Hot gases escaping from an O-ring led to the January 1986 Challenger disaster, which killed seven astronauts. In the Discovery mission, however, Mr. Mitchell said the gases didn't burn the O-ring or threaten the shuttle. NASA didn't consider the leaks significant enough to hold up the launch of Atlantis last week, he added, or to delay future launches.

"This is a damn good motor," he said.

NASA discovered the evidence of the leaks when it disassembled the Discovery boosters after the end of the mission and discovered soot between the segments that make up the booster nozzle. The booster-nozzle segments are sealed by two O-rings, which line the nozzle. As added insulation, NASA injects a rubberized caulking material into small gaps between the nozzle segments. The soot was found in the caulking material. "Gases got past a thermal barrier," Mr. Mitchell said.

Edward Price, an aerospace engineering professor at Georgia Institute of Technology, said the soot indicated the leak occurred too late in the launch to harm the O-rings. If the leaks had occurred early in the launch and the gases flowed continuously past the O-rings, he said, the gases could have destroyed the seal.

NASA "has to know why [the leaks] happened," he said. "If they don't know why, they won't know what they should be watching for." Mr. Price is a member of the National Research Council's solid-rocket-booster review panel set up after the Challenger accident to monitor the

Appendix 37: Space Shuttle Shows Problem With Boosters

This article discusses the failure of the SRM redesign effort. $500.000.000.00 later and the SRM field joints still leak and are dangerous to life, equipment and mission. Also note that even though the redesign efforts were failurcs Mr Royce Mitchell still was disillusioned and felt the SRM's were wonderful

Appendix 38

No one who was watching the news on Jan. 28, 1986 will ever forget the sight. Seventy-three seconds into its flight, the *Challenger* spacecraft blew up, killing its seven-person crew.

The cause of the tragedy soon became clear. An O-ring seal, still cold because of sub-freezing temperatures the night before lift-off, had failed to do its job, allowing flames from inside the boosters to burn through the field joint and ignite the giant external tank.

Challenger's solid rocket boosters were made at a Morton Thiokol factory about 100 miles north of Salt Lake City at the foot of the great Wasatch Mountains. Some of the employees of Thiokol's Wasatch plant were not all that surprised by the tragedy. Certainly there had been plenty of warnings.

As far back as Oct. 21, 1977, an engineer at NASA's Marshall Space Flight Center in Huntsville, Ala., handed in his evaluation of the O-ring joint designed by Morton Thiokol. His conclusion: The design was "unacceptable." But NASA paid little attention to that warning.

The engineer, Leon Ray, then took his findings to two companies that manufacture O-ring seals, Parker Seal and Precision Rubber Products. Their initial reaction, he reported, was that the "O-ring was being asked to perform beyond its intended design, and that a different type of seal should be considered." All NASA did, in 1979, was to begin investigating second sources for the solid rocket booster. But because of "budgetary priorities," funding for a competition was put on hold.

As successive shuttle launches proceeded, serious problems with the O-ring seals occurred on at least nine flights before 51-L, the *Challenger*. The first evidence of problems with the seals appeared during the *Columbia* flight in November 1981, but Thiokol ignored it. In its report to the President, the Rogers Commission, specially appointed to investigate the *Challenger* tragedy and recommend any changes that might be necessary in the space program, wrote, "Morton Thiokol, the contractor, did not accept the implications of tests early in the program that the design had a serious and unanticipated flaw." The red flag finally went up in January 1985 with flight 51-C, the *Discovery*. Roger Boisjoly was one of the first engineers at Morton Thiokol to notice real damage to the O-ring seals, before and after the 51-C flight. It was also clear by this point that low launch temperatures seriously aggravated the O-ring problem. Still, neither NASA nor Thiokol did anything more than identify the problem as critical.

FINANCIAL WORLD • JUNE 27, 1989

Appendix 38

E T H I C S

In a paper he wrote to the American Society of Mechanical Engineers, Boisjoly says, "January 1985 was the period of gross escalation of the joint seal problems. After that there were events that should have happened that would have stopped the launches, or changed the launch commit criteria to prevent launching below 53 degrees."

That summer, Boisjoly tried to form a seal-erosion task force to correct the problem, but ran into roadblocks. In his paper he wrote, "The team was frustrated from the start due to lack of management support to provide manpower and material resources.'

In October 1985, Robert Ebeling, manager of the booster ignition systems, and a member of the seal task force, wrote a memo to Thiokol management desperately trying to get attention. "HELP!" he wrote. "The seal task force is constantly being delayed by every possible means." Another letter, written by a Thiokol engineer named Scott Stein, said virtually the same thing. Of the next 10 flights to follow 51-C, there were eight instances of hot gases penetrating the field joint or nozzle joint seals. On flight 51-L the morning of Jan. 28, 1986, a seal on the aft field joint of the right solid rocket booster completely gave way, allowing a flame to penetrate the external tank. The *Challenger* exploded.

The record raises serious ethical questions: Knowing of the repeated problems with the O-rings, should Morton Thiokol's top management have taken remedial action? And having had good reason to believe the O-rings tended to malfunction at low temperatures, should they have given their approval to launch?

The top management of Morton Thiokol declined to be interviewed by FW for this article.

The Rogers Commission quickly reconstructed what happened immediately prior to launch. There had been enormous pressure to get the spacecraft airborne. Flight 51-L was already a year behind schedule. It had been postponed yet again because of high winds on Jan. 27. The night before the launch, Morton Thiokol engineers Allan McDonald, Arnie Thompson, Roger Boisjoly, Bob Ebeling and others, after learning of freezing-cold temperatures predicted for Cape Kennedy that night, pleaded with Thiokol and NASA management not to launch. An O-ring, which is rubber, loses its resiliency when it gets cold, and therefore its ability to seal the joint properly.

Originally, according to the commission's report, Thiokol managers agreed, telling NASA to scrub the launch. But a NASA manager named Larry Mulloy argued that the evidence presented by Thiokol's engineers was inconclusive. Then, Jerry Mason, senior vice president of Thiokol at the time, asked his head of engineering, Bob Lund, to "take off his engineer's cap, and put on his management cap," according to the report. Thirty minutes later, despite vehement protests from their top solid rocket booster engineers, Thiokol gave its recommendation to launch.

A defensible decision? The commission called for expert technical assistance. They ordered that veteran engineers with aerospace experience be allowed to scrutinize contractor operations in depth and report back to NASA all conceivable hazards related to the shuttle transportation system. On Sept. 22, 1988, the *Wall Street Journal* reported that Morton Thiokol had admitted to reviewing 1,027 possible problems raised by the Rogers Commission engineers. Some of the engineers themselves suggest that over 2,000 hazard reports were filed.

Morton Thiokol's Utah facility alone had to hire roughly 60 engineers to look over the solid rocket boosters.

Right from the start, the outside engineering team ran into strong opposition. Before long, several of the Rogers Commission engineers began to complain that their reports were being suppressed by Thiokol. Some may never have reached NASA, one engineer warns. From the minute they set foot in the Wasatch plant in northern Utah, their hazard reports were censored and they were treated as outcasts. In his paper to the Society of Mechanical Engineers, Boisjoly writes that Morton Thiokol's "management style would not let anything compete or interfere with the production and shipping of its boosters. The result was a program which gave the appearance of being controlled, while actually collapsing from within."

Rogers Commission engineer Steve Agee said, "We were forbidden to speak up at meetings." Kent McKinnon, a systems safety supervisor, in a letter to Joyce McDevitt, head of NASA's Safety Risk Assessment Ad Hoc Committee, said he was told to keep silent when NASA representatives were at the plant. He wrote that if he had mentioned that there were problems with management he had better "start looking for a job because management

WORLD JUNE 27, 198

Appendix 38

E T H I C S

Thiokol's Rocket Garden: ". . . no problems."

would get rid of me."

According to Agee, few of the hazard reports written by the contractors were being sent to NASA. "It didn't matter if the problem was simple or grandiose, they didn't want to tell NASA there were problems of any shape or kind."

"If a hazard couldn't be resolved, my directions were not to even submit it," said Ron Clary, another engineer commissioned to work at Thiokol in March 1987. When Clary protested, he was ordered to cite examples of how safety information was being suppressed. "That would have taken me the rest of my life," he scoffs.

Just getting the relevant information on which to base their documentation of hazards was a huge problem. "It was withheld and tended to change practically every hour," Clary says.

> NOTHING IS ILLEGAL IF A HUNDRED BUSINESSMEN DECIDE TO DO IT.
> —ATTRIBUTED TO ANDREW YOUNG

A February 1987 interoffice memo written by P.R. Dykstra, vice president of strategic planning, suggests that Thiokol management was well aware of the cover-up. Paraphrasing Pogo he writes, "We have met the enemy, and they is us." He goes on to chastise, "superficial, relatively painless solutions to complex problems are not likely to resolve the problem." In an interview with FW, Dykstra said the memo was just a joke.

Eventually, in frustration Agee went to the FBI, turning over numerous documents and once, at their suggestion, even wore a wiretap to a meeting. The FBI is currently investigating the allegations of Agee and his colleague, Tony Laine, about Morton Thiokol. In a resignation letter dated Sept. 19, 1986, Cindy Ferrari, an employee in structures design, wrote that the company as a whole has "no real concern for their employees, and their response to complaints is basically, 'If you don't like it, quit.' Employees are frustrated because their recommendations are often ignored. Life is too short to spend it working for a company which does not know what they are doing."

Laine, who has a lengthy career in aerospace and electromagnetic engineering, alleges he was fired by Thiokol for trying to examine the electromagnetic characteristics of the boosters. In a lawsuit for wrongful discharge against the company, he claims he was told not to work on electromagnetics, and only to "cut and paste" old safety reports. Thiokol claims Laine was not asked to study the electromagnetics of the booster.

Agee, before he left, had written over 200 hazard reports, many of which he marked critical. Ross Bowman, Thiokol's director of safety and quality assurance, is on record as stating Agee's work was reviewed and rejected by a Thiokol safety professional. A number of Agee's reports focused on improper grounding on railroad cars that took the booster segments down to the Kennedy Space Center, and during other segments of the booster process.

"If it had been a normal Cessna aircraft, you would have run a wire from A to B to C," Agee explains. "When there are fixes available that cost $1.98, like the grounding in certain instances, and they don't get fixed, something is wrong."

Ron Clary also remarked on the grounding, calling it a pervasive concern. "The whole time I worked there, I never saw a good ground on anything," he says.

Another problem that Agee saw during his tenure was a reckless use of "waivers and deviations." These are normal processes used in any contract situation when plans change due to unforeseen obstacles. But during his time at the Wasatch plant, he claims, "NASA bought over a thousand waivers and deviations. They needed waivers and deviations like a drug addict needs his drugs."

Agee, Boisjoly, Laine and others say they were harassed by the company in every way imaginable. "I sat next to a guy for about two months who was using a cardboard box as a desk," remarks Agee.

Clary says that when he first started working there, static-proof suits designed to permit the engineers to work in certain highly flammable areas of the plant were not given to the engineers—effectively preventing them from working in those areas. Eventually, he says, the contracted engineers began stealing them after hours so they could do their job.

Why didn't all the other Rogers Commission engineers complain? "It is a given in the contracting business to keep your head down," says Clary. "I had to get out of the field." Clary is now working in Maryland

Appendix 38

E T H I C S

for a construction company.

Clary feels that one of the reasons Thiokol's Utah plant is profitable is because it pays such low wages. But as a result, he argues, it has difficulty attracting the most talented people. In a July 1988 Ad Hoc Committee status report, the committee noted, "There is still some concern that the depth of experienced, skilled and competent people is too shallow."

In just the past five years, there have been six incidents of buildings at the Wasatch plant either blowing up or catching fire. Even in the solid fuel business, where a spark can set off 100,000 pounds of explosives in a second, such a high accident rate is unusual and suggests that major operational changes may have been necessary.

A General Accounting Office report, released in 1986, cites Thiokol's correction of safety problems as slow and incomplete. The report goes on to say that "Thiokol's implementation of actions to correct problems . . . is dismal."

In March 1986, a building at the Utah facility that was used to process fuel for Trident missiles blew up. The *Wall Street Journal* reported that in July 1987 the Air Force began withholding payments on its MX missile contract with Thiokol because of what it called poor workmanship.

That December, five workers were killed when the solid fuel in an MX missile "casting pit" ignited beneath them. Thiokol was fined $31,700 by the Utah Department of Occupational Safety and Health, and subsequently settled out of court for $14,700.

NATURE HATH PLACED MANKIND UNDER THE GOVERNANCE OF TWO SOVEREIGN MASTERS, PAIN AND PLEASURE. —JEREMY BENTHAM

A UPI article reported that in May 1988 an Air Force plant representative at Thiokol determined that the company was not meeting safety and cleanliness standards, resulting in $4 million of withheld payments once again.

Did Morton Thiokol's inattention to the

Challenger laid to rest.

O-ring problem and its decision to permit the launch of the *Challenger* violate business ethics? Michael Josephson, of the Josephson Institute for the Advancement of Ethics, reminds us that "very often people do overstate risks. An engineer will usually overemphasize risks, and a manager will use that safety margin to be less cautious." Why? "The manager is ultimately accountable," replies Josephson.

Practically speaking, the booster only had to fly for two minutes. If the booster had blown its seal just 50 seconds later, the *Challenger* would have been separated by that time and the only people to realize something had gone wrong would have been NASA and Thiokol engineers and managers. A defensible gamble?

One might argue that to the extent the space shuttle program was seriously underfunded from the outset, we all share a little of the blame. But Rex Finke, senior staff member for the Institute of Defense Analysis, has studied the budget in depth and argues that NASA got almost the exact amount that it asked for at the onset of the program. That sum was $5.25 billion, in 1971 dollars, he points out. Congress gave NASA $5.15 billion in 1972. Alex McCool, NASA's head of safety and quality assurance at Marshall Space Flight Center, remarks, "I have been here since the days of Von Braun. Even when we started to look at the system in 1972, we went with the solid rocket motors because of cost."

Finke confirms that the original plan for the shuttle system was to use liquid fuel boosters instead of solid fuel. But largely for cost reasons, NASA changed the plans and went with solid fuel boosters.

This is important because the liquid boosters are safer in some important respects. Tom Mobley, manager for shuttle applications at Martin Marietta, explains that there are several advantages to using liquid fuel boosters instead of solid fuel. "The liquids have a much higher payload capacity and the capability of being shut off if a problem occurs," he says. "Once the solid rocket boosters are turned on, there is no way to stop them from firing."

Trouble is, liquid boosters cost 25% to 50% more.

One can also argue that the unrealistic charade of competitive bidding is to blame. Morton Thiokol won the booster contract because they bid lower than the competitors. In a Dec. 12, 1973 report, NASA selection officials said Thiokol's "cost advantages were substantial and consistent throughout all areas evaluated." Ironically, the report singled out the O-ring joint design for special praise for its cost-effective use of materials. According to one engineer, the steel that is used for the massive booster cases is considered the cheapest possible that will withstand the stress.

Redesigning the boosters has cost NASA more than $500 million, just over 60% of the original estimated cost of the booster contract itself. Was it reasonable to expect Thiokol to impose unilaterally upon itself such massive cost overruns and risk losing all future contracts?

Meanwhile, not one of the Thiokol managers who made the decision to launch was fired. And in return for a $10 million reduction in award fees, Thiokol was exonerated by NASA from any legal blame for the accident. ■

I was flown to N.Y., N.Y. by Financial World Magazine and this book is the basis of the article they wrote.

Appendix 39.

RISKY BUSINESS

NASA says we're headed for the stars again. But some frightened scientists and engineers say we're heading for another disaster.

The shuttle is launched successfully, and America is back in space. We're back! America is back!
— George Bush, to a campaign rally in St. Charles, Mo., Sept. 29, 1988

If the majority of the informed speculation is correct, it's not safe enough to fly. We shouldn't be flying it. Period.
— John Pike, space policy expert for The Federation of American Scientists, in an interview, Jan. 19, 1989.

This is what you think happened:

After a week of exasperating delays, the countdown went smoothly. The Challenger gracefully lifted off the launch pad. On the ground, engineers went about their business with customary quiet confidence. All was normal, all systems Go, perfect as always, until, without warning, 72 seconds into the flight, the space shuttle Challenger and the seven lives within it disappeared in a red-blistered fireball. On the ground below, there was disbelief, then horror, then paralysis. NASA was not programmed for failure. And now, on the ground at Cape Canaveral, NASA scientists, creators of a perfect machine, were staring hollow-eyed at the skies. What could possibly have gone wrong?

That scenario, though once commonly accepted, is fiction. We know that now, because the events of the last three years — a debacle followed by unprecedented public scrutiny of a once sacrosanct national mission — have made it possible to reconstruct the final minutes of the Challenger launch. What was really happening at Cape Canaveral on that day reveals a great deal about the state of NASA at the time of the doomed launch, and about critical problems that linger.

The fact is that while public optimism is back, so is the potential for disaster.

This is what really happened:

Everyone, as usual, was scared, but this launch was scarier than most. It was colder than ever before — 24 degrees at dawn. The ice team had been worried that the thermal tiles on the outside of the spacecraft — the only protection against incineration in the searing heat of re-entry — might get damaged by icicles falling from the gantry. The entire launch area was almost entirely encased in ice. It was beautiful, and terrifying, like the ice palace in *Dr. Zhivago*. There was talk that the cold might be a problem with the solid rockets, those two gigantic explosives-packed sentinels on either side of the shuttle. But the countdown was continuing.

Maybe some of the tension came from knowledge of the disaster that had just been averted. A few weeks back, they had almost blown up a congressman, Nelson from Florida, when a piece of a sensor broke off and jammed in a fuel line.

BY DENNIS E. POWELL

I was consulted several times by James Powell in order to prepare this article. James Powell has validated about 40 persons being destroyed by MTI.

Appendix 39.

RISKY BUSINESS

Incredibly, this was discovered only by accident, and only because of bad weather: The stray piece surfaced while they were draining off the fuel after the flight was postponed.

NASA had handled that in the usual fashion: They jargoned it to sleep. "Flight rescheduled for Jan. 10 to permit removal of an obstruction in SSME #2 LOX pre-valve."

SSME means space shuttle main engine, the heart of the spacecraft. The obstruction could have blocked the flow of liquid oxygen (LOX) fuel to the main engine or continued down the fuel line to jam the engine. What NASA had not volunteered was that had the flight gone off as scheduled, the shuttle almost certainly would have blown up.

At least that had been an unanticipated problem; in any cover-your-butt bureaucracy, unanticipated problems aren't the scariest because at least there's no clear blame, no unforgiving paper trail to follow in the aftermath of a disaster. The *predictable* problems are the ones to lose sleep over, and the scariest of all of these were the main engines themselves. They'd never worked right, and they'd very nearly destroyed at least two flights already.

Ultimately, the anxiety came down to a feeling that it was only a matter of time. Disaster was inevitable; it was just a question of when.

The truth is that when Challenger lifted off at 11:38 EST that morning, there were more than a few engineers and technicians who had their fingers crossed, praying that none of the things they knew might go wrong would.

"LVLH shit hot!" astronaut Judy Resnik exclaimed into her helmet mike a quarter of a minute into the launch. NASA never explained exactly what that comment meant; the official transcript of the crew conversation, as issued by NASA, split her comment in half. At 0:14 seconds she was reported to have said, "LVLH." A second later, she was reported to have said, "[expletive deleted] hot." It sounded like technical gobbledygook followed by a pause followed by a spontaneous explosion of exuberance — the astronaut equivalent of "yee-hah!" — but some in the space program think the meaning is only too obvious, and ominous. When the tape is played, her words unmistakably run together. It sounds like one phrase: LVLH shit hot. LV and LH refer to "local vertical" and "local horizontal" — the Earth coordinates employed by the shuttle during launch. "Hot," in pilot speak, means "too much." Her words could be taken as an astronaut registering alarm that the shuttle was going too high, too fast: failing to adequately arc outward into the proper flight path.

In fact, that's what was happening. With too much thrust from one of its rockets, Challenger was off course.

The flight crew, Pilot Mike Smith and Commander Dick Scobee, may have been noticing this also. The spacecraft didn't want to go in the direction it was pointed. Something was making it yaw, making its nose want to point to the left. Four seconds after Resnik's expletive, Smith said, "Looks like we've got a lot of wind up here today." But there was no wind to speak of. NASA's weather balloons had demonstrated that before the launch; wind was one of the few things people on the ground did not have to worry about.

Within another 10 seconds the control surfaces on the trailing edge of the shuttle's wings, used to control the shuttle's descent upon re-entry to the atmosphere, began to deflect. This was very unusual, suggesting a computer-generated action to compensate for a drastic problem.

Brownish-orange smoke shot from the back of the orbiter. This type of smoke is produced by the combustion of hydrazine, the fuel in the shuttle's thrusters. The thrusters are never used except when the orbiter is actually in space. But today was terribly different.

Still there was no word from ground control.

Later, NASA's flight controller, Jay Greene, would say that the mission control consoles gave no indication anything was wrong. Later still, NASA would acknowledge that less than five seconds into the day's flight, the right solid rocket had begun producing nearly 12 percent more thrust than it was supposed to produce — an extremely serious, possibly fatal situation. To enter orbit safely, the two solid rockets must be very closely matched. This would explain the shuttle's lunge off course.

But at 70 seconds, ground control said: "Go at throttle up."

Scobee's response, "Roger, go at throttle up," did not come until a second and a half later. Play the TV videotape back, again and again, and the hesitation seems more pronounced each time, an aberration from the usually crisp back-and-forth between astronauts and ground control. It may have meant nothing at all, but some have taken it to be puzzlement at Houston's nonchalance.

No one will ever know for sure.

As the world now understands, Challenger broke apart that icy morning more than three years ago because of a failure of a simple gasket, known as an O-ring, to contain the superheated gases within the right solid-fuel rocket. A small rupture opened on the side of the rocket, loosing a blowtorch of flame that burned into and ignited the giant orange external fuel tank mounted on the underside of the shuttle orbiter.

In the anguished minutes and hours following the disaster, NASA personnel were understandably uncommunicative, and few of them to this day have given detailed accounts about their thoughts at the time. A general presumption at the Cape was that as soon as the fatal malfunction occurred, the most knowledgeable NASA engineers knew or strongly suspected what had happened: in other words, that

---

DEPARTMENT OF THE AIR FORCE
HEADQUARTERS EASTERN SPACE AND MISSILE CENTER (AFSC)
PATRICK AIR FORCE BASE, FLORIDA 32925

REPLY TO ATTN OF: PA — 28 April 1986

SUBJECT: Press Query on SRM Reliability Studies

TO: ESMC/CC

1. Atch 1 is a query from Knight-Ridder Newspapers concerning the numerous solid rocket motor reliability studies in which the USAF participated since the late 1970s.

2. A proposed response was prepared by ESMC/PA in concert with ESMC/Safety, and telecopied to SAF/PA for Air Staff coordination and response to the reporter.

3. Atch 2 is the Air Staff-approved response. Note that the data concerning the aggregated failure rate (one in 50) has been removed and is not to be used. If asked for specific failure rates, for example, for Titan SRMs, we may provide that information.

4. Since an ESMC witness before the Presidential Commission testified to the one-in-50 failure rate, that information may be released by the Commission in the future. The significance of this number, of course, is that Challenger's SRB failure occurred on the 25th mission, or 50th use of the STS SRBs.

ROBERT W. NICHOLSON, JR., Lt Col, USAF
Director of Public Affairs

2 Atch: as

cc: SAF/PAMS (Capt Wiley)
AFSC/PAM
SD/PAM
SAMTO/PA
KSC/PA (Mr. Hollinshead)
KSC/PAM (Mr. Harris)
ESMC/SE
ESMC/SEM

The Truth, But Not The Whole Truth: Memo from the Air Force ordering that statistics on solid rocket failures be withheld from the press. The censored figures showed a one in 50 likelihood of malfunction. The 50th solid rocket used was on the spaceship Challenger.

Appendix 39.

RISKY BUSINESS

never been able to carry more than 55,000 pounds, and to do that they have had to rely on overthrust — revving the engines up in excess of design capacity. The shuttle routinely throttles up to 104, and even 109 percent power, causing stresses on the engine two and four times as much as 100 percent power. The oxymoron heard on most launches, "throttle up to 109 percent power," is the most visible result of the compromises that originated in NASA's attempt to sell the shuttle program. But the biggest compromise was the decision to scrap the notion of manned rockets to deliver the orbiter into space, and instead to use unmanned solid fuel rockets.

"You can't find two reputable engineers [outside NASA] who will say solids are a good idea," says Steve Agee, an engineer brought in by Morton Thiokol to investigate the company's safety and quality assurance systems following the Challenger disaster. Solid fuel rockets are like bottle-rocket fireworks. They are explosives-packed tubes, each with a hole on one end. You light them, and stand back. Once they are lit, they cannot be unlit. If something goes wrong, they cannot be shut down. Unlike liquid fuel rockets, their output cannot be substantially altered: The thrust is constant and uncontrollable, except for very minor directional corrections.

Despite these limitations, they were attractive to NASA for reasons that were at least in part public relations: They sounded cost-effective. Unlike liquid fuel rockets, which ignite their fuel at one end only, solid rockets have the explosive power within them: Their casings must be strong enough to withstand enormous pressure, so they are hardy enough to withstand an ocean splashdown. At least in theory, solid rockets are recyclable. You can fish them out of the water and use them again.

"The decision to use solids was not made by engineers," remembers the disillusioned former NASA engineer Bill McInnis, who was at the agency at the time the decision was made. "It was made for purely political reasons. They wanted to be able to reuse something, and with solids they were pretty sure they could reuse the tubes. I doubt it would ever have saved them money or ever will — it's certainly cost them so far — but at least they could maintain this image of a reusable space plane."

NASA has repeatedly defended the use of solids as being both financially prudent and safe enough to pose no unacceptable risk. "This is a damn good motor," says Royce Mitchell, NASA's solid rocket manager at Marshall Space Flight Center.

(Whether there is in fact a practical reason to reuse the casings is a matter of debate. When there were plans for the Air Force to own and operate shuttles out of Vandenberg Air Force Base in California, the Air Force concluded there was no savings and decided it wouldn't even bother to pick up the used casings.)

The administrator of NASA at the time solids were chosen was Dr. James Fletcher, former president of the University of Utah. After NASA made the decision to use solid rockets (SRMs), the contract was awarded to Thiokol Chemical Co. of Brigham City, Utah. (The company has since been bought by the Morton salt people, and is now Morton Thiokol Inc.)

Fletcher had been on the board of an organization called Pro-Utah Inc., the purpose of which was to bring as much business to Utah as possible. And a succession of Utah's senators have been powerful in committees dealing with NASA. In fact, one of those senators, Jake Garn, actually took a ride on the shuttle.

But sound politics and sound engineering don't always coincide.

"The [solid rocket] is a flying trash can," says the consultant who investigated Thiokol's safety record. "It is an incredibly dangerous system to begin with, and Thiokol's execution makes it more dangerous still."

K— Agee.

(Agee is one of three former safety engineers who have a lawsuit pending against the rocket maker for wrongful dismissal, and for suppressing safety concerns. Thiokol maintains that Agee was an engineer with a consulting firm hired specifically to help in the solid rocket overhaul, and that he was never an employee of Thiokol per se. They declined to discuss anything further concerning the lawsuit.)

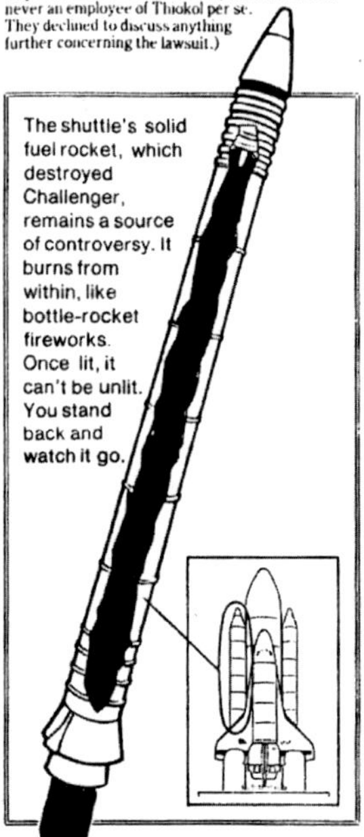

The shuttle's solid fuel rocket, which destroyed Challenger, remains a source of controversy. It burns from within, like bottle-rocket fireworks. Once lit, it can't be unlit. You stand back and watch it go.

NASA's own accounting shows that of the 25 space shuttle launches through 1986, 12 experienced some sort of solid rocket malfunction. This usually involved hot gases "blowing by" the rubbery seals between the sections of the rocket or where the nozzle is attached. But NASA decided it wasn't cause for undue alarm.

According to the report of the Rogers Commission, NASA was often accepting designs that simply didn't do what they were supposed to do. It was easier to change the specs than to change the design — and a whole lot cheaper.

It used to be, said Chief Astronaut John Young in testimony before the commission, that a piece of equipment had to be proved safe before it was allowed to be flown. But now it was the other way around: It had to be proved unsafe before it would be grounded.

A certain pre-Challenger mind-set, lambasted by the Rogers Commission, was roughly this: Well, it didn't blow up so it must be OK. NASA often uses the phrase, "within program experience," which means that it wasn't supposed to happen, it surprised us when it happened, but it didn't kill anybody so we'll live with it. Making problems go away by declaring that they aren't problems was highly criticized by the Rogers Commission, but it is something that NASA continues to do.

Russian Roulette

When NASA was contemplating the return to flight last year, one of the final tests resulted in an 18-inch crack in a solid engine nozzle skirt, the cowl around the end of the rocket. Project managers met to discuss whether to fix the problem, or to "waive" it.

According to NASA's own reports, they decided it was not desirable, but it was acceptable.

After the launch of Discovery marked America's return to space last fall, NASA announced that hot gases got into five of the 10 seals joining the casings to the nozzles, much as had happened in pre-Challenger flights. NASA said they weren't surprised, and that the problem was not of sufficient concern to cause postponement of the Atlantis flight or later launches. But the fact is, the seals were designed to prevent any breach, and the design hadn't entirely succeeded. Instead of redesigning to fix the problem, NASA decided to live with it.

"The argument that the same risk was flown before without failure is often accepted as an argument for the safety of accepting it again," wrote former Rogers Commission member Feynman nearly three years ago in an appendix to the commission report. "Because of this, obvious weaknesses are accepted again and again — sometimes without a sufficiently serious attempt to remedy them, sometimes without a flight delay because of their continued presence." He compared it to Russian roulette. "You pull the trigger and the gun doesn't go off, so it must be safe to pull the trigger again."

Another commission member, Robert Hotz, says there is an alarming distinction between the space shuttle program and any other experimental flight program, including conventional aircraft.

"This is the first time we've sent people up in something when we knew there was something wrong with it," he said. "The dangers in regular flight test are unknown dangers, but with the shuttle, we know things that are wrong with it."

Roger Boisjoly, the former Morton Thiokol engineer who tried to have the Challenger flight scrubbed because he felt certain the O-ring seals would fail in cold weather, said in an interview that this is the kind of thinking that remains at the heart of NASA's problems:

"To fix these things would shut the program down, so they're not doing it."

In April 1986, about 11 weeks after the Challenger accident, a reporter for Knight-Ridder Newspapers submitted a formal query to the Eastern Space and Missile Center (ESMC) at Patrick Air Force Base. His question: What guidance had the Air Force given NASA concerning the reliability of solid-fueled rockets?

The two-paragraph written response states that the Air Force's data was based on 1,000 uses of solid rockets over a 26-year period, and that the data dealt only with rockets that actually blew up, as opposed to those that failed for some other reason. The data had been forwarded to NASA. End of

Continued on Page 15

Appendix 39.

RISKY BUSINESS

Continued from Page 13

answer. Just what that forwarded "data" was, the response did not say.

The original answer written by the Space and Missile Center public affairs office, and obtained by Tropic, had contained an elaboration of the test results. But in a memo to his boss, ESMC director of public affairs, Lt. Col. Robert Nicholson Jr., noted:

"The data concerning the aggregated failure rate has been removed and is not to be used."

The answer was censored by Air Force brass. The statistic withheld from the media was this: One time in 50, solid rockets blew up.

"The significance of this number, of course," Nicholson's memo continued (noting that there are two solid rockets on each mission) "is that Challenger's SRB failure occurred on the 25th mission, or 50th use of the STS [shuttle] SRBs."

In other words, the demise of Challenger, rather than being some freak occurrence, was absolutely predictable. Statistically, the spacecraft exploded right on schedule.

Before Challenger, when the White House was planning to launch the unmanned nuclear-powered Galileo space probe to Jupiter, they asked NASA to supply a probable failure rate for shuttle launches. The fear was that if the shuttle blew with Galileo in the hold, the live nuclear reactor would be dumped into the Atlantic Ocean. If NASA couldn't guarantee that the risk of failure was less than one in 200, the White House wanted out.

NASA sent a reassuring report saying, yes, the risk was less than one in 200. According to the figure more commonly given by NASA, a lot less. Shuttle reliability, the space agency has claimed on more than one occasion, is such that there will be a catastrophic failure every 100,000 flights. If that were true, NASA could hope to safely launch a flight every day for nearly 273 years. That rate is 4,000 times more optimistic than the Air Force figure. NASA has never explained how it arrived at its assessment.

"That's the fundamental problem that I think I see with the shuttle right now," says John Pike of The Federation of American Scientists. "We simply don't have the first idea in the world how safe the thing is. NASA management was assuming that the incidence of failure was going to be one in 100,000. And I've seen documents to that effect. I've also talked to a lot of people who think it's more like one in a couple dozen. And if indeed the majority of the informed speculation is correct . . . it's not safe enough to fly. We shouldn't be flying it. Period."

If the catastrophic failure rate for the shuttle's solid rocket boosters is indeed one in 25 missions, then the shuttle's overall reliability is even worse. The rocket boosters are only one of many systems whose catastrophic failure rates have been calculated. A NASA-commissioned study by the California consulting firm of Pickard, Lowe and Garrick Inc. determined that in one mission in 70, all three of the shuttle's auxiliary power units (APUs) would fail. The APUs power the shuttle, including the controls that enable it to land — the flaps, the landing gear, and so forth. Without them, it would crash. The study pointed out that APUs have actually malfunctioned on at least five shuttle missions, including the November 1983 Columbia mission when two of the units failed shortly before landing.

In another NASA study a year earlier, Lockheed Co. experts determined that a malfunction of the pressurizer system for the main fuel tanks would cause a crash once every 400 missions.

And so on.

Arnold Aldrich, director of NASA's Space Transportation System, says he does not believe predicting failure rates is "meaningful." He said the important point is that "the shuttle is extremely reliable, as safe as we are able to make it after an exhaustive three-year review."

NASA plans to launch the nuclear-powered Galileo on a shuttle mission in October.

Main Engine, Main Problem

Much of the concern about the main engines focuses on the turbo pump, a fuel-handling unit that sits atop the engine. The turbo pump is a high-speed rotary device, with fins that feed thousands of gallons of liquid hydrogen and liquid oxygen to the main engines each minute. It is critical to the shuttle's safe operation. There is no backup.

The original specifications for the main engines, including the turbo pumps, required that they be reusable 55 times before being replaced. In fact, they have proved far less durable. Only one shuttle flight contained three main engines that had been used as many as five times. And that flight was the last flight of Challenger.

NASA engineers have long recognized that the design of the turbo pump is faulty, and have contracted for an improved pump to be designed and built. But, because delivery of the new unit is more than three years away, the space agency has chosen to continue launches using the older, flawed design.

The people NASA has hired to fix the problem say it isn't a safety issue at all.

"These pumps are kind of the heart of the engine and it's an area where NASA was concerned about frequent overhauls and things like that," says John Zimonis, heading up the turbo pump redesign for Pratt & Whitney. "Not a flight safety problem or not any kind of operational difficulty, but NASA thought they could perhaps improve their economics by modernizing these pumps."

But NASA's own safety director, George Rodney, belies that nonchalance: "The turbo pumps are the most critical part of the main engines. We're pushing the state of the art on turbo pumps. We have had a series of, sł all we say, a learning curve with the pumps . . . Whenever you have high-speed rotating machinery, there are failure modes that can be very serious. So, yes, you are concerned about the safety of these pumps. Whenever we find evidence of a new problem, we treat it very seriously."

NASA official Aldrich agrees there have been problems with the pumps, but says they are under control and are not serious enough to warrant scrapping launches pending delivery of the new design.

One acknowledged problem with the current design is that the pump is fitted with bearings that are insufficient to support the long central impeller shaft. This means that the shaft can resonate — vibrate — very much like a piano string. If it bends too far, the blade can scrape against the outer casing, causing chips to fly off, or, in extreme cases, loss of the blade.

"You throw a bucket — an impeller blade — and the engine comes apart quickly," says McInnis, the former NASA engineer. "This is true in all designs like that, including jet aircraft. If you have reason to think you've thrown a bucket, you punch out right now, because the whole thing's coming apart within a second or two."

The troublesome resonance takes place when the pump is spinning at about 94 percent power. The danger is minimized when the shuttle engines are at that setting only briefly, as happens when the vehicle powers through on its way to full throttle.

NASA has tried to avoid the problem by minimizing the amount of time spent at the power setting where the resonance takes place, but contingencies that could arise in the launch would make those plans meaningless.

Such was the case on Jan. 28, 1986. At one point

Some of the supposed improvements to the shuttle have provoked skepticism. The escape hatch, says Bill McInnis, is effective "only when you have nothing to escape from — when everything's working." Agrees one astronaut: "It's not going to save you from very many things."

Appendix 39.

in the Challenger launch, the spacecraft throttled down to 94 percent power. If the Challenger engines behaved as tests predicted, the turbo pump's shaft would have begun to vibrate and bend. This would have allowed its fins to strike the outer housing, and probably to chip, sending red-hot shards down the line, which is filled with propellant. It could result in a catastrophic fire.

Engineering reports show that chipping has occurred both in test and flight. Bearings and casings have also been liable to crack.

NASA says that it has made the blades stronger, less likely to crack or chip in post-Challenger overhauls. But the cracking is just one problem the turbo pumps have faced. In the Atlantis flight, a bearing along the shaft fractured — a flaw NASA attributes to a manufacturing problem. The problem was serious enough that NASA postponed the next flight one month to replace the orbiter's main engines.

"NASA has all sorts of contingency plans for one- and two-engine aborts," says McInnis. "But those plans don't make any sense, because the failure modes of the main engines are of the sort that when an engine fails, it takes out adjacent engines, too."

When the turbo pump contract was awarded in the early 1970s, Pratt & Whitney was clearly the industry leader in the design and manufacture of such units. But the Connecticut-based company had been named in a military overcharge scandal that is said to have pushed NASA away from awarding it the contract. Instead, it was decided that Rocketdyne, a Rockwell subsidiary, would build the unit.

The original Pratt & Whitney design called for additional bearings along the length of the pump's shaft. This would have prevented the harmonic problem experienced in the Rocketdyne unit.

Rocketdyne got the main engine contract in 1971. The Rocketdyne design was chosen by a NASA panel put together by Dale D. Myers, then NASA's associate administrator for Manned Space Flight. Before coming to NASA in 1970, he had been Rockwell's vice president and manager for the Space Shuttle Program. Soon after contracts for both the orbiter and its main engines were signed with Rocketdyne and Rockwell in 1971 and 1972 respectively, Myers resigned from NASA and returned to Rockwell.

In what some see as the ultimate irony, following the Challenger disaster James Fletcher, of Pro-Utah Inc. fame — the man who awarded the solid rocket contract to Thiokol and then left NASA — was called back to the agency. A new assistant administrator was also appointed. His name: Dale D. Myers.

"When Fletcher was brought back, it was a message to everyone that nothing would change. It was business as usual at NASA," says Rogers commission member Hotz.

Last summer, before the return to flight, former Assistant and Deputy Space Shuttle Director Robert [illegible] characterized NASA's new modus operandi: "No dramatic changes from the way we've done business in the past."

On the day this story went to press, Fletcher announced his resignation. His replacement had not been announced.

Too Close To The Skin

The thermal protection tiles encasing the orbiter are an engineering marvel. Spun from silicon fiber, they have astounding insulating properties. When the shuttle was being built, pictures were circulated of a technician holding a red-hot tile in his bare hand, which was unharmed, so great was the tile's ability to shield from heat.

They also have the consistency of that green plastic foam used in flower arrangements. You can stick your finger right into one.

"If we hadn't lost something like 288 tiles ferrying the first orbiter to [Kennedy Space Center] we probably would have lost the very first shuttle," remembers McInnis. "But the bonding method, the way they were attached to the shuttle, was so darn bad that they fell off just during the ferry flight. They thought they fixed it, but they weren't sure — and they never did flight tests to find out."

In early flights, large numbers of tiles would be damaged or lost. In the most recent flight of Atlantis, more than 200 tiles were lost or damaged. No flight had ever lost that many, even in the early pre-Challenger days.

"This is a Crit One issue," said one high-placed staff member of the Senate space and technology oversight committee. "Crit One" means of the highest criticality, that there is nothing to save the shuttle if the tiles are lost. "There were burns through the buffer, the tile, the insulator — and that was the orbiter sitting there. That aluminum there — that's the orbiter. That's serious."

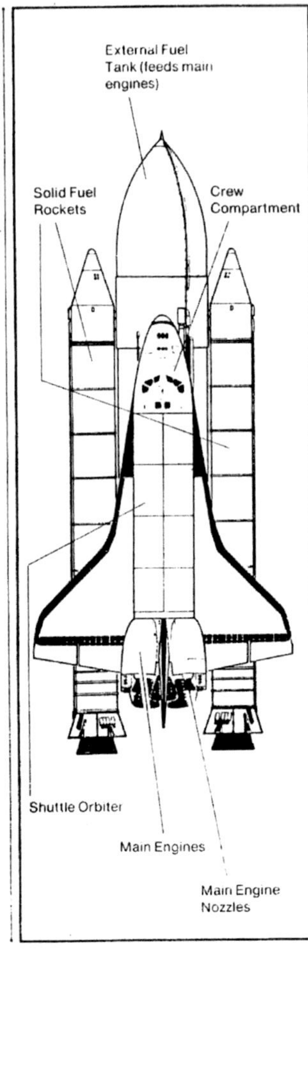

After Challenger, much was made of extensive redesigns that were taking place. The most attention, of course, was paid to the solid rockets, which were seen as the immediate cause of the disaster.

NASA has detailed hundreds of changes that have been made in the orbiter and associated hardware. The results of some of the redesign work may have lulled many into believing the shuttle is now safer than it actually is. One example is the emergency escape system installed in post-Challenger shuttles.

Several experts within and without NASA look at the system with skepticism.

"It's not going to save you from very many things," said one astronaut. "It's very limited."

"For it to be of any use," notes McInnis, "you must have perfectly working solid rocket motors, because they have to have gone through their entire burn sequence before you can cut them loose. Then you have to have perfectly working main engines, because the huge external fuel tank has to be drained down to only 5 percent full before you can cut it loose.

"Then you have to have a shuttle in full control, because it needs to be in a controlled glide before the escape system will work.

"In short, you can escape from the shuttle, really, only when you have nothing to escape from — when everything's working."

Even with the new system, the shuttle remains the only U.S. manned space vehicle with no means of escape during the most dangerous part of the flight: blastoff.

The story so far: In the full glow of the moon program, NASA proposed an elaborate space-shuttle system. Congress said no. So NASA simultaneously reduced the scope of its proposal and increased claims for its capability. Then, once the shuttle project was approved, NASA under Dr. James Fletcher, systematically closed down all other avenues to space; the entire space program, including the military program, was placed under the aegis of NASA. The shuttle, with inflated expectations and unprecedented inattention to safety, became the only show in town. All NASA's eggs were in one basket.

"Well, it actually put the Air Force's and the civilian community's eggs in that basket as well," says Dr. Bruce Murray, former director of NASA's Jet Propulsion Laboratory, the top space exploration think-tank in the United States. "That was a matter of doctrine. It really was. It was not a matter of reasoned technical alternatives, because NASA exterminated the alternatives and was able to successfully get a monopoly.

Continued on Page 21

Appendix 39.

RISKY BUSINESS

Continued from Page 19

"How could the secretary of the Air Force, the secretary of defense, and other key defense department personnel have been persuaded to let the U.S. space monitoring capability become dependent on a manned system? And the fact that their [the military's] actions since then have been to get away from that as fast as possible certainly indicates they recognize now that it was a bad thing to do."

The Air Force, already unhappy with NASA, seized upon the Challenger disaster as an opportunity to re-establish its own unmanned launch program. Other government agencies with business in outer space — even the weather bureau — followed the Air Force lead.

Part of the problem, says a second congressional staffer — one involved exclusively in monitoring space activities — is that from day to day, the people who pay attention to NASA's activities are those who have vested interests, such as space centers or concentrations of contractors in their districts. They are the ones least likely to holler, if hollering could jeopardize some aspect of some project.

"I mean, a large part of the Congress has a nice, warm, fuzzy feeling about the space program," says the staffer. "The idea that it's a good thing that the country does, and the astronauts are very brave people and gosh, you know, there's this thing out there that's gonna take pictures of Uranus and isn't that wonderful. And when you ask them what should we do with all of this, 'Well, about what we're doing now.' How much should we spend on all this? 'Well, about what we're spending now. And I think that's the dominant attitude not only in the Congress but the American public."

"In the newspaper I used to read about shuttles going up and down all the time," said former Rogers Commission member Richard Feynman, "but it bothered me a little bit that I never saw in any scientific journal any results of anything that had ever come out of the experiments on the shuttle that were supposed to be so important."

Dr. Murray says that NASA spent a lot of time and money showboating with the shuttle to justify the system by using it to do things that could have been done better by another kind of launch vehicle.

People forget, says Murray, that the principal job of Challenger was merely launching an unmanned communications satellite, which could have just as easily been launched by an unmanned rocket.

"There's absolutely no reason to have a human in the loop to launch a planetary spacecraft. There's no purpose for having a human in the loop to launch a communications satellite.

"You have to ask," Murray said, "what did we kill seven people doing?"

Last month's launch of the space shuttle Discovery was hailed by NASA officials as one of the most successful shuttle missions ever. The major accomplishments: experiments with chicken eggs and with rats with broken limbs, and the placement into orbit of the Tedrus communication satellite.

It was a duplicate of the satellite that was in Challenger's hold 3½ years ago. ■

DENNIS E. POWELL is a free-lance writer. His most recent Tropic story was on Guantanamo Naval Base in Cuba.

Appendix 40

ED MEESE

In Resigning As Attorney General, Edwin Meese Did
Nation A Favour. July 7, 1988

Insufficient evidence to indict. The words are not an endorsement. Yet Edwin Meese III waved them like a badge of honor Tuesday as he announced his long overdue resignation as U.S. attorney general.

Independent counsel James C. McKay delivered the 800-page summary of his 14- month investigation of the multiple misconduct charges against Meese to a special panel of appellate judges just hours before the attorney general revealed his plans to leave office.

While the contents of the report aren`t expected to be made public for several days, it has been known for some time that there would be no criminal charges brought against Meese. It also has been known, however, that the report will criticize the attorney general`s chronic insensitivity in ethical matters.

That would prompt an internal investigation by the Justice Department`s Office of Professional Responsibility. All of the old allegations -- of influence peddling, selling federal jobs, cronyism, financial finagling, and the manipulation of federal investigations -- would be reviewed again, this time for possible violations of federal ethics laws. By quitting now, Meese lessens the likelihood of further embarrassment for the Reagan administration.

The abrupt resignation, which will take effect within the next few weeks, brings an end to one of the least distinguished chapters in Justice Department history.

As the top law enforcement figure in the nation, Meese spent most of his time defending himself in court. His involvement in the Wedtech defense contracting scandal, his actions in a proposed $1 billion Iraqi pipeline project, his refusal to remove himself from a telephone company case at a time when he held telephone stock, all came under time-consuming scrutiny.

When he wasn`t testifying on his own behalf, Meese was criss-crossing the country, questioning the authority of the Supreme Court and attempting to undermine long established civil rights programs.

Appendix 40

During his three years on the job, the Justice Department stagnated, with key officials resigning in frustration. As Rep. Pat Shroeder, D-Colo., aptly put it, the only investigation that went anywhere on Meese`s watch was the one he didn`t know about. She referred to the massive probe of alleged fraud and bribery in Pentagon weapons procurement procedures now unfolding.

While conservative Republicans might brush off such criticism as partisan rhetoric, the man most pleased by the resignation has to be George Bush, the GOP nominee for president. With his affinity for questionable deals and associations with indicted buddies, Meese has been a sitting duck, a sleazy target for every Democrat running for office and a major distraction for the Bush campaign.

In accepting his long time friend`s resignation, President Reagan said he had no one in mind to succeed him. It won`t be easy to find a capable replacement for the short time remaining in Reagan`s second term. Considering the attorney general`s performance, an empty desk could be an improvement.

To paraphrase an old cliche, Edwin Meese has brightened up an administration just by leaving it.

Perhaps this article best focusses some of the problems that the FBI had against MTI, NASA, and others. The 'Head Cop' in America was a criminal.

Appendix 41

THOMSON REUTERS

Irene Klotz
Feb 24th 2017 12:54PM

CAPE CANAVERAL, Fla., Feb 24 (Reuters) - The Trump administration has directed NASA to study whether it is feasible to fly astronauts on the debut flight of the agency's heavy-lift rocket, a mission currently unmanned and targeted to launch in late 2018, officials said on Friday.

The study marks the Republican president's first step in shaping a vision for the National Aeronautics and Space Administration.

Under President Barack Obama, the U.S. space agency was working on the heavy-lift Space Launch System rocket and Orion deep-space capsule with the aim of sending astronauts to rendezvous with an asteroid in the mid-2020s, followed by a human expedition to Mars in the 2030s.

The request for the study from the new administration tweaks that plan by exploring whether to add a crew to an earlier test flight and perhaps setting the stage for a human return to the moon.

NASA officials said they do not feel compelled to fly the test mission with crew aboard, Bill Gerstenmaier, NASA's head of human space flight, told reporters on a conference call.

"There's not pressure to go do this," Gerstenmaier said. "I find it encouraging that we were asked to go do this feasibility study."

The study is expected to take about a month. Engineers are assessing hardware changes, schedule delays, additional costs and increased risks of flying a two-member crew on the first flight of the Space Launch System rocket, which is about four times bigger and more powerful than any current U.S. booster.

Appendix 42

THIOKOL OFFICIAL WHO LED BOOSTER REDESIGN TEAM ANNOUNCES HE'LL RESIGN

Published: May 25, 1989 12:00 a.m.

John D. Thirkill, who oversaw much of Morton Thiokol Inc.'s efforts to redesign the space shuttle booster rocket, has announced his plans to leave the aerospace company after 35 years.

Thirkill, 60, stopped working as vice president of Space Operations last week, Morton Thiokol spokesman Rocky Raab said.Although Thirkill was recently moved to the company's Aerospace Group staff, Raab said, "He has announced his retirement for the near future. He hasn't made any decision beyond retirement."

Raab said he did not know Thirkill's reason for leaving. Thirkill could not be reached at work or his South Ogden home for comment.

He became vice president of Space Operations in March 1987, when Morton Thiokol was redesigning its shuttle booster motor in response to the January 1986 Challenger explosion.

A presidential commission determined a leaking seal on one of the Challenger's two boosters failed, allowing super-hot gas to escape and ignite an explosion that destroyed the spacecraft and its seven-member crew.

Raab said that until a replacement is found, Thirkill's job will be done by his boss, Robert Lindstrom, senior vice president of Strategic and Space Operations.

Previously, Thirkill was vice president and general manager of Thiokol's Huntsville, Ala., Division.

Earlier, he was vice president of engineering at Utah's Wasatch Division; deputy director of the space shuttle program, and was involved with programs including the Minuteman, Poseidon and Trident missiles.

Thirkill began his career with Morton Thiokol in 1953 as a project engineer at the Elkton, Md., Division.

Did he resign of his own volition or was he pushed as a result of the FBI investigations? Was his Security Clearance revoked?

Appendix 43

Reporters' Relentless Pursuit Of Facts Won Times 2 Pulitzers -- Two Pulitzer Prizes

By Chuck Taylor
Seattle Times Staff Reporter

Seattle Times reporter Eric Nalder once showed up at the Federal Building in Richland to conduct an interview and to deliver a written Freedom of Information Act (FOIA) record request, known in the journalism business as a "foy-yuh."

The local chief of public relations for the U.S. Department of Energy greeted him in the lobby. Nalder handed him a letter.

"So this is your `fooey' request," the man sniffed, joking, sort of. The government isn't always chipper about disclosing public records.

"Hey," Nalder said without missing a beat, "you should just leak me some documents. You could be a working-class hero!"

A dozen years later, Nalder still is cheerfully afflicting the establishment and sweet-talking middle managers, poring over documents and ringing doorbells.

Such reporting paid off yesterday for Nalder, 51, Deborah Nelson, 44, and Alex Tizon, 37, as they shared a Pulitzer Prize in journalism for investigative reporting. Aerospace reporter Byron Acohido, 41, won the other Pulitzer for The Times, for beat reporting.

Nelson, Nalder and Tizon won for a series that exposed opportunism, nepotism and sloppy management in Native-American housing programs financed by the U.S. Department of Housing and Urban Development.

It was Nalder's second Pulitzer. He contributed to the paper's coverage of the Exxon Valdez oil spill in Alaska's Prince William Sound, which won a prize in 1990.

Acohido won for a variety of work, the most notable of which was a series of stories about how Boeing and the government have handled a puzzling problem with a rudder mechanism on 737 jetliners, a problem believed to be a possible cause in two crashes.

The four Times reporters, and hundreds like them around the country, share some characteristics that make their work special.

They are relentless, detail-oriented, likely to side with the little guy or gal and, when it suits them, to be gleefully irreverent. A common mantra might be that seen on a bumper sticker: "Question Authority."

"It must be some flaw in my personality," joked Acohido. "I want things to make sense and ring true."

Tip inspired series

Work on the tribal-housing series began about a year ago when Nelson got a telephone tip.

The tipster, from the Tulalip Indian Reservation, told of a fancy house occupied by the director of the reservation's housing authority and her husband, the authority's contracting officer, and financed with federal money intended for low-income housing.

To check out the story, Nelson tried to drive by, but the house wasn't visible from the road. She returned with photographer Greg Gilbert. They hiked through the woods, and there it was, all 5,300 square feet of it.

When Nelson called the couple, "They were vague about where the money came from," she says.

She submitted FOIA record requests to HUD and discovered an exchange of e-mail that revealed "that HUD regulations actually allowed this to happen," she said.

Word spread through Indian country in the Northwest that Nelson was nosing around, and more tips poured in.Eventually it became clear that the Tulalip house wasn't an isolated example. Nalder, a fellow records-scrounging pack rat in the next cubicle, was recruited to help Nelson.

The two had numerous cases to examine in the Northwest. HUD insisted that was the extent of the problem.

But heeding the "question authority" credo and at the urging of metropolitan editor David Boardman, Nelson and Nalder began checking housing programs at Native-American reservations nationwide, first through HUD records and then with visits.

The problems were far more widespread than HUD had said.

As the investigative pair continued their work, Tizon was added to the team. Regarded as one of the paper's most skillful writers, he was beginning a beat called American subcultures, with emphasis on Native Americans.

Tizon's first mission was to visit reservations where abuse had been found by Nelson or Nalder, and document the impoverishment - the need for the federal housing money in the first place.

Appendix 44

JOINT RUSSIA-USA PLANNING FOR MANNED MARS MISSION CONTINUES INSPITE OF ...

With the frosty relations between the USA and Russia over the USA-sponsored fascist coup in Kiev, and the resulting chaos in the eastern Ukraine, one might think that any major long-range talks of Russo-American space cooperation would be on hold, but that's not the case at all according to this article from Russia's *Sputnik* magazine from March of this year(shared by Mr. M.D.):

Russia, US to Jointly Prepare Mars, Moon Flight Road Map

When I read this article, I repeatedly asked myself why these two nations - recently at such loggerheads - could even be contemplating any sort of joint coordination of their national space programs (much less the possibility of joint missions) as is indicate in the following paragraphs:

> "
> "Russia and the United States will work together on a roadmap to send humans to Mars and the Moon, according to NASA Administrator Charles Bolden. The Russian Federal Space Agency Roscosmos and its US counterpart NASA will jointly hammer out a "road map" program on flights to Mars and the Moon, NASA Administrator Charles Bolden said on Saturday.
>
> "Bolden, who is currently on a tour of Russia's Baikonur cosmodrome in Kazakhstan, added that he had discussed joint efforts to send missions to the Red Planet with Roscosmos head Igor Komarov, including time frames and funding.
>
> "'Our area of cooperation will be Mars. *We are discussing how best to use the resources, the finance, we are setting time frames and distributing efforts in order to avoid duplication,*' Bolden said." (Emphasis added)

Now, the very idea of "avoiding duplication" falls far short of the idea of jointly sponsored manned missions, but it is suggestive. The question for the high octane speculation of the day is "why?" this need to have such talks amid the tense geopolitical situation? And my answer, highly speculative though it may be (and it is certainly *not* a speculation that exhausts all the possibilities), is that this may be a signal of a very covert agenda. As most regular readers here are aware, I am definitely in that camp that thinks there *are* significant indicators of artificial "stuff" out there on our neighboring celestial bodies. (Recall this blog from 2013: JUST WHAT THE HECK IS GOING ON WITH MARS, ANYWAY?) And it's that possibility of "stuff" out there, especially on Mars, that may be the hidden story here, for the USA may be trying to bring Russia into the Mars exploration-and-disclosure game in order to exercise influence over whatever "stuff", if any, they might find, and eventually, thus be able to influence the precise way that it might be disclosed should they confirm any "stuff" really is there.

And if that's the case, then it will be interesting to see how the other space-faring powers - China, India, Japan, Europe(i.e. Germany) - respond. It is indeed a notable fact that the European Space Agency, along with NASA, has been releasing some rather stunning pictures from Darmstadt, not only from the Red Planet, but from other celestial bodies recently, and doing so *without much comment,* almost as if a kind of limited hangout position was being put into place. And China of course released some odd pictures from the Moon(and released them in a very *odd* way on the People's Liberation Army website, to boot), and again, without much comment.

So the bottom line for me here is that there is something definitely going on with this NASA-Russia space consultation, and we can bet our bottom dollar that we're not being told the whole story. I suspect strongly that that story includes whatever pictures of Mars the Russians may have locked up in their archives, and, by careful comparison between those and current returns from the Red Planet, that quite a lot of behind the scenes discussion is being had on where a manned mission should land, the best to confirm, or deny, that there may be "stuff". And if there is "stuff", how best to handle and release the information.

See you on the flip side...

Appendix 45

Will Trump go to Mars? Nasa's nervous wait

The president-elect's priorities for the US space agency are of crucial importance for Earth as well as for future space exploration.

A booster for Nasa's Space Launch System, the rocket that might take humans to Mars, is tested in Promontory, Utah, 28June 2016.

Photograph: Bill Ingalls/AFP/Getty Images

In Nasa language it's called a pivot. It's a policy change, a U-turn or a departure from a goal set by the previous US president. Until the election of Donalds Trump, space insiders and even Nasa itself had a pretty good idea what, under a Hillary Clinton presidency, that pivot was going to be. It wasn't going to be popular but it was necessary.

People had been whispering it for more than a year. Even as the Nasa PR machine talked endlessly about "the journey to Mars", those in the know understood that it was little more than a pipe dream. Following the election, a "transition team" would be sent to take stock at Nasa, and the agency's goals would gradually pivot away from Mars and to the moon.

Returning astronauts to the moon with the cooperation of other space agencies around the world was achievable and there was a definite appetite for it. The European Space Agency and Russia were both interested in taking part and the International Space Station was seen as the blueprint for the kind of multinational cooperation that could make it happen. In essence, it was what both Hilary Clinton's Democrats and Washington Republicans were looking to make happen.

Nasa has no idea what the budget is going to be, whether the journey to Mars is going to be cancelled

"The space policies were not very different. People would joke that you could take an editorial by a Republican space person, change a couple of words here and there and then put Clinton's name on it. We were all assuming that Clinton would win. I knew who the people were that would show up at Nasa the next day to begin the transition process," says Keith Cowing, a former Nasa employee who now edits Nasawatch.com.

Then, of course, Trump won. At first it seemed no big deal, but then surprising events started to unfold. A day or two after the election, none of the expected Republican advisers were named as part of the Trump transition team for space. "They were either thrown off because they were lobbyists or had decided that they did not want to be involved," says

Cowing, who has been reporting on Nasa from Washington DC for 20 years. "Suddenly it went from what seemed to be clarity to complete mystery."

This week, however, some clarity finally emerged. Christopher Shank, who held senior policy jobs at Nasa and sat on Congress's House Science Committee, will lead the transition team. One of the first things he is going to have to look at is the rocket that Nasa are building to supposedly take astronauts to Mars: the Space Launch System (SLS).

SLS is controversial because it is so expensive. It is part of a multi-billion-dollar-per-year programme that will only fly four times between 2018 and 2026. In many ways it is Nasa's flagship programme, but it is sucking a huge amount of money out of the agency's budget on the what-if scenario of human Mars missions.

It is also looking egregiously expensive when compared with the progress being made at private companies, such as Elon Musk's SpaceX, which is developing a large rocket of its own, the Falcon Heavy.

Appendix 46:

Donalds Trump's first 100 days as president – daily updates

They wrote: "Nasa's core missions must be exploration and science – and inspirational! These are the fundamental underpinnings of a Trump civilian space programme. Nasa should be focused primarily on deep-space activities rather than Earth-centric work that is better handled by other agencies."

Appendix 47

SOME HISTORY OF THIOKOL

"Company engineers, who knew that the shuttle's boosters were not rated for operation below 40 degrees, were ignored in the executive-level decision process. Just over a minute after launch, a flare of fire from one of the boosters ignited the shuttle's external fuel tank, destroying the orbiter and killing its seven astronauts.

While Locke ordered an investigation of the accident, he failed to handle the public relations crisis that followed. At one point after the blast he told the *Wall Street Journal* that "the shuttle thing will cost us ten cents a share." While quoted out of context, Locke nonetheless came across as callous and tactless. In a demonstration of good faith, Locke conducted more than $400 million of redesign work on the boosters, working at cost.

The company suffered further damage in December 1987 when an explosion at its MX missile plant killed five employees. Furthermore, work on the boosters revealed new design flaws that had to be corrected. In 1989, despite extensive redesign work and the resumption of shuttle flights, Morton Thiokol lost a bid for a new booster design to Lockheed Corporation. At this point, Locke decided to spin off the Thiokol division.

Before dividing the companies, Locke transferred Thiokol's chemical businesses to the Morton side of the company. Morton also retained Thiokol's promising automotive airbag business. The companies were officially split on July 1, 1989, when Locke offered stockholders shares in Morton International, with options to trade them for Thiokol shares. Morton International, with $1.4 billion in sales and 8,000 employees, remained concentrated mostly in chemicals and salt, while investing nearly $100 million in its airbag business.

Independent Thiokol Diversified in the 1990s

Utah-based Thiokol Corporation emerged as a $1.1 billion company with 12,000 employees. Nevertheless, its profitable chemicals and airbag businesses had been carved away by Morton--all that remained was rockets. With greatly relaxed tensions between the United States and the Soviet Union, defence spending began to fall. Thiokol's new management team--retired U.S. Air Force general Robert T. Marsh as chairman and Edwin Garrison as president and CEO--found itself in charge of a company with a declining market and the prospect of losing the shuttle booster business to Lockheed.

Appendix 47

During this time, however, NASA was strapped for funds, and, unwilling to carry the development costs of Lockheed's trouble-plagued booster, the agency expressedinterest in extending its contract with Thiokol and its proven boosters. Thiokol subsequently signed a contract in 1991 to supply NASA with 142 solid rocket motors or the space shuttle program through 1997. Thiokol also began looking for commercial markets, finding a new customer in Motorola, whose 77-satellite Iridium system would provide seamless, worldwide cellular telephone service. Motorola hoped to use Thiokol's Castor rockets to get its system into orbit.

Despite these positive developments, the overall market for propulsion systems continued to decline as the 1990s continued. While Thiokol derived more than $1 billion in net sales from its propulsion operations in the early 1990s, by fiscal 1997this figure had been reduced to $606.1 million. Given this obvious trend, Thiokol moved decisively to make itself less dependent on its propulsion operations.

In 1991 Marsh retired and Garrison added the chairmanship of Thiokol to his duties, remaining president and CEO. Late that same year, Thiokol purchased Huck Manufacturing Company for $150 million as part of its diversification effort. Huck, based in Irvine, California, made high grade rivets, lock bolts, and other fasteners for Industrial and commercial transportation systems--notably nondefense markets. Huckwas soon renamed Huck International, Inc. and operated as a subsidiary of Thiokol. Thiokol subsequently bolstered Huck through the 1992 acquisition of a majority interest in the German-based Kamax-Aerobolt GmbH & Co. KG, a maker of fasteners for the European market; the early 1994 acquisition of the Deutsch Fastener Company, an aerospace fastener maker based in California; and the February 1995 purchase of Automatic Fastener Corporation, a Branford, Connecticut-based producer of rivets for the automotive and light truck markets. Meanwhile, in June 1993 Garrison retired as president and CEO (remaining chairman), with John R. Myers replacing him. But after Myers resigned only four months later, James R. Wilson was named president and CEO of Thiokol. In October 1995 Garrison retired as chairman, and Wilson assumed that role as well.

In December 1995 Thiokol and the Carlyle Group, a Washington D.C.-based private merchant bank, joined forces to acquire Howmet Corporation, with Thiokol holding an initial 49 percent of Howmet and Carlyle 51 percent. The Greenich, Connecticut-based Howmet, with annual revenues of about

$900 million, was a world leader in the manufacture of precision castings of super alloys and titanium, utilized primarily for jet aircraft and industrial gas turbine engine components. Thiokol had an option to purchase all of Carlyle's interest in Howmet after three years. In late 1997, however, Thiokol and Carlyle reached an agreement whereby Thiokol would purchase an additional 11 percent of Howmet, while at the same time Carlyle would sell common stock of Howmet through an initial public offering. Following the IPO, Thiokol would have a majority stake in Howmet. Beginning twoyears after the IPO, Thiokol would have the option of purchasing Carlyle's remaining shares in Howmet at market price.

Through its acquisition of Huck and its majority stake in Howmet, Thiokol had succeeded in rapidly diversifying itself. In 1991 all of the company's profits derived from propulsion systems, while by 1997 Huck and Howmet accounted for 16 percent and 39 percent, respectively, of Thiokol profits. Meanwhile, during this period, the company was restructuring its propulsion operations both to reflect an increasing emphasis on commercial markets over those of defense and to make them more efficient. During fiscal 1997, Thiokol created a single Thiokol Propulsion Groupunder which were consolidated three divisions - space, defense and launch vehicles, and science and engineering. The company named Robert Crippen, former astronaut and former director of NASA's Kennedy Space Center, to head the new group as its first president.

Approaching a new century, Thiokol was counting on its new Propulsion Group to increasingly seek out commercial customers for its boosters, as the defence market continued to look weak. The company, which had virtually erased its debt by 1997, was at the same time in a very strong position to increase its stake in Howmet and to seek out acquisitions that would further beef up Huck.

Principal Subsidiaries:

Huck International, Inc.; Howmet Corporation (60%).

What a retired UK Intelligence Officer, Paul Edwards, says about this book:

For The Right Reasons is the story of one man's struggle to do the right thing no matter what the personal cost. Steven W. Agee, is such a man, he cannot be bullied, bought or by-passed in his personal efforts to reveal the appalling state of the United States Space Program after the loss of the Space Shuttle Challenger and its 7 astronauts. Following this harrowing accident and the suspension of future shuttle launches his story starts purely by chance when Steven answers the call to work as a contractor for Morton Thiokol, the design and manufacturers of the solid rocket boosters which powered the Shuttle to orbit. His mission was to identify safety hazards and bring them to the attention of Morton Thiokol and NASA. Steven is no ordinary man, he is an elite Physicist, Engineer and Project manager educated to be the best in the US and global aerospace industry. He expected Morton Thiokol and NASA to have high quality management, excellent engineers and scientists and industry leading processes in place. What he found, was an appallingly managed and technically flawed solution for launching humans into space. This book tells of his battle with Morton Thiokol and his decision to approach the Federal Bureau of Investigation and the subsequent impacts of that decision for all involved. Steven does not pull his punches, treason has been committed, the Challenger should never have been launched, the Mormon Church and its adherents were involved in Morton Thiokol and NASA and later within the US justice system. Only the Federal Bureau of Investigation remain untarnished by corruption at the highest levels. Steven's story is compelling, a stark expose of the truth, based on engineering and scientific fact; and sadly it reveals a cover-up that severely impacts the western world still today and for some time into the future.

Paul Edwards OBE

Made in United States
Orlando, FL
04 November 2026

90205907R00328